Ordering Braids

Mathematical
Surveys
and
Monographs

Volume 148

Ordering Braids

Patrick Dehornoy
with
Ivan Dynnikov
Dale Rolfsen
Bert Wiest

American Mathematical Society
Providence, Rhode Island

The original edition of this work was published under the title *Why Are Braids Orderable?* by Patrick Dehornoy, Ivan Dynnikov, Dale Rolfsen, and Bert Wiest, 2002, by the Société Mathématique de France, Paris, France.

2000 *Mathematics Subject Classification.* Primary 20F36;
Secondary 06F05, 06F15, 20B30, 20F34, 20F38, 57M25, 57M50, 68Q25, 68Q70.

For additional information and updates on this book, visit
www.ams.org/bookpages/surv-148

Library of Congress Cataloging-in-Publication Data

Dehornoy, Patrick.
Ordering braids / Patrick Dehornoy ; with Ivan Dynnikov, Dale Rolfsen, Bert Wiest.
p. cm. — (Mathematical surveys and monographs, ISSN 0076-5376 ; v. 148)
Includes bibliographical references and index.
ISBN 978-0-8218-4431-1 (alk. paper)
1. Braid theory. 2. Linear orderings. I. Dynnikov, Ivan, 1971– II. Rolfsen, Dale. III. Wiest, Bert. IV. Title.

QA612.23.D445 2008
514′.224—dc22 2008009859

10 9 8 7 6 5 4 3 2 1 13 12 11 10 09 08

Contents

Preface

The present volume follows a book, *"Why are braids orderable?"*, written by the same authors and published in 2002 by the Société Mathématique de France in the series *Panoramas et Synthèses*. We emphasize that this is *not* a new edition of that book. Although this book contains most of the material in the previous book, it also contains a considerable amount of new material. In addition, much of the original text has been completely rewritten, with a view to making it more readable and up-to-date. We have been able not only to include ideas that were unknown in 2002, but we have also benefitted from helpful comments by colleagues and students regarding the contents of the SMF book, and we have taken their advice to heart in writing this book.

The reader is assumed to have some basic background in group theory and topology. However, we have attempted to make the ideas in this volume accessible and interesting to students and seasoned professionals alike.

In fact, the question "Why are braids orderable?" has not been answered to our satisfaction, either in the book with that title or the present volume. That is, we do not understand precisely what makes the braid groups so special that they enjoy an ordering so easy to describe, so challenging to construct and with such subtle properties as are described in these pages. The best we can offer is some insight into the easier question, "How are braids orderable?"

Patrick Dehornoy, Caen

Ivan Dynnikov, Moscow

Dale Rolfsen, Vancouver

Bert Wiest, Rennes

December 2007

Introduction

Braid theory is a beautiful subject which combines the visual appeal and insights of topology with the precision and power of algebra. It is relevant not only to algebraists and topologists, but also to scientists working in many disciplines. It even touches upon such diverse fields as polymer chemistry, molecular biology, cryptography and robotics.

The theory of braids has been an exceptionally active mathematical subject in recent decades. The field really caught fire in the mid 1980's with the revolutionary discoveries of Vaughan Jones [**115**], providing strong connections with operator theory, statistical mechanics and utilizing many ideas which originated from mathematical physics.

That braids have a natural ordering, compatible with their algebraic structure, was discovered a decade later by one of the authors (P.D.), and since then it has been intensively studied and generalized by many mathematicians, including the authors. That phenomenon is the subject of this book.

One of the exciting aspects of this work is the rich variety of mathematical techniques that come into play. In these pages, one will find subtle combinatorics, applications of hyperbolic geometry, automata theory, laminations and triangulations, dynamics, even unprovability results, in addition to the more traditional methods of topology and algebra.

A meeting of two classical subjects

It was an idea whose time was overdue—the marriage of braid theory with the theory of orderable groups. The braid groups B_n were introduced by Emil Artin [**4**] in 1925—see also [**5**]. Indeed, many of the ideas date back to the nineteenth century in the works of Hurwicz, Klein, Poincaré, Riemann, and certainly other authors. One can even find a braid sketched in the notebooks of Gauss [**97**]—see [**177**] for a discussion about Gauss and braids, including a reproduction of the picture he drew in his notebook.

The n-strand braid group B_n has a well-known presentation—other definitions will be given later:

$$B_n = \langle \sigma_1, ..., \sigma_{n-1} \mid \sigma_i\sigma_j = \sigma_j\sigma_i \text{ for } |i-j| \geqslant 2,\ \sigma_i\sigma_j\sigma_i = \sigma_j\sigma_i\sigma_j \text{ for } |i-j| = 1\rangle.$$

We use B_n^+ for the monoid with the above presentation, which is called the n-strand braid monoid. The monoid B_n^+ is included in a larger submonoid B_n^{+*} of B_n, called the dual braid monoid, which is associated with the presentation of B_n given by Birman, Ko and Lee in [**15**]—details may be found in Chapter VIII.

To each braid, there is an associated permutation of the set $\{1, ..., n\}$, with σ_i sent to $(i, i+1)$, defining a homomorphism of B_n onto the symmetric group $\mathfrak{S}_n$. The kernel of this mapping is the *pure* braid group PB_n.

The theory of ordered groups is also well over a hundred years old. One of the basic theorems of the subject is Hölder's theorem, published in 1902 [**111**], that characterizes the additive reals as the unique maximal Archimedean ordered group. It is remarkable, and somewhat puzzling, that it has taken so long for these two venerable subjects to come together as they now have.

A group or a monoid G is *left-orderable* if there exists a linear, *i.e.*, strict total, ordering $\prec$ of its elements which is left-invariant, *i.e.*, $g \prec g'$ implies $hg \prec hg'$ for all g, g', h in G. A group is right-orderable if and only if it is left-orderable, but the orderings are generally different; both choices appear in the literature with roughly equal frequency. If there is a left ordering of G which also invariant under multiplication on the right, we say that G is *orderable*, or for emphasis, *bi-orderable*.

The main theme of this book is proving and explaining the following.

THEOREM. *The Artin braid group B_n is left-orderable, by an ordering which is a well-ordering when restricted to the braid monoid B_n^+—and even to the dual braid monoid B_n^{+*}.*

Despite the high degree of interest in braid theory, the importance of the left-orderability of the braid groups, announced in 1992 [**47**], was not widely recognized at first. A possible explanation for this is that the methods of proof were rather unfamiliar to most topologists, the people most interested in braid theory. As will be seen in Chapter IV, that proof involves rather delicate combinatorial and algebraic constructions, which were partly motivated by (while being logically independent of) questions in set theory—see [**121**] for a good introduction. Subsequent combinatorial work brought new results and proposed new approaches: David Larue established in [**132, 131**] results anticipating those of [**83**], Richard Laver proved in [**137**] that the restriction of the braid ordering to B_n^+ is a well-ordering, Serge Burckel gave an effective version of the latter result in [**26, 27**]. However, these results were also not widely known for several years.

The challenge of finding a topological proof of left-orderability of B_n led to the five-author paper [**83**] giving a completely different construction of an ordering of B_n as a mapping class group. Remarkably, it leads to exactly the same ordering as [**47**]. Soon after, a new technique [**183**] was applied to yield yet another proof of orderability of the braid groups—and many other mapping class groups—using ideas of hyperbolic geometry, and moreover giving rise to many possible orderings of the braid groups. This argument, pointed out by William Thurston, uses ideas of Nielsen [**164**] from the 1920's, and it applies to many other mapping class groups. It is interesting to speculate whether Nielsen himself might have solved the problem, if asked whether braid groups are left-orderable in the following language: Does the mapping class group of an n-punctured disk act effectively on the real line by order-preserving homeomorphisms? Nielsen had laid all the groundwork for an affirmative answer.

More recently, a new topological approach using laminations was proposed in [**73**]. In common with the Mosher normal form of [**158**], it relies on using triangulations as a sort of coordinate system. Also, a combinatorial interpretation of the results of [**183**] was proposed by Jonathon Funk in [**93**], including a connection with the theory of topoi.

Braid groups are known to be automatic [**189**]. Without burdening the reader with technical details, it should be mentioned that the ordering of B_n and certain other surface mapping class groups (nonempty boundary) can be considered automatic as well, meaning roughly that it may be determined by some finite-state automaton [**182**].

Very recently [**62, 90**], the alternating decomposition—described in Chapter VII—has greatly improved our understanding of the well-ordering of the monoid B_n^+ and allowed for its extension to the dual braid monoid B_n^{+*}.

Unlike the full braid groups, the *pure* braid groups, PB_n, can be bi-ordered [**124**], by an ordering which also well-orders pure positive braids—but proves to be much simpler than the above well-ordering of B_n^+. The argument relies on a completely different approach, namely using the Magnus representation of a free group, and the fact that PB_n is a semidirect product of free groups. Subsequent work has shown that the braid groups B_n and the pure braid groups PB_n are very different from the point of view of orderability: in particular, for $n \geqslant 5$, no left-ordering of B_n can bi-order a subgroup of finite index, such as PB_n. This was proved independently in [**180**] and [**71**].

A convergence of approaches

As will be recalled in Chapter III, the orderability of a group implies various structural consequences about that group and derived objects. The fact that B_n is left-orderable implies that it is torsion-free, which had been well known. However, it also implies that the group ring $\mathbb{Z}B_n$ has no zero-divisors, which was a natural open question. Bi-orderability of PB_n shows that $\mathbb{Z}PB_n$ embeds in a skew field. In addition, it easily implies that the group PB_n has unique roots, a result proved in [**8**] by complicated combinatorial arguments, and which is definitely not true for B_n.

One may argue that such general results did not dramatically change our understanding of braid groups. The main point of interest, however, is not—or not only—the mere existence of orderings on braid groups, but the particular nature and variety of the constructions we shall present. Witness the beautiful way the ordering of PB_n is deduced from the Magnus expansion in Chapter XV, the fascinating connection between the uncountable family of orderings on B_n constructed in Chapter XIII and the Nielsen–Thurston theory, and, chiefly, the specific properties of one particular ordering on B_n. Here we refer to the ordering of B_n sometimes called the Dehornoy ordering in literature, which will be called the σ-ordering in this text.

Typically, it is the specific form of the braids greater than 1 in the σ-ordering that led to the new, efficient algorithms for the classical braid isotopy problem described in Chapters V, VII, and VIII, and motivated the further study of the algorithms described in Chapters X and XII. But what appears to be of the greatest interest here is the remarkable convergence of many approaches to one and the same object: many different points of view end up with the σ-ordering of braids, and this, in our opinion, is the main hint that this object has an intrinsic interest. Just to let the reader feel the flavour of some of the results, we state below various characterizations of the σ-ordering—the terms will be defined in the appropriate place. So, the braid β is smaller than the braid β' in the σ-ordering if and only if

- in terms of braid words—the braid $\beta^{-1}\beta'$ has a braid word representative where the generator σ_i with smallest index appears only positively (no σ_i^{-1});

• in terms of an action on a self-distributive system—for every ordered LD-system $(S, *, \prec)$, and for every sequence $\boldsymbol{x}$ in S, we have $\boldsymbol{x} \bullet \beta \prec^{\text{Lex}} \boldsymbol{x} \bullet \beta'$;

• in terms of braid word combinatorics—any sequence of handle reductions from any braid word representing $\beta^{-1}\beta'$ ends up with a σ-positive word;

• in terms of Φ-splittings, assuming that β, β' belong to B_n^+—the Φ_n-splitting of β is ShortLex-smaller than that of β';

• in terms of ϕ-splittings, assuming that β, β' belong to B_n^{+*}—the ϕ_n-splitting of β is ShortLex-smaller than that of β';

• in terms of automorphisms of a free group—for some i, the automorphism associated with $\beta^{-1}\beta'$ maps x_j to x_j for $j < i$, and it maps x_i to a word that ends with x_i^{-1};

• in terms of free group ordering—we have $\beta(z_n) \lhd \beta'(z_n)$ in $F_\infty \setminus \{1\}$;

• in terms of mapping class groups—the standardized curve diagram associated with β' first diverges from the one associated with β towards the left;

• in terms of $\mathbb{Z}^{2n}$-coordinates—the first nonzero coefficient of odd index in the sequence $(0, 1, \dots, 0, 1) \bullet \beta^{-1}\beta'$ is positive;

• in terms of Mosher's normal form—the last flip of the Mosher normal form of $\beta^{-1}\beta'$ occurs in the upper half-sphere;

• in terms of hyperbolic geometry—the endpoint of the lifting of $\beta(\Gamma_a)$ is smaller (as a real number) than the endpoint of the lifting of $\beta'(\Gamma_a)$.

Even if the various constructions of the σ-ordering depend on choosing a particular family of generators for the braid groups, namely the Artin generators σ_i, this convergence might suggest calling this ordering canonical or, at least, standard. This convergence is the very subject of this text: our aim here is not to give a complete study of any of the different approaches, but to try to let the reader feel the flavour of these different approaches. More precisely—and with the exceptions of Chapters XIII and XIV which deal with more general orderings, and of Chapter XV which deals with ordering pure braids—our aim will be to describe the σ-ordering of braids in the various possible frameworks: algebraic, combinatorial, topological, geometric, and to see which properties can be established by each technique.

As explained in Chapter II, exactly three properties of braids, called **A**, **C**, and **S** here, are crucial to prove that the σ-ordering exists and to establish its main properties. Roughly speaking, each chapter of the subsequent text—except Chapters XIII, XIV, and XV—will describe one possible approach to the question of ordering the braids, and, in each case, explain which of the properties **A**, **C**, and **S** can be proved: some approaches are relevant for establishing all three properties, while others enable us only to prove one or two of them, possibly assuming some other one already proved. We emphasize that, although these properties are established in various contexts and by very different means, there is certainly no circular reasoning involved.

However, the point of this book is not merely to prove and reprove the existence of the braid ordering. Each of the chapters gives a different viewpoint which adds new colours to our description and provides further results.

Organization of the text

Various equivalent definitions of the braid groups are described in Chapter I. The σ-ordering of braids is introduced in Chapter II, where its general properties are discussed. A number of curious examples are presented, showing that the σ-ordering has some quite unexpected properties. The well-ordering of B_n^+ is also introduced in this chapter.

Chapter III presents various applications of the braid ordering. This includes purely algebraic consequences of orderability, such as the zero-divisor conjecture, but also more specific applications following from the specific properties of the σ-ordering, such as a faithfulness criterion of representations and efficient solutions to the word problem. We point out that the braid groups provide interesting examples and counterexamples in the theory of ordered groups. In addition, we outline some applications to knot theory, the theory of pseudo-characters, and certain unprovability results arising in braid theory.

The chapters which follow contain various approaches to the orderability phenomenon. The combinatorial approaches are gathered in Chapters IV through IX, while the topological approaches are presented in Chapters X to XIV.

Chapter IV introduces left self-distributive algebraic systems (LD-systems) and the action of braids upon such systems. This is the technique whereby the orderability of braids was first demonstrated and the σ-ordering introduced. The chapter sketches a self-contained proof of left-orderability of B_n by establishing Properties **A**, **C** and **S** with arguments utilizing LD-systems. Here we consider colourings of the strands of the braids and observe that the braid relations dictate the self-distributive law among the colours. Then we can order braids by choosing orderable LD-systems as colours, a simple idea, although the existence of an orderable LD-system requires a sophisticated argument. The chapter concludes with a short discussion of the historical origins of orderable LD-systems in set theory.

A combinatorial algorithm called handle reduction is the subject of Chapter V. This procedure, which extends the idea of word reduction in a free group, is a very efficient procedure in practice for determining whether a braid word represents a braid larger than 1, and incidentally gives a rapid solution to the word problem in the braid groups. Handle reduction gives an alternative proof of Property **C**, under the assumption that Property **A** holds.

The deep structure of the braid groups discovered by Garside [**95**], and its connection with the σ-ordering, are discussed in Chapter VI. The relationship is not a simple one, but investigating the ordering of the divisors of Δ_n^d in B_n^+ leads to new insights into Solomon's descent algebras. A complete description is obtained in the case of 3-strand braids, leading to a new proof of Property **C** in that case.

Some quite recent developments in braid orderings are contained in Chapters VII and VIII, which describe normal forms for braids which are more compatible with the braid ordering than the greedy form associated with Garside theory. Chapter VII begins with an inductive scheme called the Φ_n-splitting, which yields a decomposition of every positive n-strand braid into a finite sequence of positive $(n-1)$-strand braids. The main result is that, under this decomposition, the ordering of B_n^+ is a simple lexicographic extension of the ordering of B_{n-1}^+. Most results of the chapter rely on techniques due to S. Burckel in [**27**] for encoding positive braid words by finite trees, but they are used here as a sort of black box, and their proofs are omitted.

The dual braid monoids introduced by Birman, Ko and Lee are at the heart of Chapter VIII. The monoid B_n^{+*}, which properly includes the monoid B_n^+ for $n \geqslant 3$, also has a good divisibility structure. As in Chapter VII, one introduces the notion of ϕ_n-splitting, which yields a decomposition of every braid of B_n^{+*} into a finite sequence of braids of B_{n-1}^{+*}. The main result is again that, under this decomposition, the ordering of B_n^{+*} is a lexicographic extension of the ordering of B_{n-1}^{+*}, a recent result of J. Fromentin. The consequences are similar to those of Chapter VII, including a new proof of Property **C**, and a new quadratic algorithm for comparing braids. The main interest of the approach may be that it allows for direct, elementary proofs, by contrast to Burckel's techniques, which use tricky transfinite induction arguments.

Chapter IX contains an approach to the σ-ordering using a very classical fact, that the braid groups can be realized as a certain group of automorphisms of a free group. As observed by David Larue, this method yields a quick proof of Property **A**, a partial (and not so quick) proof of Property **C**, as well as a simple criterion for recognizing whether a braid is σ-positive in terms of its action on the free group. We also outline in this chapter the interpretation developed by Jonathon Funk, in which a certain linear ordering of words in the free group is preserved under the braid automorphisms.

We begin the topological description of the σ-ordering in Chapter X. Here we realize B_n as the mapping class group of a disk with n punctures. The braid action can be visualized by use of curve diagrams which provide a canonical form for the image of the real line, if the disk is regarded as the unit complex disk. This was the first geometric argument for the left-orderability of the braid groups, and it is remarkable that the ordering described in this way is identical with the original, *i.e.*, with the σ-ordering. An advantage of this approach is that it also applies to more general mapping class groups. We emphasize that Chapters IX and X are based on very similar ideas, except that in the first one these ideas are expressed in a more algebraic language, and in the second in a more geometric one.

In Chapter XI we study in detail a topological technique which we already encountered in Chapter X for explicitly constructing σ-consistent representative braid words of any given element of the braid group, namely the method of untangling, or relaxing, curve diagrams. We discuss two examples of such algorithms: one due to Bressaud in [**19**], which has a fascinating alternative description in terms of finite state automata and word rewriting systems, and another one from [**74**] which leads to a deeper understanding of the connection between the length of a braid word and the complexity of the curve diagram of a braid. This approach can also be interpreted in terms of Teichmüller geometry [**178**].

Chapter XII continues the discussion of the σ-ordering in terms of mapping classes. However, here the geometric approach is rephrased in combinatorial terms by use of two somewhat different devices involving triangulations. The first approach, developed in [**73**], uses integral laminations. One encodes the action of a braid on the disk by counting intersections of the image of a certain triangulation with a lamination. This leads to the shortest proof of Property **A** known so far, and yet another characterization of braids larger than 1 in the σ-ordering. The second was inspired by the technique employed by Lee Mosher to establish that mapping class groups are automatic. It develops a new canonical form for braids and a method for determining σ-ordering by means of a finite state automaton.

The discussion in Chapter XIII interprets braid orderings in terms of Nielsen–Thurston theory. The key observation is that the universal cover of the punctured disk has a natural embedding in the hyperbolic plane. Thereby, braids act on a family of hyperbolic geodesics, which have a natural ordering. This point of view provides an infinitude of inequivalent orderings of braid groups and many other mapping class groups. The σ-ordering on B_n corresponds to choosing a particular geodesic in $\mathbb{H}^2$.

A recent, and quite different, topological approach to orderability is taken in Chapter XIV. Here, one considers the set of *all* left orderings of a group G, and of B_n in particular. This set is given a natural topology, forming the space $LO(G)$, which in general is compact and totally disconnected. We study the structure of $LO(B_n)$ and use this global topological approach to show that there are uncountably many essentially different left-orderings of B_n for $n \geqslant 3$, all of which provide well-orderings of B_n^+, a phenomenon noted in Chapter XIII by completely different methods.

Chapter XV is an account of an ordering of the *pure* braid groups. Unlike the full braid groups, the groups PB_n of pure braids can be given an ordering which is invariant under multiplication on both sides. The one we investigate here is defined algebraically, using the Artin combing technique, together with a specific ordering of free groups using the Magnus expansion. This ordering—which is *not* the restriction of the σ-ordering to pure braids—has the nice property that non-trivial braids in $PB_n \cap B_n^+$ are larger than 1 and well-ordered.

The final chapter contains a number of open questions related to braid orderings. Various extensions of the ideas presented in the other chapters are also discussed there.

Guidelines to the reader

An attempt has been made to keep the chapters relatively self-contained. So, apart from Chapters I, II, and XVI, all chapters are parallel one to the other rather than logically interdependent; therefore, after the first chapters, the reader can take the chapters essentially in whatever order he or she likes.

We mentioned that three properties of braids play a crucial role, namely those called **A**, **C**, and **S**, whose statements, as well as other basic definitions, are recalled at the end of the book, page 309. One of our main tasks in this text will be to prove these properties using various possible approaches. In spite of the above general remarks, it might be useful that we propose answers to the question, Which of these approaches offers the quickest, or the most elementary, proof of Properties **A**, **C** and **S**? The answer depends of course on the mathematical preferences of the reader. As for Property **A**, the shortest proofs are the one using the automorphisms of a free group in Chapter IX, and—even shorter once the curious formulas (XII.1.1) have been guessed—the one using laminations in Chapter XII.

As for Property **C**, the shortest argument is probably the one involving self-distributivity as outlined in Chapter IV, but one may prefer the approach through the handle reduction method of Chapter V, which uses nothing exotic and gives an efficient algorithm in addition, or the curve diagram approach of Chapter X, which gives a less efficient method and requires considerable effort to be made rigorous, but appeals to a natural geometric intuition.

Finally, for Property **S**, the hyperbolic geometry argument of Chapter XIII is probably the most general one, as it gives the result not only for the σ-ordering, but also for a whole family of different orderings. On the other hand, even if they may appear intricate, the combinatorial approaches of Chapters VII and VIII give the most precise and effective versions.

Although they are conceptually simple, the braid groups are very subtle non-Abelian groups which have given up their secrets only reluctantly over the years. They will undoubtedly continue to supply us with surprises and fascination, and so will in particular their orderings: despite the many approaches and results mentioned in this book, a lot of questions about braid orderings remain open today, and further developments can be expected. For the moment, we hope that this text, which involves techniques of algebra, combinatorics, hyperbolic geometry, topology, and has even a loose connection with set theory, can illuminate some facets of the ordering of braids.

Acknowledgements

We thank all colleagues and friends who suggested corrections or improvements, including Ryuji Abe, Marc Autord, Lluís Bacardit, Benjamin Beeker, Xavier Bressaud, Jérémy Chamboredon, Adam Clay, Laurent Demonet, Warren Dicks, Roger Fenn, Jean Fromentin, Jonathon Funk, Etienne Ghys, Pierre Gillibert, Eddy Godelle, John Guaschi, Christian Kassel, Gilbert Levitt, Peter Linnell, Andrey Maliutyn, Hiroshi Matsuda, Andrés Navas, Luis Paris, Andy Putman, Chloé Périn, and Hervé Sibert.

In addition, we would like to express our appreciation to the Société Mathématique de France, especially Nathalie Christiaën, and to the American Mathematical Society for their help and encouragement.

CHAPTER I

Braid Groups

In this introductory chapter, we briefly explain how the groups B_n arise in several contexts of geometry and algebra, and we mention a few basic results that will be frequently used in the sequel. Our purpose here is not to be exhaustive, and many possible approaches are not mentioned—for instance the connection with configuration spaces. All results in this chapter are classical, and we refer to textbooks for most of the proofs; see for instance [**123**], [**118**], [**14**], or [**176**].

The organization is as follows. In Section 1, we start with the Artin presentation of the group B_n in terms of generators and relations. In Section 2, we describe the connection with the geometric viewpoint of isotopy classes of families of intertwining strands. In Section 3, we address the braid group as the mapping class group of a punctured disk. Finally, in Section 4, we introduce the monoid of positive braids and mention some basic results from Garside's theory.

1. The Artin presentation

Here, we introduce the braid group B_n using the abstract presentation already mentioned in the Introduction, due to E. Artin [**4**].

1.1. Braid relations. Braid groups can be specified using a standard presentation.

DEFINITION 1.1. For $n \geqslant 2$, the *n-strand braid group* B_n is defined to be the presented group

$$\left\langle \sigma_1, \dots, \sigma_{n-1} \;\middle|\; \begin{array}{lll} \sigma_i\sigma_j = \sigma_j\sigma_i & \text{for} & |i-j| \geqslant 2 \\ \sigma_i\sigma_j\sigma_i = \sigma_j\sigma_i\sigma_j & \text{for} & |i-j| = 1 \end{array} \right\rangle. \tag{1.1}$$

The elements of B_n are called *n-strand braids*. The braid group on infinitely many strands, denoted B_∞, is defined by a presentation with infinitely many generators $\sigma_1, \sigma_2, \ldots$ subject to the same relations.

Clearly, the identity mapping on $\{\sigma_1, \dots, \sigma_{n-1}\}$ extends into a homomorphism of B_n to B_{n+1}. It can be proved easily—and it will be clear from the geometric interpretation of Section 2—that this homomorphism is injective, and, therefore, we can identify B_n with the subgroup of B_∞ generated by $\sigma_1, \dots, \sigma_{n-1}$. This is the point of view we shall always adopt in the sequel.

1.2. Braid words. According to Definition 1.1, every braid admits decompositions in terms of the generators σ_i and their inverses. A word on the letters $\sigma_1^{\pm 1}, \dots, \sigma_{n-1}^{\pm 1}$ is called an *n-strand braid word*. The *length* of a braid word w is denoted by $\ell(w)$. If the braid β is the equivalence class of the braid word w, we say that w *represents* β, or is an *expression* of β, and we write $\beta = \overline{w}$. We say that two braid words are *equivalent* if they represent the same braid, *i.e.*, if they are equivalent with respect to the least congruence that contains the relations of (1.1). As, for instance, the braid word $\sigma_1^k\sigma_1^{-k}$ represents the unit braid for every k, each braid admits infinitely many representative braid words.

2. Isotopy classes of braid diagrams

We now connect the abstract point of view of Section 1 with the concrete intuition of braids as strands that are intertwined.

2.1. Geometric braids. We denote by D^2 the unit disk with centre 0 in the plane $\mathbb{R}^2$ identified with the complex line $\mathbb{C}$, and by D_n the disk D^2 with n regularly spaced points in the real axis as distinguished points; we call these points the *puncture points* of D^2.

DEFINITION 2.1. We define an *n-strand geometric braid* to be an embedding b of the disjoint union $\coprod_{j=1}^n [0_j, 1_j]$ of n copies of the interval $[0,1]$ into the cylinder $[0,1]\times D^2$ satisfying the following properties:

• for t in $[0_j, 1_j]$, the point $b(t)$ lies in $\{t\}\times D^2$;

• the set $\{b(0_1), \dots, b(0_n)\}$ is the set of punctures of $\{0\}\times D^2$, and similarly the set $\{b(1_1), \dots, b(1_n)\}$ is the set of punctures of $\{1\}\times D^2$.

The image of each interval is called a *strand* of the braid. The idea is that, visualizing the unit interval as being horizontal, we have n strands running continuously from left to right, intertwining, but not meeting each other.

Each geometric braid b determines a permutation π in the symmetric group $\mathfrak{S}_n$ as follows. Label the punctures $P_1, \dots, P_n$. Then we take $\pi(i) = j$ if the strand of b that ends at $P_i\times\{1\}$ begins at $P_j\times\{0\}$.

A geometric braid whose permutation is trivial is called *pure*.

2.2. The group of isotopy classes. To obtain the connection with B_n, and in particular to be able to obtain a group structure, we appeal to isotopies for identifying geometric braids that are topologically equivalent.

DEFINITION 2.2. Two geometric braids b, b' are said to be *isotopic* if there is a continuous $[0,1]$-family of geometric braids b_t with $b_0 = b$ and $b_1 = b'$.

The geometric idea is that we can deform one braid into the other while holding their endpoints fixed. Note that isotopic braids induce the same permutation.

There exists a natural way of defining a product of two geometric braids using concatenation: Given two geometric braids b_1 and b_2, we squeeze the image of b_1 into the cylinder $[0,\frac{1}{2}]\times D^2$, the image of b_2 into $[\frac{1}{2},1]\times D^2$, and obtain a new, well-defined geometric braid $b_1\cdot b_2$. Clearly, this product is compatible with isotopy, hence it induces a well-defined operation on the set of isotopy classes of geometric braids.

LEMMA 2.3. *For each n, the set of isotopy classes of n-strand geometric braids equipped with the above product is a group.*

PROOF (SKETCH). The neutral element of the group is the isotopy class of the trivial geometric braid, whose strands are just straight line segments not intertwining each other. The inverse of the class of a geometric braid is the class of its reflection in the disk $\{\frac{1}{2}\}\times D^2$. □

We shall see in a moment that this group is isomorphic to the group B_n of Section 1.

Note that the function that to every braid associates the permutation it induces on the set of punctures yields a homomorphism of the group of isotopy classes of n-strand geometric braids to the symmetric group $\mathfrak{S}_n$. Our definition of the permutation associated with a braid was made precisely to obtain a homomorphism, and not an anti-homomorphism. Here we regard permutations as acting on the left, consistent with the convention for mapping classes, to be discussed in Section 3.

PROPOSITION 2.4. *The group of isotopy classes of n-strand geometric braids is isomorphic to the group B_n.*

PROOF (SKETCH). We define a homomorphism from B_n to our group of isotopy classes by sending the generator σ_i to the class of the clockwise half-twist braid involving the ith and $(i+1)$st strand indicated in Figure 1—in this picture, the cylinder has not been drawn, for simplicity, and our picture represents a side-view of the braid.

The figure also illustrates the fact that this homomorphism is well defined. Indeed, in our group of isotopy classes, we have that crossings which are far apart commute—so that we have $\sigma_i\sigma_j = \sigma_j\sigma_i$ for $|i-j| \geqslant 2$—and the Reidemeister III-type relation $\sigma_i\sigma_{i+1}\sigma_i = \sigma_{i+1}\sigma_i\sigma_{i+1}$ holds.

We leave it to the reader to verify that this homomorphism is surjective—any geometric braid can be deformed into one in which a side-view offers only finitely many crossings, all of which are transverse—and injective—our two types of relations suffice to relate any two braid diagrams representing isotopic geometric braids. The proofs, which we shall not discuss here, are similar to the proof that isotopy classes of knots are the same as knot diagrams up to Reidemeister-equivalence [**179**]. Details can be found in [**123**]. □

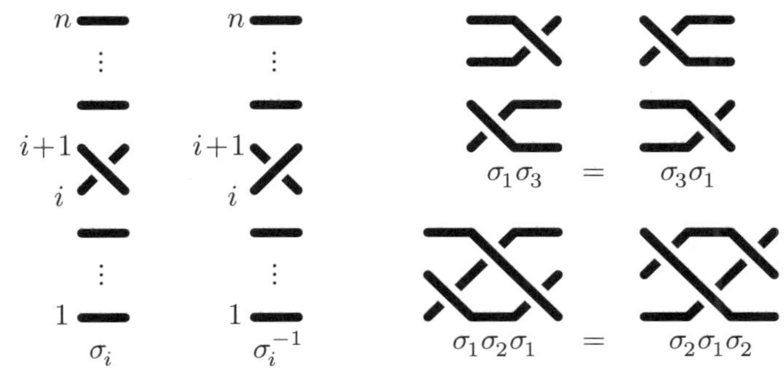

FIGURE 1. The Artin generators σ_i and σ_i^{-1} of the braid group B_n and their relations, when realized as geometric braids.

This completes the definition of B_n in terms of n-strand geometric braids. From now on, we shall no longer distinguish between B_n and the group of isotopy classes of n-strand geometric braids.

We remark that the above definition has a very natural generalization: We can replace the disk D^2 by any compact surface $\mathcal{S}$, possibly with boundary. Choosing n puncture points in $\mathcal{S}$, we can define the *n-strand braid group of the surface $\mathcal{S}$* to be the group of n-strand braids in $\mathcal{S} \times [0,1]$; see Section 3.2 of Chapter XVI.

3. Mapping class groups

We shall now identify the braid group B_n with the group of homotopy classes of self-homeomorphisms of an n-punctured disk. The idea is simply to look at braids from one end rather than from the side.

3.1. Homeomorphisms of a surface. Let $\mathcal{S}$ be an oriented compact surface, possibly with boundary, and let $\mathcal{P}$ be a finite set of distinguished interior points of $\mathcal{S}$. The most important example in this section will be the n-punctured disk D_n.

DEFINITION 3.1. The *mapping class group* $\mathcal{MCG}(\mathcal{S},\mathcal{P})$ of the surface $\mathcal{S}$ relative to $\mathcal{P}$ is the group of all isotopy classes of orientation-preserving self-homeomorphisms of $\mathcal{S}$ that fix $\partial\mathcal{S}$ pointwise and preserve $\mathcal{P}$ globally.

This means that any homeomorphism φ from $\mathcal{S}$ to itself and taking punctures to punctures represents an element of the mapping class group, provided it acts as the identity on the boundary of $\mathcal{S}$. Note that the punctures may be permuted by φ. Two homeomorphisms φ, ψ represent the same element if and only if they are isotopic through a family of boundary-fixing homeomorphisms which also fix $\mathcal{P}$. They will then induce the same permutation of the punctures.

CONVENTION 3.2. In the sequel, the product we consider on a mapping class group $\mathcal{MCG}(\mathcal{S},\mathcal{P})$ is composition: $\varphi\psi$ simply means "first apply ψ, then φ".

We remark that, in the previous paragraph, the word "isotopic" could have been replaced by the word "homotopic": by a theorem of Epstein [**76**], two homeomorphisms of a compact surface are homotopic if and only if they are isotopic.

Mapping class groups, also known as modular groups, play a prominent role in the study of the topology and geometry of surfaces, as well as in 3-dimensional topology. To illustrate the difficulty of understanding them, we note that simply proving that they admit finite—and in fact quite elegant—presentations already requires deep arguments [**105, 193**].

3.2. Connection with geometric braids. Our aim now is to sketch a proof of the following result; for details see, for instance, [**14**] or [**123**]:

PROPOSITION 3.3. *There is an isomorphism of B_n with $\mathcal{MCG}(D_n)$.*

PROOF (SKETCH). We outline a proof that $\mathcal{MCG}(D_n)$ is naturally isomorphic to the group of isotopy classes of geometric braids defined above. Let b be an n-strand geometric braid, sitting in the cylinder $[0,1]\times D^2$, whose n strands are starting at the puncture points of $\{0\}\times D_n$ and ending at the puncture points of $\{1\}\times D_n$. Then b may be considered to be the graph of the motion, as time goes from 1 to 0, of n points moving in the disk, starting and ending at the puncture points—according to Convention 3.2, letting time go from 0 to 1 would lead to an anti-isomorphism. It can be proved that this motion extends to a continuous family of homeomorphisms of the disk, starting with the identity and fixed on the boundary at all times. The end map of this isotopy is the corresponding homeomorphism $\varphi\colon D_n \to D_n$, which is well defined up to isotopy fixed on the punctures and the boundary.

Conversely, given a homeomorphism $\varphi\colon D_n \to D_n$, representing some element of the mapping class group, we want to get an n-strand geometric braid. By a well-known trick of Alexander, every homeomorphism of a disk that fixes the boundary

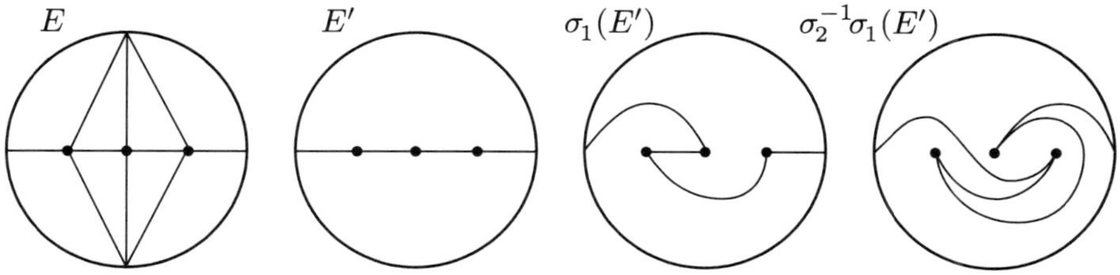

FIGURE 2. Two possible curve diagrams on D_n, and the image of one of them under the homeomorphisms σ_1 and $\sigma_2^{-1}\sigma_1$.

is isotopic to the identity, through homeomorphisms fixing the boundary. The corresponding braid is then the graph of the restriction of such an isotopy to the puncture points. Again, we must regard the isotopy parameter as going from 1 to 0 here. □

3.3. Curve diagrams. In order to visualize a mapping of a surface, it is useful to consider images of certain subsets of the surface. Let E be a diagram on the surface $\mathcal{S}$, consisting of a finite number of disjoint, properly embedded arcs, meaning the arcs terminate either on $\partial\mathcal{S}$ or in a puncture of $\mathcal{S}$. Suppose in addition that E fills $\mathcal{S}$, in the sense that the interiors of all components of the surface obtained by cutting $\mathcal{S}$ along the arcs of E are homeomorphic to open disks. Typical examples in the case $\mathcal{S} = D_n$ are the standard triangulation, as well as the collection of $n+1$ horizontal line segments indicated in Figure 2. Then the isotopy class of a homeomorphism $\varphi\colon \mathcal{S} \to \mathcal{S}$ is uniquely determined by the isotopy class of the diagram $\varphi(E)$. This fact is also illustrated in Figure 2.

It is also well known that a homeomorphism of D_n can be recovered up to homotopy from the induced isomorphism of the fundamental group $\pi_1(D_n, *)$, where $*$ is a fixed point of the boundary ∂D_n. The group $\pi_1(D_n, *)$ is a free group on n generators, say F_n. So we obtain an embedding $B_n \cong \mathcal{MCG}(D_n) \to \mathrm{Aut}(F_n)$, which can be written explicitly if we choose a base point $*$ and generators of $\pi_1(D_n, *)$. Two choices will play an important role in this text, namely when the base point $*$ is the leftmost point of the disk, and the generators are the loops $x_1, \dots, x_n$ in the first case, and $y_1, \dots, y_n$ in the second case, as shown in Figure 3.

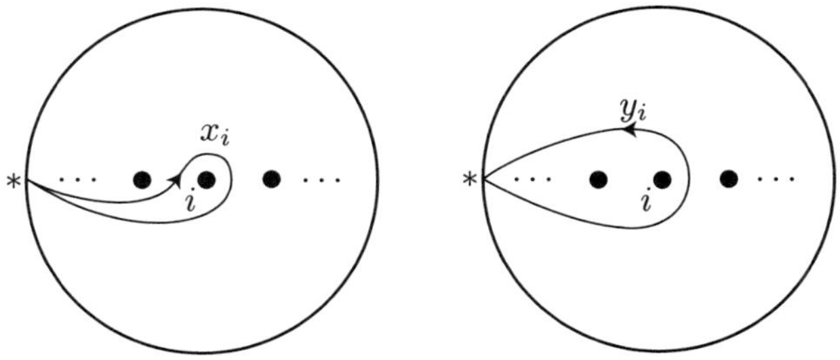

FIGURE 3. Two choices for the generators for the fundamental group of D_n.

4. Positive braids

A useful perspective is to restrict attention to a special class of braids, namely *positive* braids. This approach turns out to be extremely fruitful because the braid monoids that appear in this way turn out to have a very rich theory, based on Garside's seminal work of [**95**].

4.1. Braid monoids. As the relations of (1.1) involve no negative letter σ_i^{-1}, they also define a monoid. So, as we did in Section 1, we can introduce the following abstract definition:

DEFINITION 4.1. The *positive braid monoid* B_n^+ is defined to be the monoid that admits, as a monoid, the presentation (1.1). The elements of B_n^+ are called *positive braids.*

So the elements of B_n^+ are represented by words in the letters σ_i, but not σ_i^{-1}. Such words are called *positive.* As the relations of (1.1) preserve the word length, all positive braid words representing a given positive braid β have the same length, denoted $\ell(\beta)$—of course, the same property does not hold in the full braid group, as, there, the length two word $\sigma_1\sigma_1^{-1}$ is equivalent to the length zero empty word. In particular because of the previous seemingly trivial observation, the positive braid monoid is often easier to handle than the full braid group. Note that the length function is a morphism of the *monoid* B_n^+ to the monoid $(\mathbb{N}, +)$.

Geometrically, we may think of B_n^+ as the monoid of geometric braids with only positive crossings in their diagram, up to positive isotopy, where the isotopies are deformations through a family of braids that are again positive in the same sense.

The reason why this point of view is so fruitful is the following fundamental result due to Garside; for a proof, see, for instance, [**52**] or [**123**], or see [**95**] itself:

PROPOSITION 4.2. *The canonical mapping of B_n^+ to B_n is injective.*

Equivalently, if two positive geometric braids are isotopic, then they are isotopic through a family of positive braids. So the braid monoid embeds in the braid group, and it identifies with the subset of B_n consisting of braids which are representable by words in the letters σ_i, not using any σ_i^{-1}.

4.2. The braids δ_n and Δ_n. For each n, two particular positive n-strand braids play a fundamental role in the study of B_n^+ and, more generally, of B_n, namely the so-called *fundamental* braids δ_n and Δ_n. We also introduce the associated conjugacy automorphisms.

DEFINITION 4.3 (Figure 4). We define $\delta_1 = \Delta_1 = 1$ and, for $n \geqslant 2$,

$$\delta_n = \sigma_1\sigma_2 \ldots \sigma_{n-1} \quad \text{and} \quad \Delta_n = \delta_n\delta_{n-1} \ldots \delta_2. \tag{4.1}$$

We define the *cycling* automorphism ϕ_n to be the conjugation by δ_n, and the *flip* automorphism Φ_n to be the conjugation by Δ_n, *i.e.*, for β in B_n,

$$\phi_n : \beta \mapsto \delta_n\, \beta\, \delta_n^{-1} \quad \text{and} \quad \Phi_n : \beta \mapsto \Delta_n\, \beta\, \Delta_n^{-1}. \tag{4.2}$$

The following formulas can be read from Figure 4.

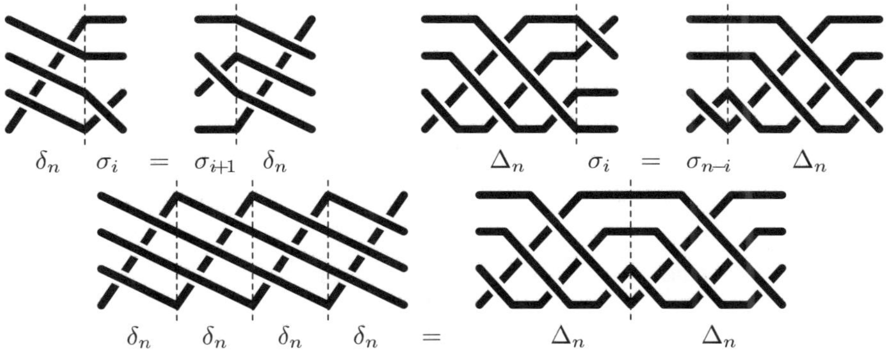

FIGURE 4. The braids δ_n, Δ_n, and their relations: δ_n corresponds to shifting the strands, while Δ_n corresponds to a global half-twist of the n strands.

LEMMA 4.4. *For each n, we have*

$$\delta_n^n = \Delta_n^2 \quad \text{and} \quad \phi_n^n = \Phi_n^2 = \mathrm{id}_{B_n}, \tag{4.3}$$

$$\phi_n(\sigma_i) = \sigma_{i+1} \quad \text{for } 1 \leqslant i \leqslant n-2, \tag{4.4}$$

$$\Phi_n(\sigma_i) = \sigma_{n-i} \quad \text{for } 1 \leqslant i \leqslant n-1. \tag{4.5}$$

Formula (4.4) implies that the generators σ_i of B_n all are pairwise conjugate, whereas (4.5) explains the name "flip automorphism" for Φ_n: in a braid diagram, Φ_n corresponds to a horizontal symmetry. Its being involutive means that the braid Δ_n^2 belongs to the centre of B_n. Actually, it is known [**36**] that, for $n \geqslant 3$, the centre of B_n consists of the powers of Δ_n^2 only. As for ϕ_n, formula (4.3) shows that it has order n, and it should rather be seen as a rotation by $2\pi/n$; see Chapter VIII for details.

4.3. Fractionary decompositions (I). The study of B_n^+ is fundamental for the understanding of B_n, because B_n is a group of fractions for B_n^+, *i.e.*, every braid is a quotient of two positive braids. Here we consider particular decompositions in which the denominator is a power of the braid Δ_n.

First, using an induction on n and playing with the braid relations, one easily obtains:

LEMMA 4.5. *For each i between 1 and $n-1$, the braid Δ_n can be expressed by a positive word that begins with σ_i.*

Then we deduce:

PROPOSITION 4.6. *Every braid β of B_n admits a decomposition $\beta = \Delta_n^{-2p}\beta'$ with $p \geqslant 0$ and $\beta' \in B_n^+$.*

PROOF. By Lemma 4.5, for each i in $\{1, \dots, n-1\}$, we can choose a positive braid word w_i such that $\sigma_i w_i$ represents Δ_n^2, *i.e.*, w_i represents $\sigma_i^{-1}\Delta_n^2$.

Let β be an arbitrary braid in B_n, and let w be an n-strand braid word representing β. Let p be the number of negative letters (letters σ_i^{-1}) in w. Let w' be the positive braid word obtained from w by replacing each letter σ_i^{-1} with w_i.

As w_i represents $\sigma_i^{-1}\Delta_n^2$ and Δ_n^2 commutes with every braid, the word w' represents $\Delta_n^{2p}\beta$, and we deduce $\beta = \Delta_n^{-2p}\beta'$, where β' is the positive braid represented by w'. □

A priori, the fractionary decomposition provided by Proposition 4.6 is not unique. Actually, we can obtain uniqueness by demanding that the exponent of Δ_n is minimal: every braid β in B_n admits a *unique* decomposition $\beta = \Delta_n^{-d}\beta'$ with $d \in \mathbb{Z}$, $\beta' \in B_n^+$ and d minimal such that such a decomposition exists.

4.4. The lattice of divisibility. In order to go further, in particular both to establish Proposition 4.2 and to improve Proposition 4.6, one needs to know more about the monoid B_n^+ and its divisibility relations, as described by Garside's theory.

DEFINITION 4.7. For β, β' in B_n, we say that β' is a *left divisor* of β, denoted $\beta' \preccurlyeq \beta$, if $\beta = \beta'\gamma$ holds for some γ in B_n^+.

Symmetrically, we say that β' is a *right* divisor of β if we have $\beta = \gamma\beta'$ for some γ in B_n^+, but we shall not introduce specific notation.

It can be noted that, for β, β' in B_n, every positive braid γ possibly satisfying $\beta = \beta'\gamma$ has to lie in B_n^+ and, therefore, the divisibility relation in B_n is the restriction of that of B_∞: there is no need to worry about the index n.

In the monoid B_n^+, no element except 1 is invertible, so the divisibility relation $\preccurlyeq$ is a partial ordering. The main result of Garside's theory [**95**] is that, for each n, the poset $(B_n^+, \preccurlyeq)$ is a lattice:

PROPOSITION 4.8 ([**77**, Chapter 9]). *Any two positive braids admit a greatest common left divisor (gcd) and a least common right multiple (lcm).*

Using Proposition 4.6, one can deduce that the poset $(B_n, \preccurlyeq)$ is also a lattice.

4.5. Fractionary decompositions (II). Proposition 4.6 gives a specific role to the powers of the braid Δ_n, and the fractionary decomposition it gives is, in general, not irreducible, in that the numerator and the denominator may have non-trivial common divisors. Garside's results enable us to improve the result. The next refinement says that, among all fractionary decompositions of a braid, there is a distinguished one through which every fractionary decomposition factorizes.

PROPOSITION 4.9. *For each braid β in B_n, there exists a unique pair of positive braids (β_1, β_2) such that $\beta = \beta_1^{-1}\beta_2$ holds and 1 is the only common left divisor of β_1 and β_2 in B_n^+. Then, for each decomposition $\beta = \gamma_1^{-1}\gamma_2$ with γ_1, γ_2 in B_n^+, we have $\gamma_1 = \gamma\beta_1$ and $\gamma_2 = \gamma\beta_2$ for some γ in B_n^+.*

PROOF. Let β belong to B_n, and let $\beta = \beta_1^{-1}\beta_2$ be a fractionary decomposition of β, *i.e.*, we assume $\beta_1, \beta_2 \in B_n^+$, such that $\ell(\beta_1) + \ell(\beta_2)$ is minimal. As the length function takes its values in $\mathbb{N}$, such a pair must exist. First, we observe that 1 is the only common left divisor of β_1 and β_2 since, if we have $\beta_1 = \gamma\beta_1'$ and $\beta_2 = \gamma\beta_2'$, then $\beta_1'^{-1}\beta_2'$ is another decomposition of β satisfying $\ell(\beta_1') + \ell(\beta_2') = \ell(\beta_1) + \ell(\beta_2) - 2\ell(\gamma)$, hence $\gamma = 1$ by the choice of β_1 and β_2.

Assume that $\beta = \gamma_1^{-1}\gamma_2$ is any fractionary decomposition of β with $\gamma_1, \gamma_2 \in B_n^+$. Using the existence of gcd's and lcm's in B_n^+ (Proposition 4.8), we shall prove that $\gamma_1^{-1}\gamma_2$ factors through $\beta_1^{-1}\beta_2$, *i.e.*, that $\gamma_1 = \gamma\beta_1$ and $\gamma_2 = \gamma\beta_2$ holds for some γ in B_n^+. The argument is illustrated on Figure 5. First, by Proposition 4.6, we can find β_0, γ_0 in B_n^+ satisfying $\gamma_0^{-1}\beta_0 = \beta_1\gamma_1^{-1}$. We deduce $\gamma_0\beta_1 = \beta_0\gamma_1$

and $\gamma_0\beta_2 = \beta_0\gamma_2$ in B_n, hence in B_n^+ by Proposition 4.2. Let γ_0' be the least common right multiple of β_0 and γ_0. By construction, we have $\beta_0 \preccurlyeq \gamma_0\beta_1$ and $\gamma_0 \preccurlyeq \gamma_0\beta_1 = \beta_0\gamma_1$, hence $\gamma_0' \preccurlyeq \gamma_0\beta_1$. Similarly, we have $\beta_0 \preccurlyeq \gamma_0\beta_2 = \beta_0\gamma_2$ and $\gamma_0 \preccurlyeq \gamma_0\beta_2$, hence $\gamma_0' \preccurlyeq \gamma_0\beta_2$. We deduce that γ_0' left-divides the left gcd of $\gamma_0\beta_1$ and $\gamma_0\beta_2$. By hypothesis, β_1 and β_2 have no nontrivial common left divisor, so the left gcd of $\gamma_0\beta_1$ and $\gamma_0\beta_2$ is γ_0, and we must have $\gamma_0' \preccurlyeq \gamma_0$, hence $\gamma_0' = \gamma_0$ since, by definition, γ_0' is a right multiple of γ_0. Hence, we have $\gamma_0 = \beta_0\gamma$ for some γ in B_n^+. As we have $\gamma_0\beta_1 = \beta_0\gamma_1$ and $\gamma_0\beta_2 = \beta_0\gamma_2$, the equality $\gamma_0 = \beta_0\gamma$ implies $\gamma_1 = \gamma\beta_1$ and, similarly, $\gamma_2 = \gamma\beta_2$. We obtained the expected factorization result.

Finally, the pair (β_1, β_2) is unique. Indeed, if $\gamma_1^{-1}\gamma_2$ is any fractionary decomposition of β satisfying $\ell(\gamma_1) + \ell(\gamma_2) = \ell(\beta_1) + \ell(\beta_2)$, the above result gives gives a braid γ satisfying $\gamma_1 = \gamma\beta_1$ and $\gamma_2 = \gamma\beta_2$ and the hypothesis on the length implies $\gamma = 1$, hence $\gamma_1 = \beta_1$ and $\gamma_2 = \beta_2$. □

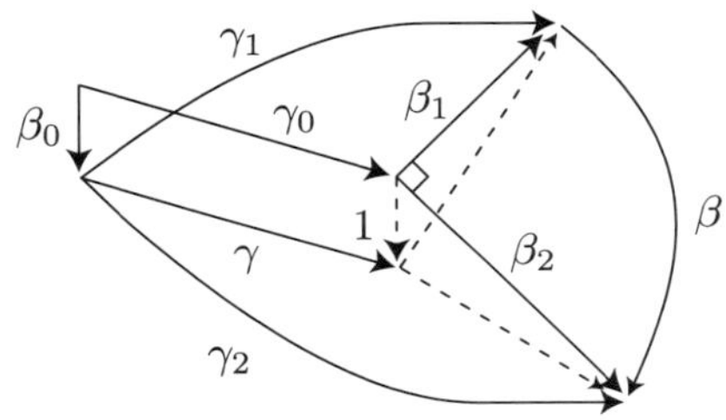

FIGURE 5. Decomposition of a braid into an irreducible fraction: the hypothesis that β_1 and β_2 have no common left divisor implies that every other fractionary decomposition $\gamma_1^{-1}\gamma_2$ factors through $\beta_1^{-1}\beta_2$.

Thus, every braid has a distinguished minimal expression as an irreducible fraction. The situation with B_n and B_n^+ is therefore similar to that of positive rational numbers and their unique expression as an irreducible fraction.

DEFINITION 4.10. For each braid β, the unique braids β_1 and β_2 associated with β by Proposition 4.9 are called the *left denominator* and the *left numerator* of β, and respectively denoted $D_L(\beta)$ and $N_L(\beta)$.

Of course, as B_n is not commutative for $n \geqslant 3$, the position of the denominator matters. Owing to the symmetry of the braid relations, each braid admits a similar irreducible decomposition $N_R(\beta)D_R(\beta)^{-1}$ in which the denominator lies on the right.

CHAPTER II

A Linear Ordering of Braids

In this chapter, we introduce the linear ordering of braids, sometimes called the *Dehornoy ordering*, that is the main subject of this book, and we list its main properties known so far. The construction starts with the notion of a σ-positive braid, and it relies on three basic properties, called **A**, **C**, and **S**, from which the σ-ordering can easily be constructed and investigated. In this chapter, we take Properties **A**, **C**, and **S** for granted, and explore their consequences. The many different proofs of these statements will be found in subsequent chapters.

The chapter is organized as follows. In Section 1, we introduce the σ-ordering and its variant the σ^{Φ}-ordering starting from Properties **A** and **C**. In Section 2, we give many examples of the sometimes surprising behaviour of the σ-ordering, and we introduce Property **S**. In Section 3, we develop global properties of the σ-ordering, involving Archimedian property, discreteness, density, and convex subgroups. Finally, in Section 4, we investigate the restriction of the σ-ordering to the monoid B_n^+ of positive braids, showing that this restriction is a well-ordering, and we give an inductive construction of the σ-ordering of B_n^+ from the σ-ordering of B_{n-1}^+.

CONVENTION. In this chapter and everywhere in this book, when we speak of positive braids, we always mean those braids that lie in the monoid B_∞^+, *i.e.*, those braids that admit at least one expression by a word containing no letter σ_i^{-1}. Such braids are sometimes called Garside positive braids, but we shall not use that name here. So the word "positive" never refers to any of the specific linear orderings we shall investigate hereafter. For the latter case, we shall introduce specific names for the braids that are larger than 1, typically σ-positive and σ^{Φ}-positive in the case of the σ-ordering and of the σ^{Φ}-ordering.

1. The σ-ordering of B_n

In this section we give a first definition of the σ-ordering of braids, based on the notion of a σ-positive braid word—many alternative definitions will be given in subsequent chapters. We explain how to construct the σ-ordering from two specific properties of braids called **A** and **C**. We also introduce a useful variant of the σ-ordering, called the σ^{Φ}-ordering, which is its image under the flip automorphism. Finally, we briefly discuss the algorithmic issues involving the σ–ordering.

1.1. Ordering a group. We start with premilinary remarks about what can be expected here. First, we recall that a *strict ordering* of a set Ω is a binary relation $\prec$ that is antireflexive ($x \prec x$ never holds) and transitive (the conjunction of $x \prec y$ and $y \prec z$ implies $x \prec z$). A strict ordering of Ω is called *linear* (or *total*) if, for all x, x' in Ω, one of $x = x'$, $x \prec x'$, $x' \prec x$ holds. Then, we recall the notion of an orderable group.

DEFINITION 1.1. (i) A *left-invariant ordering*, or *left-ordering*, of a group G is a strict linear ordering $\prec$ of G such that $g \prec h$ implies $fg \prec fh$ for all f, g, h in G. A group G is said to be *left-orderable* if there exists at least one left-invariant ordering of G.

(ii) A *bi-invariant ordering*, or *bi-ordering*, of a group G is a left-ordering of G that is also right-invariant, *i.e.*, $g \prec h$ implies $gf \prec hf$ for all f, g, h in G. A group G is said to be *bi-orderable* if there exists at least one bi-invariant ordering of G.

PROPOSITION 1.2. *For $n \geqslant 3$, the group B_n is not bi-orderable.*

PROOF. If $\prec$ is a bi-invariant ordering of a group G, then $g \prec h$ implies $\varphi(g) \prec \varphi(h)$ for each inner automorphism φ of G. Now, in the case of B_n, the inner automorphism Φ_n associated with Garside's fundamental braid Δ_n of (I.4.1) exchanges σ_i and σ_{n-i} for each i. Hence it is impossible to have $\sigma_1 \prec \sigma_{n-1}$ and $\Phi_n(\sigma_1) \prec \Phi_n(\sigma_{n-1})$ simultaneously. □

Therefore, in the best case, we shall be interested in orders that are invariant under multiplication on one side. Then, both sides play symmetric roles, as an immediate verification gives

LEMMA 1.3. *Assume that G is a group and $\prec$ is a left-invariant ordering of G. Define $g \mathrel{\tilde{\prec}} h$ to mean $g^{-1} \prec h^{-1}$. Then $\tilde{\prec}$ is a right-invariant ordering of G.*

We shall concentrate hereafter on left-invariant orderings. Specifying such an ordering is actually equivalent to specifying a subsemigroup of a certain type, called a positive cone.

DEFINITION 1.4. A subset P of a group G is called a *positive cone* on G if P is closed under multiplication and $G \setminus \{1\}$ is the disjoint union of P and P^{-1}.

LEMMA 1.5. (i) *Assume that $\prec$ is a left-invariant ordering of a group G. Then the set P of all elements in G that are larger than 1 is a positive cone on G, and $g \prec h$ is equivalent to $g^{-1}h \in P$.*

(ii) *Assume that P is a positive cone on a group G. Then the relation $g^{-1}h \in P$ is a left-invariant ordering of G, and P is then the set of all elements of G that are larger than 1.*

The verification is easy. Note that the formula $hg^{-1} \in P$ would define a right-invariant ordering.

1.2. The σ-ordering of braids. We now introduce on B_n a certain binary relation that will turn out to be a left-invariant ordering. The construction involves particular braid words defined in terms of the letters they contain.

DEFINITION 1.6. A braid word w is said to be *σ-positive* (*resp. σ-negative*) if, among the letters $\sigma_i^{\pm 1}$ that occur in w, the one with lowest index occurs positively only, *i.e.*, σ_i occurs but σ_i^{-1} does not (*resp.* negatively only, *i.e.*, σ_i^{-1} occurs but σ_i does not).

For instance, $\sigma_3\sigma_2\sigma_3^{-1}$ is a σ-positive braid word: the letter with lowest index is σ_2 (there is no $\sigma_1^{\pm 1}$), and there is one σ_2 but no σ_2^{-1}. By contrast, the word $\sigma_2^{-1}\sigma_3\sigma_2$, which is equivalent to $\sigma_3\sigma_2\sigma_3^{-1}$, is neither σ-positive nor σ-negative: the letter with lowest index is σ_2 again, but, here, both σ_2 and σ_2^{-1} appear.

DEFINITION 1.7. For β, β' in B_n, we say that $\beta <_n \beta'$ is true if $\beta^{-1}\beta'$ admits an n-strand representative word that is σ-positive.

EXAMPLE 1.8. Let $\beta = \sigma_2$ and $\beta' = \sigma_3\sigma_2$. Among the 4-strand braid words that represent the quotient $(\sigma_2)^{-1}(\sigma_3\sigma_2)$, there is the word $\sigma_2^{-1}\sigma_3\sigma_2$, which is neither σ-positive nor σ-negative, but there is also the word $\sigma_3\sigma_2\sigma_3^{-1}$—and many others. As the latter word is a 4-strand braid word that is σ-positive, $\beta <_4 \beta'$ is true.

Similarly, we have

$$\sigma_1 >_\infty \sigma_2 >_\infty \sigma_3 >_\infty \cdots \tag{1.1}$$

since, for each i, the braid word $\sigma_{i+1}^{-1}\sigma_i$ is σ-positive.

The central property is the following result of [**47**] (see Remark 1.16) which implies the first part of the theorem mentioned in the Introduction:

PROPOSITION 1.9. (i) *For $2 \leqslant n \leqslant \infty$, the relation $<_n$ is a left-invariant ordering of B_n.*

(ii) *For each n, the relation $<_n$ is the restriction of $<_\infty$ to B_n.*

Owing to (ii) above, we shall drop the subscripts and simply write $<$ for $<_n$. The order $<$ will be called the *σ-ordering* of braids, which is coherent with its definition in terms of the generators σ_i.

By definition, the relation $\beta >_n 1$ is true if and only if β admits at least one σ-positive n-strand representative word. According to Lemma 1.5, proving Proposition 1.9(i) amounts to proving that the set of all such braids is a positive cone. The latter result is a consequence of the following two statements:

Property A (Acyclicity). *A σ-positive braid word is not trivial.*

Property C (Comparison). *Every nontrivial braid of B_n admits an n-strand representative word that is σ-positive or σ-negative.*

PROOF OF PROPOSITION 1.9 FROM PROPERTIES **A** AND **C**. (i) Let P_n be the set of all n-strand braids that admit a σ-positive n-strand representative word. We shall prove that P_n is a positive cone in B_n. First, the concatenation of two σ-positive n-strand braid words is a σ-positive n-strand braid word; hence P_n is closed under multiplication.

Then, we claim that $B_n \setminus \{1\}$ is the disjoint union of P_n and P_n^{-1}. Indeed, Property **A** implies $1 \notin P_n$, and therefore $1 \notin P_n^{-1}$ as $1^{-1} = 1$ holds. So $P_n \cup P_n^{-1}$ is included in $B_n \setminus \{1\}$. Now assume $\beta \in P_n \cap P_n^{-1}$. We deduce $\beta^{-1} \in P_n$, whence

$$1 = \beta\beta^{-1} \in P_n \cdot P_n \subseteq P_n,$$

which contradicts $1 \notin P_n$. So P_n and P_n^{-1} must be disjoint. Finally, Property **C** (for B_n) means that $P_n \cup P_n^{-1}$ covers $B_n \setminus \{1\}$.

(ii) Assume $\beta, \beta' \in B_n$. Any σ-positive n-strand braid word representing $\beta^{-1}\beta'$ *a fortiori* witnesses the relation $\beta <_\infty \beta'$, so $\beta <_n \beta'$ implies $\beta <_\infty \beta'$. Conversely, assume $\beta <_\infty \beta'$. As $<_n$ is a linear ordering of B_n, one of $\beta <_n \beta'$ or $\beta \geqslant_n \beta'$ holds. In the latter case, we would deduce $\beta \geqslant_\infty \beta'$, which contradicts the hypothesis $\beta <_\infty \beta'$. So $\beta <_n \beta'$ is the only possibility. □

Property **A** has four different proofs in this text: they can be found on pages 73, 175, 190, and 224. As for Property **C**, no fewer than eight proofs are given, on pages 60, 89, 116, 148, 164, 190, 201, and 205.

In addition to being invariant under left multiplication, the σ-ordering of braids is invariant under the shift endomorphism, defined as follows.

DEFINITION 1.10. For w a braid word, the *shifting* of w is the braid word $\mathrm{sh}(w)$ obtained by replacing each letter σ_i with σ_{i+1}, and each letter σ_i^{-1} with σ_{i+1}^{-1}.

The explicit form of the braid relations implies that the shift mapping induces an endomorphism of B_∞, still denoted sh and called the *shift endomorphism.* The same argument guaranteeing that the canonical morphism of B_{n-1} into B_n is an embedding shows that the shift endomorphism of B_∞ is injective.

PROPOSITION 1.11. *For all braids β, β', the relation $\beta < \beta'$ is equivalent to* $\mathrm{sh}(\beta) < \mathrm{sh}(\beta')$.

PROOF. The shifting of a σ-positive braid word is a σ-positive braid word, so $\beta < \beta'$ implies $\mathrm{sh}(\beta) < \mathrm{sh}(\beta')$. Conversely, as $<$ is a linear ordering, the only possibility when $\mathrm{sh}(\beta) < \mathrm{sh}(\beta')$ is true is that $\beta < \beta'$ is true as well, as $\beta \geqslant \beta'$ would imply $\mathrm{sh}(\beta) \geqslant \mathrm{sh}(\beta')$. □

It is straightforward to check that, conversely, the σ-ordering is the only partial ordering on B_∞ that is invariant under multiplication on the left and under the shift endomorphism, and satisfies for all braids β, β' the inequality

$$1 < \mathrm{sh}(\beta)\, \sigma_1\, \mathrm{sh}(\beta').$$

1.3. Equivalent formulations. Before proceeding, we introduce derived notions in order to restate Properties **A** and **C** in slightly different forms. First, we can refine the notion of a σ-positive braid word by taking into account the specific index i that is involved.

DEFINITION 1.12. A braid word is said to be *σ_i-positive* if it contains at least one letter σ_i, but no σ_i^{-1} and no $\sigma_j^{\pm 1}$ with $j < i$. Similarly, it is said to be *σ_i-negative* if it contains at least one σ_i^{-1}, but no σ_i and no $\sigma_j^{\pm 1}$ with $j < i$. It is said to be *σ_i-free* if it contains no $\sigma_j^{\pm 1}$ with $j \leqslant i$.

So a braid word is σ-positive if and only if it is σ_i-positive for some i. Note that, for $i \geqslant 2$, a word w is σ_i-positive if and only if it is $\mathrm{sh}^{i-1}(w_1)$ for some σ_1-positive word w_1—we recall that sh is the shift mapping of Definition 1.10. Similarly, a braid word w is σ_i-free if and only if it is $\mathrm{sh}^{i}(w_1)$ for some w_1.

Then Properties **A** and **C** can be expressed in terms of σ_1-positive, σ_1-negative, and σ_1-free words.

PROPOSITION 1.13. *Property* **A** *is equivalent to:*
Property A (second form). *A σ_1-positive braid word is not trivial.*

PROOF. Every σ_1-positive braid word is σ-positive, so the first form of Property **A** implies the second form.

Conversely, assume the second form of Property **A**. Let w be a σ-positive word. Then w is σ_i-positive for some i. As observed above, this means that we have $w = \mathrm{sh}^{i-1}(w_1)$ for some σ_1-positive word w_1. By the second form of Property **A**, the word w_1 is not trivial, *i.e.*, it does not represent the unit braid. As the shift endomorphism of B_∞ is injective, this implies that w is not trivial either. So, the first form of Property **A** is satisfied. □

PROPOSITION 1.14. *Property* **C** *is equivalent to:*
Property C (second form). *Every braid of B_n admits an n-strand representative word that is σ_1-positive, σ_1-negative, or σ_1-free.*

PROOF. A σ-positive braid word is either σ_1-positive or σ_1-free, so the first form of Property **C** implies the second form.

Conversely, assume the second form of Property **C**. We prove the first form using induction on $n \geqslant 2$. For $n = 2$, the two forms coincide. Assume $n \geqslant 3$. Let β be a nontrivial n-strand braid. By the second form of Property **C**, we find an n-strand braid word w representing β that is σ_1-positive, σ_1-negative, or σ_1-free. In the first two cases, we are done. Otherwise, let $w_1 = \mathrm{sh}^{-1}(w)$, which makes sense as, by hypothesis, w contains no letter $\sigma_1^{\pm 1}$. As the shift endomorphism of B_∞ is injective, the word w_1 does not represent 1, so the induction hypothesis implies that w_1 is equivalent to some $(n-1)$-strand braid word w_1' that is σ-positive or σ-negative. By construction, the word $\mathrm{sh}(w_1')$ represents β and it is σ-positive or σ-negative. □

On the other hand, it will be often convenient in the sequel to have a name for the braids that admit a σ-positive word representative. So, we introduce the following natural terminology.

DEFINITION 1.15. A braid β is said to be *σ-positive inside B_n* (*resp.* *σ-negative, σ_i-positive, σ_i-negative, σ_i-free*) if, among all word representatives of β, there is at least one n-strand braid word that is σ-positive (*resp.* σ-negative, σ_i-positive, σ_i-negative, σ_i-free).

We insist that, in Definition 1.15, we only demand that there exists *at least one* word representative with the considered property. So, for instance, the braid $\sigma_2^{-1}\sigma_3\sigma_2$ is σ_2-positive since, among its many word representatives, there is one, namely $\sigma_3\sigma_2\sigma_3^{-1}$, that is σ_2-positive—there are many more: $\sigma_3\sigma_2\sigma_3^{-1}\sigma_3\sigma_3^{-1}$ is another σ_2-positive 4-strand braid word that represents the braid $\sigma_2^{-1}\sigma_3\sigma_2$.

With this terminology, $\beta <_n \beta'$ is equivalent to $\beta^{-1}\beta'$ being σ-positive inside B_n. Similarly, Property **A** means that a σ-positive braid is not trivial, and Property **C** that every nontrivial braid of B_n is σ-positive or σ-negative inside B_n.

REMARK 1.16. By Proposition 1.9(ii), a braid β of B_n satisfies $\beta >_n 1$ if and only if it satisfies $\beta >_\infty 1$, hence β is σ-positive inside B_n if and only if it is σ-positive inside B_∞. In other words, if an n-strand braid admits a word representative that is σ-positive, then it admits a word representative that is σ-positive and is an n-strand braid word, an *a priori* stronger property. Building on this result, we shall often drop the mention "inside B_n", exactly as when we write $<$ for $<_n$. However, a careful distinction has to be made when proving Property **C**. It can be mentioned that the original argument of [**47**] only leads to a proof of Property **C** in B_∞: this is enough to order every braid group B_n, but not to deduce Property **C** in B_n; see Chapter IV.

1.4. The σ^Φ-ordering of braids. If $\prec$ is an ordering of a group G and φ is an automorphism of G, then the relation $\varphi(g) \prec \varphi(h)$ defines a new ordering of G with the same invariance properties as $\prec$. In the case of B_n, the flip automorphism, *i.e.*, the inner automorphism Φ_n associated with the braid Δ_n, plays an important role, and it is natural to introduce the image of the σ-ordering under Φ_n, *i.e.*, the flipped version of the σ-ordering. As will be seen in Section 4, the new ordering so obtained has some nice properties not shared by the original version, particularly in terms of avoiding the infinite descending sequence of (1.1).

We recall from Lemma I.4.4 that Φ_n exchanges σ_i and σ_{n-i} for $1 \leqslant i < n$, thus corresponding to a symmetry in the associated braid diagrams.

DEFINITION 1.17. For $2 \leqslant n < \infty$ and β, β' in B_n, we declare that $\beta <_n^\Phi \beta'$ is true if we have $\Phi_n(\beta) < \Phi_n(\beta')$.

PROPOSITION 1.18. *The relation $<_n^\Phi$ is a left-invariant ordering of B_n. Moreover, for all β, β' in B_n, the relations $\beta <_n^\Phi \beta'$ and $\beta <_{n+1}^\Phi \beta'$ are equivalent.*

PROOF. The first part is clear as Φ_n is an automorphism of B_n.

Assume $\beta, \beta \in B_n$ and $\beta <_n^\Phi \beta'$. By definition, we have $\Phi_n(\beta) < \Phi_n(\beta')$, hence $\mathrm{sh}(\Phi_n(\beta)) < \mathrm{sh}(\Phi_n(\beta'))$ by Proposition 1.11. By construction, we have

$$\Phi_{n+1}(\beta) = \mathrm{sh}(\Phi_n(\beta)) \quad \text{and} \quad \Phi_{n+1}(\beta') = \mathrm{sh}(\Phi_n(\beta')),$$

so $\beta <_{n+1}^\Phi \beta'$ follows. As $<_n^\Phi$ is a linear ordering, this is enough to conclude that $<_n^\Phi$ coincides with the restriction of $<_{n+1}^\Phi$ to B_n. □

Owing to Proposition 1.18, we shall drop the subscripts and simply write $<^\Phi$ for the ordering of B_∞ whose restriction to B_n is $<_n^\Phi$. For instance, we have

$$1 <^\Phi \sigma_1 <^\Phi \sigma_2 <^\Phi \dots .$$

The flipped order $<^\Phi$ is easily described in terms of word representatives.

DEFINITION 1.19. (i) A braid word w is said to be *σ^Φ-positive* (*resp.* *σ^Φ-negative*) if, among the letters $\sigma_i^{\pm 1}$ that occur in w, the one with *highest* index occurs positively only (*resp.* negatively only).

(ii) A braid β is said to be *σ^Φ-positive* (*resp.* *σ^Φ-negative*) if it admits at least one braid word representative that is σ^Φ-positive (*resp.* σ^Φ-negative).

The only difference between a σ-positive and a σ^Φ-positive braid word is that, in the former case, we consider the letter σ_i with lowest index, while, in the latter case, we consider the letter σ_i with highest index.

PROPOSITION 1.20. *For all braids β, β', the relation $\beta <^\Phi \beta'$ holds if and only if $\beta^{-1}\beta'$ is σ^Φ-positive.*

PROOF. By construction, an n-strand braid word w is σ^Φ-positive if and only if the n-strand braid word $\Phi_n(w)$ is σ-positive. □

Thus the flipped order $<^\Phi$ is the counterpart of the σ-order $<$ in which the highest index replaces the lowest index, and σ^Φ-positive words replace σ-positive words. It is therefore natural to call it the *σ^Φ-ordering* of braids.

As the flip Φ_n is an automorphism of the group B_n, the properties of $<$ and $<^\Phi$ are similar. However, there are at least two reasons for considering both $<$ and $<^\Phi$. First, there is no flip on B_∞, and the two orderings differ radically on B_∞: (1.1) shows that $(B_\infty^+, <)$ has infinite descending sequences, while we shall see in Section 4.1 below that $(B_\infty^+, <^\Phi)$ is a well-ordering, and, therefore, it has no infinite descending chain. The second reason is that, in subsequent chapters, certain approaches demand that one specific version be used: the original version $<$ in Chapter IV, the flipped version $<^\Phi$ in Chapters VII and VIII.

1.5. Algorithmic aspects. The σ-ordering of braids is a complicated object. However, it is completely effective in that there exist efficient comparison algorithms. In this section (and everywhere in the sequel) we denote by $\overline{w}$ the braid represented by a braid word w—but, as usual, we use σ_i both for the letter and for the braid it represents.

PROPOSITION 1.21. *For each n, the σ-ordering of B_n has at most a quadratic complexity: there exists an algorithm that, starting with two n-strand braid words w, w' of length ℓ, runs in time $O(\ell^2)$ and decides whether $\overline{w} < \overline{w'}$ holds.*

At this early stage, we cannot yet describe the algorithms witnessing to the above upper complexity bound. It turns out that most of the proofs of Property **C** alluded to in Section 1.2 provide an effective comparison algorithm. Some of them are quite inefficient—typically the one of Chapter IV—but several lead to a quadratic complexity. This is particularly the case with those based on the Φ-normal form of Chapter VII and on the ϕ-normal form of Chapter VIII: in both cases, the normal form can be computed in quadratic time, and, then, the comparison itself can be made in linear time. This is also the case with the lamination method of Chapter XII: in this case, the coordinates of a braid can be computed in quadratic time, and the comparison (with the unit braid) can then be made in (sub)linear time. Similar results are conjectured in the case of the handle reduction method of Chapter V and the Tetris algorithm of Chapter XI—see Chapter XVI for further discussion.

Let us mention that, for a convenient definition for the RAM complexity of the input braids, the algorithm of Chapter XII even leads to a complexity upper bound which is quadratic independently of the braid index n, *i.e.*, there exists an absolute constant C so that the running time for complexity ℓ input braids in B_∞ is bounded above by $C \cdot \ell^2$.

We also point out that every comparison algorithm for the σ-ordering of braids automatically gives a solution to the braid word problem, *i.e.*, to the braid isotopy problem: indeed, we have $\overline{w} = \overline{w'}$ if and only if we have neither $\overline{w} < \overline{w'}$ nor $\overline{w} > \overline{w'}$. It also leads to a comparison for the flipped version $<^\Phi$ of the σ-ordering, as, if w, w' are n-strand braid words, $\overline{w} <^\Phi \overline{w'}$ is equivalent to $\overline{\Phi_n(w)} < \overline{\Phi_n(w')}$, and the flip automorphism Φ_n can be computed in linear time.

Another related question is that of effectively finding σ-positive representative words, *i.e.*, starting with a braid word w, finding an equivalent braid word w' that is σ-positive, σ-negative, or empty. Property **C** asserts that this is always possible. Every algorithmic solution to that problem gives a comparison algorithm as, by Property **A**, w' being σ-positive implies $\overline{w} = \overline{w'} > 1$, but, conversely, deciding $\overline{w} > 1$ does not require that we exhibit a σ-positive witness.

PROPOSITION 1.22. *The σ-positive representative problem has at most an exponential complexity: there exist a polynomial $P(n, \ell)$ and an algorithm that, starting with an n-strand braid word w of length ℓ, runs in time $2^{P(n,\ell)}$ and returns a braid word of length bounded by $2^{P(n,\ell)}$ that is equivalent to w and is σ-positive, σ-negative, or empty.*

The handle reduction approach of Chapter V gives the precise form of such a polynomial: $P(n, \ell) = n^4\ell$. From the transmission-relaxation approach of Chapter XI, an asymptotically better estimate can be extracted: $P(n, \ell) = \mathrm{const} \cdot n\ell$. However, the algorithm outlined in Chapter XI is just polynomial, but the output of the algorithm is not a braid word in the standard sense but a zipped word, this meaning that, sometimes, instead of writing one and the same subword many times, the algorithm outputs the subword once and specifies the number of repetitions. This allows us to make the size of the output bounded above by a polynomial in n and ℓ though the length of the word after unzipping is not known to be of polynomial size so far.

It is likely that the approach of Chapter VIII leads to much better results: very recently, J. Fromentin announced a new algorithm that solves the σ-positive representative problem with a quadratic time complexity and a linear space complexity, without zipping the output. We refer to Chapter XVI for further discussion.

2. Local properties of the σ-ordering

We shall now list—with or without proof—some properties of the σ-ordering of braids. In this section, we consider properties that can be called local in that they involve finitely many braids at a time.

2.1. Curious examples. We start with a series of examples, including some rather surprising ones, that illustrate the complexity of the σ-ordering. The reader should note that all examples below live in B_3. This shows that, despite its simple definition, even the σ-ordering of 3-strand braids is a quite complicated object.

The first example shows that the σ-ordering is not invariant under multiplication on the right, as was already known from Proposition 1.2.

EXAMPLE 2.1. Let $\beta = \sigma_1\sigma_2^{-1}$, and $\gamma = \sigma_1\sigma_2\sigma_1$, *i.e.*, $\gamma = \Delta_3$. The word $\sigma_1\sigma_2^{-1}$ contains one occurrence of σ_1 and no occurrence of σ_1^{-1}, so the braid β is σ-positive, and $\beta > 1$ is true. On the other hand, the braid $\gamma^{-1}\beta\gamma$ is represented by the word $\sigma_1^{-1}\sigma_2^{-1}\sigma_1^{-1}\sigma_1\sigma_2^{-1}\sigma_1\sigma_2\sigma_1$, hence also by the equivalent word $\sigma_2\sigma_1^{-1}$, as, by Lemma I.4.4, we have $\Delta_3^{-1}\sigma_i\Delta_3 = \sigma_{3-i}$ for $i = 1, 2$. The word $\sigma_2\sigma_1^{-1}$ contains one letter σ_1^{-1} and no letter σ_1. So, by definition, we have $\gamma^{-1}\beta\gamma < 1$, and, therefore, $\beta\gamma < \gamma$. So $1 < \beta$ does not imply $\gamma < \beta\gamma$.

A phenomenon connected with the noninvariance under right multiplication is that a conjugate of a braid that is larger than 1 may be smaller than 1. Example 2.1 actually gives us an illustration of this situation: in fact, in this case, the conjugate is the inverse.

EXAMPLE 2.2. Let $\beta = \sigma_1\sigma_2^{-1}$ again. Then β is σ_1-positive, hence larger than 1. By Lemma I.4.4, conjugating by Δ_3 amounts to exchanging σ_1 and σ_2. So we have $\Delta_3\beta\Delta_3^{-1} = \sigma_2\sigma_1^{-1}$, a σ_1-negative braid, hence smaller than 1, *i.e.*, we have $\beta > 1$ and $\Delta_3\beta\Delta_3^{-1} < 1$; however, we shall see in Corollary 3.7 below that the conjugates of a braid β cannot be too far from β.

An easy exercise is that every left-invariant ordering such that $g \prec h$ implies $g^{-1} \succ h^{-1}$ is also right-invariant. As the braid ordering is not right-invariant, there must exist counterexamples, *i.e.*, braids β, γ satisfying $\beta < \gamma$ and $\beta^{-1} < \gamma^{-1}$. Here are examples of this situation.

EXAMPLE 2.3. Let $\beta = \Delta_3$ and $\gamma = \sigma_2^2\sigma_1$. Then we find $\beta^{-1}\gamma = \sigma_1\sigma_2^{-1}$, a σ_1-positive word, and $\beta\gamma^{-1} = \sigma_1\sigma_2^{-1}$, again a σ_1-positive word, So, in this case, we have $1 < \beta < \gamma$ and $\beta^{-1} < \gamma^{-1}$.

EXAMPLE 2.4. Here is a stronger example. Let $\beta = \sigma_2^{-1}\sigma_1^2\sigma_2$ and $\gamma = \Delta_3$. We now find $\beta^{-1}\gamma = \sigma_1\sigma_2^{-1}\sigma_1$ (see below), a σ_1-positive word, and $\beta\gamma^{-1} = \sigma_2^{-1}\sigma_1\sigma_2^{-1}$, a σ_1-positive word. So we obtain again $1 < \beta < \gamma$ and $\beta^{-1} < \gamma^{-1}$. But there is more. We claim that $\beta^{-p}\gamma = \sigma_1\sigma_2^{-2p+1}\sigma_1$ holds for $p \geqslant 1$. Indeed, for $p = 1$, we have

$$\beta^{-1}\gamma = \sigma_2^{-1}\sigma_1^{-1} \cdot \sigma_1^{-1}\sigma_2\sigma_1\sigma_2 \cdot \sigma_1 = \sigma_2^{-1}\sigma_1^{-1} \cdot \sigma_2\sigma_1 \cdot \sigma_1 = \sigma_1\sigma_2^{-1}\sigma_1.$$

For $p \geqslant 2$, applying the induction hypothesis, we find

$$\begin{aligned}\beta^{-p}\gamma &= \sigma_2^{-1}\sigma_1^{-2}\sigma_2 \cdot \beta^{-p+1}\gamma \\ &= \sigma_2^{-1}\sigma_1^{-2}\sigma_2 \cdot \sigma_1\sigma_2^{-2p+3}\sigma_1 = \sigma_1\sigma_2^{-2} \cdot \sigma_2^{-2p+3}\sigma_1 = \sigma_1\sigma_2^{-2p+1}\sigma_1.\end{aligned}$$

As $\sigma_1\sigma_2^{-2p+1}\sigma_1$ is a σ_1-positive word for each p, we have in this case $1 < \beta^p < \gamma$ for each positive p, and $\beta^{-1} < \gamma^{-1}$.

Even more curious situations occur. Assume that β is a σ_1-positive braid. Then the sequence $1, \beta, \beta^2, \ldots$ is strictly increasing, and its entries admit expressions in which more and more letters σ_1 occur. One might therefore expect that, eventually, the braid β^p dominates σ_1, which only contains one letter σ_1. The next example shows this is not the case.

EXAMPLE 2.5. Consider $\beta = \sigma_2^{-1}\sigma_1$. Then $\beta^p < \sigma_1$ holds for each p. The inequality clearly holds for $p \leqslant 0$. For positive p, we will show that $\sigma_1^{-1}\beta^p$ is σ_1-negative. To this end, we prove the equality

$$\sigma_1^{-1}\beta^p = \sigma_2(\sigma_2\sigma_1^{-1})^{p-1}\sigma_1^{-1}\sigma_2^{-1} \tag{2.1}$$

using induction on $p \geqslant 1$. For $p = 1$, (2.1) reduces to $\sigma_1^{-1}\sigma_2^{-1}\sigma_1 = \sigma_2\sigma_1^{-1}\sigma_2^{-1}$, which directly follows from the braid relation. For $p \geqslant 2$, we find

$$\begin{aligned}\sigma_1^{-1}\beta^p &= (\sigma_1^{-1}\beta^{p-1}) \cdot \sigma_2^{-1}\sigma_1 \\ &= \sigma_2(\sigma_2\sigma_1^{-1})^{p-2}\sigma_1^{-1}\sigma_2^{-1} \cdot \sigma_2^{-1}\sigma_1 \\ &= \sigma_2(\sigma_2\sigma_1^{-1})^{p-2}\sigma_2\sigma_1^{-2}\sigma_2^{-1} = \sigma_2(\sigma_2\sigma_1^{-1})^{p-1}\sigma_1^{-1}\sigma_2^{-1},\end{aligned}$$

using the induction hypothesis and the equality $\sigma_1^{-1}\sigma_2^{-2}\sigma_1 = \sigma_2\sigma_1^{-2}\sigma_2^{-1}$.

It can be observed that, more generally, $\beta^p < \sigma_2^{-q}\sigma_1$ holds for all nonnegative p and q. So the ascending sequence β^p does not even approach σ_1, as it remains below each entry in the descending sequence $\sigma_2^{-q}\sigma_1$.

Our last example will demonstrate that the σ-ordering of B_n is not Conradian.

DEFINITION 2.6. A left-invariant ordering $\prec$ of a group G is *Conradian* if for all g, h in G that are greater than 1, there exists a positive integer p satisfying $h \prec gh^p$.

Conrad used this property in [**38**] to show that such left-ordered groups share many of the properties of bi-orderable groups; see Section XV.5 for more details.

PROPOSITION 2.7. *For $n \geqslant 3$, the σ-ordering of the braid group B_n is not Conradian.*

PROOF. Let $\beta = \sigma_2^{-1}\sigma_1$ and $\gamma = \sigma_2^{-2}\sigma_1$. Clearly, β and γ are σ_1-positive, so $\beta > 1$ and $\gamma > 1$ hold. We claim that $\gamma\beta^p < \beta$ holds for each $p \geqslant 0$. To see that, using induction on $p \geqslant 0$, we prove the equality

$$\beta^{-1}\gamma\beta^p = \sigma_2^2(\sigma_1^{-1}\sigma_2)^{p-1}\sigma_1^{-2}\sigma_2^{-1}. \tag{2.2}$$

For $p = 0$, using the braid relations, we find

$$\beta^{-1}\gamma = \sigma_1^{-1}\sigma_2\sigma_2^{-2}\sigma_1 = \sigma_1^{-1}\sigma_2^{-1}\sigma_1 = \sigma_2\sigma_1^{-1}\sigma_2^{-1} = \sigma_2^2(\sigma_1^{-1}\sigma_2)^{-1}\sigma_1^{-2}\sigma_2^{-1}.$$

For $p = 1$, we have

$$\beta^{-1}\gamma\beta = \sigma_1^{-1}\sigma_2\sigma_2^{-2}\sigma_1\sigma_2^{-1}\sigma_1 = \sigma_1^{-1}\sigma_2^{-1}\sigma_1\sigma_2^{-1}\sigma_1 = \sigma_2\sigma_1^{-1}\sigma_2^{-2}\sigma_1 = \sigma_2^2\sigma_1^{-2}\sigma_2^{-1}.$$

For $p \geqslant 1$, again using the equality $\sigma_1^{-1}\sigma_2^{-2}\sigma_1 = \sigma_2\sigma_1^{-2}\sigma_2^{-1}$ of Example 2.5, we find

$$\begin{aligned}\beta^{-1}\gamma\beta^p &= (\sigma_2^2(\sigma_1^{-1}\sigma_2)^{p-2}\sigma_1^{-2}\sigma_2^{-1})(\sigma_2^{-1}\sigma_1)\\ &= \sigma_2^2(\sigma_1^{-1}\sigma_2)^{p-2}\sigma_1^{-1}\sigma_2\sigma_1^{-2}\sigma_2^{-1} = \sigma_2^2(\sigma_1^{-1}\sigma_2)^{p-1}\sigma_1^{-2}\sigma_2^{-1}.\end{aligned}$$

For $p \geqslant 1$, the right-hand side of (2.2) is σ_1-negative, and, for $p = 0$, it is equivalent to the σ_1-negative word $\sigma_2\sigma_1^{-1}\sigma_2$, so, in each case, we obtain $\beta < \gamma\beta^p$. □

2.2. Property S. After the many counterexamples of Section 2.1, we turn to positive results.

We have seen in Example 2.1 that the σ-ordering of braids is not invariant under multiplication on the right, and, therefore, that a conjugate of a braid larger than 1 need not be larger than 1. This phenomenon cannot, however, occur with conjugates of positive braids, *i.e.*, of braids that can be expressed using the generators σ_i only, and not their inverses. The core of the question is the last of the three fundamental properties of braids we shall develop here:

Property S (Subword). *Every braid of the form $\beta^{-1}\sigma_i\beta$ is σ-positive.*

Property **S** was first proved by Richard Laver in [**137**]. In this text, proofs of Property **S** appear on pages 83, 152, 193, and 263.

Using the compatibility of $<$ with multiplication on the left and a straighforward induction, we deduce the following result, which explains our terminology:

Proposition 2.8. *Assume that β, β' are braids and some braid word representing β' is obtained by inserting positive letters σ_i in a braid word representing β. Then we have $\beta' > \beta$.*

We recall that B_∞^+ denotes the submonoid of B_∞ generated by the braids σ_i. Another consequence of Property **S** is:

Proposition 2.9. *If β belongs to B_∞^+ and is not 1, then $\beta' > 1$ is true for every conjugate β' of β. More generally, $\beta > 1$ is true for every quasi-positive braid β, the latter being defined as a braid that can be expressed as a product of conjugates of positive braids.*

Proof. Assume $\beta' = \gamma^{-1}\beta\gamma$ with $\beta \in B_\infty^+$. By definition, β is a product of finitely many braids σ_i, so, in order to prove $\beta' > 1$, it suffices to establish that $\gamma^{-1}\sigma_i\gamma > 1$ holds for each i, and this is Property **S**. □

As was noted by Stepan Orevkov [**166**], the converse implication is not true: the braid $\sigma_2^{-5}\sigma_1\sigma_2^2\sigma_1$ is a non-quasi-positive braid but every conjugate of it is σ-positive.

By applying the flip automorphism Φ_n, we immediately deduce from Property **S** that every braid of the form $\beta^{-1}\sigma_i\beta$ is also σ^Φ-positive, and that the counterpart of Proposition 2.8 involving the ordering $<^\Phi$ is true. A direct application is the following result, which is important for analysing the restriction of $<^\Phi$ to B_∞^+:

Proposition 2.10. *For each n, the set B_n^+ is the initial segment of $(B_\infty^+, <^\Phi)$ determined by σ_n, i.e., we have $B_n^+ = \{\beta \in B_\infty^+ \mid \beta <^\Phi \sigma_n\}$.*

PROOF. By definition, $\beta <^{\Phi} \sigma_n$ holds for every β in B_n^+. Indeed, if w is any n-strand braid word representing β, then $w^{-1}\sigma_n$ is a σ_n^{Φ}-positive word representing $\beta^{-1}\sigma_n$.

Conversely, assume that β is a positive braid satisfying $\beta <^{\Phi} \sigma_n$. Let w be a positive braid word representing β, and let σ_i be the generator with highest index occurring in w. By the counterpart of Proposition 2.8, we have $\beta \geqslant^{\Phi} \sigma_i$, and, therefore, $i \geqslant n$ would contradict the hypothesis $\beta <^{\Phi} \sigma_n$. □

Another application of Property **S** is the following property from [**51**]. We recall that sh denotes the shift endomorphism of B_∞ that maps σ_i to σ_{i+1} for every i.

PROPOSITION 2.11. *For each braid β, we have $\beta < \mathrm{sh}(\beta)\,\sigma_1$.*

PROOF. Let β be an arbitrary braid in B_n. We claim that the braid $\beta^{-1}\mathrm{sh}(\beta)\sigma_1$ is σ_1-positive. To see that, we write, inside B_{n+1},

$$\beta^{-1}\mathrm{sh}(\beta)\sigma_1 = \underline{(\beta^{-1}\sigma_2 \dots \sigma_n\beta)} \cdot \underline{(\sigma_n^{-1} \dots \sigma_2^{-1})} \cdot \underline{(\sigma_2 \dots \sigma_n\beta^{-1}\sigma_n^{-1} \dots \sigma_2^{-1})} \cdot \mathrm{sh}(\beta)\sigma_1.$$

The first underlined fragment is a conjugate of the positive braid $\sigma_2 \dots \sigma_n$, so, by Property **S**, it is σ-positive, hence either σ_1-positive or σ_1-free. The second underlined fragment is σ_1-free. Next, it is easy to check with a picture that the third underlined fragment is equal to $\sigma_1^{-1}\mathrm{sh}(\beta^{-1})\sigma_1$. Putting things together, we obtain

$$\beta^{-1}\mathrm{sh}(\beta)\sigma_1 = \beta' \cdot \underline{\sigma_1^{-1}\mathrm{sh}(\beta^{-1}) \cdot \sigma_1 \cdot \mathrm{sh}(\beta)\sigma_1},$$

where β' is a braid that is either σ_1-positive or σ_1-free. But, now, we see that the underlined expression is a conjugate of σ_1, so, by Property **S**, it is σ-positive, hence σ_1-positive or σ_1-free. We deduce that $\beta^{-1}\mathrm{sh}(\beta)\sigma_1$ itself is σ_1-positive or σ_1-free.

Finally, it is impossible that $\beta^{-1}\mathrm{sh}(\beta)\sigma_1$ be σ_1-free. Indeed, let π be the permutation of $\{1, \dots, n\}$ induced by β. Then the initial position of the strand that finishes at position 1 in any diagram representing $\beta^{-1}\mathrm{sh}(\beta)\sigma_1$ is $\pi^{-1}(\pi(1) + 1)$, which cannot be 1.

So the only possibility is that $\beta^{-1}\mathrm{sh}(\beta)\sigma_1$ is σ_1-positive, hence σ-positive. □

3. Global properties of the σ-ordering

We turn to more global properties, involving infinitely many braids at a time. Here we successively consider the Archimedean property, the question of density and the associated topology, and convex subgroups.

3.1. The Archimedean property. We shall show that the σ-ordering and, more generally, any left-invariant ordering of B_n fails to be Archimedean for $n \geqslant 3$. However, certain partial Archimedean properties involving the central elements Δ_n^2 are satisfied.

DEFINITION 3.1. A left-ordered group $(G, \prec)$ is said to be *Archimedean* if, for all g, h larger than 1 in G, there exists a positive integer p for which $g \prec h^p$ holds.

In other words, the powers of any nontrivial element are cofinal in the ordering. For example, an infinite cyclic group, with either of the two possible orderings, is Archimedean. On the other hand, $\mathbb{Z} \times \mathbb{Z}$ with lexicographic ordering is not Archimedean, whereas Archimedean orderings for the same group do exist, by embedding $\mathbb{Z} \times \mathbb{Z}$ in the additive real numbers, sending the generators to rationally independent numbers, and taking the induced ordering.

PROPOSITION 3.2. *The σ-ordering of B_n is not Archimedean for $n \geqslant 3$.*

PROOF. For every positive integer p, we have $1 < \sigma_2^p < \sigma_1$. □

One can say more.

PROPOSITION 3.3. *For $n \geqslant 3$, every left-invariant ordering of B_n fails to be Archimedean.*

This follows from the the fact that B_n is not Abelian for $n \geqslant 3$ and from a result of P. Conrad [**38**] generalizing the classical theorem of Hölder [**111**]: any left-invariant Archimedean ordering of a group must also be right-invariant, and the group embeds, simultaneously in the algebraic and order senses, in the additive real numbers. In particular, such a group is Abelian.

By contrast to the previous negative result, there is a partial Archimedean property involving the central element Δ_n^2, namely that every braid is dominated by some power of the braid Δ_n^2.

The results we shall establish turn out to be true not only for the σ-ordering, but also for any left-invariant ordering of B_n. So, for the rest of this section, we consider this extended framework. When $\prec$ denotes a strict ordering, $\preccurlyeq$ denotes the corresponding nonstrict ordering, *i.e.*, $x \preccurlyeq y$ stands for "$x \prec y$ or $x = y$".

LEMMA 3.4. *Assume that $\prec$ is a left-invariant ordering of B_n. Then $\Delta_n^{2p} \prec \beta$ implies $\beta^{-1} \prec \Delta_n^{-2p}$, and the conjunction of $\Delta_n^{2p} \prec \beta$ and $\Delta_n^{2q} \prec \gamma$ implies $\Delta_n^{2p+2q} \prec \beta\gamma$. The same implications hold for $\preccurlyeq$.*

PROOF. Assume $\Delta_n^{2p} \prec \beta$. Multiplying by β^{-1} on the left, we get $\beta^{-1}\Delta_n^{2p} \prec 1$, which is also $\Delta_n^{2p}\beta^{-1} \prec 1$. Multiplying by Δ_n^{-2p} on the left, we deduce $\beta^{-1} \prec \Delta_n^{-2p}$.

Now assume $\Delta_n^{2p} \prec \beta$ and $\Delta_n^{2q} \prec \gamma$. By multiplying the first inequality by Δ_n^{2q} on the left, we obtain $\Delta_n^{2p+2q} \prec \Delta_n^{2q}\beta = \beta\Delta_n^{2q}$. By multiplying the second inequality by β on the left, we obtain $\beta\Delta_n^{2q} \prec \beta\gamma$. We deduce $\Delta_n^{2p+2q} \prec \beta\gamma$. □

LEMMA 3.5. *Assume that $\prec$ is a left-invariant ordering of B_n satisfying $1 \prec \Delta_n$. Then, for each i in $\{1, \dots, n-1\}$, we have $\Delta_n^{-2} \prec \sigma_i \prec \Delta_n^2$.*

PROOF. By Lemma I.4.4, we have $\delta_n^n = \Delta_n^2$, so the hypothesis $1 \prec \Delta_n$ implies $1 \prec \Delta_n^2 = \delta_n^n$, hence $1 \prec \delta_n$, and, therefore, $1 \prec \delta_n \prec \delta_n^2 \prec \dots \prec \delta_n^n = \Delta_n^2$.

Assume that $\Delta_n^2 \preccurlyeq \sigma_i$ holds for some i. Let j be any element of $\{1, \dots, n-1\}$. By formulas (I.4.3) and (I.4.4), we can find p with $0 \leqslant p \leqslant n-1$ satisfying $\sigma_j = \delta_n^{-p}\sigma_i\delta_n^p$. Then we obtain

$$1 \prec \delta_n^{n-p} = \delta_n^{-p}\Delta_n^2 \preccurlyeq \delta_n^{-p}\sigma_i \preccurlyeq \delta_n^{-p}\sigma_i\delta_n^p = \sigma_j.$$

So $1 \prec \sigma_j$ holds for each generator σ_j. Applying Lemma 3.4, we deduce that, if a braid β can be represented by a positive braid word that contains at least one letter σ_i, then $\Delta_n^2 \preccurlyeq \beta$ holds. This applies in particular to Δ_n, and we deduce $\Delta_n^2 \preccurlyeq \Delta_n$, which contradicts the assumption $1 \prec \Delta_n$.

Similarly, assume that $\sigma_i \preccurlyeq \Delta_n^{-2}$ holds. Consider again any σ_j. If p is as above, we also have $\sigma_j = \delta_n^{n-p}\sigma_i\delta_n^{p-n}$, since δ_n^n lies in the center of B_n. Then we find

$$\sigma_j = \delta_n^{n-p}\sigma_i\delta_n^{p-n} \prec \delta_n^{n-p}\sigma_i \preccurlyeq \delta_n^{n-p}\Delta_n^{-2} = \delta_n^{-p} \preccurlyeq 1.$$

This time, $\sigma_j \prec 1$ holds for each j. As Δ_n is a positive braid, this implies $\Delta_n \prec 1$, which contradicts the assumption $1 \prec \Delta_n$. □

Gathering the results, we immediately deduce:

PROPOSITION 3.6. *Assume $\prec$ is a left-invariant ordering of B_n and $1 \prec \Delta_n$ holds. Then, for each braid β in B_n, there exists a unique integer p for which $\Delta_n^{2p} \preccurlyeq \beta \prec \Delta_n^{2p+2}$ is true. Moreover, if β can be represented by a braid word of length ℓ, we have $|p| \leqslant \ell$.*

PROOF. Lemma 3.5 implies that each generator σ_i lies in the interval $(\Delta_n^{-2}, \Delta_n^2)$. Then Lemma 3.4 implies that every braid that can be represented by a word of length ℓ lies in the interval $[\Delta_n^{-2\ell}, \Delta_n^{2\ell})$. As this interval is the disjoint union of the intervals $[\Delta_n^{2p}, \Delta_n^{2p+2})$ for $-\ell \leqslant p < \ell$, the result of the proposition follows. □

In this way, we obtain a decomposition of $(B_n, \prec)$ into a sequence of disjoint intervals of size Δ_n^2, as suggested in Figure 1.

As noted by A. Malyutin and N.Yu. Netsvetaev in [**150**], the previous result implies that the action of conjugacy cannot move a braid too far.

COROLLARY 3.7 (Figure 1). *Assume that $\prec$ is a left-invariant ordering of B_n satisfying $1 \prec \Delta_n$. Then, if β and β' are conjugate,*

$$\Delta_n^{2p} \preccurlyeq \beta \prec \Delta_n^{2p+2} \quad \textit{implies} \quad \Delta_n^{2p-2} \preccurlyeq \beta' \prec \Delta_n^{2p+4}. \tag{3.1}$$

So, in particular, $\beta\Delta_n^{-4} \prec \beta' \prec \beta\Delta_n^4$ is always true.

PROOF. Assume $\Delta_n^{2p} \preccurlyeq \beta \prec \Delta_n^{2p+2}$ and $\beta' = \gamma\beta\gamma^{-1}$. By Proposition 3.6, we have $\Delta_n^{2q} \preccurlyeq \gamma \prec \Delta_n^{2q+2}$ for some q. Lemma 3.4 first implies $\Delta_n^{-2q-2} \prec \gamma^{-1} \preccurlyeq \Delta_n^{-2q}$, and then

$$\Delta_n^{2q+2p-2q-2} \prec \gamma\beta\gamma^{-1} \prec \Delta_n^{2q+2+2p+2-2q},$$

which gives $\Delta_n^{2p-2} \prec \beta' \prec \Delta_n^{2p+4}$. □

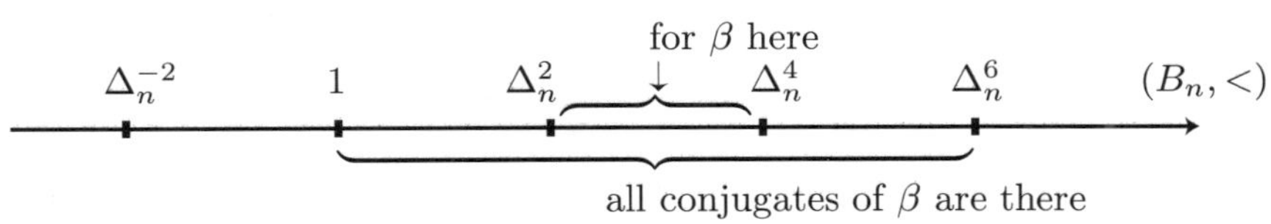

FIGURE 1. Powers of Δ_n^2 and the action of conjugacy on $(B_n, <)$.

All the previous results apply to the σ-ordering, as it is a left-invariant ordering of B_n and $1 < \Delta_n$ is satisfied. Note that, in this case, Corollary 3.7 is optimal in the sense that we cannot replace intervals of length Δ_n^2 with intervals of length Δ_n in Lemma 3.4: for instance, we have $1 < \sigma_1^2\sigma_2 < \Delta_3$ and $\Delta_3^2 < \Delta_3\sigma_1^2\sigma_2 < \Delta_3^3$.

3.2. Discreteness and density. Left-invariant orderings of a group have a sort of homogeneity—the ordering near any two group elements has similar order properties, because of invariance under left translation. In particular, there is a basic dichotomy between discrete and dense orders.

DEFINITION 3.8. A left-invariant ordering of a group is said to be *discrete* if its positive cone has a least element; it is said to be *dense* if the positive cone does not have a least element.

Equivalently, a left-invariant ordering of a group is discrete if every group element has an immediate successor and predecessor, and it is dense if between any two group elements one can find another element of the group. One verifies easily that, in a discretely left-ordered group, with least element ε larger than 1, the immediate successor of a group element g is $g\varepsilon$ and its immediate predecessor is $g\varepsilon^{-1}$.

The braid orderings display both types.

PROPOSITION 3.9. *The σ-ordering of B_n is discrete, with least σ-positive element σ_{n-1}.*

PROOF. Clearly σ_{n-1} is σ-positive. Conversely, assume that β belongs to B_n and is σ-positive. If β is σ_i-positive for some i with $i \leqslant n-2$, then $\sigma_{n-1}^{-1}\beta$ is σ_i-positive as well, so $\sigma_{n-1} < \beta$ holds. On the other hand, if β is σ_{n-1}-positive, it must be σ_{n-1}^p for some $p \geqslant 1$, and we find $\sigma_{n-1}^{-1}\beta = \sigma_{n-1}^{p-1}$, hence $\sigma_{n-1} \leqslant \beta$. □

As the flip automorphism Φ_n is an isomorphism of $(B_n, <)$ to $(B_n, <^\Phi)$, the flipped version $<^\Phi$ of the σ-ordering is also discrete on B_n, and σ_1 is the least σ^Φ-positive element. In the inclusions $B_n \subseteq B_{n+1}$, the σ^Φ-ordering has the pleasant property that the same element σ_1 is least σ-positive in each braid group. For this reason, we see a difference in the two orderings in the limit. The reader may easily verify the following.

PROPOSITION 3.10. *The σ-ordering of B_∞ is dense, whereas the σ^Φ-ordering of B_∞ is discrete, with σ_1 being the least element larger than 1.*

COROLLARY 3.11. *The ordered set $(B_\infty, <)$ is order-isomorphic to $(\mathbb{Q}, <)$.*

PROOF. A well-known result of Cantor says that any two countable linearly ordered sets that are dense—there always exists an element between any two elements—and unbounded—there is no minimal or maximal element—are isomorphic: assuming that the sets are $\{a_n \mid n \in \mathbb{N}\}$ and $\{b_n \mid n \in \mathbb{N}\}$, one alternatively defines $f(a_0)$, $f^{-1}(b_0)$, $f(a_1)$, $f^{-1}(b_1)$, etc. so as to keep f order-preserving.

Here the rationals are eligible, and the set B_∞ is countable. So, in order to apply Cantor's criterion, it suffices to prove that $(B_\infty, <)$ is dense and unbounded. The former result is Proposition 3.10. The latter is clear: for every braid β, we have $\beta\sigma_1^{-1} < \beta < \beta\sigma_1$. □

Of course, the order-isomorphism of Corollary 3.11 could not be an isomorphism in the algebraic sense, as B_∞ is non-Abelian.

Every linearly ordered set has an order topology, with open intervals forming a basis for the topology. If the ordering is discrete, as is the case for the σ-ordering of B_n for $n < \infty$, then the topology is also discrete. Since B_∞, with the σ-ordering, is order isomorphic with the rational numbers, its order topology is metrizable. In fact, it has a natural metric, as follows.

PROPOSITION 3.12. *For $\beta \neq \beta'$ in B_∞, define $d(\beta, \beta')$ to be 2^{-p} where p is the greatest integer satisfying $\beta^{-1}\beta' \in \mathrm{sh}^p(B_\infty)$, completed with $d(\beta, \beta) = 0$. Then d is a distance on B_∞, and the topology of B_∞ associated with the linear order $<$ is the topology associated with d.*

PROOF. It is routine to verify that d is a distance. The open disk of radius 2^{-p} centered at β is the left coset $\beta\,\mathrm{sh}^p(B_\infty)$, *i.e.*, the set of all braids of the form $\beta\,\mathrm{sh}^p(\gamma)$.

Assume now that β_1, β, β_2 lie in B_n and $\beta_1 < \beta < \beta_2$ holds. We will show that the open d-disk around β of radius 2^{-n+1} is included in the interval (β_1, β_2). Indeed, if $d(\beta, \gamma) < 2^{-n+1}$, then $\beta^{-1}\gamma$ belongs to $\mathrm{sh}^n(B_\infty)$. The hypothesis $\beta_1 < \beta$ implies that $\beta_1^{-1}\beta$ is σ_i-positive for some $i \leqslant n-1$. Writing $\beta_1^{-1}\gamma = (\beta_1^{-1}\beta)(\beta^{-1}\gamma)$, we see that $\beta_1^{-1}\gamma$ is also σ_i-positive and, therefore, $\beta_1 < \gamma$ is true. A similar argument gives $\gamma < \beta_2$.

Conversely, let us start with an arbitrary open d-disk $\beta\,\mathrm{sh}^p(B_\infty)$. Let β' be a braid in this disk; we have to find an open $<$-interval containing β' which lies entirely in the disk. By hypothesis, we have $\beta' = \beta\,\mathrm{sh}^p(\gamma)$ for some γ of B_∞. Let γ_1 and γ_2 be any braids satisfying $\gamma_1 < \gamma < \gamma_2$. Then the interval $(\beta\,\mathrm{sh}^p(\gamma_1), \beta\,\mathrm{sh}^p(\gamma_2))$ contains $\beta\,\mathrm{sh}^p(\gamma)$ and is included in the disk, because $\mathrm{sh}^p(B_\infty)$ is convex—see Proposition 3.17 below. This completes the proof that the topologies associated with $<$ and with d coincide. □

3.3. Dense subgroups. It is clear that densely ordered groups can have subgroups which are discretely ordered (by the same ordering)—witness $\mathbb{Z}$ in $\mathbb{Q}$. But the reverse can happen, too. For example, the lexicographic ordering on $\mathbb{Q} \times \mathbb{Z}$ is discrete—with least positive element $(0, 1)$—whereas the subgroup $\mathbb{Q} \times \{0\}$ is densely ordered. This latter phenomenon happens quite naturally also for the braid groups.

Note that, if one allows the generators σ_i to commute, the braid relation $\sigma_i\sigma_{i+1}\sigma_i = \sigma_{i+1}\sigma_i\sigma_{i+1}$ implies that σ_i and σ_{i+1} become equal. From this one sees that the Abelianization of B_n is infinite cyclic, and the Abelianization map $B_n \to \mathbb{Z}$ can be identified with the sum of the exponents of a word in the σ_i generators. The commutator subgroup $[B_n, B_n]$ consists exactly of braids expressed in the generators σ_i with exponent sum zero.

Proposition 3.13 ([**37**]). *For $n \geqslant 3$, the commutator subgroup $[B_n, B_n]$ is densely ordered under the σ-ordering.*

Proof. For simplicity, we will prove this just for $n = 3$, referring the reader to [**37**] for the general case, whose proof is similar.

For contradiction, suppose $[B_3, B_3]$ has a least σ-positive element β. We consider the braid $\beta\sigma_2\beta^{-1}$. There are three possibilities:

Case 1: $\beta\sigma_2\beta^{-1}$ is σ_1-positive. Then β must be σ_1-positive. So is $\beta\sigma_2\beta^{-1}\sigma_2^{-1}$ and we have $1 < \beta\sigma_2\beta^{-1}\sigma_2^{-1}$. On the other hand, as β is σ_1-positive, $\sigma_2\beta^{-1}\sigma_2^{-1}$ is σ_1-negative, and we have $\sigma_2\beta^{-1}\sigma_2^{-1} < 1$ and $\beta\sigma_2\beta^{-1}\sigma_2^{-1} < \beta$. So the commutator $\beta\sigma_2\beta^{-1}\sigma_2^{-1}$ is a smaller σ-positive element of $[B_3, B_3]$ than β, contradicting the hypothesis on β.

Case 2: $\beta\sigma_2\beta^{-1}$ is σ_1-negative. A similar argument gives $1 < \beta\sigma_2^{-1}\beta^{-1}\sigma_2 < \beta$, again a contradiction.

Case 3: $\beta\sigma_2\beta^{-1}$ is σ_2^p for some p. Counting the exponents, we see that the only possibility is $p = 1$, *i.e.*, β commutes with σ_2. It is shown in [**84**] that the centralizer of the subgroup of B_3 generated by σ_2 is the subgroup (isomorphic to $\mathbb{Z} \times \mathbb{Z}$) generated by σ_2 and Δ_3^2, so we must have $\beta = (\sigma_1\sigma_2\sigma_1)^{2q}\sigma_2^r$ for some integers q, r. But, since β is σ_1-positive and a commutator, we have $q > 0$ and $6q + r = 0$. Now, consider $\beta' = \sigma_1\sigma_2^{-1}$. We have $\beta' > 1$ and $\beta' \in [B_3, B_3]$, and an easy calculation gives $\beta' < \beta$, again contradicting the hypothesis on β. □

Other subgroups of B_n with $n \geqslant 3$ which are shown to be densely ordered by the σ-ordering in [**37**] include the following:

• $[PB_n, PB_n]$, the commutator subgroup of the pure braid group; but PB_n itself is discretely ordered, with least positive element σ_{n-1}^2;

• the subgroup of Brunnian braids—defined as braids such that, for every strand, its removal results in a trivial braid;

• the subgroup of homotopically trivial braids, as considered in [**99**];

• kernels of the Burau representation for those n for which this representation is unfaithful—it is known to be unfaithful for $n \geqslant 5$ and faithful for $n \leqslant 3$.

The method of proof is to identify explicitly which braids can possibly be the least σ-positive elements of a given normal subgroup of B_n.

3.4. Convex subgroups. Convex subgroups play an important role in the theory of orderable groups.

DEFINITION 3.14. If $(G, \prec)$ is a left-ordered group, a subgroup H of G is said to be *convex* if, for all h, h' in H and g in G satisfying $h \prec g \prec h'$, one has $g \in H$.

An equivalent criterion for convexity of H is the conjunction of $1 \prec g \prec h$, $g \in G$, and $h \in H$ implies $g \in H$. It is easy to verify that the collection of convex subgroups of a given group is linearly ordered by inclusion. Moreover, if N is a normal convex subgroup of the left-ordered group G, then the quotient group G/N is left-orderable by ordering cosets according to their representatives.

If the ordering of G is discrete, and H is a convex subgroup distinct from $\{1\}$, then the ordering on H is also discrete, and H contains the minimal positive element of G, which is also minimal positive in H.

We shall see that there are rather few convex subgroups in the braid groups under the σ-ordering.

PROPOSITION 3.15. *The group B_n has no proper normal convex subgroup.*

PROOF. Suppose H is a normal and convex subgroup of B_n distinct of $\{1\}$. As remarked above, the minimal positive element σ_{n-1} of B_n belongs to H by convexity. Since H is normal, σ_1 also belongs to H, as the Garside braid Δ_n conjugates it to σ_{n-1}. All the other σ_i generators are positive and less than σ_1, so they must also be in H, and therefore we have $H = B_n$, alternatively, we can observe that all generators σ_i are conjugated to σ_{n-1} in B_n, as seen in Lemma I.4.4. □

PROPOSITION 3.16. *For i in $\{1, \dots, n-1\}$, let H_i be the subgroup of B_n generated by $\sigma_i, \dots, \sigma_{n-1}$. Then each subgroup H_i is convex in B_n and these are the only nontrivial convex subgroups.*

PROOF. First, we verify that H_i is convex. Suppose $1 < \gamma < \beta$ with $\beta \in H_i$ and $\gamma \in B_n$. Note that the σ-positive elements of H_i are exactly the σ_j-positive braids in B_n with $j \geqslant i$. So β is σ_j-positive for some $j \geqslant i$. By hypothesis, γ is σ_k-positive for some k in $\{1, \dots, n-1\}$. If we had $k < j$, then $\beta^{-1}\gamma$ would be σ_j-positive, implying $\beta < \gamma$ and contradicting the hypothesis. Therefore, we have $k \geqslant j \geqslant i$ and γ lies in H_i.

It remains to show that there are no other nontrivial convex subgroups. Assume that C is a convex subgroup of B_n distinct of $\{1\}$. Let i be the least positive integer such that C contains a σ_i-positive braid, say β. We will show that $C = H_i$. First note that C contains each σ_j with $j > i$, because $\sigma_j^{-1}\beta$ is σ_i-positive and we have $1 < \sigma_j < \beta \in C$.

Now we may write $\beta = \beta_0\sigma_i\beta_1\sigma_i \dots \sigma_i\beta_m$ for some $m \geqslant 1$ and some β_i belonging to H_{i+1}, hence to C. Since C is a subgroup and β_0 belongs to C, the braid β' defined by $\beta' = \sigma_i\beta_1\sigma_i \dots \sigma_i\beta_m$ also belongs to C. In case $m > 1$, we conclude $\sigma_i^{-1}\beta'$ is also σ_i-positive and therefore we have $1 < \sigma_i < \beta'$. On the other hand, if $m = 1$ holds, we have $\beta' = \sigma_i\beta_1$. In either case, we conclude that σ_i belongs to C. We have shown that C is included in H_i. If the inclusion were proper, then C would contain a braid which is σ_j-positive for some $j < i$, contradicting our choice of i. □

Almost exactly the same argument shows the following.

PROPOSITION 3.17. *The nontrivial convex subgroups of B_∞ are exactly those of the form* $\mathrm{sh}^i(B_\infty)$. *None of these is normal.*

Finally, using the flip automorphism Φ_n, we see that, when the σ^Φ-ordering $<^\Phi$ replaces the σ-ordering, then the convex subgroups of B_n are the groups B_i with $i \leqslant n$. The same holds for B_∞.

4. The σ-ordering of positive braids

In this section, we review some results about the restriction of the orderings $<$ and $<^\Phi$ to the braid monoids B_n^+, most of which will be further developed in Chapters VII and VIII. As the many examples of Section 2.1 showed, the σ-ordering is a quite complicated ordering. By contrast, its restriction to the monoid B_n^+ is a simple ordering, namely a well-ordering. In particular, every nonempty set of positive braids has a least element, and, if it is bounded, it has a least upper bound.

We give two proofs of the well-order property for the σ-ordering of B_n^+. Due to Laver [**137**] and based on Property **S**, the first one uses Higman's subword lemma, and it is not constructive. Then, we give another argument, which is constructive and much more precise. It is based on Serge Burckel's approach in [**27**]. Here we follow the new description of [**62**], which relies on an operation called the Φ_n-splitting of a braid. It shows that the ordering of B_n^+ is a sort of lexicographical extension of the ordering of B_{n-1}^+.

Most of the properties described in this section for the monoids B_n^+ extend to the case of the so-called dual braid monoids B_n^{+*}. Introduced by Birman, Ko, and Lee in [**15**], the dual monoid B_n^{+*} is a submonoid of B_n that properly includes B_n^+. Interestingly, the proofs turn out to be easier in the case of B_n^{+*} than in the case of B_n^+. We refer to Chapter VIII for details.

4.1. The well-order property. Restricting a linear ordering to a proper subset always gives a linear ordering, but the properties of the initial ordering and of its restriction may be very different—we already saw examples in Section 3.3. This is what happens with the σ-ordering of B_n and its restriction to B_n^+. For instance, we saw in Proposition 3.9 that $(B_n, <)$ is discrete, and that every braid β has an immediate predecessor, namely $\beta\sigma_{n-1}^{-1}$. The situation is radically different with B_n^+. In particular, $(B_n^+, <)$ has limit points: for instance, in $(B_3^+, <)$, the braid σ_1 is the least upper bound of the increasing sequence $(\sigma_2^p)_{p\geqslant 0}$; see Figure 2.

We recall that a linear ordering is called a *well-ordering* if every nonempty subset has a least element, or, equivalently, provided some very weak form of the Axiom of Choice is assumed, if it admits no infinite descending sequence. A direct consequence of Property **S** is the following important result.

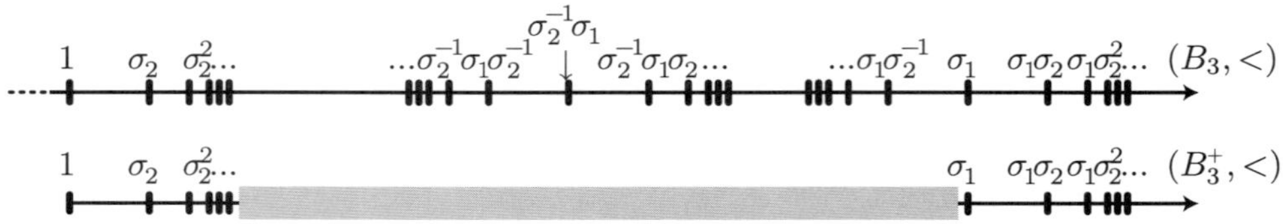

FIGURE 2. Restricting to positive braids completely changes the ordering: for instance, in $(B_3^+, <)$, the braid σ_1 is the limit of σ_2^p, whereas, in $(B_3, <)$, it is an isolated point with immediate predecessor $\sigma_2^{-1}\sigma_1$; the grey part in B_3 includes infinitely many braids, such as $\sigma_2^{-1}\sigma_1$ and its neighbours—and much more—but none of them lies in B_3^+.

PROPOSITION 4.1. *For every n, the restriction of $<$ to B_n^+ is a well-ordering.*

PROOF. A theorem of Higman [**108**], known as Higman's subword lemma, says: An infinite set of words over a finite alphabet necessarily contains two elements w, w' such that w' can be obtained from w by inserting intermediate letters (in not necessarily adjacent positions). Let $\beta_1, \beta_2, ...$ be an infinite sequence of braids in B_n^+. Our aim is to prove that this sequence is not strictly decreasing. For each p, choose a positive braid word w_p representing β_p. There are only finitely many n-strand braid words of a given length, so, for each p, there exists $p' > p$ such that $w_{p'}$ is at least as long as w_p. So, inductively, we can extract a subsequence $w_{p_1}, w_{p_2}, ...$ in which the lengths are nondecreasing. If the set $\{w_{p_1}, w_{p_2}, ...\}$ is finite, there exist k, k' such that w_{p_k} and $w_{p_{k'}}$ are equal, and then we have $\beta_{p_k} = \beta_{p_{k'}}$. Otherwise, by Higman's theorem, there exist k, k' such that w_{p_k} is a subword of $w_{p_{k'}}$, and, by construction, we must have $p_k < p_{k'}$. By Property **S**, this implies $\beta_{p_k} \leqslant \beta_{p_{k'}}$ in B_n^+. So, in any case, the sequence $\beta_1, \beta_2, ...$ is not strictly decreasing. □

The previous proof actually shows more.

PROPOSITION 4.2. *Assume that M is a submonoid of B_∞ generated by finitely many braids, each of which is a conjugate of some σ_i—hence of σ_1. Then the restriction of $<$ to M is a well-ordering.*

PROOF. In the proof of Proposition 4.1, Property **S** is used to ensure that, if a word w in the generators σ_i of B_n is a subword of another word w', then we have $\overline{w} \leqslant \overline{w'}$, where $\overline{w}$ denotes the braid represented by w. Now the same property holds for the generators of M, as each of them is a conjugate of some σ_i. Indeed, inserting a pattern of the form $v\sigma_i v^{-1}$ after w_1 in a braid word w_1w_2 amounts to inserting σ_i in the equivalent braid word $w_1vv^{-1}w_2$, and, therefore, the braid represented by $w_1 \cdot v\sigma_i v^{-1} \cdot w_2$ is larger than the braid represented by w_1w_2. □

Typically, the dual braid monoids investigated in Chapter VIII are eligible for Proposition 4.2.

REMARK 4.3. The hypothesis that the monoid M is finitely generated is crucial in Proposition 4.2. For instance, we already observed that the submonoid B_∞^+ of B_∞ is not well-ordered by the σ-ordering, as we have an infinite descending sequence $\sigma_1 > \sigma_2 > ...$. Such phenomena already occur inside B_3: for instance, the submonoid of B_3 generated by all conjugates $\sigma_2^{-p}\sigma_1\sigma_2^p$ of σ_1—and, more generally, the submonoid of all quasi-positive n-strand braids, defined to be the submonoid of B_n generated by all conjugates of $\sigma_1, \dots, \sigma_{n-1}$—contains the infinite descending sequence $\sigma_1 > \sigma_2^{-1}\sigma_1\sigma_2 > \sigma_2^{-2}\sigma_1\sigma_2^2 > \dots$.

Being a well-ordering has strong consequences. In particular, in contrast to what the examples of Section 2.1 showed, the well-order property implies the most general form of the phenomenon observed in Figure 2:

COROLLARY 4.4. *Every nonempty subset of B_n^+ is either cofinal or it has a least upper bound inside $(B_n^+, <)$.*

Indeed, for X included in B_n^+, unless X is unbounded in B_n^+, the set of all upper bounds of X is nonempty, hence it admits a least element.

4.2. The recursive construction of the ordering on B_n^+. We gave above a quick proof for Proposition 4.1, but the latter is not constructive, and it gives no direct description of the well-ordering $(B_n^+, <)$. We shall now give such a description, based on a recursive construction that connects $(B_{n-1}^+, <)$ and $(B_n^+, <)$. This approach leads in particular to considering the ordering of B_n^+ as an iterated extension of the ordering of B_2^+, *i.e.*, of the standard ordering of natural numbers.

To explain the results, it is crucial to use the flipped version of the σ-ordering, *i.e.*, the ordering $<^\Phi$ defined from σ^Φ-positive braids. The reason is that, although $(B_n^+, <)$ and $(B_n^+, <^\Phi)$ are isomorphic, the pairs $(B_n^+, B_{n-1}^+, <)$ and $(B_n^+, B_{n-1}^+, <^\Phi)$ are not, and the connection between B_n^+ and B_{n-1}^+ is more easily described in the case of $<^\Phi$.

The starting point of the approach is the following result from [**62**]. We recall that Φ_n denotes the flip automorphism (both of B_n and of B_n^+) that exchanges σ_i and σ_{n-i} for $1 \leqslant i \leqslant n-1$.

PROPOSITION 4.5. *Assume $n \geqslant 3$. Then, for each braid β in B_n^+, there exists a unique sequence $(\beta_p, \dots, \beta_1)$ in B_{n-1}^+ such that β admits the decomposition*

$$\beta = \Phi_n^{p-1}(\beta_p) \cdot \dots \cdot \Phi_n(\beta_2) \cdot \beta_1, \tag{4.1}$$

and for each r the only generator σ_i that right divides $\Phi_n^{p-r}(\beta_p) \cdot \dots \cdot \beta_r$ is σ_1. The sequence $(\beta_p, \dots, \beta_1)$ is called the Φ_n-splitting of β.

The result easily follows from the fact that every positive braid β of B_n^+ admits a unique maximal right divisor that lies in B_{n-1}^+. The unusual enumeration of the sequence from the right emphasizes that the construction starts from the right and involves right divisors.

Now, the main result says that, through the Φ_n-splitting, the ordering of B_n^+ is just a lexicographical extension of the ordering of B_{n-1}^+, more exactly a `ShortLex`-extension in the sense of [**77**], *i.e.*, the variant of the lexicographical extension in which the length is first taken into account.

PROPOSITION 4.6. *Assume $n \geqslant 3$. Let β, β' belong to B_n^+, and let $(\beta_p, \dots, \beta_1)$ and $(\beta'_{p'}, \dots, \beta'_1)$ be their Φ_n-splittings. Then $\beta <^\Phi \beta'$ holds if and only if $(\beta_p, \dots, \beta_1)$ is smaller than $(\beta'_{p'}, \dots, \beta'_1)$ for the `ShortLex`-extension of $(B_{n-1}^+, <^\Phi)$, i.e., we have either $p < p'$, or $p = p'$ and there exists $q \leqslant p$ satisfying $\beta_r = \beta'_r$ for $r > q$ and $\beta_q <^\Phi \beta'_q$.*

The result appears as Corollary VII.4.6, and it is also a consequence of Corollary VIII.3.3, with a disjoint argument.

The Φ_n-splitting of a positive braid can be computed easily, and a direct outcome of Proposition 4.6 is the existence, already mentioned in Section 1.5, of a quadratic upper bound for the complexity of the σ- and σ^{Φ}-orderings.

COROLLARY 4.7. *For each n, the orderings $<^{\Phi}$ and $<$ of B_n can be recognized in quadratic time.*

PROOF. We use induction on $n \geqslant 2$. Let w be an n-strand braid word of length ℓ. By Proposition I.4.6, we can obtain in time $O(\ell)$ two positive n-strand braid words w_1, w_2 such that w is equivalent to $w_1^{-1} w_2$. Then $\overline{w} >^{\Phi} 1$ is equivalent to $\overline{w_2} >^{\Phi} \overline{w_1}$. The Φ_n-splittings of the braids $\overline{w_1}$ and $\overline{w_2}$ can be computed in time $O(\ell^2)$; see Chapter VII. The induction hypothesis implies that the comparison of the sequences so obtained can be done in time $O(\ell^2)$ as well. The argument is similar for the σ-ordering as the shift automorphism Φ_n is computable in linear time. □

4.3. The length of $(B_n^+, <^{\Phi})$. Contrary to an arbitrary linear ordering, a well-ordering is completely determined up to isomorphism by a unique parameter, namely its length, usually specified by an ordinal number. In the case of the braid ordering on B_n^+, the length easily follows from the recursive characterization of Proposition 4.6.

We recall that ordinals are a transfinite continuation of the sequence of natural numbers: after the natural numbers comes ω, the first infinite ordinal, then $\omega + 1$, $\omega + 2$, etc. For our purposes, it is enough to know that ordinals come equipped with a well-ordering and with arithmetic operations (addition, multiplication, exponentiation) that extend those of $\mathbb{N}$. For more background information about ordinals, we refer to any textbook in set theory, for instance [**138**].

PROPOSITION 4.8. *For each n, the ordered set $(B_n^+, <^{\Phi})$ has ordinal type $\omega^{\omega^{n-2}}$.*

In other words, the length of $(B_n^+, <^{\Phi})$ is the ordinal $\omega^{\omega^{n-2}}$. The proof is an easy induction on n.

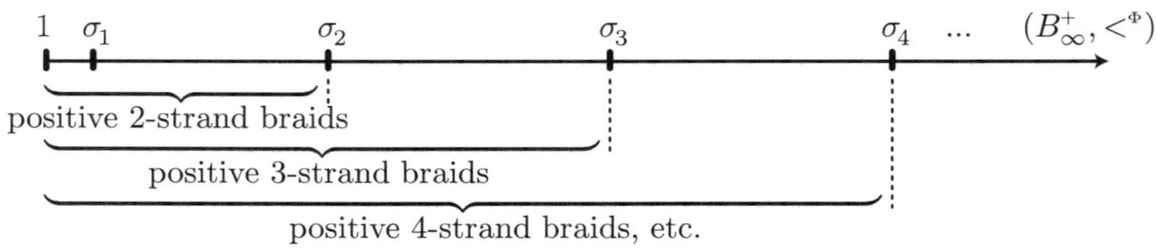

FIGURE 3. The well-ordered set $(B_\infty^+, <^{\Phi})$: an increasing union of end-extensions; for each n, the subset B_n^+ is the initial interval determined by σ_n.

By Proposition 2.10, the ordered set $(B_\infty^+, <^{\Phi})$ is the increasing union of the sets $(B_n^+, <^{\Phi})$, each set B_n^+ being an initial segment of the next one; see Figure 3. It is easy to deduce

PROPOSITION 4.9. *The ordered set $(B_\infty^+, <^{\Phi})$ is a well-ordering with ordinal type $\omega^{\omega^{\omega}}$.*

As the flip automorphism Φ_n preserves B_n^+ globally, the results about $(B_n^+, <^{\Phi})$ translate into similar results about $(B_n^+, <)$. In particular, Proposition 4.8 implies

COROLLARY 4.10. *For each n, the well-ordering $(B_n^+, <)$ has ordinal type $\omega^{\omega^{n-2}}$.*

However, we have no counterpart of Proposition 4.9 for $<$: the set B_n^+ is not an initial segment of $(B_\infty^+, <)$, and the latter is not a well-ordered set since it contains the infinite descending sequence of (1.1).

4.4. The rank of a positive braid. One of the nice features when an ordering $\prec$ of a set Ω is a well-ordering is that, for $x \in \Omega$, the position of x in $(\Omega, \prec)$ is unambiguously specified by an ordinal number, called the *rank* of x, namely the order type of the initial segment $\{y \in \Omega \mid y \prec x\}$. The rank function establishes an isomorphism between $(\Omega, \prec)$ and an initial segment of the sequence of ordinals: by construction, $x \prec x'$ is true if and only if the rank of x is smaller than the rank of x'.

So, in our current case, every positive braid β in B_n^+ is associated with a well-defined ordinal number, the rank of β, that specifies its position in $(B_n^+, <^\Phi)$. Moreover, Proposition 2.10 (or simply Figure 3) shows that the rank of β in $(B_n^+, <^\Phi)$ coincides with its rank in $(B_\infty^+, <^\Phi)$, and we can forget about the braid index.

Some values of the rank function are easily computed. For instance, the rank of the braid σ_i is the ordinal $\omega^{\omega^{i-2}}$ for $i \geqslant 2$: indeed, it is the ordinal type of the initial interval determined by σ_i. By Proposition 2.10, the latter is B_i, which, by Proposition 4.8, has ordinal type $\omega^{\omega^{i-2}}$. More values can be read in Figure 4.

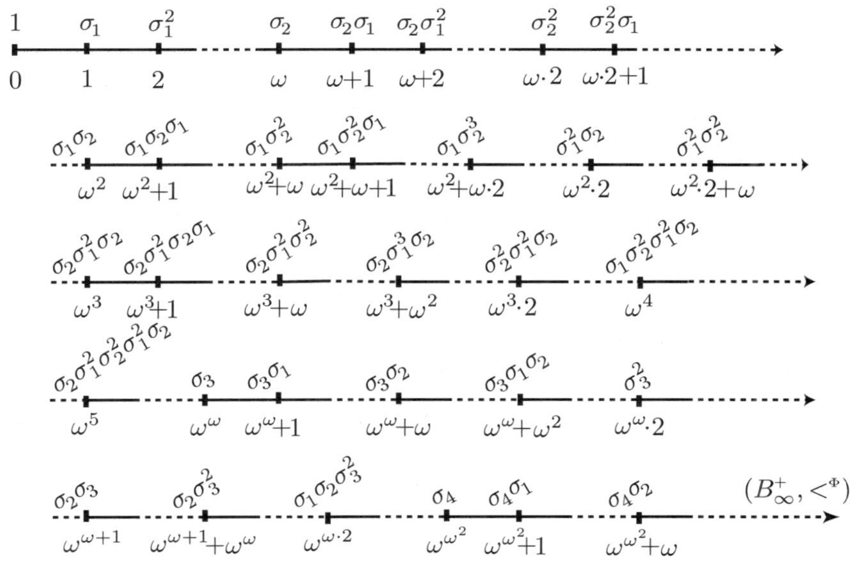

FIGURE 4. Ranks in the well-ordering $(B_\infty^+, <^\Phi)$: the position of each braid is unambiguously specified by an ordinal number that measures the length of the initial interval it determines.

REMARK 4.11. By construction, the rank mapping provides an order-isomorphism between positive braids and ordinals. Except for 2-strand braids, this mapping is *not* an algebraic homomorphism with respect to the ordinal sum: in general, the rank of $\beta_1\beta_2$ is not the sum of the ranks of β_1 and β_2. This happens to be true for $\beta_2 = \sigma_1$, which has rank 1, but, for instance, we can read in Figure 4 that the rank of σ_2 is ω, while that of $\sigma_1\sigma_2$ is ω^2, which is not $1+\omega$.

Arguably, an optimal description of $(B_\infty^+, <^\Phi)$ would consist of a closed formula explicitly computing, for each positive braid β, the rank of β, *i.e.*, determining the absolute position of β in $(B_\infty^+, <^\Phi)$. An algorithmic method has been described in [**28**], but, so far, it leads to no closed formula in the general case. However, in the case of 3-strand braids, such a formula exists. It relies on identifying distinguished word representatives called Φ-*normal*, from which the rank can be directly read.

DEFINITION 4.12. A nonempty positive 3-strand braid word $\sigma_{[p]}^{e_p}...\sigma_2^{e_2}\sigma_1^{e_1}$ is said to be Φ-*normal* if the inequalities $e_p \geqslant 1$ and $e_r \geqslant e_r^{\min}$ for $r < p$ are satisfied, where we set $e_1^{\min} = 0$, $e_2^{\min} = 1$, and $e_r^{\min} = 2$ for $r \geqslant 3$, and use $[p]$ to denote 1 for odd p, and 2 for even p.

So the criterion is that a positive 3-strand braid word w is Φ-normal if the successive blocks of letters σ_1 and σ_2 in w, enumerated from the right, and insisting that the rightmost block is a (possibly empty) block of σ_1, have a minimal legal size prescribed by the absolute numbers $e_r^{\min}$. It is easy to check that every nontrivial braid β of B_3^+ is represented by a unique Φ-normal word, naturally called its Φ-*normal form*. Then we have the following explicit formula for the rank.

PROPOSITION 4.13. *For each braid β in B_3^+, the rank of β in $(B_\infty^+, <^\Phi)$ is*

$$\omega^{p-1} \cdot e_p + \sum_{p>r\geqslant 1} \omega^{r-1} \cdot (e_r - e_r^{\min}), \tag{4.2}$$

where $\sigma_{[p]}^{e_p}...\sigma_2^{e_2}\sigma_1^{e_1}$ is the Φ-normal form of β.

This makes the description of the ordered set $(B_3^+, <^\Phi)$ complete.

EXAMPLE 4.14. The Φ-normal form of Δ_3 is $\sigma_1\sigma_2\sigma_1$, as the latter word satisfies the defining inequalities, contrary to $\sigma_2\sigma_1\sigma_2$, *i.e.*, $\sigma_2^1\sigma_1^1\sigma_2^1\sigma_1^0$, in which the third exponent from the right, namely 1, is smaller than the minimal legal value $e_3^{\min} = 2$. So, in this case, the sequence $(e_p, ..., e_1)$ is $(1, 1, 1)$, and, applying (4.2), we deduce that the rank of Δ_3 in $(B_3^+, <^\Phi)$ is $\omega^2 \cdot 1 + \omega \cdot (1-1) + 1 \cdot (1-0)$, *i.e.*, $\omega^2 + 1$. The reader can check that, more generally, the flip normal form of Δ_3^d corresponds to the length $d+2$ exponent sequence $(1, 2, ..., 2, 1, d)$, implying that the rank of Δ_3^d is the ordinal $\omega^{d+1} + d$. More values can be read in Figure 4.

4.5. Connection between positive and arbitrary braids. By Proposition I.4.6, every braid is a quotient of two positive braids. It follows that, in theory, the ordering of arbitrary braids is determined by its restriction to positive braids.

PROPOSITION 4.15. *Let $\beta_1, ..., \beta_p$ be a finite family of braids in B_n. Then, for d large enough, $\Delta_n^d\beta_1$, ..., $\Delta_n^d\beta_p$ lie in B_n^+, and the mutual positions of $\beta_1, ..., \beta_p$ in $(B_n, <)$ are the same as the mutual positions of the positive braids $\Delta_n^d\beta_1, ..., \Delta_n^d\beta_p$ in $(B_n^+, <)$.*

The result is clear, as the braid ordering $<$ is left-invariant. A similar result holds for $<^\Phi$.

However, it turns out that this result is of little help in establishing global properties of the braid ordering, and so far there is not much to say about the connection. We just mention two easy remarks involving the left numerators and denominators introduced in Proposition I.4.9 and their right counterpart.

PROPOSITION 4.16. *For each braid β, the right denominator $D_R(\beta)$ (resp. the left denominator $D_L(\beta)$) is the $<$-minimal positive braid β_1 such that $\beta\beta_1$ (resp. $\beta_1\beta$) is positive.*

PROOF. By construction, we have $\beta \cdot D_R(\beta) = N_R(\beta)$ and $D_L(\beta) \cdot \beta = N_L(\beta)$, and both $N_R(\beta)$ and $N_L(\beta)$ are positive braids.

Conversely, assume that β_1 and $\beta\beta_1$ lie in B_∞^+. Then we have $\beta = (\beta\beta_1)\beta_1^{-1}$. By the right counterpart of Proposition I.4.9, we have $\beta_1 = D_R(\beta)\gamma$ for some γ in B_∞^+. Necessarily γ is trivial or σ-positive, and, therefore, we have both $\beta_1 \geqslant D_R(\beta)$ and $\beta_1 \geqslant^\Phi D_R(\beta)$.

Symmetrically, assume that β_1 and $\beta_1\beta$ lie in B_∞^+. Then we have $\beta = \beta_1^{-1}(\beta_1\beta)$. By Proposition I.4.9, there exists γ in B_∞^+ satisfying $\beta_1 = \gamma D_L(\beta)$. As γ belongs to B_∞^+, Property **S** implies both $\beta_1 \geqslant D_L(\beta)$ and $\beta_1 \geqslant^\Phi D_L(\beta)$. □

PROPOSITION 4.17. *For each braid β, the relations $\beta > 1$ and $N_L(\beta) > D_L(\beta)$ are equivalent. Similarly, $\beta >^\Phi 1$ and $N_L(\beta) >^\Phi D_L(\beta)$ are equivalent.*

The verification is straightforward as $<$ and $<^\Phi$ are left-invariant. Note that no such relation exists with the right numerators and denominators: for instance, for $\beta = \sigma_2^{-1}\sigma_1$, we have $\beta > 1$, but $N_R(\beta) = \sigma_1\sigma_2 < D_R(\beta) = \sigma_2\sigma_1$.

The previous observations are rather trivial and do not shed much light on the structure of $(B_n, <)$. The point is that the fractionary decompositions defines two injections ι_L and ι_R of B_n into a subset of $B_n^+ \times B_n^+$, but neither of them preserves the ordered structure. On the other hand, we can easily define a well-ordering on $B_n^+ \times B_n^+$ by using a lexicographical extension of the ordering of B_n^+, and, appealing to ι_L or ι_R, deduce a well-ordering of B_n, but the latter will not be invariant under left (or right) multiplication.

CHAPTER III

Applications of the Braid Ordering

In this chapter, we gather the main applications of the braid ordering known so far. As was mentioned in the Introduction, several kinds of applications for the braid ordering can be considered, typically those that follow from the orderability of the braid groups, those that follow from the specific properties of the ordering in terms of σ-positivity, and those that more specifically involve the positive braid monoid.

The first category mainly involves results about torsion-freeness, absence of zero-divisors in group algebra, and, more generally, various results about the algebras RB_n where R is a ring. The second family includes, in particular, efficient solutions to the word problem, faithfulness criteria for representations of the braid group, as well as recent results about detection of prime knots and links, and pseudo-characters of braids. The third family contains some results of logic, namely unprovability statements that, roughly speaking, show that the braid ordering is so complicated, or so long, that certain properties cannot be established from the axioms of certain weak systems.

Although interesting, the applications described in this chapter are not so numerous—they may even appear as somehow disappointing—and they certainly do not exhaust the possibilities. We think that the braid ordering is a potentially powerful tool, and we hope for further applications. In particular, the well-order property is a very strong statement, and using it properly should lead to rich applications.

REMARK. In this chapter, we shall only mention applications of the braid ordering that directly involve braids and their orderings. Actually, there also exist results that are more indirect applications, namely results that were motivated by the investigation of braid orderings, even if they do not involve the latter directly. Typical of this family are the very recent results announced in [**109**] about descents of permutations; see Section 2.3 of Chapter XVI. Such results do not involve braids, but they answer questions that were directly inspired by the approach explained in Chapter VI, and, therefore, they can arguably be considered as applications of the σ-ordering of braids, as are most of the combinatorial results of [**60**].

More generally, the same comment applies to many results of the current book that do not involve any braid ordering, but have been inspired, at least in part, by the investigation of braid orderings. Witness, for instance, the Φ- and ϕ-normal forms of braids of Chapters VII and VIII, or the transmission-relaxation algorithm of Chapter XI.

The organization of this chapter follows the above-mentioned skeleton. In Section 1, we mention some general applications of orderability. In Section 2, we list applications that more specifically involve the σ-ordering and σ-positive braids. Finally, in Section 3, we consider applications that rely on the fact that the σ-ordering of positive braids is a well-ordering.

1. Consequences of orderability

The existence of the σ-ordering implies that the braid group B_n is left-orderable, but we have seen in Proposition II.1.2 that, for $n \geqslant 3$, it is not bi-orderable. This implies some algebraic consequences that we now review.

1.1. Torsion. In a left-orderable group, $1 \prec g$ implies $g^{-1} \prec 1$, and also $g \prec g^2 \prec g^3, \ldots$, and we conclude (with a similar argument for $g \prec 1$) that, if G is left-orderable, then G has no elements of finite order. In this way, we obtain a short proof of the following classical result.

PROPOSITION 1.1. *The braid groups are torsion-free.*

By contrast, for $n \geqslant 3$, the braid group B_n has generalized torsion, *i.e.*, a product of conjugates of a nontrivial element may be trivial, which is a sufficient reason for not being bi-orderable. Indeed, let $\beta = \sigma_1 \sigma_2 \sigma_1^{-1} \sigma_2^{-1}$. Then conjugating β by Δ_3 gives $\Delta_3 \beta \Delta_3^{-1} = \sigma_2 \sigma_1 \sigma_2^{-1} \sigma_1^{-1}$, and we find $\beta \cdot \Delta_3 \beta \Delta_3^{-1} = 1$: the braid β is not trivial, but some product of the conjugates of β is.

REMARK 1.2. The torsion-freeness of B_n follows from hypotheses that are much weaker than its orderability: the very simple argument of [**57**] shows that every group that is a group of fractions for a monoid which has no nontrivial unit and which admits least common multiples is torsion-free. The latter hypotheses are fulfilled by the braid groups as well as, much more generally, by all Garside groups of [**53**].

1.2. Group algebra. The following conjecture, dating from the first half of the twentieth century, is still unsolved. Suppose R is a ring (commutative, with unit) and G a group. The group ring RG is the free module generated by the elements of G, endowed with a multiplication in an obvious way. If G has a torsion element, say g in G has order p, then in $\mathbb{Z}G$ there are necessarily zero-divisors. For example, one calculates

$$(1-g)(1+g+g^2+\cdots+g^{p-1}) = 1-g^p = 0.$$

The zero-divisor conjecture claims that, if G is a torsion-free group and R has no zero-divisors, then the group ring RG also has no zero-divisors [**168**]. Frustrating as attempts at this conjecture have been, even for the ring $\mathbb{Z}$, the question is easily settled for left-orderable groups.

LEMMA 1.3. *The zero-divisor conjecture is true if "left-orderable" replaces "torsion-free" in the hypothesis.*

PROOF. Consider a product in RG, say

$$\Big(\sum_{i=1}^{p} r_i g_i\Big)\Big(\sum_{j=1}^{q} s_j h_j\Big) = \sum_{i,j} (r_i s_j)(g_i h_j)$$

with $h_1 \prec \ldots \prec h_q$. If $g_{i_0} h_{j_0}$ is a minimal term in the right hand side in the given ordering, use left-invariance to deduce $j_0 = 1$, and conclude that $g_{i_0} h_{j_0}$ is the unique minimal term, and, therefore, it cannot cancel with any other term. Similarly, the greatest term cannot cancel with any other term. So the product is nonzero unless all r_i's or all s_j's are zero and it is equal to 1 if and only if we have $p = q = 1$, $r_1 s_1 = 1$, and $g_1 h_1$ is the identity element of G. □

The previous argument also shows that, if G is a left-orderable group, there are no exotic units in RG: the only invertible elements are the monomials rg with r a unit of R and g in G.

In studying the braid groups B_n and their representations, the group rings $\mathbb{Z}B_n$ and $\mathbb{C}B_n$ are especially important. It was not until the proof that B_n is left-orderable that we knew the following fact.

PROPOSITION 1.4. *The rings $\mathbb{Z}B_n$ and $\mathbb{C}B_n$ have no zero-divisors, and consequently no idempotents.*

For bi-orderable groups we have a stronger conclusion, due independently to Malcev [**145**] and Neumann [**163**]: if G is bi-orderable, then $\mathbb{Z}G$ embeds in a skew field. We shall see in Chapter XV that the pure braid group PB_n is bi-orderable, so we deduce

PROPOSITION 1.5. *For every n, the group ring $\mathbb{Z}PB_n$ embeds in a skew field.*

The corresponding result for $\mathbb{Z}B_n$ has been proved by Linnell and Schicks recently [**141**].

We have observed that, with left-invariant orderings, we can have $g \prec h$ and $g' \prec h'$ but $gg' \succ hh'$; in a bi-ordered group, one easily establishes that $g \prec h$ and $g' \prec h'$ together imply $gg' \prec hh'$. In particular, $g \prec h$ implies $g^p \prec h^p$ for all positive p. So the bi-orderability of PB_n gives a new, short proof of the following:

PROPOSITION 1.6. *For every n, the group PB_n has unique roots; i.e., if β and β' are pure braids and β^p is equal to β'^p for some positive p, then β and β' are equal.*

The full braid groups B_n, with $n > 2$, certainly do not have unique roots. For instance $(\sigma_1\sigma_2)^3$ and $(\sigma_2\sigma_1)^3$ are equal in B_3 whereas $\sigma_1\sigma_2$ and $\sigma_2\sigma_1$ are distinct; they even determine distinct permutations. This example shows that a pure braid of PB_n can have multiple roots in B_n.

It was recently shown that nonisomorphic groups may have isomorphic integral group rings [**106**]. Another interesting property of orderable groups is that such a phenomenon is impossible if at least one of the groups is left-orderable [**130**]. Applying this result to the braid groups, we obtain

PROPOSITION 1.7. *Assume that G is a group and the ring $\mathbb{Z}G$ is isomorphic to $\mathbb{Z}B_n$. Then the group G is isomorphic to B_n.*

1.3. Analysis. Let G be an infinite discrete group, and let $L^2(G)$ denote the complex Hilbert space with Hilbert basis $\{g \mid g \in G\}$. The space $L^2(G)$ is the set of formal sums $\sum_{g\in G} a_g g$ with $a_g \in \mathbb{C}$ and $\sum_{g\in G} |a_g|^2 < \infty$. The group ring $\mathbb{C}G$ may be considered as the subset of $L^2(G)$ for which all but finitely many of the a_g are zero. If α, β are two elements of $L^2(G)$, say $a = \sum_{g\in G} a_g g$ and $b = \sum_{h\in G} b_h h$, the formal product defined by $ab = \sum_{g,h\in G} a_g b_h gh$ may not lie in $L^2(G)$ in general, but, if a belongs to $\mathbb{C}G$, then it does. It is conjectured that, if G is torsion-free, and a in $\mathbb{C}G$ and b in $L^2(G)$ are both nonzero, then ab is also nonzero. This is an extension of the zero divisor conjecture for group rings. Now, if G is left-orderable, and a in $\mathbb{C}G$ and b in $L^2(G)$ are both nonzero, then we can deduce $ab \neq 0$ [**140**]. In the case of braid groups, we thus obtain

PROPOSITION 1.8. *Assume $a \in \mathbb{C}B_n$ and $b \in L^2(B_n)$ with a and b nonzero. Then ab is nonzero.*

2. Applications of more specific properties

Besides the previous consequences of the fact that the braid groups are orderable, other properties follow from the specific characterization of the ordering in terms of σ-positive braid words, *i.e.*, from Properties **A** and **C**.

2.1. Faithfulness of representations. Property **C**, *i.e.*, the fact that every nontrivial braid is σ-positive or σ-negative, immediately provides the following criterion for establishing the faithfulness of a representation.

PROPOSITION 2.1. *Assume that f is a homomorphism of B_n into a group G such that the image under f of each σ-positive braid is not* 1. *Then f is injective.*

As will be seen in Chapter IX, the criterion applies to the well-known Artin representation of B_n in the automorphisms of a free group. In Proposition IX.1.6, we shall see that, if β is σ-positive, then the associated automorphism $\widehat{\beta}$ of the free group based on $\{x_1, \dots, x_n\}$ sends x_1 to a word that finishes with x_1^{-1} and, therefore, $\widehat{\beta}$ cannot be the identity. In this way, one (re)proves that the Artin representation is an embedding.

Other homomorphisms of B_n to $\mathrm{Aut}(F_n)$ have been defined by Wada in [**192**]. Using the criterion of Proposition 2.1, Shpilrain shows in [**184**] that some of them are faithful.

PROPOSITION 2.2. *Assume that F_n is the free group based on $\{x_1, \dots, x_n\}$. Then the following maps induce embeddings of B_n into* $\mathrm{Aut}(F_n)$:

- $\widehat{\sigma_i}(x_i) = x_i^p x_{i+1} x_i^{-p}$, $\widehat{\sigma_i}(x_{i+1}) = x_i$, $\widehat{\sigma_i}(x_k) = x_k$ *for* $k \neq i, i+1$, *where p is a fixed nonzero integer;*
- $\widehat{\sigma_i}(x_i) = x_i x_{i+1}^{-1} x_i$, $\widehat{\sigma_i}(x_{i+1}) = x_i$, $\widehat{\sigma_i}(x_k) = x_k$ *for* $k \neq i, i+1$;
- $\widehat{\sigma_i}(x_i) = x_i^2 x_{i+1}$, $\widehat{\sigma_i}(x_{i+1}) = x_{i+1}^{-1} x_i^{-1} x_{i+1}$, $\widehat{\sigma_i}(x_k) = x_k$ *for* $k \neq i, i+1$.

For instance, one can check that, in the last case, the image of a σ-positive braid β is an automorphism $\widehat{\beta}$ such that $\widehat{\beta}(x_1)$ begins with x_1^2—and, therefore, it cannot be the identity.

Linear representations can also be investigated from this viewpoint. In the case of the Burau representation, whose possible faithfulness is an open problem in the case of B_4, it is shown in [**61**] that the Burau image of a σ-positive 4-strand braid that admits a σ-positive word representative containing at most four letters σ_1 is not trivial—but this is far from enough to draw conclusions in the general case.

Whether the criterion applies to any of the other classical or recently discovered linear representations of the braid groups, such as the Lawrence–Krammer representation of [**128, 12**] which is known to be faithful, is an open question.

Finally, let us mention that (an extension of) Proposition 2.1 is used in [**59**] to show the faithfulness of the extension of Artin's representation to the group of so-called parenthesized braids; see Section XVI.3.6.

2.2. Efficient algorithms and cryptography. Some of the most convincing applications of the σ-ordering could be that it leads to efficient algorithms.

Using an ordering to pilot an algorithm is a natural idea. A direct realization of this vague principle is the handle reduction method that will be described in Chapter V. Indeed, both the intuition of the method and its correctness directly stem from the σ-ordering: the principle of handle reduction consists in getting rid of patterns $\sigma_i \dots \sigma_i^{-1}$ or $\sigma_i^{-1} \dots \sigma_i$ in a braid word so as to obtain a word that is σ-positive or σ-negative; hence, it is the most naive attempt to prove Property **C**. On the other hand, the convergence of the method relies on the fact that certain key subwords keep decreasing with respect to the σ-ordering when the algorithm is performed.

Handle reduction is, in practice, the most efficient solution to the braid word problem known so far. In addition, it is extremely simple to implement, and, therefore, it is relevant for possible uses of braids in applied mathematics and in cryptology.

It has been proposed to use braid groups as distinguished platform groups for developing new cryptosystems [**3, 126**]; for a survey see [**56**]. This is a very natural idea, because braid groups are neither too simple—they are non-Abelian and admit no obvious decompositions in terms of more simple groups—nor too complicated—the word problem is decidable, *i.e.*, there is no problem to unambigously specify an element of the group. However, some difficult problems quickly appear, because, for $n \geqslant 3$, the braid group B_n is not amenable, which makes it difficult to measure sets of braids and to prove probabilistic statements about braids. Anyway, several projects exist in this direction, and the subject is currently being investigated.

In addition to efficient solutions to the word problem, designing cryptographical protocols also requires hash-functions. Let us mention here that the encoding of B_n into $\mathbb{Z}^{2n}$ given by the formulas of Section XII.1 could be used to define a perfect collision-free hash-function on B_n, possibly giving another application of the σ-ordering to the subject.

Still another application of the material developed in this text to cryptography is a braid-based cryptographical protocol relying on the self-distributive operation $*$ on B_∞ defined in Section IV.1.2 [**142**].

2.3. Connection with knot theory. Braids are connected with knots and links under the closure operation. One associates with every braid β the oriented link represented by the diagram ("closed braid") $\widehat{\beta}$ obtained by connecting the output ends to the input ends as shown in Figure 1. Conversely, it is known that every oriented link, hence in particular every knot, is the closure of some braid; see for instance [**14**]. The generic question is to recognize the properties of the link represented by $\widehat{\beta}$ from those of β, typically whether it is a prime link. Following work by A.V. Malyutin and N.Yu. Netsvetaev [**150**], and by H. Matsuda, we shall see that the σ-ordering can be useful in this task: typically, a closure of a braid that is large in the σ-ordering has to represent a nontrivial link.

PROPOSITION 2.3 ([**150**]). *Assume that β is a braid in B_n satisfying $\beta < \Delta_n^{-4}$ or $\beta > \Delta_n^4$. Then the link represented by $\widehat{\beta}$ is prime, i.e., it is noncomposite, nonsplit, and nontrivial.*

The previous result is connected with the important notion of a pseudo-character.

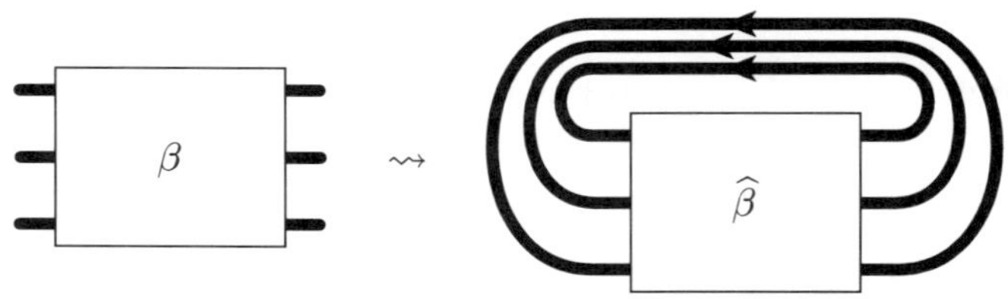

FIGURE 1. The closure of a braid, here a 3-strand braid.

DEFINITION 2.4. Assume that G is a group. A map $\chi : G \to \mathbb{R}$ is called a *pseudo-character* of G if the quantity

$$\sup_{g,h \in G} |\chi(gh) - \chi(g) - \chi(h)|, \tag{2.1}$$

called the *defect* of χ, is finite and, in addition, $\chi(g^p) = p\chi(g)$ holds for all g in G and p in $\mathbb{N}$.

It is easily seen that a pseudo-character is necessarily a conjugacy invariant, *i.e.*, it takes the same value on h and ghg^{-1} for all g, h.

In the case of braid groups, up to a multiplicative constant, the only $\mathbb{R}$-valued character, *i.e.*, the only pseudo-character with zero defect, is the exponent sum, *i.e.*, the homomorphism that takes each σ_i to 1.

It follows from deep results by Bestvina and Fujiwara about mapping class groups [**11**] that the space of all pseudo-characters on B_n is infinite dimensional, but only a few concrete examples are known. One example is the above-mentioned exponent sum. Another one is associated with the signature, *i.e.*, the function that maps β to the signature of the link represented by $\widehat{\beta}$—to obtain a pseudo-character, consider the limit of $\mathrm{sign}(\beta^p)/p$. Following J.M. Gambaudo and E. Ghys [**94**], its defect, which is related to the Meyer cocycle, is at most $2n$ for β in B_n.

The σ-ordering enables one to define one more pseudo-character on B_n.

DEFINITION 2.5. For β in B_n, denote by $\lfloor \beta \rfloor$ the unique integer r such that $\Delta_n^{2r} \leqslant \beta < \Delta_n^{2r+2}$; see Proposition II.3.6. Then the *twist* $\omega(\beta)$ of β is defined to be the limit of $\lfloor \beta^p \rfloor / p$ for $p \to \infty$.

PROPOSITION 2.6 ([**147**]). *For $n \geqslant 3$, the twist function is a pseudo-character on B_n with defect* 1. *It takes rational values, and it is the only pseudo-character on B_n that takes a nonnegative value on each σ-positive braid and takes the value* 1 *on Δ_n^2.*

We then have the following refinement of Proposition 2.3.

PROPOSITION 2.7 ([**147**]). *Assume that β is a braid in B_n satisfying $|\omega(\beta)| > 1$. Then the link represented by $\widehat{\beta}$ is prime.*

The result can be proved directly. It also follows from the stronger statement of [**148**] that, if χ is a pseudo-character of B_n that vanishes on B_{n-1} and has defect d, then $|\chi(\beta)| > d$ implies that $\widehat{\beta}$ is prime.

Other—yet related—questions involve Markov moves and their generalizations. The closures of conjugate braids are isotopic links, but, in the other direction, braids lying in different conjugacy classes may have isotopic closures. In this case, they can be connected by various transformations, such as Markov moves, flypes, or, more generally, moves associated with so-called templates.

A theorem of Birman and Menasco [**16**] states that, if β, β' are nonconjugate 3-strand braids such that $\widehat{\beta}$ and $\widehat{\beta'}$ represent the same link and the link is prime, then one can go from $\widehat{\beta}$ to $\widehat{\beta'}$ by one flype move. Another result of [**150**] is that, if β is a braid in B_3 that satisfies $\beta < \Delta_3^{-6}$ or $\beta > \Delta_3^6$, then $\widehat{\beta}$ is eligible for no flype move. We deduce the following corollary.

PROPOSITION 2.8. *Assume that β is a braid in B_3 that satisfies $\beta < \Delta_3^{-6}$ or $\beta > \Delta_3^6$. Then the link represented by $\widehat{\beta}$ corresponds to a unique conjugacy class in B_3.*

It is conjectured that a similar result holds for each B_n, *i.e.*, if β is a braid in B_n that satisfies $\beta < \Delta_n^{-2n}$ or $\beta > \Delta_n^{2n}$, then the link represented by $\widehat{\beta}$ corresponds to a unique conjugacy class in B_n. H. Matsuda (private communication, 2007) announced a proof for $n = 4$.

2.4. More braid properties. We still mention two applications in which the specific form of the σ-ordering is crucial.

The first one is an observation by Edward Formanek.

LEMMA 2.9. *Assume that β is a braid and some power of β lies in the image of the shift endomorphism. Then so does β.*

PROOF. Assume $\beta \in B_n \setminus \mathrm{sh}(B_{n-1})$. By Property **C**, the braid β is σ_1-positive, σ_1-negative, or σ_1-free. The latter is impossible, as it means that β lies in the image of sh. So β is σ_1-positive or σ_1-negative. Then, by construction, β^p is also σ_1-positive or σ_1-negative for each nonzero p. By Property **A**, this implies that β^p is not σ_1-free, *i.e.*, it does not belong to the image of sh. □

PROPOSITION 2.10. *For every n, the group B_n is isolated in B_∞; i.e., if β belongs to B_∞ and some power of β belongs to B_n, then β belongs to B_n.*

PROOF. Assume that β belongs to B_∞, and some nonzero power β^p belongs to B_n. Choose m such that β belongs to B_m and $m > n$ holds. Consider $\Phi_m(\beta)$. We recall that Φ_m denotes the flip automorphism of B_m that exchanges σ_i and σ_{m-i} for each i between 1 and $m-1$. The hypothesis that β^p belongs to B_n implies that $\Phi_m(\beta^p)$, which is $(\Phi_m(\beta))^p$, belongs to the image of the shift endomorphism. By Lemma 2.9, this implies that $\Phi_m(\beta)$ also belongs to the image of sh, hence that β belongs to B_{m-1}. Therefore, the smallest m such that β belongs to B_m is at most n. □

The second application involves the so-called palindromic braids. For each braid word w, let $\mathrm{rev}(w)$ denote the braid word obtained from w by reversing the order of the letters, *i.e.*, by reading w from right to left. As both sides of each of the braid relations of (I.1.1) are invariant under rev, the latter induces a well-defined antiautomorphism of B_n, still denoted rev, for every n.

DEFINITION 2.11. A braid β is said to be *palindromic* if $\mathrm{rev}(\beta) = \beta$ holds.

A motivation for investigating palindromic braids is that their closures are links that are invariant under the Weierstrass involution of the solid torus; see [**66**]. A simple way to obtain palindromic braids is to use the mapping π that sends every braid β to $\beta \cdot \mathrm{rev}(\beta)$. Studying the injectivity of the mapping π is the main goal of [**66**]. The σ-ordering gives an immediate answer:

PROPOSITION 2.12. *The mapping* $\pi : \beta \mapsto \beta \cdot \mathrm{rev}(\beta)$ *is injective on* B_∞.

PROOF. Assume $\beta\cdot\mathrm{rev}(\beta) = \beta'\cdot\mathrm{rev}(\beta')$. Let $\gamma = \beta^{-1}\beta'$. Then we have $\beta' = \beta\gamma$, and the hypothesis becomes $\beta \cdot \mathrm{rev}(\beta) = \beta\gamma \cdot \mathrm{rev}(\gamma)\mathrm{rev}(\beta)$, hence $\gamma \cdot \mathrm{rev}(\gamma) = 1$ by cancelling β and $\mathrm{rev}(\beta)$. By definition of the braid ordering, $\gamma > 1$ implies $\mathrm{rev}(\gamma) > 1$, hence $\gamma \cdot \mathrm{rev}(\gamma) > 1$, and, therefore, $\gamma \cdot \mathrm{rev}(\gamma) \neq 1$. Similarly, $\gamma < 1$ implies $\gamma \cdot \mathrm{rev}(\gamma) < 1$, and, therefore, $\gamma \cdot \mathrm{rev}(\gamma) \neq 1$. So the only possibility for obtaining $\gamma \cdot \mathrm{rev}(\gamma) = 1$ is $\gamma = 1$, *i.e.*, $\beta = \beta'$. □

2.5. Producing examples and counterexamples. The σ-ordering of braid groups is definitely a complicated ordering, witnessing to various nontrivial properties. So, besides the applications of the ordering to proving new properties of braids, we can also think of applications to the general theory of ordered groups, where the σ-ordering of B_n can be used to construct examples or counterexamples. This line of research seems quite promising and it might even turn out to provide the most interesting applications of the σ-ordering.

This is typically what is done by A. Navas in [**162**]: in this paper, braid groups and their orderings are mainly used as examples illustrating dynamical properties involving the set of all left-invariant orders on a group. We refer to Chapter XIV for more details about this approach.

Other examples are provided by the various specific properties of the σ-ordering. Witness all examples of Section II.2.1, which would not be easily realized in a generic left-ordered group. Among those properties for which examples may be rare, we may also think of the property that the restriction to some submonoid is a well-ordering with high ordinal type.

Further applications in the same vein may involve not the specific σ-ordering of B_n, but the whole space $LO(B_n)$ of all left-orderings on B_n. Typically, we shall see in Chapter XIV that B_n is a group such that the space $LO(B_n)$ of all left-orders on B_n has isolated points. It seems to be the only example known so far among groups with infinitely many left-orderings.

Similarly, we will see in Section XV.5 that B_4 is locally indicable but not bi-orderable, and that B_5 is left-orderable but not locally indicable. However, it has a finite index subgroup P_5 which is locally indicable, because it is bi-orderable. Not very many examples of such groups are known.

3. Application of well-orderability

The property that the restriction of the σ-ordering of braids to the braid monoids B_n^+ is a well-ordering is very strong, and we might expect striking applications. So far, not many have been identified, but we hope this will happen in the future. For the moment, we mention recent results from logic that exploit the existence of a well-ordering with high order type to deduce unprovability statements.

3.1. Distinguished elements. The well-order property asserts that every nonempty subset of B_∞^+ contains a $<^\Phi$-minimal element, and that every nonempty subset of B_n^+ contains a $<$-minimal element. This gives a very natural and powerful way to distinguish an element. For instance, we have

Proposition 3.1. *For each braid β in B_n^+, the intersection of the conjugacy class of β with B_n^+ contains a unique minimal element with respect to $<$.*

Let $\mu(\beta)$ denote the above minimal element. Note that, if we are able to algorithmically compute $\mu(\beta)$ for each β in B_n^+, then we obtain an immediate solution for the conjugacy problem in the group B_n. Indeed, if β, β' are any braids in B_n, we easily find a nonnegative integer d such that $\Delta_n^d\,\beta$ and $\Delta_n^d\,\beta'$ lie in B_n^+, and then β and β' are conjugate in B_n if and only if $\Delta_n^d\,\beta$ and $\Delta_n^d\,\beta'$ are, hence if and only if we have $\mu(\Delta_n^d\,\beta) = \mu(\Delta_n^d\,\beta')$.

This scheme is actually of little use so far, as we have as yet no way of computing the function μ. However, the very simple connection of the braid ordering with the Φ_n-splitting operation of Proposition II.4.5 may be seen as a promising sign; see Sections XVI.2.4 and XVI.2.5.

What we said for the conjugacy problem also applies to other similar problems, for instance the problem of identifying a unique distinguished braid representing each knot or link.

3.2. Unprovability statements. The σ^Φ-ordering of B_n^+ is a well-ordering with ordinal type $\omega^{\omega^{n-2}}$, and the σ^Φ-ordering of B_∞^+ is a well-ordering with ordinal type ω^{ω^ω}. These ordinals are not extremely large in the hierarchy of countable ordinals, but they are large enough to give rise to unprovability statements. The general idea is that, although the well-order property forbids that infinite descending sequences exist, there exist nevertheless finite descending sequences that are so long that their existence cannot be proved in weak logical systems.

We describe some results along this line of research, referring to [**33**] for details. In order to construct a long sequence of braids, we start with an arbitrary braid in B_3^+ and then repeat some transformation until, if ever, the trivial braid is obtained. Here, the transformation at step t will consist in removing one crossing, but, in all cases but one, introducing t new crossings. It is reminiscent of the Kirby–Paris' Hydra Game [**125**], with Hercules chopping off one head of the Hydra and the Hydra sprouting t new heads. The paradoxical result is that, contrary to what examples suggest, one always reaches the trivial braid after finitely many steps.

To make the description precise, we refer to the Φ-normal form of Definition II.4.12. Every 3-strand braid is represented by a unique Φ-normal diagram, consisting of blocks of σ_1 and σ_2, alternately. We define the *critical block* to be the rightmost block whose size exceeds the minimal legal size prescribed by the numbers $e_r^{\min}$, if such a block exists, and to be the leftmost block otherwise.

Definition 3.2 (Figure 2). For β is a nontrivial positive 3-strand braid, and t a positive integer, we define $\beta\{t\}$ to be the braid represented by the following diagram: in the Φ-normal diagram of β, we remove one crossing in the critical block, and add t crossings in the next block, if it exists, *i.e.*, if the critical block is not the final block of σ_1. The *$\mathcal{G}_3$-sequence from β* is defined by $\beta_0 = \beta$ and $\beta_t = \beta_{t-1}\{t\}$ for $t \geqslant 1$; it stops when the trivial braid 1 is possibly obtained.

It is easy to check that the $\mathcal{G}_3$-sequence from $\sigma_2^2\sigma_1^2$ has length 14: it consists of $\sigma_2^2\sigma_1^2$, $\sigma_2^2\sigma_1$, σ_2^2, $\sigma_2\sigma_1^3$, $\sigma_2\sigma_1^2$, $\sigma_2\sigma_1$, σ_2, σ_1^7, σ_1^6, σ_1^5, σ_1^4, σ_1^3, σ_1^2, σ_1, and finally 1. Similarly, the $\mathcal{G}_3$-sequence from Δ_3 has length 30. Not all examples are so easy: starting from $\sigma_1^2\sigma_2^2\sigma_1^2$, a braid with six crossings only, one does reach the trivial braid, but after no less than $90,159,953,477,630$ steps.

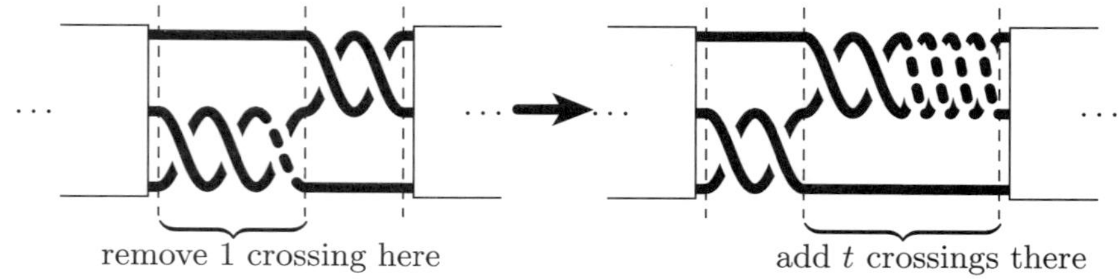

FIGURE 2. Inductive construction of the $\mathcal{G}_3$-sequence: at step t—here $t = 4$—we remove one crossing in the critical block, but add t new crossings in the next block unless the critical block is the final block of σ_1's.

PROPOSITION 3.3. *For each β in B_3^+, the $\mathcal{G}_3$-sequence from β is finite, i.e., there exists a finite number t for which we have $\beta_t = 1$.*

PROOF (SKETCH). The result follows from the conjunction of two results: the σ^Φ-ordering of B_3^+ is a well-ordering; hence, it possesses no infinite descending sequence, and every $\mathcal{G}_3$-sequence is descending with respect to $<^\Phi$. The latter is a consequence of the definition of $\beta\{t\}$ from β and of the connection between the σ^Φ-ordering of B_3^+ and the Φ-normal form. □

Although braids are not natural numbers, it should be clear that we can encode braids and their basic operations using natural numbers and the usual arithmetic operations. Therefore, it makes sense to speak of braid properties that can be proved from a certain system of arithmetical axioms: by this we mean that some reasonable encoding of braids by natural numbers has been fixed once for all and we consider the arithmetic counterpart of the braid property we have in mind.

The standard first-order Peano axiomatization of arithmetic consists of a few basic axioms involving addition and multiplication, plus the induction scheme, which asserts that, for each first-order formula $\Phi(x)$ involving $+, \times$ and $<$, the conjunction of $\Phi(0)$ and $\forall n(\Phi(n) \Rightarrow \Phi(n+1))$ implies $\forall n(\Phi(n))$. Weaker systems appear when one uses the same base axioms but restricts the induction principle to formulas of a certain type. For instance, $\mathsf{I}\Sigma_k$ denotes the subsystem of the Peano system in which the induction principle is restricted to the formulas Φ of the form $\exists x_1 \forall x_2 \exists x_3 \ldots Q x_k(\Psi)$, where Q is $\exists$ or $\forall$ according to the parity of k and Ψ is a formula that only contains bounded quantifications $\forall x < y$ and $\exists x < y$.

Most of the usual theorems involving braids turn out to be provable from the axioms of the subsystem $\mathsf{I}\Sigma_1$. By contrast, the above result about $\mathcal{G}_3$-sequences is the first result known so far that *cannot* be proved from the axioms of $\mathsf{I}\Sigma_1$.

PROPOSITION 3.4 ([**33**]). *Proposition 3.3 is an arithmetic statement that cannot be proved from the axioms of $\mathsf{I}\Sigma_1$.*

PROOF (SKETCH). Let $T(\beta)$ denote the length of the $\mathcal{G}_3$-sequence from β. Then the function $p \mapsto T(\Delta_3^p)$ grows so fast that it eventually dominates every function that can be proved to exist from the axioms of $\mathsf{I}\Sigma_1$. Technically, one uses the so-called Hardy hierarchy of fast-growing functions and the Ackermann function. □

Further results can be established. For instance, one can define $\mathcal{G}_\infty$-sequences that live in the monoid B_∞^+ and resemble $\mathcal{G}_3$-sequences in that they are both very long and descending in the braid ordering. As the order-type of $(B_\infty^+, <^\Phi)$ is larger than that of $(B_3^+, <^\Phi)$, namely ω^{ω^ω} instead of ω^ω, the sequences can be made longer, and proving their finiteness is therefore more difficult.

PROPOSITION 3.5. [**33**] *The finiteness of $\mathcal{G}_\infty$-sequences is an arithmetic statement that cannot be proved from the axioms of* $\mathsf{I}\Sigma_2$.

The previous results involve special sequences of braids, obtained by iterating some basic step. Other results involve general descending sequences of braids.

For β in B_3^+, define the *degree* $\deg(\beta)$ of β to be the least d such that β is a left divisor of Δ_3^d; see Chapter VI. For each d, there exist finitely many positive 3-strand braids of degree at most d. Hence, there exists an integer N—namely the number of 3-strand braids of degree at most d, plus 1—such that no descending sequence $(\beta_0, \dots, \beta_N)$ in $(B_3^+, <^\Phi)$ satisfies $\deg(\beta_t) \leqslant d$ for each t. Relaxing the bound on the degree leads to the following notion:

DEFINITION 3.6. For $f : \mathbb{N} \to \mathbb{N}$, we denote by $\boldsymbol{WO}_f$ the statement:

For each d, there exists N such that no descending sequence $(\beta_0, \dots, \beta_N)$ in $(B_3^+, <^\Phi)$ satisfies $\deg(\beta_t) \leqslant d + f(t)$ for each t.

So $\boldsymbol{WO}_f$ says that there is no very long descending sequence of braids with degree bounded by f. With this terminology, the above observation means that $\boldsymbol{WO}_f$ is true when f is a constant function. Actually, $(B_3^+, <^\Phi)$ being well-ordered easily implies that $\boldsymbol{WO}_f$ is true for every function f; *i.e.*, it is provable in some sufficiently strong system, for instance the full Peano system. However, using $\Box$ for the square function $x \mapsto x^2$, one can show that, if $\boldsymbol{WO}_\Box$ is provable from the axioms of some system S, then the finiteness of $\mathcal{G}_3$-sequences is also provable from these axioms. Therefore, Proposition 3.4 implies that $\boldsymbol{WO}_\Box$ cannot be proved from $\mathsf{I}\Sigma_1$. By contrast, for f constant, the principle $\boldsymbol{WO}_f$ can be proved from $\mathsf{I}\Sigma_1$. So we are led to looking for the transition between $\mathsf{I}\Sigma_1$-provability and $\mathsf{I}\Sigma_1$-unprovability.

The transition happens to be sharp. Indeed, denoting by Ack the standard Ackermann function and by Ack_r the level r approximation to Ack, and using f^{-1} for the functional inverse of f, we have

PROPOSITION 3.7 ([**33**]). *For $r \geqslant 0$, let f_r be defined by $f_r(x) = \lfloor \sqrt[\mathrm{Ack}_r^{-1}(x)]{x} \rfloor$, and f be defined by $f(x) = \lfloor \sqrt[\mathrm{Ack}^{-1}(x)]{x} \rfloor$.*

(i) *For each r, the principle $\boldsymbol{WO}_{f_r}$ is provable from the axioms of* $\mathsf{I}\Sigma_1$.

(ii) *The principle $\boldsymbol{WO}_f$ is not provable from the axioms of* $\mathsf{I}\Sigma_1$.

The functions involved in Proposition 3.7 are all of the form $x \mapsto \sqrt[g(x)]{x}$ where g is a very slowly increasing function. What is remarkable here is that a seemingly tiny change of the parameters causes the jump from provability to unprovability. The proof is a—rather sophisticated—mixture of combinatorial methods and specific results about the number of 3-strand braids satisfying some order and degree constraints, in the vein of those mentioned in Section 1 of Chapter VI.

REMARK 3.8. As was said above, braids can be encoded into natural numbers: typically, in the case of 3-strand braids, the Φ-normal form associates with every braid a finite sequence of natural numbers, namely the so-called exponent sequence. So all results in this section can be translated into results dealing with natural numbers exclusively, and one may wonder to which extent braids are really involved there. Actually, they arguably are, inasmuch as both the intuition for the definitions and the technical arguments used in the proofs directly come from the theory of braids and their specific ordering.

CHAPTER IV

Self-distributivity

This chapter presents an algebraic technique that was developed in the beginning of the 1990's, and which led to the first proof of the orderability of the braid groups [**45, 47**]. Subsequently, Richard Laver used a related method to prove that the same ordering is a well-ordering when restricted to positive braids [**137**]. The approach is complete in that it provides proofs of Properties **A**, **C**, and **S** of Chapter II—we recall that a synopsis of the properties can be found on page 309. In the current survey, the details are given only for the proof of a weak form of Property **C**, denoted $\mathbf{C}_\infty$. The argument for the latter is simple and natural in this approach, while those for Properties **A** and **S** are much more intricate, and we sketch them only.

It had been observed for many years [**20, 85**] that braids are connected with left self-distributive systems (LD-systems for short), an LD-system being defined as a set equipped with a binary operation satisfying the left self-distributivity law

$$\text{(LD)} \qquad x * (y * z) = (x * y) * (x * z).$$

When we think of $x * y$ as x acting on y, the LD-law expresses that the action preserves $*$-multiplication. In particular, David Joyce [**116**] and Sergei Matveev [**152**] introduced for every knot K a particular LD-system Q_K, the fundamental quandle of K, that characterizes the isotopy type of K up to a mirror image.

Here we use a different approach: instead of associating some particular LD-system S_β with every braid β, we choose one fixed LD-system S and use it for all braids uniformly by defining an action of B_n on n-tuples from S. It is not surprising that the action leads to an ordering on B_n when S happens to be an ordered LD-system, in a sense that will be made precise below. Most classically considered LD-systems, in particular all racks in the sense of [**85**], are connected with conjugation in a group and are strongly nonorderable. The point that made the construction described below possible is the discovery of new, orderable LD-systems at the end of the 1980's. It is worth mentioning that the first example of an orderable LD-system came from set theory and it relied on an unprovable large cardinal hypothesis. This example is not needed here, and it was never needed. The construction we shall describe below was made precisely in order to eliminate any use of this example. However, the whole story might never have happened without the hint from set theory that orderable LD-systems could exist.

This chapter is organized as follows. In Section 1, we show how to use self-distributive systems to colour the strands of a positive braid and, using a certain self-distributive operation defined on braids and the derived notion of a special braid, we deduce a proof of Property **C**. In Section 2, we study the question of extending the colourings to arbitrary braids and, at the expense of assuming the existence of an orderable LD-system, we deduce Property **A**. In Section 3, we introduce a certain group G_{LD} that captures some geometric aspects of the self-distributivity law and show how it leads to the existence of an orderable LD-system as needed to prove Property **A**. In Section 4, we describe R. Laver's approach to self-distributivity, and the way it leads to a proof of Property **S**. Finally, we mention in an appendix a connection between the self-distributive law and set theory, and the role of the latter in the discovery of the orderability of braid groups.

1. Colouring positive braids

The aim of this section is to observe that, whenever $(S, *)$ is an LD-system, there exists an action of the braid monoid B_n^+ on the n-fold product S^n. By using a specific LD-system whose elements are certain braids called special, and by resorting to some general properties of monogenerated LD-systems, we obtain a simple proof of Property **C** in B_∞, *i.e.*, of the result that every nontrivial braid admits a word representative in which the generator σ_i with lowest index i occurs only positively. The argument is effective, but it essentially takes place in B_∞: as a weird consequence, the σ-definite word representative we obtain for a braid of B_n may involve generators σ_i with i much larger than n.

1.1. The action of positive braids on an LD-system. We start with the idea of colouring the strands in a braid diagram. So, we assume that S is a fixed nonempty set and let w be a positive braid word. Then we attribute colours from S to the input ends in the diagram encoded by w, we propagate them along the strands, and we compare the sequence of output colours with the sequence of input colours.

If the colours are just pushed along the strands, the final sequence is a permutation of the initial one, and the only piece of information about our braid we obtain is its projection to the symmetric group.

Things become more interesting when we allow colours to change at crossings. Here we shall consider the case when the colour of the back strand is preserved, but the colour of the front strand may change depending on the two colours which have crossed. This amounts to using a function of $S \times S$ into S, *i.e.*, a binary operation $*$ on S, with the rule

Thus, we define a right action of n-strand positive braid words on S^n by

$$(1.1) \qquad \boldsymbol{x} \bullet \varepsilon = \boldsymbol{x}, \qquad \boldsymbol{x} \bullet \sigma_i w = (x_1, \dots, x_{i-1}, x_i * x_{i+1}, x_i, x_{i+2}, \dots, x_n) \bullet w,$$

where ε denotes the empty braid word—everywhere hereafter, when $\boldsymbol{x}$ denotes a sequence, we use $x_1, x_2, \dots$ for successive entries of $\boldsymbol{x}$.

LEMMA 1.1. *The action defined in* (1.1) *is compatible with the braid relations if and only if* $(S, *)$ *is an LD-system.*

PROOF. Compatibility with the relations $\sigma_i\sigma_j = \sigma_j\sigma_i$ with $|i-j| \geqslant 2$ is obvious. As for the relations $\sigma_i\sigma_j\sigma_i = \sigma_j\sigma_i\sigma_j$ with $|i-j| = 1$, we see in the diagrams of Figure 1 that compatibility is guaranteed if and only if the equality $x * (y * z) = (x * y) * (x * z)$ holds in S, *i.e.*, if $(S, *)$ is what we have called an LD-system. ☐

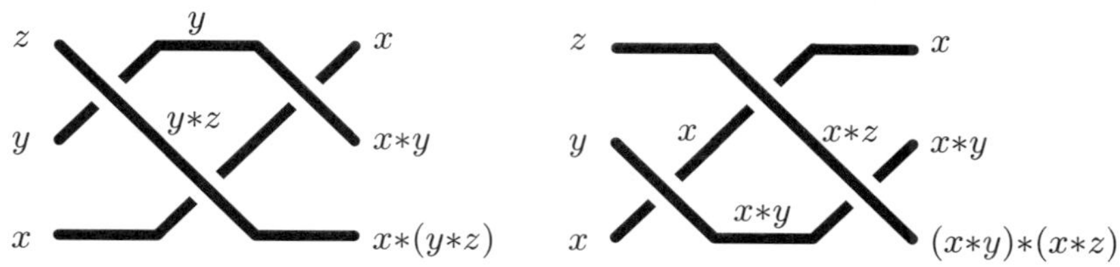

FIGURE 1. Compatibility of coloring with the braid relations: the output colours in the left and the right diagrams coincide if and only if the equality $x * (y * z) = (x * y) * (x * z)$ holds

We can therefore state:

PROPOSITION 1.2. *For every LD-system $(S, *)$ and every n, the rule* (1.1) *defines an action of B_n^+ on S^n.*

The natural question of whether the above action extends to not necessarily positive braids will be addressed in Section 2 below, but in this section it is enough to consider the case of B_n^+. Before going on, let us observe that several classical examples of LD-systems exist, and that they lead to not less classical results about braids.

EXAMPLE 1.3 (Permutation). Let S be an arbitrary set. Defining $x * y = y$ turns S into an LD-system. For $\boldsymbol{x}$ in S^n and β in B_n^+, we find

$$\boldsymbol{x} \bullet \beta = \mathrm{perm}(\beta)^{-1}(\boldsymbol{x}),$$

where $\mathrm{perm}(\beta)$ denotes the permutation induced by β. Thus the action gives the surjective homomorphism of B_n^+ onto the symmetric group $\mathfrak{S}_n$.

EXAMPLE 1.4 (Shift). Defining $x * y = y + 1$ turns $\mathbb{Z}$ into an LD-system. For $\boldsymbol{x}$ in $\mathbb{Z}^n$ and β in B_n^+, we find

$$\sum(\boldsymbol{x} \bullet \beta) = \ell(\beta) + \sum \boldsymbol{x}, \tag{1.2}$$

where $\sum \boldsymbol{x}$ denotes $x_1 + \cdots + x_n$, and $\ell(\beta)$ denotes the length of any positive braid word representing β.

EXAMPLE 1.5 (Mean). Let E be a $\mathbb{Z}[t]$-module. Defining $x * y = (1-t)x + ty$ turns E into an LD-system. For $\boldsymbol{x}$ in E^n and β in B_n^+, the entries of $\boldsymbol{x} \bullet \beta$ are linear combinations of those of $\boldsymbol{x}$, *i.e.*, we have

$$\boldsymbol{x} \bullet \beta = \boldsymbol{x} \cdot \rho(\beta)$$

for some $n \times n$-matrix $\rho(\beta)$. The mapping ρ is a linear representation of B_n^+, namely the (unreduced) Burau representation.

EXAMPLE 1.6 (Conjugation). Let G be a group. Defining $x * y = xyx^{-1}$ turns G into an LD-system. In particular, let F_n be the free group based on $x_1, \ldots, x_n$. For β in B_n^+, define elements $y_1, \ldots, y_n$ of F_n by

$$(y_1, \ldots, y_n) = (x_1, \ldots, x_n) \bullet \beta,$$

and let $\varphi(\beta)$ be the endomorphism of F_n that maps x_i to y_i for every i. Then φ is a homomorphism of B_n^+ into $\mathrm{End}(F_n)$, and its image is actually included in $\mathrm{Aut}(F_n)$. The action is the Artin representation of B_n^+ in $\mathrm{Aut}(F_n)$, to which we shall return in Chapter IX.

1.2. A self-distributive operation on B_∞. Hereafter, as in [**47**], we appeal to still another LD-system, namely one whose elements are braids. To this end, we introduce a self-distributive operation on B_∞ which is a sort of twisted conjugacy. We recall that sh denotes the shift endomorphism of B_∞, which maps σ_i to σ_{i+1} for every i.

DEFINITION 1.7. For β_1, β_2 in B_∞, we put

$$\beta_1 * \beta_2 = \beta_1 \cdot \mathrm{sh}(\beta_2) \cdot \sigma_1 \cdot \mathrm{sh}(\beta_1^{-1}), \tag{1.3}$$

where $\cdot$ refers to usual braid multiplication.

For instance, we find $1 * 1 = \sigma_1$, $1 * \sigma_1 = \sigma_2\sigma_1$, $\sigma_1 * 1 = \sigma_1^2\sigma_2^{-1}$, which shows that the operation $*$ is neither commutative nor idempotent. Note that, because of the shift operator in (1.3), the operation $*$ is defined on B_∞ only, and it induces no well-defined operation on B_n for any finite n.

At this point, formula (1.3) comes as a rabbit out of a hat. Several justifications can be given. In particular, we shall see in Remark 3.11 how (1.3) is connected with some natural operation arising on an extension of B_∞ and explaining all its properties. At this point, we shall content ourselves with direct verifications.

LEMMA 1.8. *The system $(B_\infty, *)$ is a left cancellative LD-system; i.e., the operation $*$ obeys the self-distributivity law and, in addition, $\gamma * \beta = \gamma * \beta'$ implies $\beta = \beta'$.*

PROOF. Let $\beta_1, \beta_2, \beta_3$ be any braids. Expanding the definition of (1.3), we find

$$\begin{aligned}
\beta_1 * (\beta_2 * \beta_3) &= \beta_1 \cdot \mathrm{sh}\beta_2 \cdot \mathrm{sh}^2\beta_3 \cdot \sigma_2 \cdot \mathrm{sh}^2\beta_2^{-1} \cdot \sigma_1 \cdot \mathrm{sh}\beta_1^{-1},\\
(\beta_1 * \beta_2) * (\beta_1 * \beta_3) &= (\beta_1 \cdot \mathrm{sh}\beta_2 \cdot \sigma_1 \cdot \mathrm{sh}\beta_1^{-1}) * (\beta_1 \cdot \mathrm{sh}\beta_3 \cdot \sigma_1 \cdot \mathrm{sh}\beta_1^{-1})\\
&= (\beta_1 \cdot \mathrm{sh}\beta_2 \cdot \sigma_1 \cdot \mathrm{sh}\beta_1^{-1}) \cdot \mathrm{sh}(\beta_1 \cdot \mathrm{sh}\beta_3 \cdot \sigma_1 \cdot \mathrm{sh}\beta_1^{-1})\\
&\qquad \cdot \sigma_1 \cdot \mathrm{sh}(\beta_1 \cdot \mathrm{sh}\beta_2 \cdot \sigma_1 \cdot \mathrm{sh}\beta_1^{-1})^{-1}\\
&= \beta_1 \cdot \mathrm{sh}\beta_2 \cdot \sigma_1 \cdot \mathrm{sh}\beta_1^{-1} \cdot \mathrm{sh}\beta_1 \cdot \mathrm{sh}^2\beta_3 \cdot \sigma_2 \cdot \mathrm{sh}^2\beta_1^{-1}\\
&\qquad \cdot \sigma_1 \cdot \mathrm{sh}^2\beta_1 \cdot \sigma_2^{-1} \cdot \mathrm{sh}^2\beta_2^{-1} \cdot \mathrm{sh}\beta^{-1}\\
&= \beta_1 \cdot \mathrm{sh}\beta_2 \cdot \sigma_1 \cdot \mathrm{sh}^2\beta_3 \cdot \sigma_2 \cdot \mathrm{sh}^2\beta_1^{-1}\\
&\qquad \cdot \sigma_1 \cdot \mathrm{sh}^2\beta_1 \cdot \sigma_2^{-1} \cdot \mathrm{sh}^2\beta_2^{-1} \cdot \mathrm{sh}\beta_1^{-1}.
\end{aligned}$$

As σ_1 commutes with every braid in the image of sh^2 and $\sigma_1\sigma_2\sigma_1\sigma_2^{-1} = \sigma_2\sigma_1$ holds, both $\beta_1 * (\beta_2 * \beta_3)$ and $(\beta_1 * \beta_2) * (\beta_1 * \beta_3)$ equal $\beta_1 \cdot \mathrm{sh}\beta_2 \cdot \mathrm{sh}^2\beta_3 \cdot \sigma_2\sigma_1 \cdot \mathrm{sh}^2\beta_2^{-1} \cdot \mathrm{sh}\beta_1^{-1}$, and the LD-law is satisfied.

Then expanding $\gamma * \beta = \gamma * \beta'$ gives $\gamma \cdot \mathrm{sh}\beta \cdot \sigma_1 \cdot \mathrm{sh}\gamma^{-1} = \gamma \cdot \mathrm{sh}\beta' \cdot \sigma_1 \cdot \mathrm{sh}\gamma^{-1}$, whence $\mathrm{sh}\beta = \mathrm{sh}\beta'$ as cancellation is legal in the group B_∞. This implies $\beta = \beta'$, as the shift endomorphism is injective. □

As it is an LD-system, $(B_\infty, *)$ is eligible for colouring positive braids: in this case, we use braids to color braid diagrams. The key point here is that the external action of (positive) braids on a sequence of braids can be connected with an internal multiplication inside B_∞.

NOTATION 1.9. For $(\beta_1, \dots, \beta_n)$ a sequence in B_∞^n, we write

$$\prod^{\mathrm{sh}}(\beta_1, \dots, \beta_n) = \beta_1 \cdot \mathrm{sh}(\beta_2) \cdot \ldots \cdot \mathrm{sh}^{n-1}(\beta_n).$$

LEMMA 1.10. *For every sequence* $(\beta_1, \ldots, \beta_n)$ *in* B_∞^n *and* β *in* B_n^+, *we have*

$$\prod^{\mathrm{sh}}((\beta_1, \ldots, \beta_n) \bullet \beta) = \prod^{\mathrm{sh}}(\beta_1, \ldots, \beta_n) \cdot \beta. \tag{1.4}$$

PROOF. We use induction on the length of β. The result is true for $\beta = 1$. Assume $\beta = \sigma_i\, \beta'$. First, we find

$$\begin{aligned}
\prod^{\mathrm{sh}}((\beta_1, \ldots, \beta_n) \bullet \sigma_i) &= \prod^{\mathrm{sh}}(\beta_1, \ldots, \beta_{i-1}, \beta_i * \beta_{i+1}, \beta_i, \ldots, \beta_n)\\
&= \beta_1 \cdot \mathrm{sh}\beta_2 \cdot \ldots \cdot \mathrm{sh}^{i-2}\beta_{i-1} \cdot \mathrm{sh}^{i-1}(\beta_i * \beta_{i+1}) \cdot \mathrm{sh}^i\beta_i \cdot \mathrm{sh}^{i+1}\beta_{i+2} \cdot \ldots \cdot \mathrm{sh}^{n-1}\beta_n\\
&= \beta_1 \cdot \mathrm{sh}\beta_2 \cdot \ldots \cdot \mathrm{sh}^{i-2}\beta_{i-1} \cdot (\mathrm{sh}^{i-1}\beta_i \cdot \mathrm{sh}^i\beta_{i+1} \cdot \sigma_i \cdot \mathrm{sh}^i\beta_i^{-1}) \cdot \mathrm{sh}^i\beta_i\\
&\qquad\qquad \cdot \mathrm{sh}^{i+1}\beta_{i+2} \cdot \ldots \cdot \mathrm{sh}^{n-1}\beta_n\\
&= \beta_1 \cdot \mathrm{sh}\beta_2 \cdot \ldots \cdot \mathrm{sh}^{i-2}\beta_{i-1} \cdot \mathrm{sh}^{i-1}\beta_i \cdot \mathrm{sh}^i\beta_{i+1} \cdot \sigma_i \cdot \mathrm{sh}^{i+1}\beta_{i+2} \cdot \ldots \cdot \mathrm{sh}^{n-1}\beta_n\\
&= \beta_1 \cdot \mathrm{sh}\beta_2 \cdot \ldots \cdot \mathrm{sh}^{i-2}\beta_{i-1} \cdot \mathrm{sh}^{i-1}\beta_i \cdot \mathrm{sh}^i\beta_{i+1} \cdot \ldots \cdot \mathrm{sh}^{n-1}\beta_n \cdot \sigma_i\\
&= \prod^{\mathrm{sh}}(\beta_1, \ldots, \beta_n) \cdot \sigma_i,
\end{aligned}$$

as σ_i commutes with $\mathrm{sh}^k\beta$ for $k \geqslant i+1$. Applying the induction hypothesis and using (1.1), we deduce

$$\begin{aligned}
\prod^{\mathrm{sh}}((\beta_1, \ldots, \beta_n) \bullet \beta) &= \prod^{\mathrm{sh}}(((\beta_1, \ldots, \beta_n) \bullet \sigma_i) \bullet \beta')\\
&= \prod^{\mathrm{sh}}((\beta_1, \ldots, \beta_n) \bullet \sigma_i) \cdot \beta'\\
&= (\prod^{\mathrm{sh}}(\beta_1, \ldots, \beta_n) \cdot \sigma_i) \cdot \beta' = \prod^{\mathrm{sh}}(\beta_1, \ldots, \beta_n) \cdot \beta,
\end{aligned}$$

which is (1.4). □

1.3. Special braids. Besides the full LD-system $(B_\infty, *)$, we shall also consider the sub-LD-system of $(B_\infty, *)$ generated by the unit braid 1. The braids lying in this subsystem are called special. They will play an important role what follows.

DEFINITION 1.11. We denote by B_{sp} the closure of $\{1\}$ in $(B_\infty, *)$. The elements of B_{sp} are called *special* braids.

A braid is special if it admits an expression that exclusively involves the braid 1 and the operation $*$. For instance, 1, σ_1, $\sigma_2\sigma_1$, $\sigma_1^2\sigma_2^{-1}$ are special braids, as we have $\sigma_1 = 1 * 1$, $\sigma_2\sigma_1 = 1 * (1 * 1)$, $\sigma_1^2\sigma_2^{-1} = (1 * 1) * 1$.

By definition, the set B_{sp} is closed under operation $*$, and, therefore B_{sp} equipped with (the restriction of) $*$ is an LD-system, and is therefore eligible for colouring positive braid diagrams. Applying Lemma 1.10, we deduce.

PROPOSITION 1.12. (i) *Every braid in* B_n^+ *can be expressed as*

$$\beta_1 \cdot \mathrm{sh}(\beta_2) \cdot \ldots \cdot \mathrm{sh}^{n-1}(\beta_n), \tag{1.5}$$

where $\beta_1, \ldots, \beta_n$ *are special braids.*

(ii) *Every braid in* B_n *can be expressed as*

$$\mathrm{sh}^{n-1}(\beta_n^{-1}) \cdot \ldots \cdot \mathrm{sh}(\beta_2^{-1}) \cdot \beta_1{}^{-1} \cdot \beta_1' \cdot \mathrm{sh}(\beta_2') \cdot \ldots \cdot \mathrm{sh}^{n-1}(\beta_n'), \tag{1.6}$$

where $\beta_1, \ldots, \beta_n, \beta_1', \ldots, \beta_n'$ *are special braids.*

PROOF. (i) Assume $\beta \in B_n^+$. Define

$$(\beta_1, \ldots, \beta_n) = (1, \ldots, 1) \bullet \beta.$$

As the input sequence $(1, \ldots, 1)$ consists of special braids and B_{sp} is closed under operation $*$, all braids involved in coloring a positive diagram associated with β are special, so, in particular, the output colours $\beta_1, \dots, \beta_n$ are special. Then, we have $\prod^{\mathrm{sh}}(1, \dots, 1) = 1$, so applying (1.4) directly gives (1.5).

By Proposition I.4.6, every braid in B_n can be expressed as a quotient $\beta^{-1} \cdot \beta'$ with β, β' in B_n^+. Then (ii) follows from applying (i) to β and β'. □

Thus we have obtained for every braid a decomposition in terms of special braids. It can be shown that, for β a positive braid, the expression of β given in (1.5) is unique, but we shall not need this result here.

EXAMPLE 1.13. Consider the braid $\sigma_1^{-2}\sigma_2\sigma_1$, which is $\beta^{-1}\beta'$ with $\beta = \sigma_1^2$ and $\beta' = \sigma_2\sigma_1$. Applying (1.1), we find

$$(1, 1, 1) \bullet \beta = (\sigma_1^2\sigma_2^{-1}, \sigma_1, 1) \quad \text{and} \quad (1, 1, 1) \bullet \beta' = (\sigma_2\sigma_1, 1, 1).$$

We deduce for $\sigma_1^{-2}\sigma_2\sigma_1$ the expression

$$\mathrm{sh}^2(1)^{-1} \cdot \mathrm{sh}(\sigma_1)^{-1} \cdot (\sigma_1^2\sigma_2^{-1})^{-1} \cdot (\sigma_2\sigma_1) \cdot \mathrm{sh}(1) \cdot \mathrm{sh}^2(1),$$

or, using the trivial braid 1 and the operations $*$, $^{-1}$ and sh exclusively,

$$\mathrm{sh}^2(1)^{-1} \cdot \mathrm{sh}(1 * 1)^{-1} \cdot ((1 * 1) * 1)^{-1} \cdot (1 * (1 * 1)) \cdot \mathrm{sh}(1) \cdot \mathrm{sh}^2(1), \tag{1.7}$$

i.e., when trivial terms are removed, $\mathrm{sh}(1 * 1)^{-1} \cdot ((1 * 1) * 1)^{-1} \cdot (1 * (1 * 1))$.

1.4. The Comparison Property. Our aim is to establish:

PROPOSITION 1.14 (Property $\mathbf{C}_\infty$). *Every braid is σ_1-positive, σ_1-negative, or σ_1-free.*

Equivalently, every braid admits a word representative in which the letters σ_1 and σ_1^{-1} do not both appear. Property $\mathbf{C}_\infty$ is Property $\mathbf{C}$ inside B_∞: the difference with $\mathbf{C}$ is that, in $\mathbf{C}_\infty$, we do not demand that for each n and each n-strand braid word there exists an equivalent n-strand braid word that is σ_1-positive, σ_1-negative, or σ_1-free. Note that Property $\mathbf{C}_\infty$ is sufficient to construct the σ-ordering of B_∞, and, from there, of each B_n by restriction.

When we look at (1.7) or, more generally, at (1.6), we see that possible letters $\sigma_1^{\pm 1}$ can come from the central factors β_1 and β'_1 only, as all other factors appear with a shift. We are thus left with the question of establishing Property $\mathbf{C}_\infty$ in the particular case of a braid that is the quotient of two *special* braids. We shall see now that, in such a case, the result directly follows from the specific definition of the operation $*$ and from general results about monogenerated LD-systems. The key role is played by the iterated left-divisibility relation.

DEFINITION 1.15. (i) For $(S, *)$ a binary system, and x, y in S, we say that x is an *iterated left divisor* of y and write $x \sqsubset y$ if there exists a positive integer p and elements $z_1, \dots, z_p$ in S satisfying

$$y = (\dots((x * z_1) * z_2) * \dots) * z_p. \tag{1.8}$$

(ii) We say that $(S, *)$ has the *Comparison Property* if any two distinct elements of S are comparable for $\sqsubset$, in the sense that at least one of $x \sqsubset y$ and $y \sqsubset x$ is true.

In a semigroup, the relation of iterated left-divisibility coincides with the standard left-divisibility relation, as $(x * z_1) * z_2$ is also $x * (z_1 * z_2)$. By contrast, in a general LD-system, there is no reason why the left-divisibility relation should be transitive, and that is why we consider iterated left-divisibility.

Then, the following general result holds in every monogenerated LD-system, *i.e.*, in every LD-system that is generated by a single element.

PROPOSITION 1.16. *Every monogenerated LD-system has the Comparison Property.*

A proof will be given in Section 1.5 below. For the moment, we shall see that Proposition 1.16 easily implies Property $\mathbf{C}_\infty$. The point is that the iterated left-divisibility relation in $(B_{\mathrm{sp}}, *)$ is directly connected with occurrences of σ_1 and σ_1^{-1}.

LEMMA 1.17. *Assume that β, β' are special braids satisfying $\beta \sqsubset \beta'$. Then the braid $\beta^{-1}\beta'$ is σ_1-positive.*

PROOF. By definition, saying that $\beta \sqsubset \beta'$ is true means that there exist special braids $\beta_1, \ldots, \beta_p$ satisfying

$$\beta' = (\ldots((\beta * \beta_1) * \beta_2) * \ldots) * \beta_p. \tag{1.9}$$

Expanding the latter product gives

$$\beta' = \beta \cdot \mathrm{sh}(\beta'_1) \cdot \sigma_1 \cdot \mathrm{sh}(\beta'_2) \cdot \sigma_1 \cdot \ldots \cdot \sigma_1 \cdot \mathrm{sh}(\beta'_{p+1}),$$

with $\beta'_1 = \beta_1$, $\beta'_2 = \beta^{-1}\beta_2$, $\beta'_q = ((\ldots((\beta*\beta_1)*\beta_2)*\ldots)*\beta_{q-1})^{-1}\beta_q$ for $3 \leqslant q \leqslant p$, and $\beta'_{p+1} = ((\ldots((\beta*\beta_1)*\beta_2)*\ldots)*\beta_p)^{-1}$. Hence the braid $\beta^{-1}\beta'$ admits a representative braid word containing p letters σ_1 and no letter σ_1^{-1}. □

PROPOSITION 1.18. *Proposition 1.16 implies Property $\mathbf{C}_\infty$: every braid in B_∞ is σ_1-positive, σ_1-negative, or σ_1-free.*

PROOF. Let $\beta \in B_\infty$. By Proposition 1.12(ii), we have a decomposition

$$\beta = \mathrm{sh}(\beta_2) \cdot \beta_1^{-1}\,\beta'_1 \cdot \mathrm{sh}(\beta'_2),$$

where β_1 and β'_1 are special braids. By construction, special braids form a monogenerated LD-system, namely the one generated by the braid 1. Then, by Proposition 1.16, at least one of the following three cases occur:

(i) $\beta_1 \sqsubset \beta'_1$: then, by Lemma 1.17, $\beta_1^{-1}\beta'_1$ is σ_1-positive, and, therefore, so is β, as the factors $\mathrm{sh}(\beta_2)$ and $\mathrm{sh}(\beta'_2)$ cannot destroy σ_1-positivity;

(ii) $\beta_1 = \beta'_1$: then β belongs to the image of sh, so it is σ_1-free;

(iii) $\beta'_1 \sqsubset \beta_1$: then, as in (i), β^{-1} is σ_1-positive and, therefore, β is σ_1-negative.

Thus Property $\mathbf{C}_\infty$ is proved. □

EXAMPLE 1.19. Let us consider the braid $\sigma_1^{-2}\sigma_2\sigma_1$ of Example 1.13 again. The point is to compare the braids $\beta_1 = (1 * 1) * 1$ and $\beta'_1 = 1 * (1 * 1)$ involved in (1.7) with respect to iterated left divisibility in $(B_{\mathrm{sp}}, *)$. In the current case, it is easy to see that $\beta_1 * \beta_1 = \beta'_1$ holds by applying left self-distributivity twice:

$$1 * (1 * 1) = (1 * 1) * (1 * 1) = ((1 * 1) * 1) * ((1 * 1) * 1).$$

So we have $\beta_1 \sqsubset \beta'_1$, and the computation of Lemma 1.17 then gives

$$\beta_1^{-1}\beta'_1 = \mathrm{sh}(\beta_1) \cdot \sigma_1 \cdot \mathrm{sh}(\beta_1)^{-1} = (\sigma_2^2\sigma_3^{-1}) \cdot (\sigma_1) \cdot (\sigma_3\sigma_2^{-2}),$$

an expression that contains one σ_1 and no σ_1^{-1}. Introducing these values in (1.7), we finally obtain

$$\sigma_1^{-2}\sigma_2\sigma_1 = (\sigma_2^{-1}) \cdot (\sigma_2^2\sigma_3^{-1}) \cdot (\sigma_1) \cdot (\sigma_3\sigma_2^{-2}) = \sigma_2^{-1}\sigma_2^2\sigma_3^{-1}\sigma_1\sigma_3\sigma_2^{-2},$$

which contains no σ_1^{-1}, contrary to the initial expression $\sigma_1^{-2}\sigma_2\sigma_1$. Of course, the latter expression can be shortened into $\sigma_2\sigma_1\sigma_2^{-2}$ by applying further reduction and braid relations, but our point is to show that the method of Proposition 1.18 is effective. Actually, our algorithm is still incomplete, as we did not show how to effectively obtain the special braids witnessing the relation $\beta_1 \sqsubset \beta_1'$: in the case above, we just guessed the values; finding them systematically requires the methods explained in Section 3.

REMARK 1.20. The previous argument takes place in B_∞, but not in any particular B_n: in Example 1.19, we started with a braid in B_3, but obtained a σ_1-positive expression that involves the letter σ_3. In general, the braid representative provided by Proposition 1.18 depends on the special braids β_q witnessing the iterated left-divisibility relation of (1.9). The techniques of Section 3 give an effective upper bound for the letters σ_i possibly occurring in their decompositions, but a huge one. Let us mention that Richard Laver, using normal forms methods in free LD-systems analogous to those mentioned in Section 4 below, has given a proof of Property **C** for each fixed group B_n in an unpublished work written around 1994.

1.5. The free monogenerated LD-system. It remains to establish Proposition 1.16, *i.e.*, to prove that every monogenerated LD-system has the Comparison Property. To this end, we follow the argument of [**44**], building on results of [**43**] about free LD-systems.

LEMMA 1.21. *Assume that S is an LD-system that has the Comparison Property and S' is a homomorphic image of S. Then S' has the Comparison Property as well.*

PROOF. Let f be a homomorphism of S onto S'. Assume $x \sqsubset y$ in S. Then, for some $p \geqslant 1$ and $z_1, \dots, z_p$ in S, we have $y = (...((x * z_1) * z_2)...) * z_p$ in S. As f is a homomorphism, we deduce $f(y) = (...((f(x) * f(z_1)) * f(z_2))...) * f(z_p)$, hence $f(x) \sqsubset f(y)$ in S'.

Let x', y' be any two elements in S'. By hypothesis, there exist x, y in S satisfying $x' = f(x)$ and $y' = f(y)$. As S has the Comparison Property, at least one of $x = y$, $x \sqsubset y$, $y \sqsubset x$ holds in S. Therefore, at least one of $x' = y'$, $x' \sqsubset y'$, $y' \sqsubset x'$ holds in S', and S' has the Comparison Property. □

For trivial reasons, there exists a most general monogenerated LD-system with the universal property that every monogenerated LD-system is a homomorphic image of it, namely the *free* monogenerated LD-system. Lemma 1.21 shows that, in order to establish Proposition 1.16, it is sufficient to establish

PROPOSITION 1.22. *The free monogenerated LD-system has the Comparison Property.*

The free monogenerated LD-system is the counterpart of the additive semigroup $\mathbb{Z}_+$ of positive integers when self-distributivity replaces associativity. Note that the counterpart of Proposition 1.22 is then true: in this case, iterated divisibility is the relation $\exists z(y = x + z)$, *i.e.*, the standard ordering, and the Comparison Property is the assertion that any two positive integers are comparable. So our aim is to prove a counterpart of the latter property when the LD-law replaces associativity. As the LD-law is more complicated than associativity, the proof will be less easy, but, at least, the property itself should appear natural.

In order to prove Proposition 1.22, we need to start from a description of free LD-systems. When we deal with associative structures like groups or monoids, the elements of a system generated by a set X can be described as the evaluation at X of words, *i.e.*, of finite sequences of letters. In particular, the free monoid generated by X can be described as the family of all words built from X, and the free group generated by X can be described as the quotient of the set of all words built from X and a copy X^{-1} of X obtained by collapsing all pairs xx^{-1} and $x^{-1}x$. When we deal with nonassociative structures like LD-systems, we can no longer start from words as, in general, the position of the brackets matters. Instead, we shall use *terms*, which are formal expressions involving letters, brackets, and the symbol $*$.

DEFINITION 1.23. For $n \geqslant 1$, we denote by T_n the set of all well-formed expressions constructed using letters among $x_1, \dots, x_n$ and the binary operator $*$, namely the closure of $\{x_1, \dots, x_n\}$ under the operation $(t_1, t_2) \mapsto t_1 * t_2$. The elements of T_n are called *terms*.

Thus, x_2, $x_1 * x_3$, $(x_1 * x_1) * x_1$ are terms, while $*x_2x_1**$ is not a term. The system $(T_n, *)$ is sometimes called the *absolutely free system*—or the free magma in [**17**]—generated by $x_1, \dots, x_n$. It has the universal property that every binary system generated by n elements $a_1, \dots, a_n$ is a homomorphic image of T_n under the evaluation mapping that takes x_n to a_n. A term t involving $x_1, \dots, x_n$ can be adequately compared with an n variable polynomial. In our case, it is often useful to see terms as labelled binary trees: every term that is not a single letter has well defined left and right subterms, and we inductively define the tree associated with a term t to be the binary tree consisting of a root with two successors, namely a left subtree which is the tree associated with the left subterm of t, and a right subtree which is the tree associated with the right subterm of t. For instance, the trees associated with the terms above, respectively, are

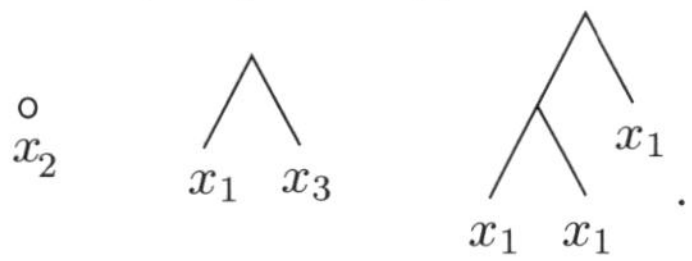

As said above, every binary system generated by $x_1, \dots, x_n$ is a quotient of $(T_n, *)$. For instance, the free monoid generated by $x_1, \dots, x_n$ is the quotient of T_n under the equivalence relation that identifies two terms t, t' if and only if one can go from t to t' by applying associativity, *i.e.*, by moving brackets: thus the equivalence class of t is fully characterized by the word obtained from t by forgetting all brackets, and we recover the description of the free monoid generated by $x_1, \dots, x_n$ as the collection of all words on $\{x_1, \dots, x_n\}$. In the case of the LD-law, we have a similar description.

DEFINITION 1.24. We say that two terms t, t' in T_n are *LD-equivalent*, denoted $t =_{LD} t'$, if we can go from t to t' by applying finitely many transformations consisting in replacing a subterm of the form $t_1 * (t_2 * t_3)$ with the corresponding term $(t_1 * t_2) * (t_1 * t_3)$ or vice versa; see Figure 2.

For instance, we have $x * (y * z) =_{LD} ((x * y) * x) * ((x * y) * z)$—we write $x, y, \dots$ for $x_1, x_2, \dots$—because we can go from the first term to the second by applying (LD) twice:

$$x * (y * z) =_{LD} (x * y) * (x * z) =_{LD} ((x * y) * x) * ((x * y) * z): \tag{1.10}$$

in the first step, we distribute x to y and z, in the second step, we distribute $x * y$ to x and z.

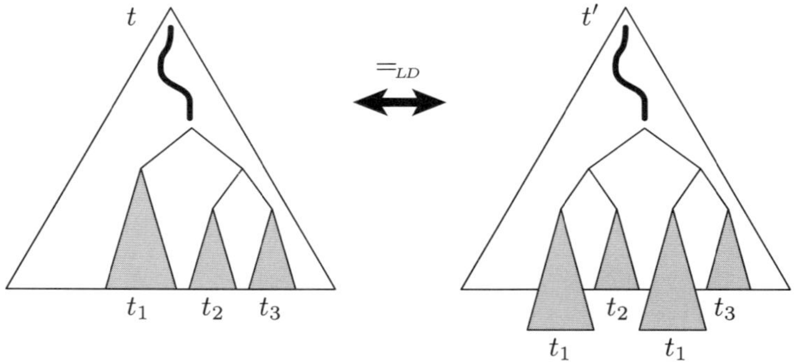

FIGURE 2. Applying the LD-law to a term viewed as a tree: some subterm which can be expressed as $t_1 * (t_2 * t_3)$ is replaced with the corresponding term $(t_1 * t_2) * (t_1 * t_3)$, or *vice versa*; note that the size of the tree (number of leaves) necessarily changes as t_1 is duplicated, and that the height may change as well—contrary to the case of the associativity law.

LEMMA 1.25. *For every n, LD-equivalence is an equivalence relation on T_n compatible with the operation $*$. The quotient-system $T_n/{=_{LD}}$ is a free LD-system of rank n—i.e., every LD-system generated by at most n elements is a homomorphic image of $T_n/{=_{LD}}$.*

PROOF. The relation $=_{LD}$ is an equivalence relation by construction, and it is compatible with the operation $*$ because we insisted that (LD) can be applied to any subterm of the considered term. To see that the quotient-system $T_n/{=_{LD}}$ is a free LD-system based on $\{x_1, \dots, x_n\}$, it suffices to check the expected universal property. So, assume that $(S, *)$ is an LD-system and $a_1, \dots, a_n$ are arbitrary elements of S. Then there exists a unique homomorphism f of T_n to S mapping x_i to a_i for each i, namely the evaluation mapping. The hypothesis that $(S, *)$ is an LD-system implies that f is constant on each LD-equivalence class, hence f induces a well-defined homomorphism of $T_n/{=_{LD}}$ to S. □

The above result is trivial, and, at this point, we still have no concrete realization of free LD-systems as, contrary to the case of associativity, we have no simple, effective description of LD-equivalence classes. Our method in the rest of this section consists of analyzing LD-equivalence by privileging one of the possible orientations of the LD-law.

DEFINITION 1.26. Assume $t, t' \in T_n$. We say that t' is an *LD-expansion* of t if we can go from t to t' by applying finitely many transformations consisting in replacing a subterm $t_1 * (t_2 * t_3)$ with the corresponding term $(t_1 * t_2) * (t_1 * t_3)$.

Thus the only difference between LD-equivalence and LD-expansion is that in the latter we only apply the LD-law in the expanding direction, *i.e.*, going from left to right in Figure 2. For instance, each step of (1.10) is an LD-expansion, so the term $((x * y) * x) * ((x * y) * z)$ is an LD-expansion of the term $x * (y * z)$—but the converse is not true: being an LD-expansion is an antisymmetric relation.

The proof that the free monogenerated LD-system has the Comparison Property relies on three specific properties of the LD-law. The first, and most important one, is as follows. By construction, LD-equivalence is the equivalence relation generated by LD-expansions, so, if two terms t, t' are LD-equivalent, there exists a sequence $t = t_0, t_1, \ldots, t_{2p} = t'$ such that t_i is an LD-expansion of t_{i-1} and t_{i+1} for each odd i. The point is that this zigzag may always be assumed to have length two.

LEMMA 1.27. *Two LD-equivalent terms admit a common LD-expansion.*

PROOF (SKETCH, SEE ALSO LEMMA 3.5). The point is to prove that, if t' and t'' are any two LD-expansions of some term t, then t' and t'' admit a common LD-expansion. To prove this, let us say that t' is a k-expansion of t if t' is obtained from t by applying the LD-law at most k times in the expanding direction. Then, for every term t, one can explicitly define a certain LD-expansion ∂t of t that is a common LD-expansion of all 1-expansions of t, and check that, if t' is an LD-expansion of t, then $\partial t'$ is an LD-expansion of ∂t. Then, using an induction, one shows that for every k the term $\partial^k t$ is an LD-expansion of all k-expansions of t. It follows that, if t' and t'' are any two LD-expansions of some term t, then t' and t'' admit common LD-expansions, namely all terms $\partial^k t$ with k sufficiently large.

The definition of the term ∂t is inductive: ∂t equals t when t is a single letter, and $\partial(t_1 * t_2)$ is obtained by replacing each letter x_i in ∂t_2 with the term $\partial t_1 * x_i$. For instance, the reader can check that one has

$$\partial(x * (x * (x * x))) = ((x{*}x){*}(x{*}x)){*}((x{*}x){*}(x{*}x))$$

—and that $\partial^2(x * (x * (x * x)))$ is a complicated term with 42 letters x. $\square$

The second property is specific to the case of terms in T_1, *i.e.*, to trees where all leaves are given the same label x. Hereafter, we use a distinguished sequence of such terms, recursively defined by $x^{[1]} = x$, $x^{[k+1]} = x * x^{[k]}$. Pictorially, $x^{[k]}$ corresponds to a right comb of length k.

LEMMA 1.28. *For each t in T_1, we have $x^{[k+1]} =_{LD} t * x^{[k]}$ for k sufficiently large.*

PROOF. Define the size of a term t of T_1 to be the number of occurrences of x in t. By definition, any term in T_1 is either the single letter x or it is $t_1 * t_2$, where t_1 and t_2 have smaller size than t. So we can argue using induction on the size of t. For $t = x$, we have $x^{[k+1]} = x * x^{[k]}$ for every k, by definition. Assume now $t = t_1 * t_2$. Assuming the result is true for t_1 and t_2, we obtain for k sufficiently large,

$$\begin{aligned} x^{[k+1]} &=_{LD} t_1 * x^{[k]} =_{LD} t_1 * (t_2 * x^{[k-1]}) \\ &\qquad =_{LD} (t_1 * t_2) * (t_1 * x^{[k-1]}) =_{LD} (t_1 * t_2) * x^{[k]} = t * x^{[k]}, \end{aligned}$$

which is the result for t. $\square$

The third property needed for the proof of Proposition 1.22 involves iterated left subterms. For t a term that is not a single letter, we denote by $\mathrm{left}(t)$ the left subterm of t; then, we use $\mathrm{left}^2(t)$ for the left subterm of the left subterm of t, etc. For a given term t, the term $\mathrm{left}^q(t)$ exists for q at most the length of the leftmost branch of t viewed as a tree. For instance, for $t = (x * y) * (x * z)$, we have $\mathrm{left}(t) = x * y$, $\mathrm{left}^2(t) = x$, and $\mathrm{left}^3(t)$ is not defined.

LEMMA 1.29. *If t is a term that is not a single letter, and if t' is an LD-expansion of t, then there exists q such that* $\mathrm{left}^q(t')$ *is an LD-expansion of* $\mathrm{left}(t)$.

Proof. For an induction, it suffices to prove the result when t' is obtained from t by applying the LD-law to some subterm t_0 of t, in the expanding direction. We consider the various possible positions of t_0 inside t. If t_0 is t itself, then, by definition of the LD-law, $\mathrm{left}^2(t')$ is equal to $\mathrm{left}(t)$. If t_0 is the subterm $\mathrm{left}^r(t)$ with $r \geqslant 1$, then $\mathrm{left}(t')$ is an LD-expansion of $\mathrm{left}(t)$. In all other cases, we have $\mathrm{left}(t') = \mathrm{left}(t)$. □

We are now ready to conclude.

Proof of Proposition 1.22. (See Figure 3.) Our aim is to prove that the Comparison Property is true in the free monogenerated LD-system, *i.e.*, in the quotient-system $T_1/ =_{LD}$. So we start with arbitrary terms t, t' in T_1, and aim at proving that the $=_{LD}$-classes of t and t' are comparable with respect to $\sqsubset$. By Lemma 1.28, the terms $t * x^{[k]}$ and $t' * x^{[k]}$ are LD-equivalent for k large enough, as both are LD-equivalent to $x^{[k+1]}$. Hence, by Lemma 1.27, $t * x^{[k]}$ and $t' * x^{[k]}$ admit a common LD-expansion, say t_0. Now, by Lemma 1.29, there exist nonnegative integers q, q' such that $\mathrm{left}^q(t_0)$ is an LD-expansion of $\mathrm{left}(t * x^{[k]})$, *i.e.*, of t, and $\mathrm{left}^{q'}(t_0)$ is an LD-expansion of $\mathrm{left}(t' * x^{[k]})$, *i.e.*, of t'. Thus we have $t =_{LD} \mathrm{left}^q(t_0)$, and $t' =_{LD} \mathrm{left}^{q'}(t_0)$. If q and q' are equal, we deduce

$$t =_{LD} \mathrm{left}^q(t_0) = \mathrm{left}^{q'}(t_0) =_{LD} t',$$

i.e., using $\overline{t}$ for the LD-equivalence class of t, we have $\overline{t} = \overline{t'}$.

Assume now $q = q' + p$ with $p \geqslant 1$. Then we have $\mathrm{left}^q(t_0) = \mathrm{left}^p(\mathrm{left}^{q'}(t_0))$, which, by definition, means that we have

$$\mathrm{left}^{q'}(t_0') = (...((\mathrm{left}^q(t_0) * t_1) * t_2)...) * t_p$$

for some $t_1, \dots, t_p$, which implies $t' =_{LD} (...((t * t_1) * t_2)...) * t_p$. Hence, we obtain now $\overline{t'} = (...((\overline{t} * \overline{t_1}) * \overline{t_2})...) * \overline{t_p}$, *i.e.*, $\overline{t} \sqsubset \overline{t'}$.

The argument is symmetric for $q' < q$, leading to $\overline{t'} \sqsubset \overline{t}$. □

So our proof of the Comparison Property for the free monogenerated LD-system is complete. Lemma 1.21 then implies that every monogenerated LD-system has the Comparison Property, (Proposition 1.16). So does in particular the LD-system $(B_{\mathrm{sp}}, *)$, and thus our proof of Property **C**—more exactly $\mathbf{C}_\infty$—is complete.

Remark 1.30. The key point in the previous argument is Lemma 1.27. The latter has much in common with the property that any two braids in the monoid B_n^+ admit a least common right multiple—which actually can be deduced from Lemma 1.27 using the techniques of Section 3. Here the operator ∂ plays the role of Garside's fundamental braid Δ_n, or, more exactly, of the inner automorphism Φ_n of B_n^+ associated with Δ_n. The framework of Garside categories [**129, 67**] enables one to give a unified treatment for these seemingly remote situations.

2. Colouring arbitrary braids

In Section 1, we investigated the action of positive braids on the powers of an LD-system, and, by choosing a convenient LD-system consisting of special braids, we proved Property **C**—more exactly, Property $\mathbf{C}_\infty$. We shall now discuss the extension of this action to arbitrary, not necessarily positive braids. There are two ways of performing this extension: if we insist on extending the action of B_n^+ into a

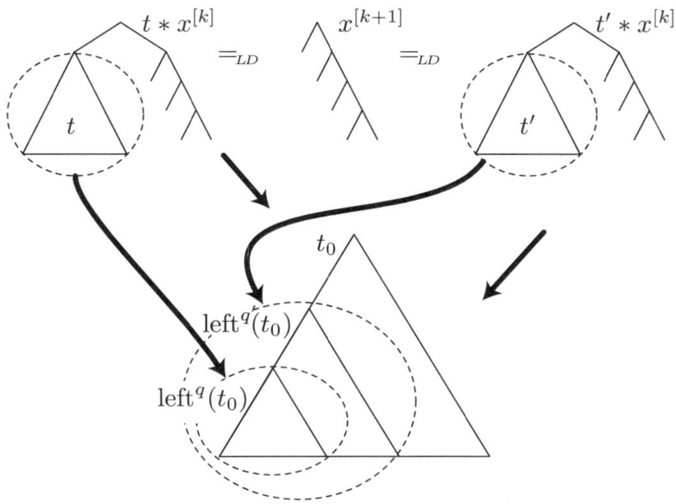

FIGURE 3. Proof of Proposition 1.22: For any two terms t, t', we can find a common LD-expansion t_0 of $t * x^k$ and $t' * x^k$ for k large enough; then some iterated left subterm of t_0 is an LD-expansion of t, and some iterated left subterm of t_0 is an LD-expansion of t'; now, two iterated left subterms of a given term always are comparable with respect to the iterated left subterm relation: for instance, in the picture, q is larger than q' and, therefore, $\mathrm{left}^q(t_0)$ is an iterated left subterm of $\mathrm{left}^{q'}(t_0)$.

full action of B_n, then we must drastically reduce the eligible LD-systems to what are called racks, and no application is to be expected in terms of braid orderings; on the other hand, if we accept lowering our ambitions and content ourselves with what will be called a partial action, then many LD-systems remain eligible. In particular, using the partial action on an orderable LD-system leads to a very short proof of Property **A**, *i.e.*, of the fact that a σ-positive braid is never trivial.

2.1. The action of braids on a rack. Coming back to the context of Section 1.1, let us try to extend our colourings to arbitrary braid diagrams. So, we have to colour negative crossings. In order to find the most flexible definition, let us first assume that the set of colours S is equipped with two more binary operations, say $\circ$ and $\bar{*}$, and consider the rule

This amounts to extending the action of braid words on sequences of colours by

$$\boldsymbol{x} \bullet \sigma_i^{-1} = (x_1, \dots, x_{i-1}, x_i \circ x_{i+1}, x_{i+1} \bar{*} x_i, x_{i+2}, \dots, x_n). \tag{2.1}$$

The reader will easily check that the action defined by (1.1) and (2.1) is compatible with the relations $\sigma_i\sigma_i^{-1} = \sigma_i^{-1}\sigma_i = 1$ if and only if the following laws are satisfied:

$$y \circ x = x, \qquad x * (x \bar{*} y) = x \bar{*} (x * y) = y. \tag{2.2}$$

Thus, the operation $\circ$ has to be trivial, while $\bar{*}$ must be chosen so that the left translations associated with $\bar{*}$ are inverses of the left translations associated with $*$.

So $\bar{*}$ exists if and only if the left translations associated with $*$ are bijective, in which case we necessarily have

$$x \mathbin{\bar{*}} y = \text{ the unique } z \text{ satisfying } x * z = y. \tag{2.3}$$

Therefore, we can state

PROPOSITION 2.1. *Define a rack to be an LD-system $(S, *)$ in which all left translations are bijective, i.e., we have $\forall x, y \exists! z (x * z = y)$. Then, for every rack $(S, *)$ and every n, the rules (1.1), (2.1), and (2.3) (with $x \circ y = y$) define an action of B_n on S^n.*

As in Section 1.1, for $\boldsymbol{x}$ a sequence of elements of S, and β a braid, we shall write $\boldsymbol{x} \bullet \beta$ for the result of applying β to $\boldsymbol{x}$, *i.e.*, for $\boldsymbol{x} \bullet w$ where w is an arbitrary braid word representing β.

It is easily checked that all LD-systems mentioned in Examples 1.3 through 1.6 are racks, and, therefore, there exists a well-defined action of braids on the powers of each of these LD-systems. On the other hand, the braid LD-system $(B_\infty, *)$ is not a rack: we observed that left translations in $(B_\infty, *)$ are injective, but they are not surjective in general. More precisely, there exists a braid γ satisfying $\beta * \gamma = \beta'$ if and only if the braid $\beta^{-1} \cdot \beta' \cdot \mathrm{sh}(\beta) \cdot \sigma_1^{-1}$ belongs to the image of the shift endomorphism, which is not the case in general: for instance, $\sigma_2 \sigma_1^{-1}$ does not belong to the image of sh, and, therefore, there exists no braid γ satisfying $1 * \gamma = \sigma_2$.

Racks are very interesting objects, and they turn out to be quite useful in topology [**85**]. However, they are not suitable for the order applications we have in mind, because a rack is always connected with the conjugacy of a group. Indeed, assume that $(Q, *)$ is a rack, and let L be the function from Q to the permutation group $\mathfrak{S}_Q$ that maps x of Q to the associated left translation $L(x)$, *i.e.*, to the map $y \mapsto x * y$. Then we have $L(x * y) = L(x) L(y) L(x)^{-1}$, *i.e.*, L carries the $*$-operation of Q to the conjugacy operation of $\mathfrak{S}_Q$. The mapping L need not be injective in general, but, nevertheless, its existence implies that the operation on Q is close to the conjugacy of a group in many aspects.

As we are interested in ordering braids, it is not surprising that using LD-systems equipped with an ordering can be useful. The following notion appears natural in this context.

DEFINITION 2.2. We say that $(S, *, \prec)$ is an *ordered* LD-system if $(S, *)$ is an LD-system, and $\prec$ is a (strict) linear ordering on S such that $x \prec x * y$ is always true, and $y \prec z$ implies $x * y \prec x * z$.

We say that an LD-system $(S, *)$ is *orderable* if there exists at least one linear ordering $\prec$ on S such that $(S, *, \prec)$ is an ordered LD-system. By definition, the ordering in an ordered LD-system extends the left-divisibility relation, hence its iterated version $\sqsubset$ of Definition 1.15, and, therefore, the latter admits no cycle in an orderable LD-system. Note that the connection between the ordering and the operation in an ordered LD-system is the same as the connection between the standard ordering of positive integers and their addition.

Now, the bad news follows:

PROPOSITION 2.3. *An orderable LD-system is never a rack.*

PROOF. Assume that $(S, *, \prec)$ is an ordered LD-system. Then, by definition, we have $x * x \prec (x * x) * y$ for all x, y in S, hence, in particular,

$$x * x \neq (x * x) * x. \tag{2.4}$$

On the other hand, we claim that every rack obeys the law $x * y = (x * x) * y$. Indeed, if $(S, *)$ is a rack, then, for all x, y in S, we find

$$(x * x) * y = (x * x) * (x * (x \mathbin{\overline{*}} y)) = x * (x * (x \mathbin{\overline{*}} y)) = x * y,$$

where $\overline{*}$ is the symmetric operation of (2.3). $\square$

Therefore, the approach of using orderable LD-systems to colour the strands of arbitrary braids cannot work: the only LD-systems that can colour arbitrary braids are racks, and those are never orderable.

However, we observe

LEMMA 2.4. *Every orderable LD-system is left cancellative, i.e., $x * y = x * z$ implies $y = z$.*

PROOF. Assume that $(S, *, \prec)$ is an ordered LD-system and y, z are distinct elements of S. Then $y \prec z$ or $z \prec y$ holds. By definition, this implies $x * y \prec x * z$, or $x * z \prec x * y$, respectively. In either case, we deduce $x * y \neq x * z$. $\square$

2.2. The partial action of braids on a left-cancellative LD-system. In view of Proposition 2.3 and Lemma 2.4, we are led to considering the possible extension of the braid action to LD-systems that are not racks, but, at least, are left-cancellative: this amounts to saying that the left translations are not necessarily bijective—as is the case for a rack—but, at least, are injective. This is what we shall do now. The price to pay for the extension is that we shall obtain a partial action only: it will no longer be true that every set of input colours can be propagated, but the result is that, for each given braid, there exists at least one sequence of admissible initial colours, and the corresponding output colours do not depend on the choice of the braid word representative. Precisely, we shall establish

PROPOSITION 2.5. *Assume that $(S, *)$ is a left-cancellative LD-system. Then the rules* (1.1), (2.1), *and* (2.3)*—considered as a partial, not necessarily everywhere defined operation—define a partial action of n-strand braid words on S^n that has the following properties:*

(i) *If $w_1, \dots, w_p$ are n-strand braid words, there exists at least one sequence $\boldsymbol{x}$ in S^n such that each of $\boldsymbol{x} \bullet w_1, \dots, \boldsymbol{x} \bullet w_p$ is defined;*

(ii) *If w, w' are equivalent n-strand braid words and both $\boldsymbol{x} \bullet w$ and $\boldsymbol{x} \bullet w'$ are defined, they are equal.*

For the rest of this section, we assume that $(S, *)$ is a left-cancellative LD-system—some of the results extend to arbitrary LD-systems, but we shall not consider that case here. This means that, for all x, z in S, there exists at most one y satisfying $x * y = z$, and, therefore, the braid colourings associated with S are unique when they exist. Indeed, the rules for colour propagation are

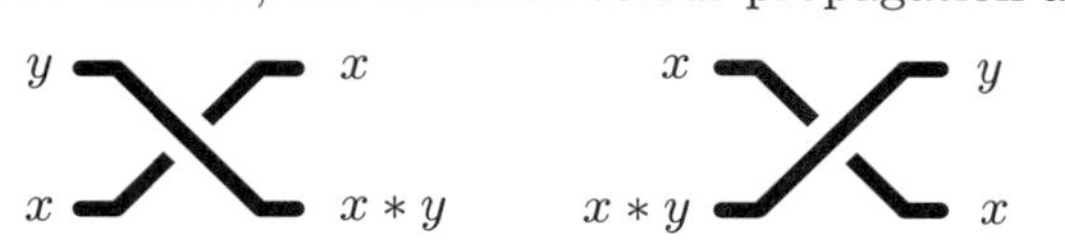

and, for each initial sequence $\boldsymbol{x}$ in S^n and each n-strand braid word w, either the initial colours can be propagated throughout the diagram encoded by w and there is exactly one output sequence which will be denoted by $\boldsymbol{x} \bullet w$, or there exists at least one negative crossing where the division is impossible and then $\boldsymbol{x} \bullet w$ does not exist. Note that, if w is uv, then $\boldsymbol{x} \bullet w$ exists if and only if $\boldsymbol{x} \bullet u$ and $(\boldsymbol{x} \bullet u) \bullet v$ exist, and, in this case, we have the expected equality. In other words, all rules for an action are obeyed, provided that one does not get stuck at some negative crossing.

The first, obvious observation is that we can always obtain colourings for diagrams that are encoded by words in which all negative letters precede all positive letters.

LEMMA 2.6. *Assume that u, v are positive n-strand braid words. Let $\boldsymbol{y}$ be an arbitrary sequence in S^n, and let $\boldsymbol{x} = \boldsymbol{y} \bullet u$. Then the sequence $\boldsymbol{x} \bullet u^{-1}v$ exists.*

PROOF. We apply the colours $\boldsymbol{y}$ in the middle of the diagram encoded by $u^{-1}v$ and propagate them to the left through u^{-1} and to the right through v, as below,

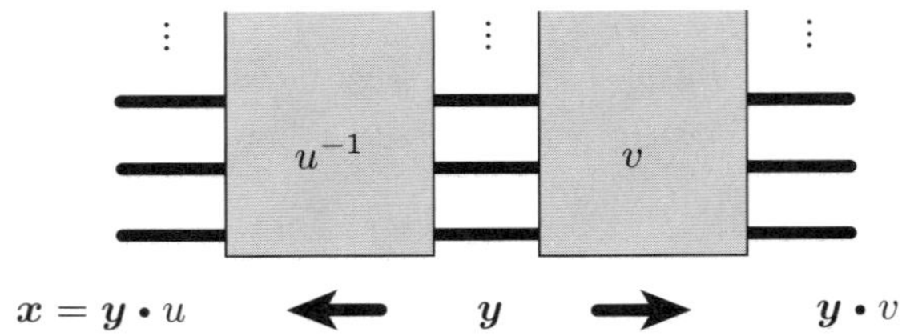

By construction, the colouring rules are obeyed everywhere, and we find $\boldsymbol{x} \bullet u^{-1} = \boldsymbol{y}$, so $\boldsymbol{x} \bullet u^{-1}v$ exists and is equal to $\boldsymbol{y} \bullet v$. □

By Proposition I.4.6, every braid word w is equivalent to a braid word of the form $u^{-1}v$ with u, v positive. This, however, is *not* sufficient to deduce that the existence of $\boldsymbol{x} \bullet u^{-1}v$ implies the existence of $\boldsymbol{x} \bullet w$: for instance, for $w = \sigma_1^{-1}\sigma_1$, then w is equivalent to $u^{-1}v$, with $u = v = \varepsilon$ (the empty word); now, $(1, 1) \bullet u^{-1}v$ is defined, whereas $(1, 1) \bullet w$ is not. So we have to be more careful. What we shall prove is that each braid word w can be transformed into an equivalent fractionary word of the form $u^{-1}v$ using only certain transformations that preserve the action in some convenient way, namely *(left) subword-reversing* transformations—or simply left reversing. We refer the reader to [**50, 55**] and [**53**] for more precision and further developments of the subword-reversing technique.

DEFINITION 2.7. For braid words w, w', we say that w is *left reversible* to w' if w' can be obtained from w by iteratively

- deleting a subword $\sigma_i\sigma_i^{-1}$, or
- replacing a subword $\sigma_i\sigma_j^{-1}$ with $|i - j| \geqslant 2$ by $\sigma_j^{-1}\sigma_i$, or
- replacing a subword $\sigma_i\sigma_j^{-1}$ with $|i - j| = 1$ by $\sigma_j^{-1}\sigma_i^{-1}\sigma_j\sigma_i$.

EXAMPLE 2.8. Consider $w = \sigma_1\sigma_2^{-1}\sigma_2\sigma_3^{-1}$. Then w contains the factor $\sigma_1\sigma_2^{-1}$, so it is left reversible to $w_1 = \sigma_2^{-1}\sigma_1^{-1}\sigma_2\sigma_1\sigma_2\sigma_3^{-1}$—note that w also contains the factor $\sigma_2\sigma_3^{-1}$, so it is left reversible to $\sigma_1\sigma_2^{-1}\sigma_3^{-1}\sigma_2^{-1}\sigma_3\sigma_2$ as well. Then w_1 contains the factor $\sigma_2\sigma_3^{-1}$, so it is left reversible to $w_2 = \sigma_2^{-1}\sigma_1^{-1}\sigma_2\sigma_1\sigma_3^{-1}\sigma_2^{-1}\sigma_3\sigma_2$, etc. The reader can check that all sequences of subword-reversing from w end in seven steps with the word $\sigma_2^{-1}\sigma_1^{-1}\sigma_3^{-1}\sigma_2^{-1}\sigma_1^{-1}\sigma_3\sigma_2\sigma_1\sigma_3\sigma_2$. The latter word can no longer be reversed, for it contains no more factor of the form $\sigma_i\sigma_j^{-1}$.

It should be clear that, if w is left reversible to w', then w and w' are equivalent, since each elementary reversing step consists in replacing a subword by an equivalent subword. Conversely, we note that, for a given word w, it is false that every word equivalent to w can be obtained by left reversing from w: for instance, starting from the word $\sigma_1^{-1}\sigma_1$, we cannot reach the empty word—actually, we can reach no word other than $\sigma_1^{-1}\sigma_1$ since the latter contains no subword of the form $\sigma_i\sigma_j^{-1}$. We shall appeal to the following fundamental result, which is one of the many consequences of the Garside theory of braid monoids.

LEMMA 2.9. *For each braid word w, there exist positive words u, v such that w is left reversible to $u^{-1}v$.*

PROOF (SKETCH). The words of the form $u^{-1}v$ with u, v positive are those words that are terminal with respect to left reversing, and the problem is to prove that, starting from a word w, at least one sequence of reversing steps terminates in a finite number of steps with a word $u^{-1}v$ as above.

So the main problem is to prove termination. It is not hard to see that it is sufficient to do it when we start with a word of the form uv^{-1}, with positive u, v. Then, the point is that, in the braid monoid B_n^+, the braids represented by u and v admit a least left common multiple β and that every word w occurring in the left reversing of uv^{-1} is drawn in the set of divisors of β—in the sense of Definition V.2.2 below. There are finitely many such divisors, and one can deduce that no infinite reversing sequence exists. □

Let us return to S-colourings. We observed above that, if w, w' are equivalent braid words, and $\boldsymbol{x}$ is a sequence of colours from S, then the existence of $\boldsymbol{x} \bullet w$ does not guarantee that of $\boldsymbol{x} \bullet w'$ in general. The interest of left reversing is to prevent such problems.

LEMMA 2.10. *Assume that w is left reversible to w', and that $\boldsymbol{x} \bullet w'$ exists. Then $\boldsymbol{x} \bullet w$ exists as well, and we have $\boldsymbol{x} \bullet w = \boldsymbol{x} \bullet w'$.*

PROOF. (See Figure 4.) It suffices to prove the result when w is left reversible to w' in one step. If w' has been obtained from w by deleting a factor $\sigma_i\sigma_i^{-1}$ or replacing $\sigma_i\sigma_j^{-1}$ with $\sigma_j^{-1}\sigma_i$ in the case $|i-j| \geqslant 2$, the result is obvious. Assume that w' has been obtained from w by replacing a factor $\sigma_i\sigma_j^{-1}$ with $|i-j| = 1$ by $\sigma_j^{-1}\sigma_i^{-1}\sigma_j\sigma_i$. Without loss of generality, we can assume $i = 1$ and $j = 2$. Our aim is to prove that, if $(x, y, z) \bullet \sigma_2^{-1}\sigma_1^{-1}\sigma_2\sigma_1$ exists, so does $(x, y, z) \bullet \sigma_1\sigma_2^{-1}$. Now, the hypothesis that $(x, y, z) \bullet \sigma_2^{-1}$ exists implies that there exists y' in S satisfying $y = z * y'$. Similarly, the hypothesis that $(x, z, y') \bullet \sigma_1^{-1}$ exists implies that there exists x' in S satisfying $x = z * x'$. But, then, $(x, y, z) \bullet \sigma_1^{-1}\sigma_2$ exists, and, because of

$$z * (y' * x') = (z * y') * (z * x') = x * y,$$

we find $(x, y, z) \bullet \sigma_2^{-1}\sigma_1^{-1}\sigma_2\sigma_1 = (x, y, z) \bullet \sigma_1\sigma_2^{-1}$. □

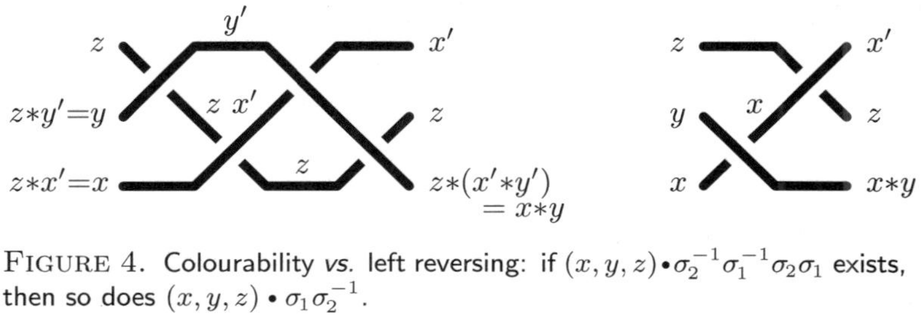

FIGURE 4. Colourability *vs.* left reversing: if $(x, y, z)\bullet\sigma_2^{-1}\sigma_1^{-1}\sigma_2\sigma_1$ exists, then so does $(x, y, z) \bullet \sigma_1\sigma_2^{-1}$.

We can now establish the existence part of Proposition 2.5.

PROOF OF PROPOSITION 2.5(i). Assume first $p = 1$, *i.e.*, we consider one single braid word w. By Lemma 2.9, there exist positive words u, v such that w is left reversible to $u^{-1}v$. By Lemma 2.6, there exists a sequence $\boldsymbol{x}$ in S^n such that $\boldsymbol{x} \bullet u^{-1}v$ exists. By Lemma 2.10, $\boldsymbol{x} \bullet w$ exists as well.

Assume now $p \geqslant 2$, *i.e.*, we wish to colour several braid words $w_1, \dots, w_p$ simultaneously—that case is needed for the proof of Proposition 2.19 only. For each k, there exist positive words u_k, v_k such that w_k is left reversible to $u_k^{-1}v_k$. Then, Proposition I.4.8 implies that the braids represented by $u_1, \dots, u_p$ admit a common left multiple in the monoid B_n^+, *i.e.*, we can find positive braid words $u'_1, \dots, u'_p$ such that $u'_1u_1, \dots, u'_pu_p$ are pairwise equivalent. As in the proof of Lemma 2.6, let $\boldsymbol{y}$ be any sequence of colours from S. By Proposition 1.2, the value of $\boldsymbol{y} \bullet (u'_ku_k)$ does not depend on k. If $\boldsymbol{x}$ is the latter sequence, then, by construction, $\boldsymbol{x} \bullet u_k^{-1}$ is defined for every k, and so is $\boldsymbol{x} \bullet (u_k^{-1}v_k)$. Applying Lemma 2.10 as above, we conclude that $\boldsymbol{x} \bullet w_k$ exists as well for every k. □

As for the uniqueness part of Proposition 2.5, we shall appeal to the right counterpart of left subword-reversing.

DEFINITION 2.11. For braid words w, w', we say that w is *right reversible* to w' if w' can be obtained from w by iteratively

- deleting a subword $\sigma_i^{-1}\sigma_i$, or
- replacing a subword $\sigma_i^{-1}\sigma_j$ with $|i - j| \geqslant 2$ by $\sigma_j\sigma_i^{-1}$, or
- replacing a subword $\sigma_i^{-1}\sigma_j$ with $|i - j| = 1$ by $\sigma_j\sigma_i\sigma_j^{-1}\sigma_i^{-1}$.

As the braid relations are symmetric, right reversing is completely similar to left reversing. The words that are terminal with respect to right reversing are those of the form uv^{-1} with u, v positive, and an argument symmetric to that for Lemma 2.9 gives

LEMMA 2.12. *For each braid word w, there exist positive words u, v such that w is right reversible to uv^{-1}.*

As for colourings, we also obtain a connection with right reversing, but one should notice that the orientation is reversed:

LEMMA 2.13. *Assume that w is right reversible to w', and that $\boldsymbol{x} \bullet w$ exists. Then $\boldsymbol{x} \bullet w'$ exists as well, and we have $\boldsymbol{x} \bullet w' = \boldsymbol{x} \bullet w$.*

PROOF. (See Figure 5.) The critical step is to prove that, if $(x, y, z) \bullet \sigma_1^{-1}\sigma_2$ exists, so does $(x, y, z) \bullet \sigma_2\sigma_1\sigma_2^{-1}\sigma_1^{-1}$. The existence of $(x, y, z) \bullet \sigma_1^{-1}$ implies that $x = y{*}x'$ holds for some x', and, then, using the relation $y{*}(x'{*}z) = (y{*}x'){*}(y{*}z) = x * (y * z)$, we obtain the expected equality. □

We can then complete the proof of Proposition 2.5.

PROOF OF PROPOSITION 2.5(ii). Let β be the braid represented by w. By (the right counterpart of) Proposition I.4.9, there exists a unique decomposition $\beta = N_R(\beta)D_R(\beta)^{-1}$ where $N_R(\beta)$ and $D_R(\beta)$ are positive braids with no common right divisor in B_n^+. Let u_0, v_0 be positive braid words representing $N_R(\beta)$ and $D_R(\beta)$, respectively. We claim that, if $\boldsymbol{x} \bullet w$ is defined, then so is $\boldsymbol{x} \bullet u_0v_0^{-1}$, and these two

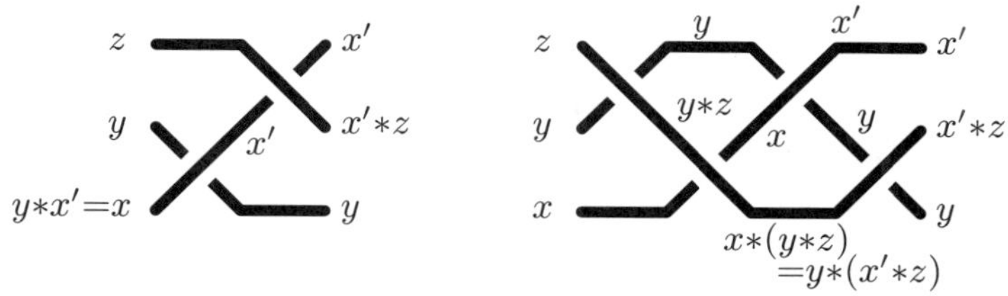

FIGURE 5. Colourability *vs.* right reversing: if $(x, y, z) \bullet \sigma_1\sigma_2^{-1}$ exists, then so does $(x, y, z) \bullet \sigma_2\sigma_1\sigma_2^{-1}\sigma_1^{-1}$.

sequences are equal. The claim implies the proposition for, if w' is equivalent to w, it also represents the braid β, and we then find $\boldsymbol{x} \bullet w = \boldsymbol{x} \bullet u_0v_0^{-1} = \boldsymbol{x} \bullet w'$.

So assume that $\boldsymbol{x} \bullet w$ is defined. By Lemma 2.12, there exist positive words u, v such that w is right reversible to uv^{-1}. Then Lemma 2.13 implies that $\boldsymbol{x} \bullet uv^{-1}$ is defined and $\boldsymbol{x} \bullet w = \boldsymbol{x} \bullet uv^{-1}$ holds. The right counterpart of Proposition I.4.9 implies that the fractionary decomposition of β associated with uv^{-1} factorizes through the one associated with $N_R(\beta)D_R(\beta)^{-1}$, *i.e.*, with $u_0v_0^{-1}$. Hence, we must have $u \equiv u_0w_0$ and $v \equiv v_0w_0$ for some positive braid word w_0. By Proposition 1.2, this implies that $\boldsymbol{x} \bullet (u_0w_0)(w_0^{-1}v_0^{-1})$ exists and is equal to $\boldsymbol{x} \bullet uv^{-1}$, hence to $\boldsymbol{x} \bullet w$. Now, by construction, $\boldsymbol{x} \bullet u_0w_0w_0^{-1}v_0^{-1}$ is equal to $\boldsymbol{x} \bullet u_0v_0^{-1}$ as $\boldsymbol{y} \bullet w_0w_0^{-1}$ exists and is equal to $\boldsymbol{y}$ for each $\boldsymbol{y}$. □

2.3. A proof of Property A. We thus have extended the action of braids to all left-cancellative LD-systems, at the expense of having a partial action only, *i.e.*, one that need not be defined everywhere. We shall now use this partial action in the case of ordered LD-systems, as introduced in Definition 2.2. By Lemma 2.4, the latter are eligible for our current approach. Then, provided we take for granted the result from self-distributive algebra that orderable LD-systems do exist, we shall obtain a very easy and natural proof of Property **A**, *i.e.*, of the fact that a σ_1-positive braid is never trivial.

PROPOSITION 2.14. **(Property A)** *Assume that ordered LD-systems exist. Then a σ_1-positive braid word never represents* 1.

PROOF. (See Figure 6.) Let $(S, *, \prec)$ be an ordered LD-system, and w be an n-strand braid word containing p letters σ_1 and no letter σ_1^{-1}. By Proposition 2.5(i), there exists a sequence of colours $\boldsymbol{x}$ in S^n such that $\boldsymbol{x} \bullet w$ is defined. For $0 \leqslant q \leqslant p$ let z_q be the colour of the bottom strand after the qth σ_1-crossing. By construction, we have $z_1 = z_0 * y_1$ for some y_1, hence $z_0 \prec z_1$ by definition of an ordered LD-system. Similarly, we have $z_2 = z_1 * y_2$ for some y_2, hence $z_1 \prec z_2$, etc. So, the sequence of z_k's is increasing, and we have $z_p > z_0$. Now, if w were equivalent to the empty word ε, as $\boldsymbol{x} \bullet \varepsilon$ certainly exists, Proposition 2.5(ii) would imply $\boldsymbol{x} \bullet w = \boldsymbol{x} \bullet \varepsilon = \boldsymbol{x}$, and, in particular, we would have $z_p = z_0$, contradicting $z_p > z_0$. So w is not equivalent to ε, *i.e.*, the braid word w does not represent the trivial braid 1. □

REMARK 2.15. Let us call Property $\mathbf{A}_i$ the statement that a braid that admits a representative braid word containing at least one letter σ_i but no letter σ_i^{-1} is not trivial, *i.e.*, a braid word where σ_i occurs but σ_i^{-1} does not represent the unit braid. Then Property **A** is just Property $\mathbf{A}_1$. The connection of Property $\mathbf{A}_2$, or, more generally, $\mathbf{A}_i$ for $i \geqslant 2$, with Property **A** is not clear: indeed, using the shift

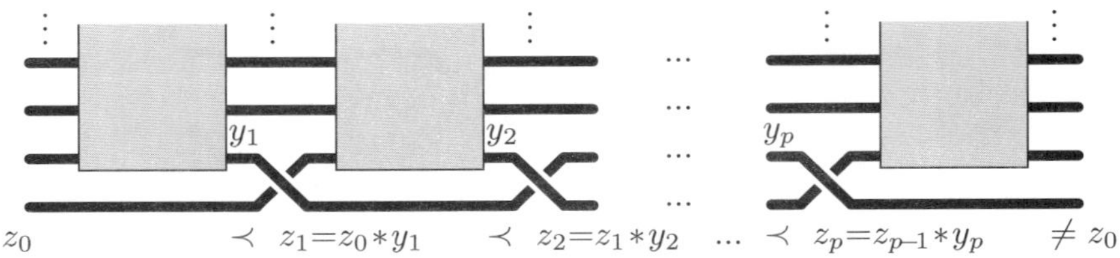

FIGURE 6. A σ_1-positive braid word is never trivial: use an ordered LD-system to colour the strands; then the colours of the first position make an increasing sequence, so, in particular, the last one cannot be the first one.

endomorphism and Property **A**, we deduce that a σ_2-positive braid is not trivial. But Property $\mathbf{A}_2$ claims more, as it involves all braid words with at least one σ_2 and no σ_2^{-1}, regardless of whether they also contain σ_1 and σ_1^{-1}. It turns out that the argument used above for Property **A** can be adapted to prove Property $\mathbf{A}_i$ for each i. Assuming that S is an ordered LD-system generated by a single element g, one can show that the quantity

$$x_1 * (x_2 * (...(x_{i-1} * (x_i * g))...))$$

increases at each letter σ_i and is preserved under each letter $\sigma_j^{\pm 1}$ with $j \neq i$. We refer to [**52**, Proposition 2.18] for details.

Before we justify the existence of orderable LD-systems in Section 3, let us observe that this existence is actually directly equivalent to Property **A**.

PROPOSITION 2.16. *Assume Property* **A**. *Then* $(B_{\mathrm{sp}}, *, \sqsubset)$ *is an ordered LD-system.*

PROOF. First, we claim that the relation $\sqsubset$ is a linear ordering on special braids. Indeed, we have seen in Lemma 1.17 that, if β, β' are special braids and $\beta \sqsubset \beta'$ holds, then the quotient-braid $\beta^{-1}\beta'$ is σ_1-positive. By Property **A**, such a word is not trivial, so $\beta \sqsubset \beta'$ implies $\beta \neq \beta'$, *i.e.*, $\sqsubset$ is an antireflexive relation. As, by construction, it is transitive, it is a strict ordering. Moreover, as $(B_{\mathrm{sp}}, *)$ is monogenerated, it satisfies the Comparison Property, so $\sqsubset$ is a linear ordering on B_{sp}. Finally, $\beta \sqsubset \beta * \beta_1$ is true by definition, and $\beta_1 \sqsubset \beta_2$ implies $\beta * \beta_1 \sqsubset \beta * \beta_2$ in every LD-system. So $(B_{\mathrm{sp}}, *, \sqsubset)$ is an ordered LD-system. □

Thus Property **A** is just another way of asserting the existence of an orderable LD-system. The equivalence can be used in both directions. If we have a cheap proof of Property **A**, such as those given in Chapters IX or XII below, then we deduce an equally cheap proof for the existence of an orderable LD-system. In the other direction, if we have a direct proof that orderable LD-systems exist, such as the one that will be sketched in the next section, then we obtain a proof of Property **A**.

2.4. A realization of the free LD-system of rank 1. We conclude this section with the observation that special braids provide a realization of the free monogenerated LD-system inside the braid group B_∞, which in turn implies a simple characterization of the braid ordering in terms of colourings by means of an ordered LD-system.

PROPOSITION 2.17. *Every monogenerated orderable LD-system is free, and then the iterated divisibility relation $\sqsubset$ is the unique relation that makes it an ordered LD-system.*

PROOF. Assume that $(S, *, \prec)$ is an ordered LD-system generated by a single element g. Let π denote the canonical projection of the free LD-system $T_1/{=_{LD}}$ onto S that maps x to g. We claim that π is injective. Indeed, let t, t' be terms in T_1 that are not LD-equivalent. By Proposition 1.16 (Comparison Property) there exist $p \geqslant 1$ and terms $t_1, \ldots, t_p$ satisfying $t' = (\ldots((t * t_1) * t_2)\ldots) * t_p$ or *vice versa*. Applying the homomorphism π, we deduce that $\pi(t) \sqsubset \pi(t')$ holds in S. By definition of an ordered LD-system, $x \sqsubset y$ implies $x \prec y$, hence $x \neq y$ in S. So we have $\pi(t) \neq \pi(t')$, and π is injective, hence bijective. Therefore, $(S, *)$ is free.

Next, we claim that $\sqsubset$ is a strict linear ordering on S. Indeed, it is transitive by construction, and any two distinct elements of S are $\sqsubset$-comparable by the Comparison Property. So it remains to prove that $x \sqsubset x$ never holds. But that would mean that there exists $p \geqslant 1$ and elements $y_1, \ldots, y_p$ in S satisfying $x = (\ldots((x * y_1) * y_2) * \ldots) * y_p$. Now, as in the proof of Proposition 2.16, this would imply $x \prec x * y_1 \prec (x * y_1) * y_2 \prec \ldots \prec x$, contradicting the hypothesis that $\prec$ is a strict ordering.

At this point, we know that $\sqsubset$ is included in $\prec$, that $\prec$ is a strict ordering, and that $\sqsubset$ is a strict linear ordering: this is enough to conclude that $\sqsubset$ and $\prec$ coincide. ☐

Applying Proposition 2.16, we deduce

COROLLARY 2.18. *Special braids equipped with $*$ form a free LD-system of rank* 1.

Thus we obtained inside B_∞ a realization for the free LD-system with one generator. An application of the previous result is the second characterization of the braid ordering mentioned in the Introduction (see Figure 7).

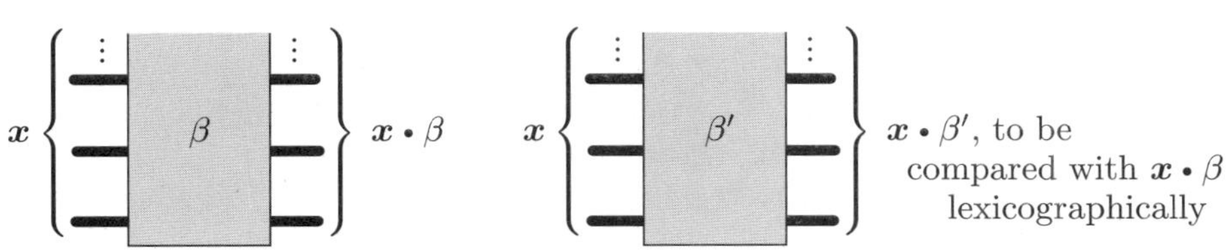

FIGURE 7. The σ-ordering of B_∞ in terms of colourings: if $(S, \prec)$ is an ordered LD-system, then β is smaller than β' if, when we apply the same input colours from S to β and β', then the output colours from β are smaller than those from β' with respect to the lexicographical ordering on sequences of colours.

PROPOSITION 2.19. *Assume that $(S, *, \prec)$ is an ordered LD-system. Denote by $\prec^{\text{Lex}}$ the lexicographical extension of $\prec$ to S^n. Then, for all β, β' in B_n, the relation $\beta < \beta'$ is true if and only if, for some sequence $\boldsymbol{x}$ in S^n, both $\boldsymbol{x} \bullet \beta$ and $\boldsymbol{x} \bullet \beta'$ are defined and we have $\boldsymbol{x} \bullet \beta \prec^{\text{Lex}} \boldsymbol{x} \bullet \beta'$, if and only if $\boldsymbol{x} \bullet \beta \prec^{\text{Lex}} \boldsymbol{x} \bullet \beta'$ holds for any sequence $\boldsymbol{x}$ such that both $\boldsymbol{x} \bullet \beta$ and $\boldsymbol{x} \bullet \beta'$ are defined.*

PROOF. As the braid relation $<$ is a linear ordering and, by Proposition 2.5, we know that there exist sequences $\boldsymbol{x}$ such that both $\boldsymbol{x} \bullet \beta$ and $\boldsymbol{x} \bullet \beta'$ are defined, it suffices to show that, if $\boldsymbol{x} \bullet \beta \prec^{\text{Lex}} \boldsymbol{x} \bullet \beta'$ is satisfied for at least one sequence $\boldsymbol{x}$, then we have $\beta < \beta'$.

Here we shall only consider the case when S is monogenerated, and refer to [**52**] for the general case. Then, by Corollary 2.18, we may assume that S is the set B_{sp} of special braids equipped with the operation of (1.3), *i.e.*, we take for $\boldsymbol{x}$ a sequence of special braids. Let $(\beta_1, \dots, \beta_n) = \boldsymbol{x} \bullet \beta$ and $(\beta'_1, \dots, \beta'_n) = (\beta'_1, \dots, \beta'_n) \bullet \beta'$. Then Lemma 1.10 gives

$$\begin{aligned}\beta^{-1}\beta' &= \left(\prod^{\text{sh}}(\boldsymbol{x} \bullet \beta)\right)^{-1} \cdot \prod^{\text{sh}}(\boldsymbol{x} \bullet \beta')\\ &= \left(\prod^{\text{sh}}(\beta_1, \dots, \beta_n)\right)^{-1} \cdot \prod^{\text{sh}}(\beta'_1, \dots, \beta'_n)\\ &= \text{sh}^{n-1}\beta_n^{-1} \cdot \dots \cdot \text{sh}\beta_2^{-1} \cdot \beta_1^{-1} \cdot \beta'_1 \cdot \text{sh}\beta'_2 \cdot \dots \cdot \text{sh}^{n-1}\beta'_n,\end{aligned}$$

and the hypothesis $\boldsymbol{x} \bullet \beta \prec^{\text{Lex}} \boldsymbol{x} \bullet \beta'$ together with Lemma 1.17 implies that the latter braid is σ-positive. Hence $\beta < \beta'$ holds. □

3. The group of left self-distributivity

In Section 2.3 we somehow cheated the reader in that our proof of Property **A** uses the so far unproven result that there exists an orderable LD-system. This result cannot be considered standard, and we shall now outline a proof. The argument turns out to be rather delicate, but it is nicely conceptual, relying on a precise analysis of the LD-law by means of an object which is interesting in its own right, namely a certain group that plays for the LD-law the role that the famous Thompson's groups F and V play for associativity and for associativity together with commutativity.

3.1. The geometry of the LD-law. Our aim in this section is to give a direct proof of the existence of an orderable LD-system. Actually, we shall prove:

PROPOSITION 3.1. *The free monogenerated LD-system is orderable.*

By definition, every sub-LD-system of an orderable LD-system is orderable, so, owing to Propositions 2.16 and 2.17, proving that there exists at least one orderable LD-system or proving that the above specific LD-system, namely the free LD-system of rank 1, is orderable are actually equivalent tasks.

Hereafter, we come back to the formalism of Section 1.5. We recall that T_1 denotes the collection of all well-formed terms constructed using one letter x and the operator $*$, and that $=_{LD}$ denotes the smallest congruence on terms that contains all instances of the LD-law. Introducing the following notation is convenient.

DEFINITION 3.2. If t, t' are terms, we say that $t \sqsubset_{LD} t'$ holds if there exists $p \geqslant 1$ and terms $t_1, \dots, t_p$ satisfying $t' =_{LD} (\dots((t * t_1) * t_2)\dots) * t_p$.

Thus $t \sqsubset_{LD} t'$ is true if and only if there exist terms t_0, t'_0 such that $t_0 =_{LD} t$ and $t'_0 =_{LD} t'$ hold and so does $t_0 \sqsubset t'_0$, *i.e.*, t_0 is an iterated left subterm of t'_0. What we did in the proof of Proposition 1.22 was precisely proving that, for all terms t, t' in T_1, at least one of $t =_{LD} t'$, $t \sqsubset_{LD} t'$, $t' \sqsubset_{LD} t$ is true.

LEMMA 3.3. *In order to prove Proposition 3.1, it is sufficient to prove that the relations $=_{LD}$ and $\sqsubset_{LD}$ on T_1 are disjoint.*

PROOF. By Lemma 1.25, the free LD-system of rank 1 can be realized as the quotient-structure $T_1/{=_{LD}}$. Write $\overline{t}$ for the LD-class of a term t. Then, by construction, the iterated left-divisibility relation $\sqsubset$ on $T_1/{=_{LD}}$ is the projection of the relation $\sqsubset_{LD}$: for all terms t, t' in T_1, the relations $\overline{t} \sqsubset \overline{t'}$ and $t \sqsubset_{LD} t'$ are equivalent. Hence, assuming that $=_{LD}$ and $\sqsubset_{LD}$ are disjoint just means that $\sqsubset$ is antireflexive in $T_1/{=_{LD}}$. In this case, as it is transitive by construction, it is a strict ordering, and even a strict linear ordering by Proposition 1.22. Finally, $t \sqsubset_{LD} t * t_1$ is always true by definition, and $t \sqsubset_{LD} t'$ implies $t_0 * t \sqsubset_{LD} t_0 * t'$ by using the LD-law. So $T_1/{=_{LD}}$ is an ordered LD-system—note that the last part of the argument is the same as for Proposition 2.16. □

So, from now on, our aim will be to prove that the relations $=_{LD}$ and $\sqsubset_{LD}$ on T_1 are disjoint. To this end, we shall analyse what can be called the *geometry* of the LD-law. By definition, saying that two terms t, t' are LD-equivalent means that we can transform t into t' by repeatedly replacing a subterm of the form $t_1 * (t_2 * t_3)$ with the corresponding term $(t_1 * t_2) * (t_1 * t_3)$, or vice versa, *i.e.*, by applying the LD-law. The idea will be to take into account the *position* where the law is applied—and that is why we speak of geometry here. To this end, we fix a system of addresses for the subterms of a term. A simple system is obtained by viewing terms as binary trees and describing the path that goes from the root of the tree to the root of the considered subterm, by using 0, for instance, for forking to the left and 1 for forking to the right. This allows us to speak of the αth subterm of a given term t. Notice that, for each term t, the αth subterm of t is defined for finitely many addresses α only. For instance, if t is the term $(x * x) * x$, the 0th subterm of t is the term $x * x$, and the set of all addresses α for which the αth subterm of t is defined consists of the five addresses $\varnothing$ (the empty address, *i.e.*, the address of the root), $0, 1, 00$, and 01.

DEFINITION 3.4. For α an address, we denote by LD_α the (partial) operator on terms that corresponds to applying the LD-law to the αth subterm in the expanding direction. We denote by $\mathcal{G}_{LD}$ the monoid generated by all operators LD_α and their inverses using reversed composition; we think of $\mathcal{G}_{LD}$ as acting on terms on the right, writing $t \bullet f$ for the result of applying f to t.

The operator LD_α is a partial operator: the term t belongs to the domain of LD_α if and only if the αth subterm of t exists, and it can be decomposed as $t_1 * (t_2 * t_3)$, in which case $t \bullet LD_\alpha$ is the term obtained by replacing the above subterm with $(t_1 * t_2) * (t_1 * t_3)$ in t. The operator LD_α is injective, and its inverse LD_α^{-1} corresponds to applying the LD-law at α in the contracting direction. For every term t, the set of those addresses α's such that $t \bullet LD_\alpha$ exists is finite. For instance, if t is the term $x_1 * (x_2 * (x_3 * x_4))$, then $t \bullet LD_\varnothing$ and $t \bullet LD_1$ only are defined, and the values are $(x_1 * x_2) * (x_1 * (x_3 * x_4))$ and $x_1 * ((x_2 * x_3) * (x_2 * x_4))$, respectively.

By construction, two terms t, t' are LD-equivalent if and only if some element of the monoid $\mathcal{G}_{LD}$ maps t to t'.

Now, we would like to replace the monoid $\mathcal{G}_{LD}$ with a group. Because it consists of injective partial operators, the monoid $\mathcal{G}_{LD}$ is an inverse monoid, but not a group. Moreover, it may happen that the composition of two operators in $\mathcal{G}_{LD}$ is just empty, *i.e.*, it applies to no term at all: for instance, the branches 00... and 10... must have the same length in any term belonging to the image of $LD_\varnothing$, *i.e.*, to the domain of $LD_\varnothing^{-1}$, and this is never the case for a term in the image of $LD_\varnothing \cdot LD_1$, so the operator $LD_\varnothing \cdot LD_1 \cdot LD_\varnothing^{-1}$ is just empty. It follows that there is no way to quotient $\mathcal{G}_{LD}$ into a nontrivial group without distorting it completely. So, we resort to an indirect approach, namely guessing a presentation of $\mathcal{G}_{LD}$, and then introducing the group G_{LD} defined by this presentation: the idea is that, if we guessed the presentation correctly, then the group G_{LD} should resemble the monoid $\mathcal{G}_{LD}$, and all results about the action of $\mathcal{G}_{LD}$ on terms should admit purely syntactic counterparts in G_{LD}. The first step is therefore to find relations between the various operators LD_α.

LEMMA 3.5. *For all addresses α, β, γ, the following relations hold in $\mathcal{G}_{LD}$:*

$$LD_{\alpha 0\beta} \cdot LD_{\alpha 1\gamma} = LD_{\alpha 1\gamma} \cdot LD_{\alpha 0\beta},$$
$$LD_{\alpha 0\beta} \cdot LD_{\alpha} = LD_{\alpha} \cdot LD_{\alpha 00\beta} \cdot LD_{\alpha 10\beta},$$
$$LD_{\alpha 10\beta} \cdot LD_{\alpha} = LD_{\alpha} \cdot LD_{\alpha 01\beta},$$
$$LD_{\alpha 11\beta} \cdot LD_{\alpha} = LD_{\alpha} \cdot LD_{\alpha 11\beta},$$
$$LD_{\alpha 1} \cdot LD_{\alpha} \cdot LD_{\alpha 1} \cdot LD_{\alpha 0} = LD_{\alpha} \cdot LD_{\alpha 1} \cdot LD_{\alpha}.$$

Verification is easy. The above relations are quite natural: they are nothing but a syntactic counterpart to Lemma 1.27. Notice that, for each pair of addresses α, β, there exists exactly one relation in the list above taking the form $LD_\alpha ... = LD_\beta ...$, *i.e.*, explaining how to obtain a common LD-expansion for $t \bullet LD_\alpha$ and $t \bullet LD_\beta$. According to the strategy sketched above, we introduce

DEFINITION 3.6. We denote by G_{LD} the group generated by an infinite sequence of generators τ_α indexed by addresses, *i.e.*, by finite sequences of 0's and 1's, subject to the relations of Lemma 3.5, *i.e.*, $\tau_{\alpha 0\beta} \cdot \tau_{\alpha 1\gamma} = \tau_{\alpha 1\gamma} \cdot \tau_{\alpha 0\beta}$, etc.

Let us denote by $\mathcal{G}_{LD}^+$ (*resp.* G_{LD}^+) the submonoid of $\mathcal{G}_{LD}$ (*resp.* G_{LD}) generated by the elements LD_α (*resp.* τ_α), *i.e.*, we forbid inverses. Lemma 3.5 implies that $\mathcal{G}_{LD}^+$ is a quotient of G_{LD}^+, so the action of $\mathcal{G}_{LD}^+$ on terms factors through an action of G_{LD}^+—we shall denote by $t \bullet g$ the result of letting g act on t. This action however does not extend to the group G_{LD}, as $\mathcal{G}_{LD}$ is not a quotient of G_{LD}, due to the fact that the composition of two operators in $\mathcal{G}_{LD}$ may be the empty operator, *i.e.*, an operator whose domain is empty.

REMARK 3.7. The previous approach applies to every algebraic law, and, more generally, to every family of algebraic laws [**46, 48, 54**]: in each case, some monoid describes the associated geometry, and, in good cases, a group appears. In the case of associativity, the group involved happens to be Richard Thompson's group F investigated in [**155, 32**]; similarly, the group corresponding to associativity plus commutativity is Thompson's group V [**58**].

3.2. The blueprint of a term. The core of the argument for proving Proposition 3.1 is the following observation: Lemma 1.28 tells us that, for every term t in T_1, the relation $x^{[k+1]} =_{LD} t * x^{[k]}$ holds for k sufficiently large. By construction, this means that some element f of the monoid $\mathcal{G}_{LD}$—depending on t and, *a priori*, on k—maps $x^{[k+1]}$ to $t * x^{[k]}$, *i.e.*, in some sense, constructs the term t from the universal term $x^{[k+1]}$. Moreover, the inductive proof of Lemma 1.28 gives us an explicit definition of such an element f. Indeed, assume that t is $t_1 * t_2$, that f_1 maps $x^{[k+1]}$ to $t_1 * x^{[k]}$, and that f_2 maps $x^{[k]}$ to $t_2 * x^{[k-1]}$. For g in $\mathcal{G}_{LD}$, let $\mathrm{sh}_1(g)$ denote the shifted version of g consisting of applying g to the right subterm of its argument. Then we read on the sequence

$$x^{[k+1]} \xrightarrow{f_1} t_1 * x^{[k]} \xrightarrow{\mathrm{sh}_1(f_2)} t_1 * (t_2 * x^{[k-1]}) \xrightarrow{LD_\varnothing} (t_1 * t_2) * (t_1 * x^{[k-1]}) \xrightarrow{\mathrm{sh}_1(f_1^{-1})} t * x^{[k]}$$

that the operator

$$f_1 \cdot \mathrm{sh}_1(f_2) \cdot LD_\varnothing \cdot \mathrm{sh}_1(f_1^{-1})$$

maps $x^{[k+1]}$ to $t * x^{[k]}$. In this way, we obtain

LEMMA 3.8. *For t in T_1, define χ_t in $\mathcal{G}_{LD}$ inductively by $\chi_x = \mathrm{id}$ and*

$$\chi_{t_1 * t_2} = \chi_{t_1} \cdot \mathrm{sh}_1(\chi_{t_2}) \cdot LD_\varnothing \cdot \mathrm{sh}_1(\chi_{t_1}^{-1}). \tag{3.1}$$

Then, for every t and for k sufficiently large, the operator χ_t maps $x^{[k+1]}$ to $t * x^{[k]}$.

According to our strategy, we introduce the counterpart $[\![t]\!]$ of the operator χ_t in G_{LD}: $[\![t]\!]$ should be seen as a sort of copy of t inside G_{LD}. We denote by sh_1 (*resp.* sh_0) the left shift endomorphism of G_{LD} that maps τ_α to $\tau_{1\alpha}$ (*resp.* $\tau_{0\alpha}$) for every α. We recall that T_1 is the set of all terms constructed using the letter x.

DEFINITION 3.9. For t in T_1, the *blueprint* of t is the element $[\![t]\!]$ of G_{LD} inductively defined by $[\![x]\!] = 1$ and

$$[\![t_1 * t_2]\!] = [\![t_1]\!] \cdot \mathrm{sh}_1([\![t_2]\!]) \cdot \tau_\varnothing \cdot \mathrm{sh}_1([\![t_1]\!]^{-1}). \tag{3.2}$$

We recall that our aim is to prove that the relations $=_{LD}$ and $\sqsubset_{LD}$ exclude each other in T_1. To see that, we translate them to G_{LD} using $[\![t]\!]$ as a counterpart to t.

LEMMA 3.10. (i) *If $t =_{LD} t'$ is satisfied, then $[\![t]\!]^{-1} \cdot [\![t']\!]$ lies in the subgroup of G_{LD} generated by the elements $\tau_{0\alpha}$.*

(ii) *If $t \sqsubset_{LD} t'$ is satisfied, then $[\![t]\!]^{-1} \cdot [\![t']\!]$ admits an expression where $\tau_\varnothing$ occurs, but $\tau_\varnothing^{-1}$ does not.*

PROOF (SKETCH). (i) First, observe that the statement is natural. Indeed, we know that, if $t =_{LD} t'$ holds, then some element f of $\mathcal{G}_{LD}$ maps t to t', and, therefore, $\mathrm{sh}_0(f)$ maps $t * x^{[k]}$ to $t' * x^{[k]}$. Now, by construction, the operator $\chi_t^{-1} \cdot \chi_{t'}$ also maps $t * x^{[k]}$ to $t' * x^{[k]}$. This means that the operator $\chi_t^{-1} \cdot \chi_{t'}$ coincides with some operator $\mathrm{sh}_0(f)$ on at least one term, which, by substitution arguments, implies that it does on every term where defined. If our axiomatization is correct, it should therefore be true that the element $[\![t]\!]^{-1} \cdot [\![t']\!]$ of G_{LD} coincides with some element of the form $\mathrm{sh}_0(a)$, *i.e.*, belongs to the subgroup $\mathrm{sh}_0(G_{LD})$ of G_{LD}. The point is that, if the latter statement is true, it must be provable by a direct verification. This is exactly what happens: for instance, the verification corresponding to $f = LD_\varnothing$ is the relation

$$[\![(t_1 * t_2) * (t_1 * t_3)]\!] = [\![t_1 * (t_2 * t_3)]\!] \cdot \tau_\varnothing,$$

which follows from the definition of the blueprint and the defining relations of G_{LD}.

(ii) By (i), it is sufficient to prove the relation for $t \sqsubset t'$, and even for $t' = t * t_0$, as an induction then gives the result. Now, the definition gives

$$[\![t]\!]^{-1} \cdot [\![t']\!] = \mathrm{sh}_1([\![t_0]\!]) \cdot \tau_\varnothing \cdot \mathrm{sh}_1([\![t]\!]^{-1}),$$

which has the desired form: $\tau_\varnothing$ occurs, but $\tau_\varnothing^{-1}$ does not. □

REMARK 3.11. The braid group B_∞ is a quotient of the group G_{LD}: indeed, mapping τ_{1^n} to σ_{n-1} and collapsing all generators τ_α such that the address α contains at least one 0 defines a surjective homomorphism, as can be seen by comparing the relations of Lemma 3.5 with the braid relations. The existence of this homomorphism is the core of the connection between braids and left self-distributivity. The reader may observe that, when one projects the formula (3.2), itself a mere translation of the easy proof of Lemma 1.28, to B_∞, one obtains the braid operation $*$: this is the way the latter naturally appears.

3.3. A preordering on G_{LD}. It therefore remains to *separate* the elements of G_{LD} according to whether they can be expressed with or without $\tau_\varnothing$ and $\tau_\varnothing^{-1}$. To this end, we use an order argument in G_{LD} relying on a study of where left subterms are mapped.

By definition, a term t' in T_∞ is an LD-expansion of another term t if some element of $\mathcal{G}_{LD}^+$ maps t to t'. In this framework, Lemma 1.27 can be rephrased as the result that every element in $\mathcal{G}_{LD}$ can be expressed as a fraction fg^{-1} with f, g in $\mathcal{G}_{LD}^+$. The proof uses the relations of Lemma 3.5 only, so it applies to the group G_{LD}:

LEMMA 3.12. *Every element of the group G_{LD} can be expressed as ab^{-1} with a, b in G_{LD}^+.*

Then, Lemma 1.29 tells us that, if t' is an LD-expansion of t, then, for every k such that t has a kth iterated left subterm, there exists k' such that the k'th iterated left subterm of t' is an LD-expansion of the kth iterated left subterm of t. It is easy to see that the index k' only depends on k and on the element f of $\mathcal{G}_{LD}^+$ that describes the passage from t to t'. More precisely, there exist two maps $\delta : \mathbb{N} \times \mathcal{G}_{LD}^+ \to \mathbb{N}$ and $\pi : \mathcal{G}_{LD}^+ \times \mathbb{N} \to \mathcal{G}_{LD}^+$ such that, for every f in $\mathcal{G}_{LD}^+$, every term t, and every k sufficiently small, we have

$$\text{left}^{\delta(k,f)}(t \bullet f) = \text{left}^k(t) \bullet \pi(f,k). \tag{3.3}$$

This once again can be proved using the relations of Lemma 3.5 exclusively, so we can obtain a counterpart in G_{LD}^+ (once again, if the property is true—and it is!—its proof is a simple verification):

LEMMA 3.13. *There exists two maps $d : \mathbb{N} \times G_{LD}^+ \to \mathbb{N}$ and $p : G_{LD}^+ \times \mathbb{N} \to G_{LD}^+$ such that, for every a in G_{LD}^+, every term t, and every k sufficiently small, we have*

$$\text{left}^{d(k,a)}(t \bullet a) = \text{left}^k(t) \bullet p(a,k). \tag{3.4}$$

We are ready to conclude the argument.

PROOF OF PROPOSITION 3.1 (SKETCH). Our aim is to prove that the free LD-system on one generator is orderable. By Lemma 3.3, this amounts to proving that the relations $=_{LD}$ and $\sqsubset_{LD}$ are disjoint on T_1. Now, by Lemma 3.10, this amounts to proving that an element of the group G_{LD} cannot simultaneously belong to the subgroup generated by the elements $\tau_{0\alpha}$, and admit an expression where $\tau_\varnothing$ occurs but $\tau_\varnothing^{-1}$ does not. So the problem is to separate the considered subsets of G_{LD}.

Now, let us introduce two subsets of G_{LD} as follows: We say that an element c of G_{LD} belongs to $P_<$ (*resp.* $P_=$) if there exists a decomposition $c = ab^{-1}$ with a, b in G_{LD}^+ satisfying

$$d(1,b) < d(1,a) \quad (\textit{resp.} =). \tag{3.5}$$

It is not hard to prove that the sets $P_<$ and $P_=$ are disjoint and that they are closed under product. The point is to prove that, if an element c of G_{LD} decomposes into $c = ab^{-1} = a'b'^{-1}$ with a, b, a', b' in G^+_{LD} and we have, say, $d(1, b) < d(1, a)$, then we necessarily have $d(1, b') < d(1, a')$. Now, under the above hypotheses, we have $ag = a'g'$ and $bg = b'g'$ for some g, g' in G^+_{LD}, which implies $d(1, ag) = d(1, a'g')$, *i.e.*,

$$d(d(1, a), g) = d(d(1, a'), g'),$$

and, similarly,

$$d(d(1, b), g) = d(d(1, b'), g').$$

As the mappings $d(\cdot, g)$ and $d(\cdot, g')$ are increasing—this is the point, and it is a natural result since its geometric counterpart for the map δ is obvious—it is clear that $d(1, b) < d(1, a)$ is equivalent to $d(1, b') < d(1, a')$.

Then one can easily show that $P_=$ is closed under inverse and prove the inclusions $P_= \cdot P_< \subseteq P_<$, and $P_< \cdot P_= \subseteq P_<$—thus the relation $c^{-1}c' \in P_< \cup P_=$ defines a preordering on G_{LD}, *i.e.*, a reflexive and transitive relation, that is compatible with multiplication on the left, and $P_=$ is the associated equivalence relation. Let us consider the generators τ_α. First, we decompose $\tau_\varnothing$ as the fraction with trivial denominator $\tau_\varnothing = \tau_\varnothing \cdot 1^{-1}$, and we find

$$d(1, 1) = 1 < d(1, \tau_\varnothing) = 2,$$

hence $\tau_\varnothing$ belongs to $P_<$. On the other hand, we have for each α

$$d(1, 1) = 1 = d(1, \tau_{0\alpha}) = d(1, \tau_{1\alpha}),$$

hence $\tau_{0\alpha}$ and $\tau_{1\alpha}$ belong to $P_=$. Now, if c belongs to the subgroup $\mathrm{sh}_0(G_{LD})$ of G_{LD}, then, by the above computation, c belongs to $P_=$. On the other hand, if c admits an expression where $\tau_\varnothing$ occurs but $\tau_\varnothing^{-1}$ does not, c belongs to $P_<$. We conclude that the two cases exclude each other. $\square$

This completes the proof of Proposition 3.1, *i.e.*, the proof that the free monogenerated LD-system is orderable—and, therefore, this also completes a proof of Property **A**.

4. Normal forms in free LD-systems

Here we sketch the argument developed by Richard Laver in [**137**] to prove Property **S** by using left self-distributivity.

The problem is as follows. We have to prove that for each braid β and each i the inequality $\beta < \sigma_i\beta$ is satisfied. To this end, owing to Proposition 2.19, we could use a convenient ordered LD-system, and prove $\boldsymbol{x} \bullet \beta <^{\text{Lex}} \boldsymbol{x} \bullet \sigma_i\beta$ for some sequence $\boldsymbol{x}$. We could, for instance, try to use the ordered LD-systems $(B_\infty, *, <)$ and $(B_{\text{sp}}, *, <)$, which are eligible. The problem is that, if S is any of these ordered LD-systems, the action of B_n on S^n is not order preserving: for instance, we have $(1, \sigma_1) <^{\text{Lex}} (\sigma_1, 1)$, but

$$(1, \sigma_1) \bullet \sigma_1 = (\sigma_2\sigma_1, 1) >^{\text{Lex}} (\sigma_1, 1) \bullet \sigma_1 = (\sigma_1^2\sigma_2^{-1}, \sigma_1),$$

as $\sigma_2\sigma_1 > \sigma_1^2\sigma_2^{-1}$ is true since the braid $\sigma_1^{-1}\sigma_2^{-1}\sigma_1^2\sigma_2^{-1}$ admits the σ_1-negative expression $\sigma_2^2\sigma_1^{-1}\sigma_2^{-2}$. So, when using the ordered LD-system $(B_\infty, *, <)$, we certainly have $\boldsymbol{x} <^{\text{Lex}} \boldsymbol{x} \bullet \sigma_i$ for every $\boldsymbol{x}$, but it is not clear how to deduce $\boldsymbol{x} \bullet \beta <^{\text{Lex}} \boldsymbol{x} \bullet \sigma_i\beta$ in general.

The solution proposed by Laver consists in working with the free LD-system on countably many generators to have more space, and to use a partial action of B_n on some proper subset D of (some superset of) that free LD-system, one whose elements are sparse enough to obtain an order-preserving action.

4.1. LD-monoids. Actually, the argument requires using a still larger structure, namely a free *LD-monoid* with countably many generators.

DEFINITION 4.1. An *LD-monoid* is defined to be a monoid $(M, \cdot, 1)$ equipped with a second binary operation $*$ so that the following mixed laws are satisfied:

$$x \cdot y = (x * y) \cdot x, \tag{4.1}$$

$$(x \cdot y) * z = x * (y * z), \tag{4.2}$$

$$x * (y \cdot z) = (x * y) \cdot (x * z), \tag{4.3}$$

$$1 * x = x, \quad x * 1 = 1. \tag{4.4}$$

Observe that every LD-monoid is an LD-system, as the second operation $*$ must be left self-distributive:

$$x * (y * z) = (x \cdot y) * z = ((x * y) \cdot x) * z = (x * y) * (x * z).$$

LD-monoids occur naturally in the study of LD-systems [**42, 133**]. Many examples of LD-systems are in fact LD-monoids: for instance, if G is a group, G equipped with its group multiplication and with conjugation is an LD-monoid. Moreover, there exists an easy uniform way for embedding a given LD-system S into an LD-monoid $\widehat{S}$ built from the free monoid generated by S—a construction closely connected to that used in [**78, 143**] to study the set-theoretical Yang–Baxter equation. In particular, the completion of the free LD-system of rank n is a free LD-monoid of rank n. Hereafter, for $1 \leqslant n \leqslant \infty$, we denote by F_n^{LD} the free LD-system of rank n, and by F_n^{LDM} the free LD-monoid of rank n. Such structures are eligible for our approach, as we have

PROPOSITION 4.2. *For every n, $1 \leqslant n \leqslant \infty$, the free LD-monoid F_n^{LDM} is a left-cancellative LD-system. There exists a unique ordering $\prec$ on F_n^{LDM} such that $x \prec x * y$ and $x \preccurlyeq x \cdot y$ always hold and we have $x_1 \succ x_2 \succ ... \succ x_n$.*

4.2. Decreasing division form. The problem with the action on say F_1^{LD} is that we lack space for separating the elements: typically, Lemma 1.28 shows that any two elements of F_1^{LD} become equal when multiplied on the right by some sufficiently large power of the generator. To avoid such phenomena, which discard the possibility of an order-preserving braid action, we consider a convenient subset D of F_∞^{LDM}.

The construction of D is rather delicate, and it appeals to the normal form results established in [**134, 135**]. The elements of F_∞^{LDM} can be represented as equivalence classes of terms constructed using an infinite series of letters x_1, x_2, ... and two binary operators $*$ and $\cdot$, with respect to the congruence $=_{LDM}$ corresponding to identities (4.1)–(4.4) together with the usual laws of a monoid, *i.e.*, associativity and neutral element. Defining a normal form means selecting in each equivalence class of $=_{LDM}$ a distinguished term. We use $\sqsubseteq_{LD}$ for the union of $\sqsubset_{LD}$ and $=_{LD}$.

DEFINITION 4.3. We say that the term t is in *division normal form*, or, for short, is *normal*, if it has the form

$$((...((x_i * t_1) * t_2)...) * t_{n-1}) \odot t_n, \tag{4.5}$$

where $\odot$ is either $*$ or $\cdot$, $t_1, \dots, t_n$ are normal, and we have

$$t_{k+2} \sqsubseteq_{LD} (...((x_i * t_1) * t_2)...) * t_k$$

for $k \leqslant n-2$, and, if $\odot$ is $\cdot$, $t_n \sqsubset_{LD} (...((x_i * t_1) * t_2)...) * t_{n-2}$.

Normal terms make distinguished representatives for all $=_{LDM}$-classes:

PROPOSITION 4.4. *Every element of F_∞^{LDM} is represented by a unique normal term; moreover, we have $x \prec y$ with respect to the ordering of Proposition 4.2 if and only if the normal term representing x precedes the normal term representing y in the lexicographical extension $<^{\text{Lex}}$ of the ordering $x_1 > x_2 > \dots$.*

Let us now introduce a subset of F_∞^{LDM}.

DEFINITION 4.5. We say that a normal term t is *decreasing* if, for each subterm $t_1 * t_2$ or $t_1 \cdot t_2$ of t, we have $t_2 <^{\text{Lex}} t_1$. We denote by D the subset of F_∞^{LDM} consisting of those elements whose normal form is decreasing.

It is easy to see that D is a proper subset of F_∞^{LDM}: for instance, the element $x_2 * x_1$ does not belong to D, as its normal form is the nondecreasing normal term $x_2 * x_1$.

The main result is then the existence of an action of B_∞ on D—*not* on $D^{\mathbb{N}}$: the action is not that of Section 1, and it should rather be thought of as an analog of the action of B_∞ on a free group considered in Chapter IX.

PROPOSITION 4.6. (i) *The formulas*

$$x_{i-1} \bullet \sigma_i = x_{i-1} * x_i, \quad x_i \bullet \sigma_i = x_{i-1}, \quad x_j \bullet \sigma_i = x_j \text{ for } j \neq i-1, i \tag{4.6}$$

induce a well-defined and faithful partial action of B_∞ on D in the following sense: for any two distinct braids β, β' in B_∞, there exists at least one element x of D such that $x \bullet \beta'$ and $x \bullet \beta'$ are defined and distinct.

(ii) *Moreover, for every x in D, we have $x \preccurlyeq x \bullet \sigma_i$, and, for all x, y in D, the relation $x \prec y$ is equivalent to $x \bullet \sigma_i \prec y \bullet \sigma_i$, whenever these expressions are defined.*

We skip the proof, which is an intricate induction on normal terms—the argument for the existence of at least one element x in D, such that $x \bullet \beta$ and $x \bullet \beta'$ are defined, is the same as the one of Section 2.2 for the partial action on the powers of an LD-system. It is then easy to conclude.

PROPOSITION 4.7 (Property **S**). *Every braid of the form $\beta^{-1}\sigma_i\beta$ is σ-positive.*

PROOF (SKETCH). By definition of the action of B_∞ on D, if β, β' belong to the image of sh^i, *i.e.*, can be represented without using $\sigma_1, \dots, \sigma_i$, then $x \bullet \beta \prec x \bullet \sigma_i\beta'$ is true—whenever the terms are defined—and one can deduce that any inequality $\beta < \beta'$ in B_∞ implies $x \bullet \beta' \prec x \bullet \beta'$ in D when the terms are defined.

Now, let β be an arbitrary braid. By Proposition 4.6(i), there exists x in D such that $x \bullet \beta$ and $x \bullet \sigma_i\beta$ are defined and distinct. By Proposition 4.6(ii), we obtain $x \preccurlyeq x \bullet \sigma_i$, and then, inductively, $x \bullet \beta \preccurlyeq x \bullet \sigma_i\beta$, hence $x \bullet \beta \prec x \bullet \sigma_i\beta$ as we assumed $x \bullet \beta \neq x \bullet \sigma_i\beta$. Two cases are possible *a priori*, namely $\beta < \sigma_i\beta$ and $\beta > \sigma_i\beta$. By the remark above, the latter would imply $x \bullet \sigma_i\beta \prec x \bullet \beta$, so it is impossible. Hence we have $\beta < \sigma_i\beta$, *i.e.*, the braid $\beta^{-1}\sigma_i\beta$ is σ-positive. $\square$

5. Appendix: Iterations of elementary embeddings in set theory

It might be of interest to mention the connection between the results described in this chapter and some questions in set theory centered around Proposition 3.1.

Contrary to algebraic systems containing an operation that is self-distributive both on the left and on the right, which had been studied in the 1960's and 1970's by Belousov in Kichinev, and Jaroslav Ježek, Tomáš Kepka, Petr Nemeč *et al.* in Prague, LD-systems, in particular free LD-systems, had received rather little attention until the beginning of the 1980's. Then set theory provided a new, puzzling example involving the iterations of an elementary embedding of a self-similar rank. An elementary embedding is a sort of strong homomorphism; see, for instance, [**119**]—and a rank is a set with the special property that every mapping of R to R can be seen as an element of R. It follows that, if i, j are two mappings of a rank into itself, then i may be *applied* to j—as j is an element of R. Playing with this situation, it is easy to see that this application operation, denoted $*$ hereafter, satisfies the left self-distributivity law, *i.e.*, the set I of all elementary embeddings of a rank R into itself equipped with this operation $*$ is an LD-system. Early results, in particular in [**133, 43**], showed that the LD-system $(I, *)$ had complicated and presumably deep properties, motivating further investigation. In 1989, the following results were proved independently:

PROPOSITION 5.1 ([**44**]). *Assume that Proposition* 3.1 *is true, i.e., there exists an orderable LD-system. Then the word problem of the LD-law is solvable, i.e., there exists an algorithm to decide for arbitrary terms* t, t' *whether* $t =_{LD} t'$ *holds.*

The result is based on Proposition 1.22. Starting with two terms t, t', we can enumerate all terms LD-equivalent to t and t'. After finitely many steps, we shall obtain a proof of $t =_{LD} t'$, $t \sqsubset_{LD} t'$, or $t' \sqsubset_{LD} t$. If Proposition 3.1 is true, we can conclude in the last two cases that $t =_{LD} t'$ is false.

PROPOSITION 5.2 ([**134**]). *Every monogenerated subsystem of the set theoretical LD-system* $(I, *)$ *is orderable—if it exists.*

The conjunction of the above two results seems to give a proof of the decidability of the word problem for the LD-law, but *it does not.* Indeed, the existence of the system $(I, *)$ is an unprovable statement, one whose logical status is to remain open: the construction of $(I, *)$ requires starting with a very large rank, called *self-similar*, and, like the existence of an inaccessible cardinal or of a measurable cardinal, the existence of a self-similar rank cannot be deduced from the usual axioms of set theory—and it cannot even be proved to be noncontradictory. Thus, the only consequence one could deduce from Propositions 5.1 and 5.2 is:

COROLLARY 5.3. *If there exists a self-similar rank, then the word problem for the LD-law is decidable.*

This was a quite paradoxical situation, as the existence of a connection between a syntactic, finitistic question such as the word problem of an algebraic law and huge objects of set theory appears to be unlikely—though not *a priori* impossible. Between 1989 and 1992, two possible conclusions were possible: either Proposition 3.1 is inevitably connected with some strong logical axiom—as are certain combinatorial properties of the integers studied by Harvey Friedman in his reverse mathematics program [**88, 89**]—or there exists a new, direct proof of Proposition 3.1

that does not require using any weird logical assumption. The latter happened: by studying free LD-systems along the lines described in Sections 1 through 3—which entirely take place in an ordinary mathematical framework—one could build the desired proof of Proposition 3.1, with the additional benefit of introducing braids in the picture and deducing unexpected braid orderability results.

It is not clear that proving Proposition 3.1 would have been considered an interesting challenge if set theory had not given some strong hint that this property should be true. So all further developments about braid orderings can be seen as *applications* of set theory. However, it should be stressed that these are applications of a particular type, as they precisely appeared in the process of removing set theory from some earlier results. We can compare the role of set theory here with the role of physics when it gives evidence for some formulas that remain then to be proved rigorously: in some sense, adding an unprovable logical statement is not so far from, say, liberally using diverging series or infinite integrals.

Let us mention that in [**136**] Laver investigated some finite quotients of the set-theoretical LD-system $(I, *)$. Under the hypothesis that a self-similar rank exists, he deduced several combinatorial properties of these finite LD-systems. The puzzling point is that, contrary to the case of Proposition 3.1, no alternative proof avoiding set theoretical hypotheses has been found so far, nor has it either been proved that these hypotheses are inevitable, despite strong attempts by Randall Dougherty [**68**], Thomas Jech [**69**], and Aleš Drápal [**70**]; see Chapter XIII of [**52**].

CHAPTER V

Handle Reduction

Handle reduction is a combinatorial method, with a natural geometrical content, that gives a proof of Property **C**, provided Property **A** is known. It was developed in [**49**]. The main interest of the method is that it does not only give a proof that every nontrivial braid word admits a representative that is σ-positive or σ-negative, but it also gives an algorithm for finding such a representative, *i.e.*, for transforming an arbitrary braid word into an equivalent σ-positive or σ-negative braid word, and, therefore, for comparing braids with respect to the σ-ordering. Moreover, this algorithm turns out to be extremely efficient in pratice, yet no theoretical confirmation of that efficiency has been found so far.

The techniques in this chapter mainly belong to combinatorial group theory, and, in particular, the Cayley graph of the braid group plays a central role.

The chapter is organized as follows. In Section 1, we describe handle reduction, state the main convergence result, and briefly discuss practical implementations of the method. In Section 2, we prove the convergence of handle reduction. Finally, in Section 3, we study the specific case of 3-strand braids, and mention a few variants of handle reduction.

1. Description of handle reduction

To decide whether a word w represents 1 in a free group, it suffices to freely reduce it, *i.e.*, to iteratively delete all subwords of the form xx^{-1} and $x^{-1}x$; then w represents 1 if and only if the final word is empty. In the case of a nonfree group, this result is no longer true: for instance, the word $\sigma_1\sigma_2\sigma_1\sigma_2^{-1}\sigma_1^{-1}\sigma_2^{-1}$ represents 1 in the braid group B_n, yet it contains no subword $\sigma_i\sigma_i^{-1}$ or $\sigma_i^{-1}\sigma_i$.

Handle reduction is a generalization of free reduction, which it extends properly: as in the case of a free group, the result will be that a braid word represents 1 in the braid group if and only if it reduces to the empty word.

The principle of handle reduction is simple—even naive—and it directly stems from Property **C**. If a nonempty braid word w is neither σ-positive nor σ-negative, then, by definition, the letter σ_i with minimal index appears both positively and negatively in w; so, necessarily, w contains a subword of the form $\sigma_i v \sigma_i^{-1}$ or $\sigma_i^{-1} v \sigma_i$ where all letters in v are letters $\sigma_k^{\pm 1}$ with $k > i$. Such a subword will be called a *handle*, and the basic observation is that there exists a natural way to transform such a handle into an equivalent braid word in which the first and the last letters $\sigma_i^{\pm 1}$ have been deleted: this operation is called *handle reduction*. The idea is then to iterate handle reduction until no more handle is left.

1.1. Handles. We recall that sh denotes the word homomorphism that maps every letter $\sigma_i^{\pm 1}$ to $\sigma_{i+1}^{\pm 1}$, as well as the induced endomorphism of B_∞.

DEFINITION 1.1. We say that a braid word is a *σ_i-handle* if it has the form $\sigma_i^e v \sigma_i^{-e}$, where e is ±1 and all letters in v are of the form $\sigma_k^{\pm1}$ with $k > i$.

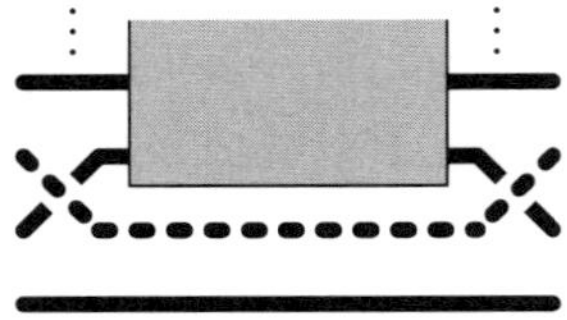

FIGURE 1. A σ_i-handle, here a σ_2-handle: the dotted strand is the handle, and the grey box is the suitcase; we draw the diagram as unbounded, because we prefer to think of B_∞ rather than of any specified B_n.

Thus, every braid word that is neither σ-positive nor σ-negative must contain a σ_i-handle for some i. The name refers to the handle formed by the $(i+1)$st strand in the associated braid diagram, as shown in Figure 1.

Now, the scheme of Figure 2 shows that every handle can be transformed into an equivalent braid word so that the initial and the final crossings have been eliminated. The principle is to call such a transformation the *reduction* of the handle, and to iterate it until no handle is left: if the process converges, the final word contains no handle, so it is either σ-positive, σ-negative, or empty.

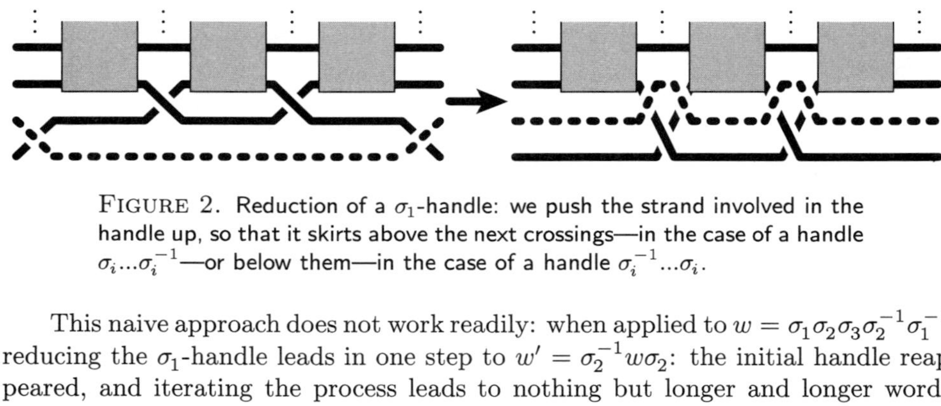

FIGURE 2. Reduction of a σ_1-handle: we push the strand involved in the handle up, so that it skirts above the next crossings—in the case of a handle $\sigma_i ... \sigma_i^{-1}$—or below them—in the case of a handle $\sigma_i^{-1} ... \sigma_i$.

This naive approach does not work readily: when applied to $w = \sigma_1\sigma_2\sigma_3\sigma_2^{-1}\sigma_1^{-1}$, reducing the σ_1-handle leads in one step to $w' = \sigma_2^{-1} w \sigma_2$: the initial handle reappeared, and iterating the process leads to nothing but longer and longer words. Now, the handle in w' is not the original handle of w, but it comes from the σ_2-handle $\sigma_2\sigma_3\sigma_2^{-1}$ of w. If we reduce this σ_2-handle into $\sigma_3^{-1}\sigma_2\sigma_3$ before reducing the σ_1-handle of w, *i.e.*, if we first go from w to $w'' = \sigma_1\sigma_3^{-1}\sigma_2\sigma_3\sigma_1^{-1}$, then applying handle reduction yields $\sigma_3^{-1}\sigma_2^{-1}\sigma_1\sigma_2\sigma_3$, a σ-positive word equivalent to w. We shall see in what follows that the previous obstruction, namely the existence of nested handles, is the only possible problem: provided we reduce nested handles first, handle reduction always comes to an end in a finite number of steps.

DEFINITION 1.2. A handle $\sigma_i^e v \sigma_i^{-e}$ is said to be *permitted* if the word v includes no σ_{i+1}-handle, *i.e.*, if all letters $\sigma_{i+1}^{\pm1}$ occurring in v, if any, have the same sign. If w is a permitted handle, say $w = \sigma_i^e v \sigma_i^{-e}$, we define the *reduct* of w to be the word obtained from v by replacing each letter $\sigma_{i+1}^{\pm1}$ with $\sigma_{i+1}^{-e}\sigma_i^{\pm1}\sigma_{i+1}^{e}$. Finally, we say that w' is obtained from w by *handle reduction*—or, simply, the w is *reducible* to w'—if w' is obtained by replacing a subword of w that is a permitted handle with its reduct.

The general form of a σ_i-handle is

$$\sigma_i^e\, v_0\, \sigma_{i+1}^{d_1}\, v_1\, \sigma_{i+1}^{d_2}\, \dots\, \sigma_{i+1}^{d_k}\, v_k\, \sigma_i^{-e}$$

with $d_j = \pm 1$ and $v_j \in \mathrm{sh}^{i+1}(B_\infty)$. Saying that this handle is permitted amounts to saying that all exponents d_j have a common value d. Then, reducing the handle means replacing it with

$$v_0\, \sigma_{i+1}^{-e}\sigma_i^d\sigma_{i+1}^e\, v_1\, \sigma_{i+1}^{-e}\sigma_i^d\sigma_{i+1}^e\, \dots\, \sigma_{i+1}^{-e}\sigma_i^d\sigma_{i+1}^e\, v_k :$$

we remove the initial and final $\sigma_i^{\pm 1}$ and replace each σ_{i+1}^d with $\sigma_{i+1}^{-e}\sigma_i^d\sigma_{i+1}^e$. In the particular case of a handle $\sigma_i^e\sigma_i^{-e}$, which is always permitted, reducing the handle just means deleting it: handle reduction extends free reduction.

EXAMPLE 1.3. Let us start with $w = \sigma_1\sigma_2\sigma_3\sigma_2\sigma_1^{-1}\sigma_2^{-1}\sigma_1^{-1}\sigma_2^{-1}\sigma_3^{-1}\sigma_2^{-1}\sigma_1^{-1}$. It will be convenient to use a for σ_1, b for σ_2, etc., and A for σ_1^{-1}, B for σ_2, etc. Choosing to reduce at each step the current leftmost handle—in the sense of Definition 1.9 below—and underlying it, we successively obtain (in columns):

$\mathtt{\underline{abcbA}BABCBA}$	$\mathtt{Ba\underline{bcB}BCBA}$	$\mathtt{BaCC\underline{bB}A}$
$\mathtt{BabcBa\underline{bB}ABCBA}$	$\mathtt{BaC\underline{bcB}CBA}$	$\mathtt{B\underline{aCCA}}$
$\mathtt{BabcB\underline{aA}BCBA}$	$\mathtt{BaCCb\underline{cC}BA}$	$\mathtt{BCC}$.

The final word contains no handle, it is σ_2-negative, and handle reduction has been successful in this case. We may observe in this simple example that the length of the braid words may increase when handle reduction is performed, and that neither the position of the first letter nor that of the last letter in the leftmost handle—the underlined subword in each word above—keeps increasing or decreasing: at first, handle reduction appears to be a chaotic process.

The following results are clear from the definition:

LEMMA 1.4. (i) *Handle reduction transforms a word into an equivalent word;*
(ii) *If a nonempty braid word w is terminal w.r.t. handle reduction, i.e., if w contains no handle, then w is σ-positive or σ-negative.*

1.2. The main result. The main result is that handle reduction converges.

PROPOSITION 1.5. *Assume that w is an n-strand braid word of length ℓ. Then every sequence of handle reduction steps from w converges in at most $2^{n^4\ell}$ steps.*

COROLLARY 1.6 (Property **C**). *Every n-strand braid word is equivalent to some n-strand braid word that is σ-positive, σ-negative, or empty.*

Indeed, Propostion 1.5 says that every n-strand braid word w is equivalent to some braid word w' that contains no handle: by definition, such a braid word w' is either empty, σ-positive, or σ-negative.

Observe that handle reduction gives a solution both for the word problem of B_∞ and for the decision problem of the linear ordering on B_∞.

PROPOSITION 1.7. *Let w be a braid word, and β be the braid represented by w.*
(i) *We have $\beta = 1$ in B_∞ if and only if w is reducible to the empty word;*
(ii) *We have $\beta > 1$ in B_∞ if w is reducible to some σ-positive braid word.*

PROOF. The only problem is that a given braid word may contain several handles, and reduction need not be confluent in general, *i.e.*, various sequences of reductions from a given word may lead to distinct final words. However, Property **A** tells us that the cases in Proposition 1.7 do not depend on the considered sequence of reductions: all words obtained by reduction from w are pairwise equivalent, and Property **A** asserts that no equivalence class may contain the empty word and a σ-positive word at the same time. □

We deduce the third equivalent definition of the braid ordering mentioned in the Introduction:

COROLLARY 1.8. *For all braids β, β', the relation $\beta < \beta'$ is true if and only if any sequence of handle reductions from any word representing $\beta^{-1}\beta'$ ends up with a σ-positive word.*

1.3. Left handle reduction. As it stands, handle reduction is not an algorithm: a given braid word may contain several handles, and, in order to obtain a deterministic method, we have to fix a strategy that chooses which handle is to be reduced first. Several choices are natural, and we shall briefly discuss them here.

There are at least two reasons for doing that. Firstly, for simplicity, we shall concentrate in Section 2 on the convergence of a particular reduction strategy, and we need to define it first. Secondly, handle reduction turns out to be in practice the most efficient solution to the braid word problem known so far, although no proven complexity upper bound explains it. So the question of finding efficient implementations is of concrete interest, particularly in view of the possible cryptographical implementations mentioned in Section III.2.2.

The most simple reduction strategy consists in systematically considering the leftmost handle, leading to an algorithm that can be called *left* handle reduction.

Hereafter, a braid word w of length ℓ is viewed as a length ℓ sequence of letters. For $1 \leqslant p \leqslant q \leqslant \ell$, the word obtained from w by deleting all letters before position p and after position q is called the (p,q)-subword of w. A prefix of w is a $(1,q)$-subword of w, *i.e.*, a subword that starts at the first letter of w.

DEFINITION 1.9. We say that v is the *leftmost* handle in a braid word w if v is a handle, there exist p, q such that v is the (p,q)-subword of w, and there exist no p', q' with $q' < q$ such that the (p', q')-subword of w is a handle.

Then the leftmost handle is always eligible for reduction:

LEMMA 1.10. *Assume that w is a braid word containing at least one handle. Then the leftmost handle in w is permitted.*

PROOF. Let q be the smallest integer such that the length q prefix w' of w contains a handle. By hypothesis, there exists p such that the (p,q)-subword of w is a handle, say $\sigma_i^e v \sigma_i^{-e}$, and, by construction, this handle is the leftmost handle in w. We claim that it is permitted. Indeed, the contrary would mean that there exist $p', q' < q$ such that the (p', q')-subword of w is a σ_{i-1}-handle, which implies that the length q' prefix of w contains a handle and contradicts the choice of q. □

In this way, we obtain a simple deterministic braid transformation.

DEFINITION 1.11. For w a braid word containing at least one handle, we define $\mathrm{red}(w)$ to be the braid word obtained from w by reducing the leftmost handle in w; we say that $\mathrm{red}(w)$ is obtained by *left handle reduction* from w.

Thus, left handle reduction consists in iterating the transformation red until a word that contains no handle is obtained. This algorithm, which was used in Example 1.3, is the one we shall investigate in Section 2 below.

1.4. Better reduction strategies. Other algorithmic implementations of handle reductions are possible. We briefly mention four variations that result in significantly improving the practical efficiency of the method.

Firstly, it is advisable to systematically perform all possible free reductions rather than waiting for the handles $\sigma_i^e \sigma_i^{-e}$ to become the leftmost handle of the current word.

Secondly, instead of reducing all handles, one can reduce only the unavoidable handles, namely the σ_1-handles together with the nested subhandles. In this way, instead of ending with a word that contains no more handles, we end with a word that is σ-positive, σ-negative, or empty, but, for instance, if the word is σ_1-positive, it may still contain σ_2-handles. This of course is not a problem if we wish only to compare the braid represented by that word with 1.

Thirdly, reduction steps may be grouped. As is clear in Figure 2, reducing a σ_1-handle may result in creating one or more σ_2-handles. This happens whenever there is more than one letter σ_2 in the σ_1-handle that is reduced. It is not hard to predict what the result is of reducing these new σ_2-handles, and, inductively, to define a sort of compound reduction that directly reduces all successive handles that may appear in this way.

Finally, the most efficient versions are obtained by using, in addition to the previous ingredients, the classical divide-and-conquer strategy: in order to reduce a word w, we decompose w into $w_1 w_2$, where the length of w_1 and w_2 are approximately equal, we reduce w_1 and w_2 separately, using the same method iteratively, and, having found reduced words w_1', w_2' equivalent to w_1 and w_2, we finally reduce $w_1' w_2'$. If w_1' and w_2' happen to be both σ-positive, or both σ-negative, thus with probability $1/2$, the last step vanishes.

With the above improvements, handle reduction turns out to be very efficient. Table 1 compares the overall computation times needed to reduce random words and to compute their greedy normal form (see Chapter VI and [**77**, Chapter 9]). Handle reduction is always more efficient: the average time for reducing a braid word of length 4,000 is always below one second, and reduction is 10 times faster than greedy normal form in the case of 4 or 16 strands, and much more when the number of strands increases. Table 2 provides additional information, namely the length of the final braid words obtained in the handle reduction algorithm.

These data suggest that the exponential upper bound following from Proposition 1.5 is *very* far from optimal—and they immediately lead to the conjectures stated in Section 2.2 of Chapter XVI, namely that the there exists a quadratic upper bound on the number of reduction steps, and a linear upper bound on the length of all words obtained in the process.

TABLE 1. Handle reduction *vs.* normal form: comparison of average CPU times in millisec. on an AMD Duron processor at 750 MHz; samples of 1,000 random braid words; C++ implementation by Hervé Sibert. In the case of handle reduction, the divide-and-conquer trick is applied until the length reaches 4 times the braid index.

	4 strands	16 strands	64 strands
64 crossings	0.20 *vs.* 5.36	0.03 *vs.* 8.65	0.016 *vs.* 23.1
256 crossings	2.71 *vs.* 77.4	0.45 *vs.* 105	0.14 *vs.* 194
1,024 crossings	54.5 *vs.* 1,526	10.2 *vs.* 1,378	1.56 *vs.* 1,899
4,096 crossings	1,560 *vs.* 29,900	1,635 *vs.* 21,990	33 *vs.* 23,640

TABLE 2. Length of the final word obtained in handle reduction: average case and (bracketed) worst case; samples of 1,000 random braid words.

	4 strands	16 strands	64 strands
64 crossings	73 (134)	62.6 (90)	63.6 (70)
256 crossings	308 (448)	258 (366)	254 (282)
1,024 crossings	1,257 (1,566)	1,126 (2,422)	1,023 (1,122)
4,096 crossings	5,034 (5,614)	5,745 (14,682)	4,169 (5,302)

2. Convergence of handle reduction

The aim of this section is to prove the convergence of handle reduction. This is a nontrivial task, as no very simple proof has been found so far. The problem is that handle reduction may increase the length of the braid word it is applied to, and, contrary to the case of free reduction, it is not clear that some parameter monotonically decreases when the reduction is applied. Although handle reduction is defined as a syntactic transformation, the arguments used below to prove its convergence are deeply geometric in nature—in particular, they make an essential use of the specific properties of braids.

2.1. Convergence of left handle reduction. In order to keep the argument as simple as possible but, on the other hand, to make it rigorous (which may be a good idea as handle reduction arguably provides the most satisfactory proof of Property **C**), we shall complete the proof in the case of left handle reduction only. Also, we shall forget the complexity issues and concentrate on the qualitative viewpoint. So, the result we will prove is

PROPOSITION 2.1. *For each braid word w, there exists an integer m such that the word* $\mathrm{red}^m(w)$ *contains no handle.*

In other words, left handle reduction, as introduced in Definition 1.11, always converges. Since w and $\mathrm{red}^m(w)$ are equivalent, and a braid word that contains no handle is either empty, σ-positive, or σ-negative, Proposition 2.1 implies Corollary 1.6 and its applications.

The proof of Proposition 2.1 consists of two steps. The first one is a boundedness result that relies on Garside's theory of positive braids. The second one is a monotonicity result that relies on Property **A**, which is taken here as an hypothesis.

2.2. A boundedness result. Our first task for proving convergence will be to show that all words obtained from a word w using handle reduction remain locked in some finite region of the Cayley graph of B_∞ depending on w only. We recall that, if v is a braid word, then $\overline{v}$ denotes the braid represented by v.

DEFINITION 2.2. Assume $X \subseteq B_\infty$, and $\beta_* \in X$. We say that a braid word w is *drawn from* β_* *in* X if, for each prefix v of w, the braid $\beta_*\overline{v}$ belongs to X.

The *Cayley graph* of the group B_∞ relative to the Artin generators σ_i is the labeled oriented graph whose vertices are the elements of B_∞ and there is a σ_i-labeled edge from the vertex β to the vertex β' if and only if the equality $\beta' = \beta\sigma_i$ holds in B_∞. Braid words correspond to paths in the Cayley graph, and saying that w is drawn from β_* in X means that the path starting from the vertex β_* and labeled by w only visits vertices that lie in X. Observe that, even if X is finite, arbitrarily long words may be drawn in X: for instance, for every k, the word $(\sigma_1\sigma_1^{-1})^k$ is drawn from 1 in $\{1, \sigma_1\}$.

In what follows, we shall consider braid words that are drawn in particular sets consisting of all left divisors of some positive braid, typically a power of Garside's fundamental braid Δ_n.

We recall from Chapter I that, if β, β' are braids, we say that β' is a *left divisor* of β, denoted $\beta \preccurlyeq \beta'$, if $\beta = \beta'\gamma$ holds for some braid γ belonging to B_∞^+.

DEFINITION 2.3. For β in B_∞^+, we denote by $\mathrm{Div}(\beta)$ the family of all left divisors of β in B_∞^+, *i.e.*, the set of all braids β' satisfying $1 \preccurlyeq \beta' \preccurlyeq \beta$.

By Proposition I.4.8, we know that, for each positive braid β, the set $\mathrm{Div}(b)$ equipped with $\preccurlyeq$ is a lattive. The aim of this section is to prove the following result:

PROPOSITION 2.4. *For each n-strand braid word w, there exist two positive braids β_*, β such that w, as well as every word obtained from w by handle reduction, is drawn from β_* in* $\mathrm{Div}(\beta)$. *(See Figure 3.)*

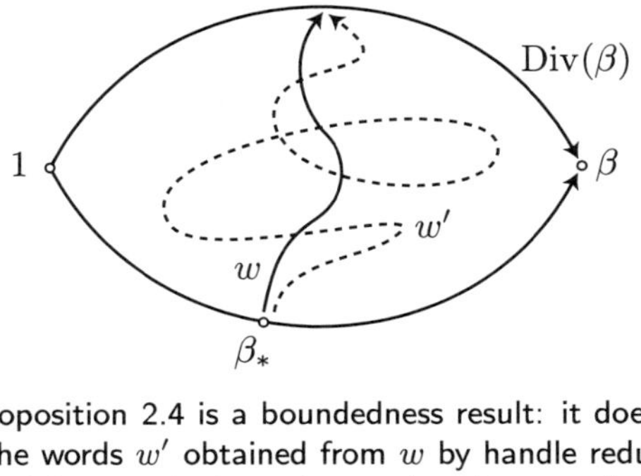

FIGURE 3. Proposition 2.4 is a boundedness result: it does not say that the lengths of the words w' obtained from w by handle reduction admit a finite upper bound, but, at the least, it says that these words cannot diverge too far from the initial word and must remain in the *finite* fragment of the Cayley graph consisting of the braids that lie between 1 and β in the sense of divibility.

Proposition 2.4 follows from two results. The first one is easy.

LEMMA 2.5. *For each n-strand braid word w, there exist positive braids β_* and β such that w is drawn from β_* in* $\mathrm{Div}(\beta)$.

PROOF. Assume that w has length ℓ and, for $0 \leqslant r \leqslant \ell$, let v_r be the length r prefix of w. The properties of Δ_n imply that, for each r, there exists $d_r \geqslant 0$ such that the braid $\Delta_n^{d_r}\overline{v_r}$ is positive, and that there exists $e_r \geqslant 0$ such that $\Delta_n^{d_r}\overline{v_r}$ is a left divisor of $\Delta_n^{d_r+e_r}$, *i.e.*, we have $1 \preccurlyeq \Delta_n^{d_r}\overline{v_r} \preccurlyeq \Delta_n^{d_r+e_r}$. Now, let $\beta_* = \Delta_n^d$ with $d = \max\{d_0, \dots, d_\ell\}$, and $\beta = \Delta_n^{d+e}$ with $e = \max\{e_0, \dots, e_\ell\}$. Then, for each r, we find $1 \preccurlyeq \beta_*\overline{v_r} \preccurlyeq \beta$. Owing to Definitions 2.2 and 2.3, this means that w is drawn from β_* in $\mathrm{Div}(\beta)$. □

The second result is a closure lemma.

LEMMA 2.6. *Assume that w is drawn from β_* in* $\mathrm{Div}(\beta)$, *and w' is obtained from w by handle reduction. Then w' too is drawn from β_* in* $\mathrm{Div}(\beta)$.

2.3. Special transformations. The proof of Lemma 2.6 consists in decomposing handle reduction into more elementary transformations and showing that the words drawn from β_* in $\mathrm{Div}(\beta)$ are closed under these elementary transformations.

DEFINITION 2.7. Let w, w' be braid words. We say that w' is obtained from w by a *special transformation* if we have $w = w_1 v w_2$ and $w' = w_1 v' w_2$ for some words w_1, w_2, and (v, v') is one of the following pairs:
type 1: $(\sigma_i\sigma_j, \sigma_j\sigma_i)$ with $|i-j| \geqslant 2$;
type 2: $(\sigma_i^{-1}\sigma_j^{-1}, \sigma_j^{-1}\sigma_i^{-1})$ with $|i-j| \geqslant 2$;
type 3: $(\sigma_i^{-1}\sigma_j, \sigma_j\sigma_i^{-1})$ with $|i-j| \geqslant 2$, $(\sigma_i^{-1}\sigma_j, \sigma_j\sigma_i\sigma_j^{-1}\sigma_i^{-1})$ with $|i-j| = 1$, $(\sigma_i^{-1}\sigma_i, \varepsilon)$;
type 4: $(\sigma_i\sigma_j^{-1}, \sigma_j^{-1}\sigma_i)$ with $|i-j| \geqslant 2$, $(\sigma_i\sigma_j^{-1}, \sigma_j^{-1}\sigma_i^{-1}\sigma_j\sigma_i)$ with $|i-j| = 1$, $(\sigma_i\sigma_i^{-1}, \varepsilon)$.

Observe that special transformations of type 3 are right reversings as defined in Chapter IV, and, symmetrically, type 4 are left reversings. Also, note that the braid transformations $\sigma_i\sigma_j\sigma_i \mapsto \sigma_j\sigma_i\sigma_j$ with $|i-j| = 1$ are not defined to be special.
Lemma 2.6 follows from the next two results.

LEMMA 2.8. *If w' is obtained from w by handle reduction, then w' can be optained from w by finitely many special transformations.*

LEMMA 2.9. *Assume that w is drawn from β_* in* $\mathrm{Div}(\beta)$, *and w' is obtained from w by a special transformation. Then w' is drawn from β_* in* $\mathrm{Div}(\beta)$.

PROOF OF LEMMA 2.8. The point is to prove that, if v is a permitted handle, and v' is its reduct, then we can go from v to v' by using special transformations. By definition, there exist exponents $e, d = \pm 1$ such that v has the form

$$v = \sigma_i^e \; u_0 \; \sigma_{i+1}^d \; u_1 \; \dots \; u_{r-1} \; \sigma_{i+1}^d \; u_r \; \sigma_i^{-e}, \tag{2.1}$$

where $u_0, \dots, u_r$ contain only letters $\sigma_k^{\pm 1}$ with $k \geqslant i+2$, and then we have

$$v' = \; u_0 \; \sigma_{i+1}^{-e}\sigma_i^d\sigma_{i+1}^e \; u_1 \; \dots \; u_{r-1} \; \sigma_{i+1}^{-e}\sigma_i^d\sigma_{i+1}^e \; u_r. \tag{2.2}$$

Assume first $d = 1$, $e = -1$. The involved words are

$$\begin{array}{lccccccccc} v = & \underline{\sigma_i^{-1}} & u_0 & \sigma_{i+1} & u_1 & \dots & u_{r-1} & \sigma_{i+1} & u_r & \sigma_i, \\ v' = & & u_0 & \sigma_{i+1}\sigma_i\sigma_{i+1}^{-1} & u_1 & \dots & u_{r-1} & \sigma_{i+1}\sigma_i\sigma_{i+1}^{-1} & u_r & . \end{array}$$

The principle is to use type 2 and 3 transformations to let the initial (underlined) letter σ_i^{-1} in v migrate to the right until it reaches to the final letter σ_i. First, σ_i^{-1} crosses u_0 using type 3 transformations for the positive letters in u_0, and type 2 transformations for the negative ones. In this way, we reach the word

$$u_0 \, \underline{\sigma_i^{-1}} \, \sigma_{i+1} \, u_1 \, \ldots \, u_{r-1} \, \sigma_{i+1} \, u_r \, \sigma_i .$$

One more type 3 transformation lets σ_i^{-1} cross σ_{i+1}, resulting in the word

$$u_0 \, \sigma_{i+1}\sigma_i\sigma_{i+1}^{-1}\underline{\sigma_i^{-1}} \, u_1 \, \ldots \, u_{r-1} \, \sigma_{i+1} \, u_r \, \sigma_i .$$

The same process lets σ_i^{-1} cross u_1, and the next σ_{i+1}, and, after r such steps, we reach the word

$$u_0 \, \sigma_{i+1}\sigma_i\sigma_{i+1}^{-1} \, u_1 \, \ldots \, u_{r-1} \, \sigma_{i+1}\sigma_i\sigma_{i+1}^{-1} \, u_r \, \underline{\sigma_i^{-1}}\sigma_i ,$$

and a final type 3 transformation leads to the expected word v'.

The argument for the case $d = -1$, $e = +1$ is similar, with transformations of type 1 and 4 instead of 2 and 3. For the case $d = 1$, $e = 1$, the argument is symmetric, *i.e.*, we start with the final letter σ_i^{-1} and let it migrate to the left, using transformations of type 2 and 4. Finally, the case $d = e = -1$ is similar, with transformations of type 1 and 3 instead of 2 and 4. □

We shall now complete the proof of Lemma 2.9. To do that, we shall use the basic result of Garside's theory of braid monoids, namely Proposition I.4.8: any two elements of B_∞^+ admit a lower bound (greatest common left divisor) and an upper bound (least common right multiple) with respect to $\preccurlyeq$.

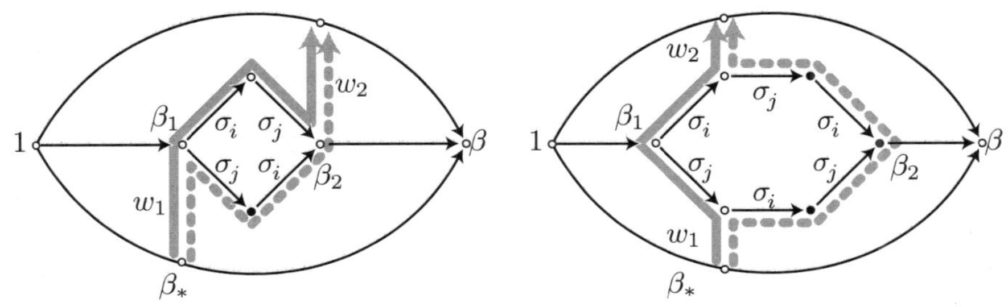

FIGURE 4. Closure of words drawn from β_* in $\mathrm{Div}(\beta)$ under special transformations. On the left, type 1: by hypothesis, w (grey plain path) is drawn from β_*, and the question is whether w' (grey dotted path) is drawn as well; the only new vertex visited is $\beta_1\sigma_j$ (in black), and it belongs to $\mathrm{Div}(\beta)$ because it lies between β_1 and β_2. On the right, type 3: with the same conventions, three new vertices are visited, and they belong to $\mathrm{Div}(\beta)$ because they lie between β_1 and β_2, while β_2 lies in $\mathrm{Div}(\beta)$ because the latter is closed under the least common right multiple operation.

PROOF OF LEMMA 2.9. We assume that w is drawn from β_* in $\mathrm{Div}(\beta)$, and that w' is obtained from w by one special transformation (see Figure 4). We have to prove that, for each prefix v of w', the braid $\beta_*\overline{v}$ lies in $\mathrm{Div}(\beta)$. By hypothesis, the result is true for each prefix of w' that is also a prefix of w, so we only have to consider those prefixes of w' that are not prefixes of w. As the divisibility relation $\preccurlyeq$ is transitive, the existence of β_1, β_2 in $\mathrm{Div}(\beta)$ satisfying

$$\beta_1 \preccurlyeq \beta_* \overline{v} \preccurlyeq \beta_2 \tag{2.3}$$

implies $\beta_* \overline{v} \in \mathrm{Div}(\beta)$, so our aim in what follows will be to exhibit, for each prefix v of w' that is not a prefix of w, two braids β_1, β_2 in $\mathrm{Div}(\beta)$ satisfying (2.3).

Assume first that w' is obtained from w by a type 1 transformation. This means that, for some w_1, w_2 and i, j with $|i - j| \geqslant 2$, we have $w = w_1\ \sigma_i \sigma_j\ w_2$ and $w' = w_1\ \sigma_j \sigma_i\ w_2$. By construction, the only prefix of w' that is not a prefix of w is $w_1 \sigma_j$. Now, let $\beta_1 = \beta_* \overline{w_1}$ and $\beta_2 = \beta_* \overline{w_1} \sigma_i \sigma_j$. As w_1 and $w_1 \sigma_i \sigma_j$ are prefixes of w and w is drawn from β_* in $\mathrm{Div}(\beta)$, the braids β_1 and β_2 lie in $\mathrm{Div}(\beta)$. Now, by construction, we have $\beta_1 \preccurlyeq \beta_* \overline{w_1} \sigma_j \preccurlyeq \beta_2$, as expected for (2.3).

Type 2 is similar. We start from $w = w_1\ \sigma_i^{-1} \sigma_j^{-1}\ w_2$ and $w' = w_1\ \sigma_j^{-1} \sigma_i^{-1}\ w_2$ with $|i - j| \geqslant 2$. The only prefix of w' that is not a prefix of w is $w_1 \sigma_j^{-1}$. Let $\beta_1 = \beta_* \overline{w_1} \sigma_i^{-1} \sigma_j^{-1}$ and $\beta_2 = \beta_* \overline{w_1}$. As above β_1 and β_2 lie in $\mathrm{Div}(\beta)$ and we have $\beta_1 \preccurlyeq \beta_* \overline{w_1} \sigma_j^{-1} \preccurlyeq \beta$, as expected for (2.3).

For type 3, we consider $w = w_1\ \sigma_i^{-1} \sigma_j\ w_2$ and $w' = w_1\ \sigma_j \sigma_i \sigma_j^{-1} \sigma_i^{-1}\ w_2$ with $|i - j| = 1$. The cases $|i - j| \geqslant 2$ and $i = j$ are similar and easier. Three prefixes of w' are not prefixes of w, namely $w_1 \sigma_j$, $w_1 \sigma_j \sigma_i$, and $w_1 \sigma_j \sigma_i \sigma_j^{-1}$. Let $\beta_1 = \beta_* \overline{w_1} \sigma_i^{-1}$, and $\beta_2 = \beta_* \overline{w_1} \sigma_j \sigma_i$. First, $w_1 \sigma_i^{-1}$ is a prefix of w, so β_1 lies in $\mathrm{Div}(\beta)$. Next, w_1 and $w_1 \sigma_i^{-1} \sigma_j$ are prefixes of w, so we have $\overline{w_1}$ and $\overline{w_1} \sigma_i^{-1} \sigma_j$, *i.e.*, $\beta_1 \sigma_i$ and $\beta_1 \sigma_j$, lie in $\mathrm{Div}(\beta)$. So, in particular, we have $\beta_1 \sigma_i \preccurlyeq \beta$ and $\beta_1 \sigma_j \preccurlyeq \beta$. Now, β_2, which is $\beta_1 \sigma_i \sigma_j \sigma_i$, is the upper bound of $\beta_1 \sigma_i$ and $\beta_1 \sigma_j$ in the divisibility lattice, so we deduce $\beta_2 \preccurlyeq \beta$, and β_2 lies in $\mathrm{Div}(\beta)$ too. Then, by construction, we have $\beta_1 \preccurlyeq \beta_* \overline{w_1} \sigma_j \preccurlyeq \beta_2$, $\beta_1 \preccurlyeq \beta_* \overline{w_1} \sigma_j \sigma_i \preccurlyeq \beta_2$, and $\beta_1 \preccurlyeq \beta_* \overline{w_1} \sigma_j \sigma_i \sigma_j^{-1} \preccurlyeq \beta_2$, as expected for (2.3).

As can be expected, type 4 is exactly similar to case 3, the existence of greatest lower bounds for the relation $\preccurlyeq$ now replacing the lowest upper bound. □

Thus the proof of Lemma 2.6, and, therefore, of Proposition 2.4 is complete.

2.4. The main prefix. The boundedness result of Proposition 2.4 is not sufficient for proving that handle reduction always converges. In particular, loops might occur. We shall see now that this is impossible, because some braid parameter attached with the reduction sequence is monotonically increasing or decreasing during the reduction process. This step of the proof is where Property **A** is used.

An outline of the argument we shall use to prove Proposition 2.1 is as follows. From now on, we restrict ourselves to the case of left handle reduction, which provides the technical advantage of guaranteeing the uniqueness of the reduction sequence from a given braid word. So, let w be a braid word, and let w_m be $\mathrm{red}^m(w)$, the braid word obtained after m steps of left handle reduction.

• The first observation is that the number of σ_1-handles in w_m is not larger than the number of σ_1-handles in w.

• The second observation is that, if we define the *main prefix* $\pi(w_m)$ of w_m to be the prefix that ends after the first letter of the leftmost σ_1-handle and if the leftmost σ_1-handle in w is positive, then the braids represented by $\pi(w_0)$, $\pi(w_1)$,... make a sequence that is nonincreasing with respect to the σ-ordering and is even strictly decreasing at each reduction step that involves a σ_1-handle. Moreover—and this is the point—there exists a σ_1-negative witness-word that is drawn inside some fixed subset of the braid group and contains one letter σ_1^{-1} for each reduction step involving a σ_1-handle.

• The third observation is that Property **A** guarantees that this cannot happen infinitely many times, and the convergence of the reduction process easily follows.

We shall now make the above sketch formal. The main step consists in analysing what happens when the leftmost handle of a braid word is reduced. As several cases are possible, the analysis requires some care, but it is easy.

DEFINITION 2.10. For w a braid word, we write $h(w)$ for the number of σ_1-handles in w; if $h(w) \geqslant 1$ holds, we define the *main prefix* $\pi(w)$ of w to be the prefix that ends after the initial letter of the leftmost σ_1-handle of w, and $e(w)$ to be the sign of this handle.

We recall from Definition 1.11 that, if w is a braid word containing at least one handle, then $\mathrm{red}(w)$ denotes the word obtained from w by reducing the leftmost handle in w—whether this handle is a σ_1-handle or not. The key technical lemma is as follows.

LEMMA 2.11. *Assume that w is a braid word drawn from β_* in $\mathrm{Div}(\beta)$ and containing at least one handle. Let $w' = \mathrm{red}(w)$. Then three cases are possible:*

• *Case 1: $h(w') = h(w) = 0$;*

• *Case 2: $h(w') < h(w)$;*

• *Case 3: $h(w') = h(w) \geqslant 1$; in this case, we have $e(w') = e(w)$, and there exists a word $\gamma(w)$ drawn from $\beta_*\pi(w)$ in $\mathrm{Div}(\beta)$ satisfying $\pi(w') \equiv \pi(w)\gamma(w)$ and such that, if the leftmost handle in w is a σ_i-handle with $i \geqslant 2$, the word $\gamma(w)$ is empty, while, if the leftmost handle in w is a σ_1-handle, the word $\gamma(w)$ contains one letter $\sigma_1^{-e(w)}$ and no letter $\sigma_1^{e(w)}$.*

PROOF. If $h(w) = 0$ holds, *i.e.*, if w contains no σ_1-handle, one goes from w to w' by reducing some σ_i-handle with $i \geqslant 2$, and w' contains no σ_1-handle either. So we are in Case 1.

We assume now $h(w) \geqslant 1$ and write e for $e(w)$. Then there exist numbers $p, q \geqslant 1$ such that, starting from the left, the letters $\sigma_1^{\pm 1}$ in w consist of p letters σ_1^e, then q letters σ_1^{-e}. So w has the form

$$v_0\ \sigma_1^e\ v_1\ \sigma_1^e\ \ldots\ v_{p-2}\ \sigma_1^e\ v_{p-1}\ \underline{\sigma_1^e\ v_p\ \sigma_1^{-e}}\ v_{p+1}\ \sigma_1^{-e}\ \ldots\ v_{p+q-1}\ \sigma_1^{-e}\ v_{p+q}\ v, \tag{2.4}$$

where each v_k contains no $\sigma_1^{\pm 1}$ and v is either empty—case $h(w) = 1$—or starts with σ_1^e—case $h(w) \geqslant 2$. The leftmost σ_1-handle in w is the underlined subword, and the main prefix $\pi(w)$ is $v_0\ \sigma_1^e\ \ldots\ v_{p-1}\ \sigma_1^e$.

Assume first that the leftmost handle in w is a σ_i-handle with $i \geqslant 2$ (see Figure 5). Then the reduction from w to w' occurs inside one of the words $v_0, \ldots, v_p$, *i.e.*, it consists in replacing some subword v_k with the corresponding word $\mathrm{red}(v_k)$. In this case, we have $h(w') = h(w)$ and $e(w') = e(w)$. Moreover, if $k = p$ holds, we have $\pi(w') = \pi(w)$, while, if $k < p$ holds, $\pi(w')$ is obtained from $\pi(w)$ by replacing the subword v_k with $\mathrm{red}(v_k)$. In all cases, we find $\pi(w') \equiv \pi(w)$, and we are in Case 3 with $\gamma(w) = \varepsilon$ (the empty word).

Assume now that w' is obtained from w by reducing the underlined σ_1-handle of (2.4). The hypothesis that $\sigma_1^e v_p \sigma_1^{-e}$ is a permitted handle implies that σ_2 and

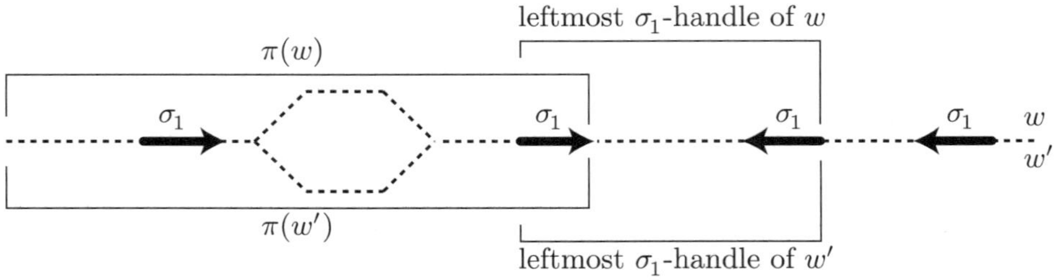

FIGURE 5. Change in the main prefix, case when the leftmost handle is not the leftmost σ_1-handle: reduction occurs in an intermediate subword, and $\pi(w')$ is equivalent to $\pi(w)$.

σ_2^{-1} do not simultaneously occur in v_p, and, therefore, v_p can be written as

$$u_0\ \sigma_2^d\ u_1\ \sigma_2^d\ \dots\ u_{r-1}\ \sigma_2^d\ u_r$$

for some $r \geqslant 0$, $d = \pm 1$, and each word u_k containing no $\sigma_1^{\pm 1}$ or $\sigma_2^{\pm 1}$.

Assume first $r = 0$, *i.e.*, v_p contains no $\sigma_2^{\pm 1}$ (see Figure 6). Then the reduct of $\sigma_1^e v_p \sigma_1^{-e}$ is v_p, so here reduction amounts to deleting the underlined letters σ_1^e and σ_1^{-e} of (2.4). If we have $p = 1$ or $q = 1$, then we obtain $h(w') < h(w)$, and we are in Case 2. Otherwise, *i.e.*, for $p, q \geqslant 2$, we have $h(w') = h(w)$, with

$$w' = v_0\ \sigma_1^e\ v_1\ \sigma_1^e\ \dots\ v_{p-2}\ \underline{\sigma_1^e\ v_{p-1}\qquad v_p\qquad v_{p+1}\ \sigma_1^{-e}}\ \dots, \tag{2.5}$$

in which the new leftmost σ_1-handle is underlined. We read on (2.5) the relations $e(w') = e(w) = e$ and $\pi(w') = v_0\sigma_1^e \dots v_{p-2}\sigma_1^e$, whence

$$\pi(w) = \pi(w')\ v_{p-1}\sigma_1^e.$$

We deduce $\pi(w') \equiv \pi(w)\ \sigma_1^{-e} v_{p-1}^{-1}$, corresponding to Case 3 with $\gamma(w) = \sigma_1^{-e} v_{p-1}^{-1}$. Indeed, by construction, the word $\gamma(w)$ is drawn from $\beta_* \overline{\pi(w)}$ in $\mathrm{Div}(\beta)$ since $v_{p-1}\sigma_1^e$ is a suffix of $\pi(w)$, which by hypothesis is drawn from β_* in $\mathrm{Div}(\beta)$.

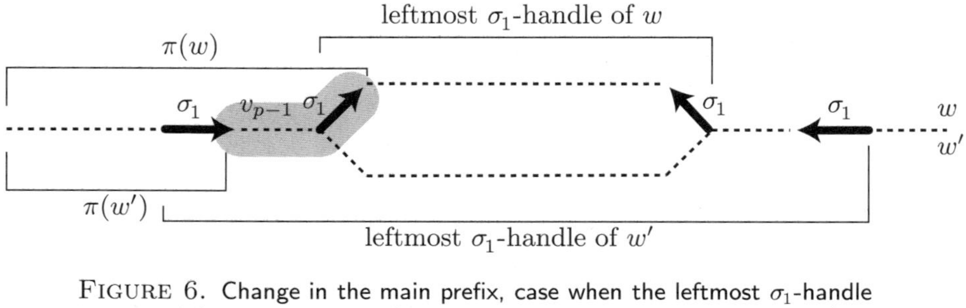

FIGURE 6. Change in the main prefix, case when the leftmost σ_1-handle is reduced (here with $e = +1$), and there is no σ_2 in the handle: $\pi(w')$ is equivalent to $\pi(w)\gamma(w)$, with $\gamma(w) = \sigma_1^{-1} v_{p-1}^{-1}$ (in grey).

Assume now $r \geqslant 1$ with $d = -e$, *i.e.*, the letter σ_2^{-e} occurs in v_p (see Figure 7). Then each σ_2^{-e} in v_p gives rise to a σ_1^{-e} in the reduct of v_p, hence in w'. If we have $p = 1$, then we obtain $h(w') < h(w)$, and we are in Case 2. Otherwise, *i.e.*, for $p \geqslant 2$, we have $h(w') = h(w)$, with

$$w' = v_0\ \sigma_1^e\ v_1\ \sigma_1^e\ \dots\ v_{p-2}\ \underline{\sigma_1^e\ v_{p-1}\qquad u_0\ \sigma_2^{-e}\sigma_1^{-e}}\sigma_2^e\ u_1\ \dots, \tag{2.6}$$

in which the new leftmost σ_1-handle is underlined. We read on (2.6) the relations $e(w') = e(w) = e$ and $\pi(w') = v_0\sigma_1^e ... v_{p-2}\sigma_1^e$, hence $\pi(w) = \pi(w')v_{p-1}\sigma_1^e$ as above, so we are in Case 3 and we conclude exactly as above.

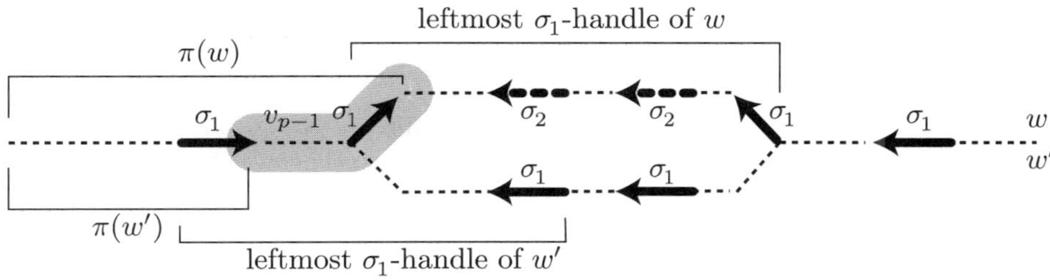

FIGURE 7. Change in the main prefix, case when the leftmost σ_1-handle is reduced (here with $e = +1$), and σ_2^{-1} occurs in the handle: each σ_2^{-1} in the handle gives a σ_1^{-1} in w', and, here again, $\pi(w')$ is equivalent to $\pi(w)\gamma(w)$, with $\gamma(w) = \sigma_1^{-1}v_{p-1}^{-1}$ (in grey).

Finally, assume $r \geqslant 1$ with $d = e$, *i.e.*, the letter σ_2^e occurs in v_p (see Figure 8). Each σ_2^e in v_p gives rise to a σ_1^{-e} in the reduct of v_p, hence in w'. If we have $q = 1$, then we obtain $h(w') < h(w)$, and we are in Case 2. Otherwise, *i.e.*, for $q \geqslant 2$, we have $h(w') = h(w)$, with

$$w' = v\ u_0\ \sigma_2^{-e}\sigma_1^e\sigma_2^e\ u_1\ ...\ u_{r-1}\ \sigma_2^{-e}\underline{\sigma_1^e\sigma_2^e\ u_r\quad v_{p+1}\ \sigma_1^{-e}}\ ... \tag{2.7}$$

in which the new leftmost σ_1-handle is underlined. We read on (2.7) the relation $e(w') = e(w) = e$. Moreover, with our notations, we have $\pi(w) = v\sigma_1^e$, and (2.7) gives

$$\pi(w)\ v_p\sigma_1^{-e} \equiv \pi(w')\ \sigma_2^e u_r.$$

We deduce $\pi(w') \equiv \pi(w)\ v_p\sigma_1^{-e}u_r^{-1}\sigma_2^{-e}$, corresponding to Case 3 with $\gamma(w) = v_p\sigma_1^{-e}u_r^{-1}\sigma_2^{-e}$, because the word $\gamma(w)$ is drawn from $\beta_*\overline{\pi(w)}$ in $\mathrm{Div}(\beta)$. Indeed, w is drawn from β_* in $\mathrm{Div}(\beta)$ by hypothesis and $\pi(w)v_p\sigma_1^{-e}$ is a prefix of w, hence $v_p\sigma_1^{-e}$ is drawn from $\beta_*\overline{\pi(w)}$ in $\mathrm{Div}(\beta)$; on the other hand, by Proposition 2.4, w' is drawn from β_* in $\mathrm{Div}(\beta)$ too, and $\pi(w')\sigma_2^e u_r$ is a prefix of w', hence u_r^{-1} is drawn from $\beta_*\overline{\pi(w')\sigma_2^e u_r}$, which is also $\beta_*\overline{\pi(w)v_p\sigma_1^{-e}}$, in $\mathrm{Div}(\beta)$. So $\gamma(w)$ is drawn from $\beta_*\overline{\pi(w)}$ in $\mathrm{Div}(\beta)$, and the proof is complete. □

2.5. Using Property A. The way Property **A** enters the argument is the following observation.

LEMMA 2.12. *Assume that β_*, β are positive braids and w is a σ_1-positive braid word drawn from β_* in* $\mathrm{Div}(\beta)$. *Then the number of occurrences of σ_1 in w is at most the cardinality of* $\mathrm{Div}(\beta)$.

PROOF. Assume that σ_1 occurs r times in w. Let $v_1, ..., v_r$ be the prefixes of w such that v_p finishes just before the pth letter σ_1 in w. By hypothesis, all braids $\beta_*\overline{v_p}$ belong to $\mathrm{Div}(\beta)$. Now $p < q$ implies $\beta_*\overline{v_p} \neq \beta_*\overline{v_q}$: indeed, by construction, we have $v_q = v_p v$, where v contains at least one letter σ_1, and no letter σ_1^{-1}, so, by Property **A**, the braid $\overline{v}$ is not 1. Hence $\beta_*\overline{v_1}, ..., \beta_*\overline{v_r}$ are pairwise distinct elements of $\mathrm{Div}(\beta)$. Therefore, r is at most the cardinality of $\mathrm{Div}(\beta)$. □

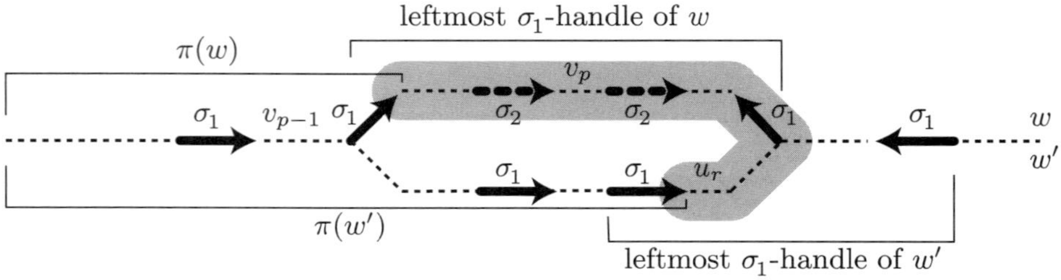

FIGURE 8. Change in the main prefix, case when the leftmost σ_1-handle is reduced (here with $e = +1$), and σ_2 occurs in the handle: each σ_2 gives a σ_1 in w', and, now $\pi(w')$ is equivalent to $\pi(w)\gamma(w)$, with $\gamma(w) = v_p\sigma_1^{-1}u_r^{-1}$ (in grey).

We are now ready to conclude, *i.e.*, to prove Proposition 2.1.

PROOF OF PROPOSITION 2.1. Using induction on $n \geqslant 2$, we prove the following.

For each n-strand braid word w, there exists m such that $\mathrm{red}^m(w)$ contains no handle—and therefore $\mathrm{red}^{m+1}(w)$ does not exist.

For $n = 2$, the only letters in w are σ_1 and σ_1^{-1}, handle reduction coincides with free reduction, and the result is clear, with m at most half the length of w.

Assume $n \geqslant 3$, and assume for a contradiction that w is an n-strand braid word such that $\mathrm{red}^m(w)$ exists for every m. We write w_m for $\mathrm{red}^m(w)$. By Lemma 2.11, the sequence $h(w_0)$, $h(w_1)$,... is nonincreasing, hence it is eventually constant. So, at the expense of possibly deleting the first w_m's, we can assume that there exists h such that $h(w_m) = h$ holds for every m.

By hypothesis, w_{m+1} is obtained from w_m by reducing its leftmost handle, which is either a σ_1-handle, or a σ_i-handle for some $i \geqslant 2$. Let M be the set of all m's such that the leftmost handle in w_m is a σ_1-handle.

Firstly, we claim that M is infinite. Indeed, let m be an arbitrary number. By hypothesis, w_m exists, and we can write

$$w_m = \mathrm{sh}(v_0)\ \sigma_1^e\ \mathrm{sh}(v_1)\ \sigma_1^e\ ...\ \sigma_1^e\ \mathrm{sh}(v_p)\ v,$$

where each word v_k is an $(n-1)$-strand braid word and v either begins with σ_1^{-e} (case $h > 0$) or is empty (case $h = 0$). By induction hypothesis, there exists for each k an integer m_k such that $\mathrm{red}^{m_k}(v_k)$ contains no handle. Let $m' = m + m_0 + \cdots + m_p$. Then, by construction, we have

$$w_{m'} = \mathrm{sh}(\mathrm{red}^{m_0}(v_0))\ \sigma_1^e\ \mathrm{sh}(\mathrm{red}^{m_1}(v_1))\ \sigma_1^e\ ...\ \sigma_1^e\ \mathrm{sh}(\mathrm{red}^{m_p}(v_p))\ v.$$

If v were empty, $w_{m'}$ would contain no handle, contradicting our hypothesis that the sequence $(w_m)_{m\geqslant 0}$ is infinite. Hence v begins with σ_1^{-e}, and the leftmost handle in $w_{m'}$ is a σ_1-handle. Thus we found an element m' of M which is at least equal to m, and M is infinite.

On the other hand, we claim that M is finite, thus getting the expected contradiction. Indeed, let β_*, β be positive braids such that w, and hence by Proposition 2.4, all words w_m are drawn from β_* in $\mathrm{Div}(\beta)$. We apply Lemma 2.11 to each word w_m. By hypothesis, we are always in Case 3 of that lemma. Let e be the common value of $e(w_m)$ for all m, and let u be the (infinite) word $\gamma(w_0)\gamma(w_1)\ldots$.

By construction, u is drawn from $\beta_* \overline{\pi(w)}$ in $\mathrm{Div}(\beta)$, it contains no letter σ_1^e, and it contains exactly one letter σ_1^{-e} for each m in M. By Lemma 2.12, the number of such letters, and therefore the cardinal of M, is bounded above by the cardinal of $\mathrm{Div}(\beta)$. In particular M is finite.

So the existence of an n-strand braid word w such that $\mathrm{red}^m(w)$ exists for every m is a contradictory assumption, and the proof is complete. □

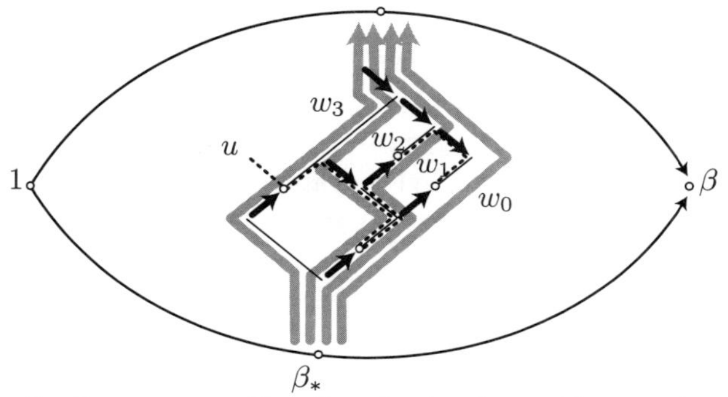

FIGURE 9. Convergence of handle reduction, here with $e = +1$: consider the paths associated with the successive words $w_0, w_1, \ldots$ a reduction sequence (in grey); the witness word u (dotted line) connects the ends of the main prefixes (circled), and it contains one letter σ_1^{-1} (bold arrow) for each reduction of the leftmost σ_1-handle.

Adapting the previous argument to the case of general handle reduction, *i.e.*, proving that *every* sequence of handle reduction converges, in whatever order the handles are reduced, is easy: instead of keeping track of the leftmost σ_1-handle, we should simultaneously keep track of each σ_1-handle; for each of them, the counterpart of Lemma 2.11 is true, and the associated prefix has the same monotonous behaviour as what was called the main prefix above.

2.6. The σ_1-content of a positive braid. The previous argument is effective, and it can easily be converted into the upper bound of complexity stated in Proposition 1.5. As the latter seems to be far from optimal, we shall discuss it briefly only.

For an induction, the point is to get an upper bound on the number of reduction steps involving the leftmost σ_1-handle—called *σ_1-reduction steps* hereafter. Let us introduce

DEFINITION 2.13. For β a positive braid, we define the *σ_1-content* $c_1(\beta)$ of β to be the maximal number of letters σ_1 in a σ-positive braid word drawn in $\mathrm{Div}(\beta)$.

If w is an n-strand word of length ℓ, then w is drawn in $\mathrm{Div}(\Delta_n^\ell)$ and, using Lemma 2.11, it is not hard to prove that the number of σ_1-reduction steps from w is bounded above by the number $c_1(\Delta_n^\ell)$, and we are left with the evaluation of the latter number. By Lemma 2.12, a coarse upper bound is the cardinality of $\mathrm{Div}(\Delta_n^\ell)$. As each element in the latter set is represented by a positive braid word of length at most ℓ times the length of Δ_n, *i.e.*, $\ell n(n-1)/2$, and there are $n-1$ possible letters, we obtain the upper bound

$$c_1(\Delta_n^\ell) \leqslant (n-1)^{\ell n(n-1)/2}, \tag{2.8}$$

from which the bound of Proposition 1.5 follows.

There are two natural ways for improving the upper bound of Proposition 1.5: obtaining a better estimate than $c_1(\Delta_n^\ell)$ for the number of σ_1-reduction steps, and obtaining for $c_1(\Delta_n^\ell)$ a better estimate than the one in (2.8). The second approach fails: we shall see in Section VI.3.6 that the value of $c_1(\Delta_n^\ell)$ is exponential in ℓ, even for $n = 3$. Thus, in order to really improve the results, the point seems to identify further constraints for the transversal word denoted u in the proof of Proposition 1.5—or to find a new convergence proof.

3. Special cases and variants

We conclude the chapter with some results about the special case of 3-strand braids, where a specific argument exists. Also, we establish for further use in Chapter VI a closure property for the Cayley graph that follows from the convergence of handle reduction, and we describe some variants of handle reduction.

3.1. The case of three strands. The case of 3-strand braids is particular in many aspects. Using a specific argument, we shall now prove the convergence of handle reduction without appealing to Property **A**, and we shall also establish a polynomial upper bound of complexity. The main step consists in considering a 3-strand braid word that contains one σ_1-handle, and to construct some planar rectangular diagram that controls its left handle reduction and induces a quadratic complexity bound.

LEMMA 3.1. *Assume that w is a 3-strand braid word of length ℓ satisfying $h(w) \leqslant 1$. Then there exists $m \leqslant \ell^2/4$ such that $\mathrm{red}^m(w)$ contains no handle.*

PROOF. (See Figure 10.) The result is obvious if w contains no σ_1-handle. So we assume now that w contains one σ_1-handle. Up to a symmetry, we may assume that the σ_1-handle in w is positive, *i.e.*, that all letters σ_1 precede all letters σ_1^{-1}. Moreover, we first assume that w contains no negative σ_2-handle, *i.e.*, no subword $\sigma_2^{-1}\sigma_2$.

We now inductively construct a sequence of numbers $m_0 < m_1 < \dots$ and a sequence of planar diagrams $D_0 \subset D_1 \subset \dots$ so that the following conditions are satisfied for each p:

(i) The diagram D_p is drawn inside the rectangular grid $\{0, \dots, \ell^+\} \times \{0, \dots, \ell^-\}$, where ℓ^+ (*resp.* ℓ^-) is the number of positive (*resp.* negative) letters in w, and it consists of arrows that are labeled σ_1, σ_2, or ε, and that connect some vertex (x, y) with integral coordinates to the vertex $(x+1, y)$—horizontal arrow—or to the vertex $(x, y-1)$—vertical arrow.

(ii) For each $q \leqslant p$, the word $\mathrm{red}^{m_q}(w)$ is defined, it contains no negative σ_2-handle, *i.e.*, no subword $\sigma_2^{-1}\sigma_2$, and it is drawn in D_p in the sense that there is a connected path starting from $(0, 0)$ whose labels form the considered word, adopting the convention that crossing a σ_i-labeled arrow from target to source contributes σ_i^{-1}; moreover, $\mathrm{red}^{m_q}(w)$ corresponds to the leftmost path in D_p.

(iii) The number m_p is at most the area of D_p, defined to be the number of squares included in D_p.

We start with $m_0 = 0$. The diagram D_0 consists of a single connected path $\Pi(w)$ starting from $(0, 0)$, constructed using induction on the length of w: assuming $w = w'\sigma_i^e$, the path $\Pi(w)$ consists of $\Pi(w')$ plus a horizontal σ_i-labeled arrow starting from the endpoint of $\Pi(w')$ if e is positive—*resp.* a vertical σ_i-labeled arrow arriving at the endpoint of $\Pi(w')$ if e is negative. So, by construction, the labels of D_0 make the word w, and the conditions (i), (ii), and (iii) are satisfied.

Assume now that m_p and D_p have been constructed, and let $w_p = \mathrm{red}^{m_p}(w)$. Assume that w_p contains at least one handle; otherwise, the construction stops. By Lemma 2.11, or by a direct induction, w_p contains no negative σ_1-handle, and, by (ii), it contains no negative σ_2-handle either. Hence, the leftmost handle in w_p is either $\sigma_i\sigma_i^{-1}$ with $i = 1$ or 2 (Case 1), or $\sigma_1\sigma_2^d\sigma_1^{-1}$ for some nonzero integer d (Case 2). By (ii), this handle corresponds in D_p to a pattern of the type

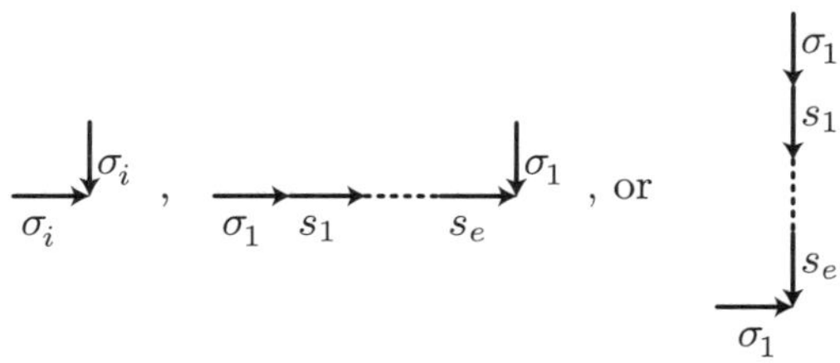

where d letters s_k are σ_2 and the remaining $e - d$ ones are ε.

In Case 1, we put $m_{p+1} = m_p + 1$ and define D_{p+1} to be the diagram obtained from D_p by completing 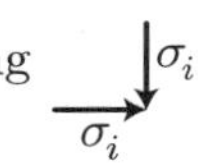 into 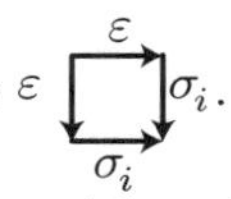.

In Case 2, we put $m_{p+1} = m_p + |d|$ and define D_{p+1} to be the diagram obtained from D_p by completing 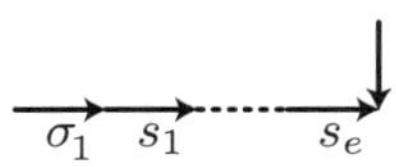 into 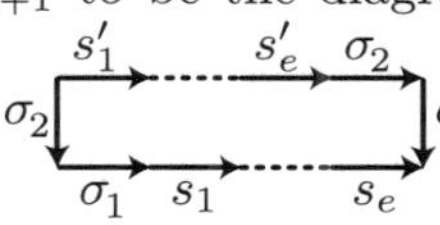, and 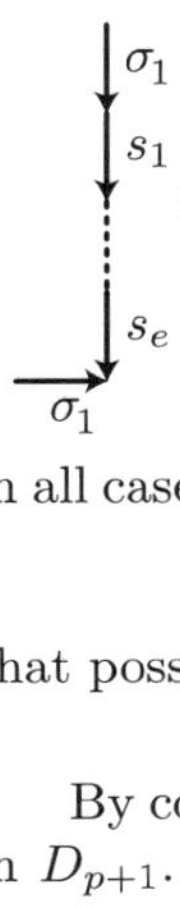 into 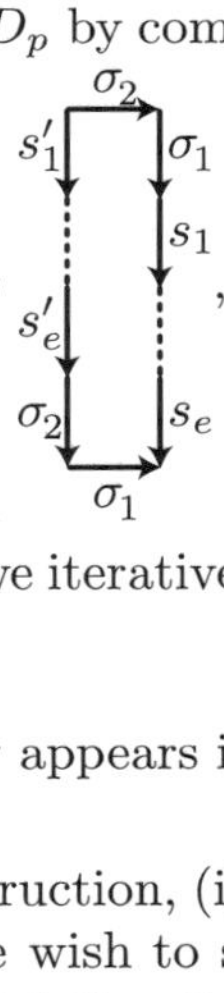, where s'_k is σ_1 for $s_k = \sigma_2$, and $s'_k = \varepsilon$ for $s_k = \varepsilon$. Moreover, in all cases, we iteratively complete each pattern 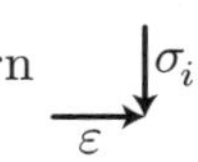 (*resp.* 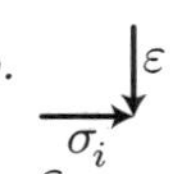, *resp.* 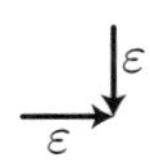) that possibly appears into 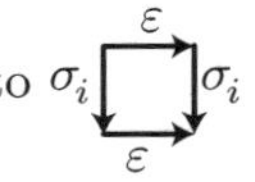 (*resp.* 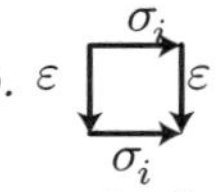, *resp.* 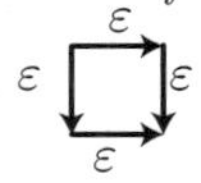).

By construction, (i) is satisfied by D_{p+1}. As for (ii), let w' be the leftmost word in D_{p+1}. We wish to show $w' = \mathrm{red}^{m_{p+1}}(w)$. In Case 1, the word w' is obtained from w_p by deleting the handle $\sigma_i\sigma_i^{-1}$, and we find

$$w' = \mathrm{red}(w_p) = \mathrm{red}(\mathrm{red}^{m_p}(w)) = \mathrm{red}^{m_{p+1}}(w).$$

In Case 2, the word w' is obtained from w_p by replacing $\sigma_1\sigma_2^d\sigma_1^{-1}$ with $\sigma_2^{-1}\sigma_1^d\sigma_2$; now, a direct verification gives $\sigma_2^{-1}\sigma_1^d\sigma_2 = \mathrm{red}^{|d|}(\sigma_1\sigma_2^d\sigma_1^{-1})$, and we deduce

$$w' = \mathrm{red}^{|d|}(w_p) = \mathrm{red}(\mathrm{red}^{m_p}(w)) = \mathrm{red}^{m_{p+1}}(w).$$

Finally, condition (iii) is maintained, as the area of D_{p+1} is the area of D_p augmented by at least 1 in Case 1, and by at least e with $e \geqslant d$ in Case 2.

Now, since all diagrams D_p are included in the grid $\{0, \dots, \ell^+\} \times \{0, \dots, \ell^-\}$, the iterative construction must stop, which means that there exists p such that the word $\mathrm{red}^{m_p}(w)$ contains no handle. Moreover, condition (iii) gives

$$m_p \leqslant \mathrm{area}(D_p) \leqslant \ell^+ \times \ell^- \leqslant \ell^2/4.$$

Finally, if the initial w contains some negative σ_2-handles, *i.e.*, subwords of the form $\sigma_2^{-1}\sigma_2$, we let w' be the word obtained from w by iteratively replacing $\sigma_2^{-1}\sigma_2$ with $\sigma_2\sigma_2^{-1}$. We then observe that the left handle reductions of w and w' require the same number of steps, because σ_2-handles are always reduced before they could interfere with the letters $\sigma_1^{\pm 1}$. □

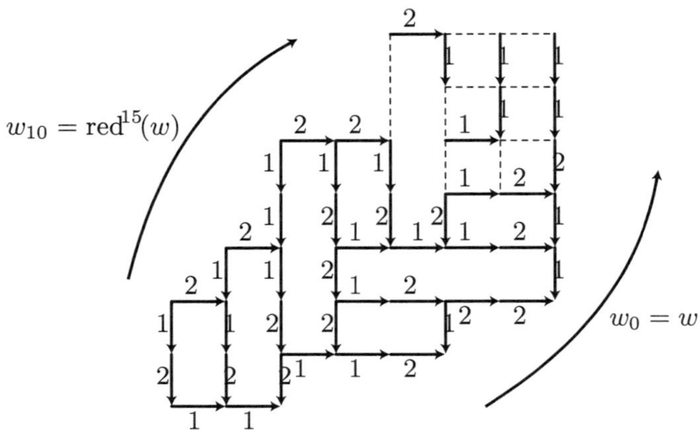

FIGURE 10. Diagram D_{10} for $w = \sigma_1^2\sigma_2^{-1}\sigma_1^2\sigma_2\sigma_1^{-1}\sigma_2^2\sigma_1^{-2}\sigma_2^{-1}\sigma_1^{-2}$; the ε-labeled arrows are represented by dotted lines, and we write i for σ_i; we start with $w_0 = w$, which corresponds to the diagonal path closest to the bottom right corner and, after 10 induction steps, we finish with the σ_1-negative word $w' = \sigma_2^{-1}\sigma_1^{-1}\sigma_2\sigma_1^{-1}\sigma_2\sigma_1^{-2}\sigma_2^3$, which corresponds to the diagonal path closest to the top left corner; in this example, the numbers m_p are $0, 1, 4, 5, 6, 7, 8, 9, 12, 14, 15$, so there are 15 steps in the left handle reduction of w—to be compared with the area of D_{10}, which is 29, and that of the framework, which is 49.

PROPOSITION 3.2. *Let w be a 3-strand braid word of length ℓ, and let h be the number of σ_1-handles in w. Then there exists $m \leqslant h\ell^2/4$ such that $\mathrm{red}^m(w)$ contains no handle.*

PROOF. We use induction on h. For $h \leqslant 1$, Lemma 3.1 gives the result. Assume $h \geqslant 2$, and write $w = w_0w'$, where w_0 is the longest prefix of w satisfying $h(w_0) = 1$. By Lemma 3.1, there exists $m_0 \leqslant \ell^2/4$ such that $\mathrm{red}^{m_0}(w_0)$ contains no handle. Now, we have $\mathrm{red}^m(w) = (\mathrm{red}^m(w)_0)w'$ for $m \leqslant m_0$, because the leftmost handle in $(\mathrm{red}^m(w)_0)w'$ lies in the prefix $\mathrm{red}^m(w)_0$. Thus we find $\mathrm{red}^{m_0}(w) = (\mathrm{red}^{m_0}(w_0))w'$. By construction, we have $h((\mathrm{red}^{m_0}(w_0))w') = h - 1$, *i.e.*, $h(\mathrm{red}^{m_0}(w)) = h - 1$, and the length of $\mathrm{red}^{m_0}(w)$ is at most ℓ. So, by induction hypothesis, there exists $m' \leqslant (h-1)\ell^2/4$ such that $\mathrm{red}^{m'}(\mathrm{red}^{m_0}(w))$ contains no handle. As we have $m_0 + m' \leqslant h\ell^2/4$, we are done. □

While the quadratic upper bound of Lemma 3.1 is easily seen to be sharp, that of Proposition 3.2 is probably not optimal: as the only upper bound on the number of σ_1-handles is the length of the initial word, we obtain a final cubical upper bound on the number of reduction steps, while a quadratic upper bound seems likely. In any case, the argument fails to extend to $n \geqslant 3$: even for words of the form uv^{-1} with u, v positive, the commutation relations $\sigma_1\sigma_3 = \sigma_3\sigma_1$ make the construction of a diagram similar to that used in the proof of Lemma 3.1 problematic.

3.2. An application. Besides its practical interest as an efficient solution to the braid word problem and its theoretical interest as a simple way of establishing Property **C**, the convergence of handle reduction may lead to results of independent interest. We now mention one such result that will be subsequently used in Chapter VI.

By definition, saying that $\beta < \beta'$ holds means that the quotient braid $\beta^{-1}\beta'$ admits at least one σ-positive word representative, *i.e.*, in the language of the Cayley graph, that there exists at least one σ-positive path going from β to β' inside the Cayley graph of B_∞. The following result states that, if β and β' lie in some set $\mathrm{Div}(\gamma)$, then, among all σ-positive paths as above, at least one is entirely drawn inside $\mathrm{Div}(\gamma)$; hence, in particular, inside B_n if β and β' are n-strand braids.

PROPOSITION 3.3. *Let γ be a positive braid. Then, for all β, β' in $\mathrm{Div}(\gamma)$, the following are equivalent:*

(i) *The relation $\beta < \beta'$ holds;*

(ii) *There exists a σ-positive word representing $\beta^{-1}\beta'$ drawn from β in B_n;*

(iii) *There exists a σ-positive word representing $\beta^{-1}\beta'$ drawn from β in $\mathrm{Div}(\gamma)$.*

PROOF. Clearly (iii) implies (ii), which implies (i), since the latter means that there exists a σ-positive word representing $\beta^{-1}\beta'$ drawn from β in B_∞. We shall prove that (i) implies (iii) by resorting to handle reduction. By doing so, we also reprove that (i) implies (ii), a result that was first proved in [**131**].

So, we assume that β, β' lie in $\mathrm{Div}(\gamma)$ and $\beta < \beta'$ holds. The problem is to prove that, among all σ-positive paths connecting β to β' in the Cayley graph of B_∞, at least one is drawn inside $\mathrm{Div}(\gamma)$. Now, let w, w' be positive words representing β and β', and let w_0 be the word $w^{-1}w'$. Then w_0 represents $\beta^{-1}\beta'$, and, by construction, it is drawn in $\mathrm{Div}(\gamma)$ from β. Now, Lemma 2.6 implies that each word obtained from w_0 by handle reduction is also drawn in $\mathrm{Div}(\gamma)$ from β, and Proposition 1.5 implies that, among these words, at least one, say w_1, contains no handle, hence it is either empty, or σ-positive, or σ-negative. We claim that w_1 is σ-positive, which gives the expected result. Indeed, the hypothesis $\beta < \beta'$ implies that $\beta^{-1}\beta'$ admits a σ-positive representative, and Property **A** then forbids that it admits another representative that is empty or σ-negative. □

3.3. Variants. We conclude with two variants of handle reduction. First, we have defined a σ_i-handle to be a braid word of the form $\sigma_i^{\pm 1} v \sigma_i^{\mp 1}$ where $\sigma_1^{\pm 1}, \dots, \sigma_i^{\pm 1}$ do not occur in v. Let us define a *generalized* σ_i-handle to be a similar braid word $\sigma_i^{\pm 1} v \sigma_i^{\mp 1}$ where only $\sigma_{i-1}^{\pm 1}$ and $\sigma_i^{\pm 1}$ are forbiden in v. Then the results for generalized handle reduction are the same as for handle reduction: in particular, the convergence result of Proposition 1.5 extends without change, at the expense of using Property $\mathbf{A}_i$ of Remark IV.2.15 rather than Property **A**, *i.e.*, $\mathbf{A}_1$.

The last variant is the so-called coarse reduction: here, one comes back to the standard notion of handle, but, instead of reducing a σ_i-handle by pushing the

$(i+1)$st strand over the next crossings as in Definition 1.2, we would also push it systematically to the nth position, if we are working with B_n. As no nontrivial result has been proved about this variant so far, we keep it for Chapter XVI.

CHAPTER VI

Connection with the Garside Structure

The aim of this chapter is to investigate the connection between the σ-ordering of braids and Garside's theory of the braid positive monoids B_n^+. It is mainly based on the references [**61, 60**].

Initiated with Garside's seminal paper [**95**] and subsequently developed by many other authors, Garside's theory is arguably the best understood part in the algebraic study of braids. It leads in particular to algorithmic solutions to the word and conjugacy problems, and, much more generally, it provides the most efficient way for addressing a number of questions about braids. Moreover, Garside's theory has now been extended to a very wide family of groups and groupoids far beyond braid groups themselves. It is therefore a natural task to study the connection between braid ordering and Garside's theory and the notions it leads to, in particular the so-called greedy normal form.

Let us say it immediately: the connection is not simple. For instance, if β and β' are positive braids, there is no easy way for recognizing whether $\beta < \beta'$ holds by inspecting the greedy normal forms of β and β'. There exists such a way in the case of divisors of Δ_n—called simple braids, or permutation braids—but, for more complicated braids, the connection becomes intricate.

However, though not so easy, the study of the connection between the Garside structure and the σ-ordering of braids provides several nontrivial results of a combinatorial nature. In particular, it leads to addressing various counting problems involving braids, in connection with the symmetric group and the Solomon descent algebra. Such questions have not been much considered so far, and one may expect further developments.

As in the case of each of the approaches to the σ-ordering developed in the various chapters of this text, two types of results may be expected, namely (re)-proving the existence of the ordering by providing new proofs of Properties **A** and/or **C**, and proving new results about the braid ordering once we know the latter exists. Here, we shall establish results of both kinds: For the first kind, we describe one more scheme for proving Property **C**, which actually is completed in the case of B_3 only. For the second kind, we give a fairly complete description of the σ-ordering of divisors of Δ_n^d, that in particular includes evaluating the number of so-called σ_i-jumps.

The organization of this chapter is as follows. In Section 1, we recall the construction of the greedy normal form of positive braids and state a few results about the number of n-strand braids of degree at most d, *i.e.*, about the number of divisors of Δ_n^d. In Section 2, we show how to deduce from such counting arguments a scheme for proving Property **C** when Property **A** is known. In Section 3, we define $S_{n,d}$ to be the $<$-increasing enumeration of the divisors of Δ_n^d and investigate the structure of this sequence. The main result is that $S_{n,d}$ is a concatenation of translated copies of fragments of $S_{n-1,d}$, the number of which can be effectively computed.

1. The degree of a positive braid

In this introductory section, we recall the construction of the greedy normal form for positive braids—this is classic, see [**95, 77, 1, 75**]—and deduce some counting results about the number of braids with a given degree—this is less classic.

We recall that B_n^+ denotes the monoid of all positive n-strand braids, *i.e.*, those n-strand braids that admit word representatives containing no letter σ_i^{-1}; note that the trivial braid 1 is called positive. Then B_∞^+ denotes the monoid of all positive braids.

1.1. Simple braids. In the theory we describe here, a key role is played by those positive n-strand braids that are divisors of the Garside fundamental braid Δ_n, usually called simple braids or permutation braids, and we first state a few basic facts about these braids.

We recall from Section I.4 that, if β, β' are (positive) braids, we say that β' is a *left divisor* of β, or, equivalently, that β is a *right multiple* of β', denoted $\beta' \preccurlyeq \beta$, if $\beta = \beta'\gamma$ holds for some positive braid γ, *i.e.*, for some braid γ belonging to B_∞^+. As in Chapter V, if β is a positive braid, we denote by $\mathrm{Div}(\beta)$ the set of all positive braids that left divide β, *i.e.*, the set $\{\gamma \mid 1 \preccurlyeq \gamma \preccurlyeq \beta\}$. According to Proposition I.4.8, for each positive braid β, the poset $(\mathrm{Div}(\beta), \preccurlyeq)$ is a lattice.

In this chapter, we restrict ourselves to positive braids, *i.e.*, to those braids β that satisfy $1 \preccurlyeq \beta$, and we specially consider the divisors of the braid Δ_n and its powers. We recall from Definition I.4.3 that Δ_n is recursively defined by

$$\Delta_1 = 1, \qquad \Delta_n = \delta_n\, \Delta_{n-1} = \sigma_1 \sigma_2 \dots \sigma_{n-1}\, \Delta_{n-1}. \tag{1.1}$$

It is easily seen that each generator σ_i is a left and a right divisor of Δ_n, and (less easily) that Δ_n is the left (and the right) lcm of $\sigma_1, \dots, \sigma_{n-1}$. Our first important notion is

Definition 1.1. A positive n-strand braid is called *simple* if it is a left divisor of Δ_n, *i.e.*, if it belongs to the set $\mathrm{Div}(\Delta_n)$.

Lemma 1.2 ([**77**, Chapter 9]). *For β a positive braid, the following are equivalent:*

(i) *The braid β is a left divisor of Δ_n;*
(ii) *The braid β is a right divisor of Δ_n;*
(iii) *In any positive diagram representing β, any two strands cross at most once.*

As explained in Chapter I, each n-strand braid β determines a permutation of $\{1, \dots, n\}$. Here we shall denote this permutation by $\mathrm{perm}(\beta)$. So, for each i with $1 \leqslant i \leqslant n$, $\mathrm{perm}(\beta)(i)$ is the initial position of the strand that finishes at position i in any diagram representing β. Then perm is a surjective homorphism of B_n to the symmetric group $\mathfrak{S}_n$.

Lemma 1.3 ([**77**, Chapter 9]). *The restriction of the mapping* perm *to simple n-strand braids is a bijection to the symmetric group $\mathfrak{S}_n$: for each permutation π in $\mathfrak{S}_n$, there exists a unique simple braid β satisfying* $\mathrm{perm}(\beta) = \pi$.

Thus there exist exactly $n!$ simple n-strand braids.

EXAMPLE 1.4. There are six simple 3-strand braids, namely 1, σ_1, σ_2, $\sigma_2\sigma_1$, $\sigma_1\sigma_2$, and Δ_3. In examples, as in previous chapters, we often use the shorter notation $\mathtt{a}$ for σ_1, $\mathtt{b}$ for σ_2, etc. Thus, we also write $\mathrm{Div}(\Delta_3) = \{1, \mathtt{a}, \mathtt{b}, \mathtt{ba}, \mathtt{ab}, \mathtt{aba}\}$.

1.2. The greedy normal form. By definition, for each positive n-strand braid β different from 1, the gcd β_1 of β and Δ_n is the maximal simple left divisor of β, and we obtain a distinguished decomposition $\beta = \beta_1\beta'$ with β_1 simple. By decomposing β' in the same way and iterating, we obtain the so-called (left) greedy normal decomposition of β.

PROPOSITION 1.5 ([**75**],[**77**, Chapter 9]). *Every nontrivial positive braid β admits a unique decomposition $\beta = \beta_1\beta_2...\beta_d$ satisfying $\beta_d \neq 1$ and, for each $r < d$,*

$$\beta_r = \mathrm{gcd}(\Delta_n, \beta_r...\beta_d). \tag{1.2}$$

DEFINITION 1.6. A sequence $(\beta_1, ..., \beta_d)$ satisfying the conditions of (1.2) is said to be *(greedy) normal.* For β a nontrivial positive braid, the unique normal sequence $(\beta_1, ..., \beta_d)$ for which $\beta_d \neq 1$ and $\beta = \beta_1\beta_2...\beta_d$ hold is called the *(left greedy) normal form* of β; then d is called the *degree* of β, and β_r is called the rth *factor* of β.

It is coherent to extend the above conventions by defining the degree of the trivial braid 1 to be 0, and the rth factor of a degree d braid to be 1 for $r > d$. Also, adding a trivial entry 1 at the right of a normal sequence still gives a normal sequence. At the expense of adding such trivial entries, we shall consider normal sequences as unbounded on the right; in this way, we can speak of the rth entry of a normal sequence for every r. Then the degree of a nontrivial braid β is the maximal index r such that the rth entry in its normal sequence is not 1.

Because every n-strand braid that divides Δ_{n+1} has to divide Δ_n, the normal form of a positive n-strand braid computed in B_n^+ and B_{n+1}^+ coincide, so we can forget about the braid index.

We shall use the following two properties of the normal form:

LEMMA 1.7 ([**75**], [**77**, Chapter 9]). *Assume that $(\beta_1, ..., \beta_d)$ is a sequence of positive braids. Then the following are equivalent:*

(i) *The sequence $(\beta_1, ..., \beta_d)$ is normal;*

(ii) *For each $r < d$, the subsequence (β_r, β_{r+1}) is normal;*

(iii) *For each $r < d$, every σ_i that divides β_{r+1} on the left divides β_r on the right.*

An important consequence of Lemma 1.7 is that the normal form of braids, being characterized by the purely local criterion of (ii) above, is connected with a so-called automatic structure for the braid group B_n—see [**77**, Chapter 9] or, for some basic definitions, Section 1.2 of Chapter XI.

LEMMA 1.8 ([**75**],[**77**, Chapter 9]). *For each positive n-strand braid β, the following are equivalent:*

(i) *The braid β belongs to* $\mathrm{Div}(\Delta_n^d)$, *i.e., it is a left—or right—divisor of Δ_n^d;*

(ii) *The degree of β is at most d.*

So speaking of a degree here is coherent: by Lemma 1.8, if positive braids β and β' have degree d and d' respectively, the degree of $\beta\beta'$ is at most $d + d'$.

EXAMPLE 1.9. There are nineteen divisors of Δ_3^2, *i.e.*, nineteen positive 3-strand braids of degree at most two. With the usual notational convention, using a dot to separate the factors of the normal form, we find the list 1, b, b.b, a, ab, ab.b, a.a, a.ab, ba, bab, bab.b, ba.a, ba.ab, b.ba, bab.a, bab.ab, ab.ba, bab.ba, bab.bab—here in an order that may look strange, but should become familiar soon, as it is the $<$-increasing one.

REMARK 1.10. We described above the left version of the greedy normal form that involves the maximal simple braid which is a left divisor of the considered braid. Naturally, a symmetric, right version involving the maximal simple braid that is a right divisor also exists, and results are similar. Note that, because the left and the right divisors of Δ_n coincide, and, therefore, so do those of Δ_n^d for each d, there is only one notion of degree: for each positive braid β, the length of the left and the right greedy normal forms of β are equal.

1.3. Counting the divisors of Δ_n^d. In what follows, we need to count the divisors of Δ_n^d in B_n^+. By Lemma 1.3, these divisors are in one-to-one correspondence with the normal sequences of length at most d, and, therefore, their number is bounded above by $(n!)^d$ for all n and d. Determining the exact value is not very difficult: owing to the local characterization of Lemma 1.7, this amounts to solving a linear recursion involving an explicit adjacency matrix.

NOTATION 1.11. For $n \geqslant 2$, $d \geqslant 0$, and β a simple n-strand braid, we define $b_{n,d}(\beta)$ to be the number of length d normal sequences in B_n^+ whose last entry is β.

So, $b_{n,d}(\beta)$ is the number of normal sequences of the form $(\beta_1, \dots, \beta_{d-1}, \beta)$. The following result is easy.

LEMMA 1.12. *For all n, d, we have*

$$\#\mathrm{Div}(\Delta_n^d) = \sum_{\beta \text{ simple}} b_{n,d}(\beta) = b_{n,d+1}(1). \tag{1.3}$$

Indeed, it suffices to observe that, for each simple braid β, the sequence $(\beta, 1)$ is normal. So $(\beta_1, \dots, \beta_d)$ is normal if and only if $(\beta_1, \dots, \beta_d, 1)$ is normal.

The general principle for computing the numbers $b_{n,d}(\beta)$ for some fixed n is to introduce the adjacency matrix that describes normal pairs of simple braids.

LEMMA 1.13. *For $n \geqslant 1$, let M_n be the $n! \times n!$ matrix with entries indexed by simple n-strand braids such that the (β, β')-entry in M_n is 1 if (β, β') is normal, and is 0 otherwise. Then, for all n, d, and β, the number $b_{n,d}(\beta)$ is the β-entry in the row matrix $(1, 1, \dots, 1) \cdot (M_n)^{d-1}$.*

PROOF. The result is easily proved using induction on d and Lemma 1.7: $(\beta_1, \dots, \beta_{d-1}, \beta)$ is normal if and only if $(\beta_1, \dots, \beta_{d-1})$ and (β_{d-1}, β) are normal, so, writing $m_{\gamma,\beta}$ for the (γ, β)-entry of M_n, we obtain

$$b_{n,d}(\beta) = \sum_{(\gamma,\beta) \text{ normal}} b_{n,d-1}(\gamma) = \sum_{\gamma \text{ simple}} b_{n,d-1}(\gamma) \cdot m_{\gamma,\beta}. \qquad \square$$

EXAMPLE 1.14. The matrix M_1 is (1), corresponding to $b_{1,d}(1) = 1$ for each d. For $n = 2$, using the enumeration $(1, \sigma_1)$ of simple braids, we find $M_2 = \begin{pmatrix} 1 & 0 \\ 1 & 1 \end{pmatrix}$, leading to $b_{2,d}(1) = d$ and $b_{2,d}(\sigma_1) = 1$, as could be expected: there are $d + 1$

TABLE 1. First values of $b_{n,d}(\Delta_i)$ for $1 \leqslant i < n$—the value is 1 for $i \geqslant n$. For instance, we read that $b_{3,3}(1)$, the number of 3-strand braids of degree at most 3 whose third factor is 1—hence the number of positive 3-strand braids of degree at most 2— is 19, as was seen in Example 1.9.

d	1	2	3	4	5	6	7
$b_{2,d}(\Delta_1)$	1	2	3	4	5	6	7
$b_{3,d}(\Delta_1)$	1	6	19	48	109	234	487
$b_{3,d}(\Delta_2)$	1	3	7	15	31	63	127
$b_{4,d}(\Delta_1)$	1	24	211	1,380	8,077	45,252	249,223
$b_{4,d}(\Delta_2)$	1	12	83	492	2,765	15,240	83,399
$b_{4,d}(\Delta_3)$	1	4	15	64	309	1,600	8,547
$b_{5,d}(\Delta_1)$	1	120	3,651	79,140	1,548,701	29,375,460	551,997,751
$b_{5,d}(\Delta_2)$	1	60	1,501	30,540	585,811	11,044,080	207,154,921
$b_{5,d}(\Delta_3)$	1	20	311	5,260	94,881	1,755,360	32,741,851
$b_{5,d}(\Delta_4)$	1	5	31	325	4,931	86,565	1,590,231

braids of degree at most d, namely the braids σ_1^e with $e < d$, whose dth factor is 1, and σ_1^d, whose dth factor is Δ_2, *i.e.*, σ_1. For $n = 3$, using the enumeration $(1, \sigma_1, \sigma_2, \sigma_2\sigma_1, \sigma_1\sigma_2, \Delta_3)$ of simple 3-strand braids, we obtain

$$M_3 = \begin{pmatrix} 1 & 0 & 0 & 0 & 0 & 0 \\ 1 & 1 & 0 & 0 & 1 & 0 \\ 1 & 0 & 1 & 1 & 0 & 0 \\ 1 & 1 & 0 & 0 & 1 & 0 \\ 1 & 0 & 1 & 1 & 0 & 0 \\ 1 & 1 & 1 & 1 & 1 & 1 \end{pmatrix},$$

from which we deduce, for instance, $\#\mathrm{Div}(\Delta_3^2) = b_{3,3}(1) = 19$.

Lemma 1.13 immediately implies

PROPOSITION 1.15. *Assume $n \geqslant 2$. Let $\rho_1, \dots, \rho_r$ be the nonzero eigenvalues of the matrix M_n, and $m_1, \dots, m_r$ be their respective multiplicities. Then there exists for each r a degree m_r polynomial P_r such that for each d we have*

$$\#\mathrm{Div}(\Delta_n^d) = P_1(d)\rho_1^d + \cdots + P_r(d)\rho_r^d. \tag{1.4}$$

Moreover, for each n, the generating function of the numbers $\#\mathrm{Div}(\Delta_n^d)$ is rational.

Similar results hold for all numbers $b_{n,d}(\beta)$. As the matrix M_n is an $n! \times n!$ matrix, completing the computation is not so easy, even for small values of n. Actually, M_n is highly redundant, with many columns repeated, and it has a zero eigenvalue with high multiplicity. In **[60]** it is shown how to replace M_n with a smaller matrix $\widehat{M_n}$ of size $p(n) \times p(n)$, where $p(n)$ is the number of partitions of n, *i.e.*, the number of finite nonincreasing sequences $(n_1, \dots, n_k)$ satisfying $n_1 + \cdots + n_k = n$. With such methods, one easily obtains the values listed in Table 1.

Also, using the reduced matrices $\widehat{M_3} = \begin{pmatrix} 1 & 0 & 0 \\ 4 & 2 & 0 \\ 1 & 1 & 1 \end{pmatrix}$ and $\widehat{M_4} = \begin{pmatrix} 1 & 0 & 0 & 0 & 0 \\ 11 & 4 & 1 & 0 & 0 \\ 5 & 3 & 2 & 1 & 0 \\ 6 & 4 & 2 & 2 & 0 \\ 1 & 1 & 1 & 1 & 1 \end{pmatrix}$, one obtains the following explicit form for (1.4) in terms of the nonzero eigenvalues of M_3, namely 1 (double) and 2, and those of M_4, namely 1 (double), 2, and $3 \pm \sqrt{6}$:

PROPOSITION 1.16. *For $d \geqslant 0$, we have*

$$\#\mathrm{Div}(\Delta_3^d) = 8 \cdot 2^d - 3d - 7,$$

$$\#\mathrm{Div}(\Delta_4^d) = \sum_{\pm} \frac{3}{20}(32 \pm 13\sqrt{6})(3 \pm \sqrt{6})^d - \frac{128}{5} \cdot 2^d + 6d + 17.$$

The main interest of the above formulas is to show that each of the involved parameters has an exponential growth with respect to d, in $O(2^d)$ for $n = 3$, and in $O((3 + \sqrt{6})^d)$ for $n = 4$. One easily deduces

COROLLARY 1.17. *For $d \geqslant 0$, we have*

$$\#\mathrm{Div}(\Delta_3^d) = 2 \cdot \#\mathrm{Div}(\Delta_3^{d-1}) + 3d + 1, \tag{1.5}$$

$$\#\mathrm{Div}(\Delta_4^d) = 6 \cdot \#\mathrm{Div}(\Delta_4^{d-1}) - 3 \cdot \#\mathrm{Div}(\Delta_3^{d-2}) + 32 \cdot 2^d - 12d - 34, \tag{1.6}$$

with initial values $\#\mathrm{Div}(\Delta_3^0) = \#\mathrm{Div}(\Delta_4^0) = 1$, *and* $\#\mathrm{Div}(\Delta_4) = 24$.

1.4. Connection with the symmetric group. The previous section contains all results needed for the subsequent investigation of the connection between the braid ordering and the Garside structure in Sections 2 and 3 below. However, before turning to this study, we briefly mention here a connection of the adjacency matrix M_n considered above and the so-called descents of a permutation.

By Lemma 1.3, there is a one-to-one correspondence between simple n-strand braids and permutations of $\{1, \dots, n\}$. Under this correspondence, the normality condition of Lemma 1.7(iii) easily translates in the language of permutations.

DEFINITION 1.18. For π a permutation of $\{1, \dots, n\}$ and $1 \leqslant i < n$, we say that i is a *descent* of π if $\pi(i) > \pi(i+1)$ holds.

LEMMA 1.19. *Assume that β, β' are simple n-strand braids associated with the permutations π and π', respectively. Then the sequence (β, β') is normal if and only if every descent of ${\pi'}^{-1}$ is a descent of π.*

PROOF. One easily checks that σ_i is a left divisor of a simple braid β if and only if the strands starting at positions i and $i+1$ cross in any positive braid diagram representing β, and, therefore, if and only if we have $\pi^{-1}(i) > \pi^{-1}(i+1)$, *i.e.*, i is a descent of π^{-1}. Symmetrically, σ_i is a right divisor of β if and only if the strands finishing at positions i and $i+1$ cross, hence if and only if we have $\pi(i) > \pi(i+1)$, *i.e.*, i is a descent of π. □

Hence, another way of introducing the matrix M_n of Lemma 1.13 is to view it as a matrix whose rows and columns are indexed by permutations of $\{1, \dots, n\}$, and the (π, π')-entry of M_n is 1 if and only if all descents of ${\pi'}^{-1}$ (also called the *recoils* of π') are descents of π. This makes the matrix M_n a quite natural object in the combinatorial theory of the symmetric group $\mathfrak{S}_n$.

Using this language, the above mentioned result—that the nonsingular part of the size $n!$ matrix M_n is a size $p(n)$ matrix $\widehat{M_n}$—turns out to be a counterpart of classical results by Solomon about the descent algebra [**188**]. As will be explained in Section 2.3 of Chapter XVI, this approach led to further questions about descents of permutations, and, in particular, to very recent results announced in [**109**].

2. Proving Property C using a counting argument

We come back now to the braid ordering, and apply the combinatorial results of Section 1 to study the connection between this ordering and the finite sets $\mathrm{Div}(\Delta_n^d)$, *i.e.*, the set of all n-strand braids of degree at most d.

In this section, we show how understanding the above connection may lead to reproving the existence of the braid ordering. At the moment, no idea leading to a proof of Property **A** is known, but there exists a simple scheme for establishing Property **C**, at least when Property **A** is known. So far, that scheme has been completed in the case $n = 3$ only, and this is the case we shall now consider.

2.1. The principle of the argument. Assume that X is a finite subset of B_n and that $(\beta_1, \ldots, \beta_N)$ is a sequence of elements of X such that each quotient-braid $\beta_r^{-1}\beta_{r+1}$ is σ-positive. Then, Property **A** guarantees that the braids β_r are pairwise distinct, so, if we can prove that N is the cardinality of X, we obtain a proof of Property **C** restricted to the set $X^{-1}X$, *i.e.*, a proof that every nontrivial quotient of elements of X is σ-positive or σ-negative.

By Garside's theory, every braid in B_n is a quotient of braids in $\mathrm{Div}(\Delta_n^d)$ for d large enough. So if, for some n, we are able

- to directly construct for each d a sequence Σ_d consisting of elements of $\mathrm{Div}(\Delta_n^d)$ with the property that the quotient of any two distinct entries is σ-positive or σ-negative, and
- to prove that the length of the sequence Σ_d is the cardinality of $\mathrm{Div}(\Delta_n^d)$,

then we obtain a proof of Property **C** for B_n.

Notice that a sequence Σ_d as above is unique, as it necessarily consists of the $<$-increasing enumeration of the considered set X. So the actual problem is not to prove an existence statement, but just to guess an explicit construction of the sequence Σ_d, without using Property **C**, after which the rest should be a simple verification.

2.2. A Pascal-like triangle. We shall now realize the previous scheme in the case of 3-strand braids. In order to guess a recursive definition of the $<$-increasing enumeration of the set $\mathrm{Div}(\Delta_3^d)$, the counting formulas of Section 1 are essential: in particular, the specific form of (1.5) suggests that the enumeration of $\mathrm{Div}(\Delta_3^d)$ might involve two copies of the enumeration of $\mathrm{Div}(\Delta_3^{d-1})$, and it is therefore not surprising that the construction we shall describe below is reminiscent of a Pascal triangle.

Our aim is to construct, for each d, a certain sequence of 3-strand braids of degree at most d. We shall often appeal to a distinguished series of 3-strand braids, that are connected with the powers of Δ_3, and we first introduce them.

DEFINITION 2.1. For $d \geqslant 0$, we define θ_d to be (the braid represented by) the length d suffix of the left infinite braid word $\ldots\sigma_2^2\sigma_1^2\sigma_2^2\sigma_1$.

Thus, we have $\theta_1 = \sigma_1$, $\theta_2 = \sigma_2\sigma_1$, $\theta_3 = \sigma_2^2\sigma_1$, etc. Easy verifications give

$$\theta_0\sigma_2^{[0]}$$

$$\theta_0\sigma_2^{[1]}+\underbrace{(\Sigma_{1,1})+\theta_1\sigma_2^{[2]}+(\Sigma_{1,2})}+\theta_2\sigma_2^{[3]}$$

$$\sigma_2\cdot\swarrow \qquad \searrow\sigma_1\sigma_2\cdot$$

$$\theta_0\sigma_2^{[2]}+\underbrace{(\Sigma_{2,1})+\theta_1\sigma_2^{[2]}+\Sigma_{2,2}}+\theta_2\sigma_2^{[2]}+\underbrace{\Sigma_{2,3}+\theta_3\sigma_2^{[2]}+(\Sigma_{2,4})}+\theta_4\sigma_2^{[2]}$$

$$\sigma_2\cdot\swarrow \qquad \searrow\sigma_1\sigma_2\cdot \qquad\qquad \sigma_1\cdot\swarrow \qquad \searrow\sigma_2\sigma_1\cdot$$

$$\theta_0\sigma_2^{[3]}+\underbrace{(\Sigma_{3,1})+\theta_1\sigma_2^{[3]}+\Sigma_{3,2}}+\theta_2\sigma_2^{[3]}+\underbrace{\Sigma_{3,3}+\theta_3\sigma_2^{[3]}+\Sigma_{3,4}}+\theta_4\sigma_2^{[3]}+\underbrace{\Sigma_{3,5}+\theta_4\sigma_2^{[3]}+(\Sigma_{3,6})}+\theta_6+\sigma_2^{[3]}$$

$$\sigma_2\cdot\swarrow \qquad \searrow\sigma_1\sigma_2\cdot \qquad\qquad \sigma_1\cdot\swarrow \qquad \searrow\sigma_2\sigma_1\cdot \qquad\qquad \sigma_2\cdot\swarrow \qquad \searrow\sigma_1\sigma_2\cdot$$

$$\cdots \qquad \cdots \qquad\qquad \cdots \qquad \cdots \qquad\qquad \cdots \qquad \cdots$$

FIGURE 1. Inductive construction of Σ_d as a Pascal triangle: the top line gives the decomposition of Σ_0, the next line gives that of Σ_1, then that of Σ_2, etc. The subsequence $\Sigma_{d,r}$ is obtained by (translating and) concatenating the subsequences $\Sigma_{d-1,r-1}$ and $\Sigma_{d-1,r}$, or $\Sigma_{d-1,r-2}$ and $\Sigma_{d-1,r-1}$, depending on the parity of r. The parenthesized sequences are empty, but are included to emphasize the pattern. If we forget about the subsequences $\theta_q\sigma_2^{[d]}$, we have a Pascal triangle.

LEMMA 2.2. *For each d, we have $\theta_{2d}\sigma_2^d = \Delta_3^d$, and $\theta_{2d}\sigma_2 = \sigma_i\theta_{2d}$, with $i = 1$ if d is odd, and $i = 2$ if d is even.*

According to the scheme described in Section 2.1, we introduce now, for each d, a certain sequence of 3-strand braids called Σ_d. It will eventually turn out that Σ_d is the $<$-increasing enumeration of the divisors of Δ_3^d.

NOTATION 2.3. If Σ, Σ' are sequences (of braids), we denote by $\Sigma + \Sigma'$ the concatenation of Σ and Σ', *i.e.*, the sequence obtained by appending Σ' after Σ. If Σ is a sequence of braids and β is a braid, we denote by $\beta\Sigma$ the translated sequence obtained by left multiplying each entry in Σ by β.

NOTATION 2.4. For $d \geqslant 0$, we write $\sigma_2^{[d]}$ for $(1, \sigma_2, \dots, \sigma_2^d)$, and define Σ_d by

$$\Sigma_d = \theta_0\sigma_2^{[d]} + \Sigma_{d,1} + \theta_1\sigma_2^{[d]} + \Sigma_{d,2} + \cdots + \theta_{2d-1}\sigma_2^{[d]} + \Sigma_{d,2d} + \theta_{2d}\sigma_2^{[d]}, \tag{2.1}$$

where $\Sigma_{d,1}, \dots, \Sigma_{d,2d}$ are defined by $\Sigma_{d,1} = \Sigma_{d,2d} = \emptyset$ and, for $2 \leqslant r \leqslant 2d-1$,

$$\Sigma_{d,r} = \begin{cases} \sigma_1(\Sigma_{d-1,r-1} + \theta_{r-1}\sigma_2^{[d-1]} + \Sigma_{d-1,r}) & \text{for } r = 0 \pmod 4,\\ \sigma_2\sigma_1(\Sigma_{d-1,r-2} + \theta_{r-1}\sigma_2^{[d-1]} + \Sigma_{d-1,r-1}) & \text{for } r = 1 \pmod 4,\\ \sigma_2(\Sigma_{d-1,r-1} + \theta_{r-1}\sigma_2^{[d-1]} + \Sigma_{d-1,r}) & \text{for } r = 2 \pmod 4,\\ \sigma_1\sigma_2(\Sigma_{d-1,r-2} + \theta_{r-1}\sigma_2^{[d-1]} + \Sigma_{d-1,r-1}) & \text{for } r = 3 \pmod 4. \end{cases}$$

The general scheme is illustrated in Figure 1: the sequence Σ_d is constructed by starting with $2d+1$ copies of $\sigma_2^{[d]}$ translated by $\theta_0, \dots, \theta_{2d}$ and inserting (translated copies of) fragments of the previous sequence Σ_{d-1}.

EXAMPLE 2.5. The first values are $\Sigma_0 = \theta_0\sigma_2^{[0]} = (1)$, then

$$\begin{aligned}\Sigma_1 &= \theta_0\sigma_2^{[1]} + \Sigma_{1,1} + \theta_1\sigma_2^{[1]} + \Sigma_{1,2} + \theta_2\sigma_2^{[1]}\\ &= (1, \mathtt{b}) + \emptyset + \mathtt{a}(1, \mathtt{b}) + \emptyset + \mathtt{ba}(1, \mathtt{b}) = (1, \mathtt{b}, \mathtt{a}, \mathtt{ab}, \mathtt{ba}, \mathtt{bab}),\end{aligned}$$

$$\begin{aligned}\Sigma_2 &= \theta_0\sigma_2^{[2]} + \Sigma_{2,1} + \theta_1\sigma_2^{[2]} + \Sigma_{2,2} + \theta_2\sigma_2^{[2]} + \Sigma_{2,3} + \theta_3\sigma_2^{[2]} + \Sigma_{2,4} + \theta_4\sigma_2^{[2]}\\
&= (1, \mathtt{b}, \mathtt{bb}) + \emptyset + \mathtt{a}(1, \mathtt{b}, \mathtt{bb}) + \mathtt{a}(\mathtt{a}, \mathtt{ab}) + \mathtt{ba}(1, \mathtt{b}, \mathtt{bb})\\
&\qquad + \mathtt{ba}(\mathtt{a}, \mathtt{ab}) + \mathtt{bba}(1, \mathtt{b}, \mathtt{bb}) + \emptyset + \mathtt{abba}(1, \mathtt{b}, \mathtt{bb})\\
&= (1, \mathtt{b}, \mathtt{bb}, \mathtt{a}, \mathtt{ab}, \mathtt{abb}, \mathtt{aa}, \mathtt{aab}, \mathtt{ba}, \mathtt{bab}, \mathtt{babb}, \mathtt{baa}, \mathtt{baab}, \mathtt{bba},\\
&\qquad \mathtt{bbab}, \mathtt{bbabb}, \mathtt{abba}, \mathtt{abbab}, \mathtt{abbabb}).\end{aligned}$$

One can check directly that the sequence Σ_d provides the $<$-increasing enumeration of $\mathrm{Div}(\Delta_3^d)$ for $d = 0, 1, 2$. We shall now prove

PROPOSITION 2.6. *For each d, the sequence Σ_d*
(i) *consists of divisors of Δ_3^d,*
(ii) *is such that the quotient of any two entries is σ-positive or σ-negative,*
(iii) *has length equal to the cardinality of* $\mathrm{Div}(\Delta_3^d)$.

When this is established, the scheme described in Section 2.1 will have been completed for the sets $\mathrm{Div}(\Delta_3^d)$. We now sketch the main steps in the proof of Proposition 2.6. Point (i) is easy:

PROOF OF PROPOSITION 2.6(i). The result is true for $d = 0$. Assume $d \geqslant 1$. By construction, each entry in Σ_d either is of the form $\theta_q\sigma_2^e$ with $0 \leqslant q \leqslant 2d$ and $0 \leqslant e \leqslant d$, or it belongs to some subsequence $\Sigma_{d,r}$ with $2 \leqslant r \leqslant 2d-1$. In the first case, $\theta_q\sigma_2^e$ is a right divisor of $\theta_{2d}\sigma_2^e$, which itself is a left divisor of $\theta_{2d}\sigma_2^d$, hence, by Lemma 2.2, of Δ_3^d. So each entry $\theta_q\sigma_2^e$ is a divisor of Δ_3^d. As for the entries coming from some subsequence $\Sigma_{d,r}$, by definition they are of the form $\beta\gamma$ with β in $\{\sigma_2, \sigma_1\sigma_2, \sigma_1, \sigma_2\sigma_1\}$ and γ an entry in Σ_{d-1}. Then β is a divisor of Δ_3, while, by induction hypothesis, γ is a divisor of Δ_3^{d-1}. So, by Lemma 1.8, $\beta\gamma$ is a divisor of Δ_3^d. □

2.3. A quotient-sequence for Σ_d. We turn to point (ii) in Proposition 2.6. To establish it, we explicitly construct a sequence of braid words witnessing the σ-positivity of the quotients of adjacent entries. These words involve the three letters σ_1, σ_2, and σ_2^{-1}.

NOTATION 2.7. For $d \geqslant 0$, we write $(\sigma_2)^d$ for the sequence $(\sigma_2, \dots, \sigma_2)$ with σ_2 repeated d times, and define W_d by $W_0 = \emptyset$ and, for $d \geqslant 1$,

$$\begin{aligned}W_d = (\sigma_2)^d &+ (\sigma_2^{-d}\sigma_1) + (\sigma_2)^d + (\sigma_2^{-d}\sigma_1) + W_{d,2} + (\sigma_1\sigma_2^{-d})\\
&+ (\sigma_2)^d + (\sigma_2^{-d}\sigma_1) + W_{d,3} + (\sigma_1\sigma_2^{-d}) + \cdots\\
&+ (\sigma_2)^d + (\sigma_2^{-d}\sigma_1) + W_{d,2d-1} + (\sigma_1\sigma_2^{-d}) + (\sigma_2)^d + (\sigma_1\sigma_2^{-d}) + (\sigma_2)^d,\end{aligned}$$

with, for $d \geqslant 2$,

$$W_{d,2} = W_{d,3} = (\sigma_2)^{d-1} + (\sigma_1\sigma_2^{-d+1}) + W_{d-1,2},$$

and, for $d \geqslant 3$ and $4 \leqslant 2r \leqslant 2d-4$,

$$\begin{aligned}W_{d,2r} = W_{d,2r+1} &= W_{d-1,2r-1} + (\sigma_2^{-d+1}\sigma_1) + (\sigma_2)^{d-1} + (\sigma_1\sigma_2^{-d+1}) + W_{d-1,2r},\\
W_{d,2d-2} = W_{d,2d-1} &= W_{d-1,2d-3} + (\sigma_2^{-d+1}\sigma_1) + (\sigma_2)^{d-1}.\end{aligned}$$

For instance, using $\mathtt{B}$ for σ_2^{-1}, we find $W_1 = (\mathtt{b})^1 + (\mathtt{Ba}) + (\mathtt{b})^1 + (\mathtt{aB}) + (\mathtt{b})^1$, whence

$$W_1 = (\mathtt{b}, \mathtt{Ba}, \mathtt{b}, \mathtt{aB}, \mathtt{b}).$$

Then, similarly, we find $W_2 = (\mathtt{b})^2 + (\mathtt{BBa}) + (\mathtt{b})^2 + (\mathtt{BBa}) + W_{2,2} + (\mathtt{aBB}) + (\mathtt{b})^2 + (\mathtt{BBa}) + W_{2,3} + (\mathtt{aBB}) + (\mathtt{b})^2 + (\mathtt{aBB}) + (\mathtt{b})^2$ with $W_{2,2} = W_{2,3} = (\mathtt{b})^1 = (\mathtt{b})$, whence

$$W_2 = (\mathtt{b}, \mathtt{b}, \mathtt{BBa}, \mathtt{b}, \mathtt{b}, \mathtt{BBa}, \mathtt{b}, \mathtt{aBB}, \mathtt{b}, \mathtt{b}, \mathtt{BBa}, \mathtt{b}, \mathtt{aBB}, \mathtt{b}, \mathtt{b}, \mathtt{aBB}, \mathtt{b}, \mathtt{b}).$$

The pattern becomes more complicated for $d \geqslant 3$ because nontrivial sequences $W_{d,r}$ appear only then.

LEMMA 2.8. *For each d, the sequence W_d is a quotient-sequence for Σ_d, i.e., the rth entry in W_d is a word representative of the quotient $\beta^{-1}\beta'$, where β and β' are the rth and $(r+1)$st entries in Σ_d.*

We skip the proof, which consists in inductively proving that W_d is a quotient-sequence for Σ_d, and, moreover, that the subsequence $W_{d,r}$ is a quotient-sequence for the subsequence $\Sigma_{d,r}$. The details use the formulas of Lemma 2.2 and require some care, but the general idea should be clear. Once this is done, it is easy to complete our argument.

PROOF OF PROPOSITION 2.6(ii). By definition, all entries in the sequence W_d are σ-positive words, as they all are of the form σ_2, $\sigma_1\sigma_2^{-e}$, or $\sigma_2^{-e}\sigma_1$ with $e \leqslant d$. By Lemma 2.8, the quotient between an entry and the next one in Σ_d is σ-positive. As the product of two or more σ-positive words is σ-positive, the same holds for the quotient between an entry and any entry that occurs after it in Σ_d. □

2.4. A proof of Property C for B_3. The last part of Proposition 2.6 that remains to be proved is point (iii), which claims that for each d the length of the sequence Σ_d equals the cardinality of $\mathrm{Div}(\Delta_3^d)$.

PROOF OF PROPOSITION 2.6(iii). Let N_d denote the length of Σ_d. Computing N_d is not very difficult. However, there is no need to do this. Indeed, we saw that the cardinality of $\mathrm{Div}(\Delta_3^d)$ obeys the inductive rule (1.5). Therefore, in order to prove the expected equality, it is sufficient to check that N_d obeys the same rule

$$N_d = 2N_{d-1} + 3d + 1, \tag{2.2}$$

and starts from the same initial value $N_1 = 6$ (or $N_0 = 1$). The latter point was checked in Example 2.5. As for the induction rule, Figure 1 shows that most entries in Σ_{d-1} give rise to two entries in Σ_d. Precisely, each entry of Σ_{d-1} not belonging to a factor of the form $\theta_{2q}\sigma_2^{[d-1]}$ gives rise to two entries in Σ_d, and, conversely, each entry in Σ_d not belonging to a factor $\theta_{2q}\sigma_2^{[d]}$ comes from such an entry in Σ_{d-1}. There are d factors $\theta_{2q}\sigma_2^{[d-1]}$ in Σ_{d-1}, each of length d, and $2d+1$ factors $\theta_{2q}\sigma_2^{[d]}$ in Σ_d, each of length $d+1$. So we obtain

$$N_d - (2d+1)(d+1) = 2(N_{d-1} - d^2),$$

and (2.2) follows. □

We thus have completed the scheme of Section 2.1 in the case of the sets $\mathrm{Div}(\Delta_3^d)$. We deduce

PROPOSITION 2.9 (Property **C** for B_3). *Each nontrivial 3-strand braid is σ-positive or σ-negative.*

PROOF. By Property **A**, the sequence Σ_d consists of pairwise distinct braids, since the quotient of any two distinct entries is σ-positive or σ-negative. As the length of Σ_d equals the cardinality of $\mathrm{Div}(\Delta_3^d)$, every element of $\mathrm{Div}(\Delta_3^d)$ must occur in Σ_d, and, therefore, every element of B_3^+ must occur in some sequence Σ_d. Let β be an arbitrary 3-strand braid. By Proposition I.4.6, there exist positive 3-strand braids β_1, β_2 such that $\beta = \beta_1^{-1}\beta_2$ holds. For d large enough, the braids β_1 and β_2 occur in the sequence Σ_d and, therefore, their quotient β is either σ-positive or σ-negative. □

2.5. A normal form. We can deduce from the above results more information about the σ-ordering of 3-strand braids. However, we shall give a sketch only, because all results below will be reproved and extended in Chapter VII.

So far, we introduced the sequence Σ_d as a sequence of braids. The explicit rule of Notation 2.4 specifies not only a braid but also a distinguished word representative for that braid.

NOTATION 2.10. We denote by $\underline{\Sigma}_d$ the sequence of braid words defined by the recursive rule of Notation 2.4.

So for instance $\underline{\Sigma}_1$ is the sequence of braid words $(1, \mathtt{b}, \mathtt{a}, \mathtt{ab}, \mathtt{ba}, \mathtt{bab})$. It is easy to inductively check that, for $d \leqslant d'$, all words occurring in $\underline{\Sigma}_d$ occur in $\underline{\Sigma}_{d'}$. Moreover, the enumeration orders are compatible: if w occurs before w' in $\underline{\Sigma}_d$, it also occurs before w' in $\underline{\Sigma}_{d'}$. So, for $d \leqslant d'$, there exists an injection $f_{d,d'}$ of $\underline{\Sigma}_d$ into $\underline{\Sigma}_{d'}$, and it is natural to introduce the direct limit $\underline{\Sigma}$ of the system $(\underline{\Sigma}_d, f_{d,d'})$: the sequence $\underline{\Sigma}$ is an infinite sequence of 3-strand braid words that is indexed by ordinal numbers smaller than ω^ω and starts with $\varepsilon, \mathtt{b}, \mathtt{bb}, \mathtt{bbb}, \dots, \mathtt{a}, \mathtt{ab}, \mathtt{abb}, \dots, \mathtt{aa}, \mathtt{aab}, \dots, \mathtt{ba}, \mathtt{bab}, \dots$. By construction, each braid in B_3^+ has a unique word representative in $\underline{\Sigma}$, *i.e.*, $\underline{\Sigma}$ defines a normal form in B_3^+. Moreover, by construction, the order in which words appear in $\underline{\Sigma}$ corresponds to the position of the braids they represent in the σ-ordering of B_3^+.

The point is that both the normal words and their ordering can be easily described: as an inspection of the recursive construction shows, the words that appear in $\underline{\Sigma}$ can be characterized by conditions on the sizes of the blocks of σ_1's and σ_2's they contain, and they turn out to be (up to exchanging σ_1 and σ_2) the Φ-normal words that will be introduced in Definition VII.2.2, while their order is (up to exchanging σ_1 and σ_2) the ordering $<^+$ of Definition VII.2.14.

So the current approach is another way of proving that every 3-strand braid admits a unique Φ-normal word representative, and that the ordering of such braids corresponds to a `ShortLex`-ordering of the normal forms. Further results, like Property **S** for B_3, *i.e.*, the result that $\beta\sigma_1 > \beta$ and $\beta\sigma_2 > \beta$ hold for each β in B_3^+, and the fact that the restriction of the σ-ordering to B_3^+ is a well-ordering of ordinal type ω^ω can be then read on the sequences Σ_d, but we shall not go into detail here.

3. The increasing enumeration of $\mathrm{Div}(\Delta_n^d)$

Apart from possibly providing a new construction of the σ-ordering of braids, our current combinatorial approach can be used to establish new properties of the ordering once its existence is known. Our ultimate goal would be to give a complete description for the σ-ordering of the positive n-strand braids of degree at most d. The task is easy for $d = 1$, *i.e.*, when we consider simple braids but, starting with degree two, things become much more complicated—as the results of Section 2 showed. What we shall do here is establish general results about the structure of the chains $(\mathrm{Div}(\Delta_n^d), <)$. The main results are Propositions 3.11 and 3.14, which connect $(\mathrm{Div}(\Delta_n^d), <)$ and $(\mathrm{Div}(\Delta_{n-1}^d), <)$. These results explain the constructions of Section 2 and should make them more easily understandable.

3.1. The case of simple braids. In the case of simple braids, *i.e.*, when we restrict ourselves to positive braids that divide some Δ_n, completely describing the braid ordering is easy.

PROPOSITION 3.1. *Assume that β, β' are simple n-strand braids. Let π and π' be the permutations associated with β and β'. Then $\beta < \beta'$ holds if and only if the sequence $(\pi(1), \dots, \pi(n))$ is lexicographically smaller than $(\pi'(1), \dots, \pi'(n))$.*

PROOF. The proof, which uses an induction on $n \geqslant 2$, is an easy exercise. For $n = 2$, the only braids we have to consider are 1 and σ_1. The associated permutations are the identity permutation and the transposition $(1, 2)$, and the result is clear.

Assume $n \geqslant 3$. For $r \geqslant 1$, define $\sigma_{r,1} = \sigma_{r-1} \dots \sigma_2 \sigma_1$. Let $r = \pi(1)$, and let π_1 be the permutation of $\{1, \dots, n-1\}$ defined by $\pi_1(i) = \pi(i+1)$ for $\pi(i+1) < r$ and $\pi_1(i) = \pi(i+1) - 1$ for $\pi(i+1) > r$. Finally, let β_1 be the simple $(n-1)$-strand braid associated with the permutation π_1. By the criterion of Lemma 1.2(iii), the braid $\sigma_{r,1}\mathrm{sh}(\beta_1)$ is simple, and the associated permutation is π. Hence, by Lemma 1.3, we have $\beta = \sigma_{r,1}\mathrm{sh}(\beta_1)$. Similarly, starting with $r' = \pi'(1)$, we find $\beta' = \sigma_{r',1}\mathrm{sh}(\beta'_1)$ for some simple braid β'_1 associated with the permutation π'_1 analogous to π_1.

Assume first $r = r'$. Then we obtain $\beta^{-1}\beta' = \mathrm{sh}(\beta_1^{-1}\beta'_1)$. By induction hypothesis, $\beta_1 < \beta'_1$ holds if and only if the sequence $(\pi_1(1), \dots, \pi_1(n-1))$ is lexicographically smaller than the sequence $(\pi'_1(1), \dots, \pi'_1(n-1))$. By construction, this happens if and only if the sequence $(\pi(2), \dots, \pi(n))$ is smaller than $(\pi'(2), \dots, \pi'(n))$, and this is the expected result.

Assume now $r \neq r'$, say $r < r'$. Then we obtain

$$\beta^{-1} \cdot \beta' = \mathrm{sh}(\beta_1^{-1}) \cdot \sigma_{r,1}^{-1} \cdot \sigma_{r',1} \cdot \mathrm{sh}(\beta'_1). \tag{3.1}$$

As shown in Figure 2, we have

$$\sigma_{r,1}^{-1} \cdot \sigma_{r',1} = \sigma_{r',1} \cdot \mathrm{sh}(\sigma_{r,1}^{-1}) = \mathrm{sh}(\sigma_{r'-1,1}) \cdot \sigma_1 \cdot \mathrm{sh}(\sigma_{r,1}^{-1}).$$

Inserting the last expression in (3.1) gives a σ_1-positive word representing $\beta^{-1}\beta'$. The case $r > r'$ is symmetric, leading to a σ_1-negative expression for $\beta^{-1}\beta'$. □

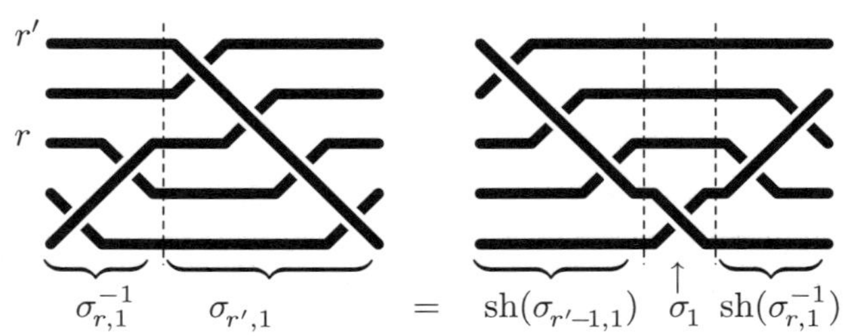

FIGURE 2. The relation $\sigma_{r,1}^{-1} \cdot \sigma_{r',1} = \sigma_{r',1} \cdot \mathrm{sh}(\sigma_{r,1}^{-1})$ for $r < r'$—here with $r = 3$ and $r' = 5$.

As soon as we consider degree two and higher, the connection becomes problematic. The greedy normal form provides for each degree two braid a distinguished decomposition as a product of two simple braids, hence a pair of permutations, and, after Proposition 3.1, one might expect that the braid associated with a pair of permutations (π_1, π_2) is smaller than the braid associated with (π'_1, π'_2) if and only if the pair (π_1, π_2) is lexicographically smaller than the pair (π'_1, π'_2), or something similar. This would show that the ordering of degree two braids is a lexicographical extension of the ordering of simple braids, and provide a complete and satisfactory description. Such a simple connection does *not* exist. For instance, we have $\sigma_1\sigma_2^2 < \sigma_1^2$, and the greedy normal forms are $(\sigma_1\sigma_2, \sigma_2)$ and (σ_1, σ_1): as $\sigma_1\sigma_2 > \sigma_1$ holds, the first sequence is larger than the second. Considering a reversed lexicographical ordering would not help: for $\sigma_1^2\sigma_2 < \sigma_2\sigma_1$, the greedy normal forms are $(\sigma_1, \sigma_1\sigma_2)$ and $(\sigma_2\sigma_1, 1)$, and now the second coordinate decreases. The compatibility is not better with the right version of the greedy normal form mentioned in Remark 1.10: $\Delta_3\sigma_1 > \Delta_3\sigma_2$ holds, but, in any reasonable sense, the right normal form of $\Delta_3\sigma_1$, namely (σ_2, Δ_3), is smaller than that of $\Delta_3\sigma_2$, namely (σ_1, Δ_3).

3.2. The sequences $S_{n,d}$. The previous counterexample suggests that, for $d \geqslant 2$, the restriction of the σ-ordering to braids of degree at most d has no simple description. What we shall do from now on is to establish partial results in view of such a description. These results involve the $<$-increasing enumeration of the set $\mathrm{Div}(\Delta_n^d)$, *i.e.*, the finite sequence obtained when all n-strand braids of degree at most d are listed in increasing order. We shall use the following notation.

NOTATION 3.2. For $n \geqslant 2$ and $d \geqslant 0$, we denote by $S_{n,d}$ the $<$-increasing enumeration of $\mathrm{Div}(\Delta_n^d)$.

As announced above, our aim is to give a description of the sequences $S_{n,d}$ that is as complete as possible. The main results are Proposition 3.11, which describes the sequence $S_{n,d}$ in terms of a certain number $h_1(\Delta_n^d)$ of shifted and translated copies of $S_{n-1,d}$, and Proposition 3.14, which determines the number $h_1(\Delta_n^d)$ in terms of the matrix M_n of Section 1.

It follows from the results of Section 2 that, for each d, the sequence $S_{3,d}$ is the sequence Σ_d of Notation 2.4. So, in that specific case, $S_{n,d}$ admits a recursive construction in terms of $S_{n,d-1}$ and of the sequences denoted $\sigma_2^{[e]}$, which are the sequences $S_{2,e}$, *i.e.*, $S_{n-1,e}$. We shall establish a general version of such results for every n.

It is useful to think of $\mathrm{Div}(\Delta_n^d)$ as a finite subset of the Cayley graph of B_n—more precisely, of B_n^+—relative to the generators σ_i. We recall that the latter is the labeled oriented graph with vertex set B_n such that there exists a σ_i-labeled edge from β to β' if and only if $\beta' = \beta\sigma_i$ holds in B_n. Then, the sequence $S_{n,d}$ is a distinguished path through (the Cayley graph of) $\mathrm{Div}(\Delta_n^d)$. For instance, we display in Figure 3 (left) the increasing enumeration of $\mathrm{Div}(\Delta_3^2)$.

The above representation can be improved. By definition, the sequence $S_{n,d}$ is $<$-increasing. Hence, for any two adjacent entries β, β', there exists a σ-positive word representing $\beta^{-1}\beta'$, whence a σ-positive path connecting β to β' in the Cayley graph of B_n. Proposition V.3.3 says more: as both β and β' are divisors of Δ_n^d, there must exist a σ-positive word representing $\beta^{-1}\beta'$ that is drawn from β in $\mathrm{Div}(\Delta_n^d)$, *i.e.*, a σ-positive path from β to β' that entirely remains inside the Cayley graph

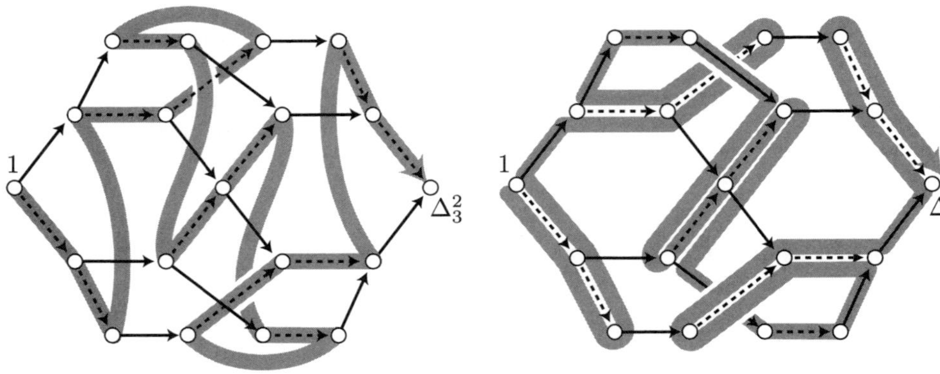

FIGURE 3. Two representations of the sequence $S_{3,2}$ as a path visiting each vertex in the Cayley graph of $\mathrm{Div}(\Delta_3^2)$; plain arrows represent σ_1, dotted arrows represent σ_2; on the left, we do not care about the quotient between successive entries and just visit each vertex once; on the right, we exhibit σ-positive words drawn inside $\mathrm{Div}(\Delta_3^2)$ that witness the transitions; the resulting path is bbBBabbBBabaBBbbBBabaBBbbaBBbb.

of $\mathrm{Div}(\Delta_n^d)$. In the case $n = 3$, the property can be checked directly from the explicit construction of the sequence W_d, and it is illustrated for Δ_3^2 on Figure 3 (right): all edges visited by the grey path belong to our graph—while *a priori* it might well happen that, in order to obtain σ-positive connections, we have to go outside of the restricted graph. This situation is general: Proposition V.3.3 guarantees that for all n and d there exists a path as above.

3.3. The partial orders $<_i$. Our analysis of the sequences $S_{n,d}$ consists in introducing a filtration of the σ-ordering in terms of partial orderings.

If β, β' are braids, then by definition $\beta < \beta'$ holds if the quotient $\beta^{-1}\beta'$ is σ_i-positive for some i . We obtain a natural refinement by bounding the index i.

DEFINITION 3.3. For β, β' in B_∞ and $i \geqslant 1$, we say that $\beta <_i \beta'$ holds or, equivalently, that (β, β') is a *σ_i-jump*, if $\beta^{-1}\beta'$ is σ_j-positive for some $j \leqslant i$.

LEMMA 3.4. *For each i, the relation $<_i$ is a strict partial ordering of B_∞ that refines $<$. For $i \leqslant j$, the relation $<_i$ refines $<_j$, and $<$ is the union of all $<_i$.*

PROOF. The relation $<_i$ is transitive because the concatenation of a σ_i-positive word and a σ_j-positive word is a $\sigma_{\min(i,j)}$-positive word. It is antireflexive, and therefore antisymmetric, because of Property **A**. □

EXAMPLE 3.5. The $<$-increasing enumerations of the divisors of Δ_3 and of Δ_3^2, *i.e.*, the sequences $S_{3,1}$ and $S_{3,2}$, have been shown in Example 2.5. When we indicate for each step the height of the corresponding jump, we find for $S_{3,1}$

$$1 <_2 \mathtt{b} <_1 \mathtt{a} <_2 \mathtt{ab} <_1 \mathtt{ba} <_2 \Delta_3,$$

where we recall $\mathtt{a}, \mathtt{b}, \ldots$ stand for $\sigma_1, \sigma_2, \ldots$. For instance, $(\mathtt{ab}, \mathtt{ba})$ is a σ_1-jump, because we have $(\mathtt{ab})^{-1}(\mathtt{ba}) = \mathtt{BAba} = \mathtt{BbaB} = \mathtt{aB}$, a σ_1-positive expression. Similarly, we obtain for $S_{3,2}$:

$$1 <_2 \mathtt{b} <_2 \mathtt{bb} <_1 \mathtt{a} <_2 \mathtt{ab} <_2 \mathtt{abb} <_1 \mathtt{aa} <_2 \mathtt{aab} <_1 \mathtt{ba} <_2 \mathtt{bab} <_2 \mathtt{babb} <_1 \mathtt{baa}$$
$$<_2 \mathtt{baab} <_1 \mathtt{bba} <_2 \mathtt{bbab} <_2 \mathtt{bbabb} <_1 \mathtt{abba} <_2 \mathtt{abbab} <_2 \mathtt{abbabb}.$$

It is then natural to count the various σ_i-jumps in the sequences $S_{n,d}$.

DEFINITION 3.6. For β a positive braid and $i \geqslant 1$, we define the *σ_i-height $h_i(\beta)$* of β to be the number of σ_i-jumps in the $<$-increasing enumeration of $\mathrm{Div}(\beta)$, augmented by 1.

For instance, we can read in Example 3.5 the values $h_1(\Delta_3) = 3$, $h_2(\Delta_3) = 6$, $h_1(\Delta_3^2) = 7$, $h_2(\Delta_3^2) = 19$. Note that by definition a σ_1-jump is a σ_2-jump, and that is why, for instance, $h_2(\Delta_3)$ is the total number of divisors of Δ_3. It is not hard to check that the σ_i-height of β can equivalently be defined as the maximal length of a $<_i$-chain included in $\mathrm{Div}(\beta)$. The following basic observations are easy.

PROPOSITION 3.7. (i) *For every positive braid β in B_n^+, we have*

$$h_1(\beta) \leqslant h_2(\beta) \leqslant \dots \leqslant h_{n-1}(\beta) = \#\mathrm{Div}(\beta). \tag{3.2}$$

(ii) *For all positive braids β, β' and $i \geqslant 1$, we have*

$$h_i(\beta\beta') \geqslant h_i(\beta) + h_i(\beta') - 1. \tag{3.3}$$

PROOF. The inequalities of (3.2) directly follow from the implications of Lemma 3.4. As for the last equality, by definition, every $<$-chain included in B_n^+ is a $<_{n-1}$-chain, hence the maximal $<_{n-1}$-chain in $\mathrm{Div}(\beta)$ is $\mathrm{Div}(\beta)$ itself, and $h_{n-1}(\beta)$ is the cardinality of $\mathrm{Div}(\beta)$. As for (3.3), the result is obvious since the concatenation of two $<_i$-chains is a $<_i$-chain. □

Using (3.3) and the coarse upper bound $\#\mathrm{Div}(\Delta_n^d) \leqslant (n!)^d$ that follows from Lemma 1.8, we deduce

$$d \cdot h_i(\Delta_n) - d + 1 \leqslant h_i(\Delta_n^d) \leqslant (n!)^d \tag{3.4}$$

for all i, n, d. The numbers $h_i(\Delta_n)$ will turn out to be important parameters for describing the sequences $S_{n,d}$, and better estimates will be given below.

3.4. The structure of sh^i-classes. For each i with $1 \leqslant i < n$, the chain $S_{n,d}$ —or, more generally, every chain $(\mathrm{Div}(\beta), <)$ for β a positive n-strand braid—can be decomposed into intervals containing no σ_i-jump and separated by σ_i-jumps. We shall now describe the structure of each of these intervals more precisely, namely show that they are lattices with respect to left-divisibility.

For each i, the image of B_∞ under sh^i is a subgroup of B_∞, so $\beta^{-1}\beta' \in \mathrm{sh}^i(B_\infty)$ defines an equivalence relation on arbitrary braids, hence on positive braids as well.

NOTATION 3.8. For $i \geqslant 0$ and β, β' positive braids, we say that $\beta \equiv_i \beta'$ holds if $\beta^{-1}\beta'$ belongs to the image of sh^i.

Hereafter, we use $\equiv_i$ to partition $\mathrm{Div}(\beta)$ into subsets, naturally called *sh^i-classes*. By definition, $\beta \equiv_0 \beta'$ holds for all β, β', while, for β, β' in B_n^+, the relation $\beta \equiv_n \beta'$, and, more generally, $\beta \equiv_i \beta'$ with $i \geqslant n$, holds only for $\beta = \beta'$. So the interesting cases are $1 \leqslant i \leqslant n-1$ only. Figure 4 shows that $\mathrm{Div}(\Delta_3)$ contains three sh^1-classes, while $\mathrm{Div}(\Delta_3^2)$ contains seven of them.

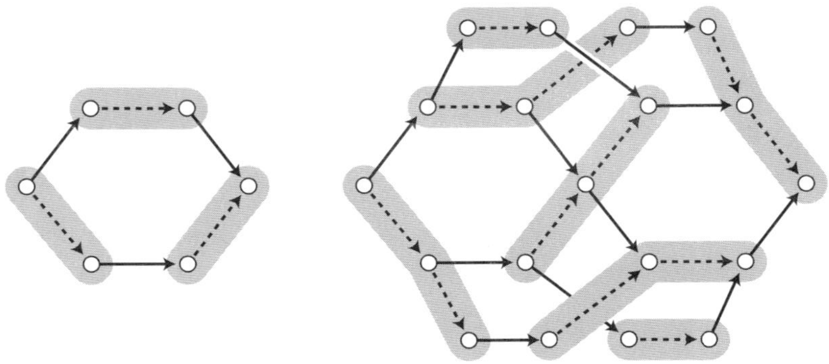

FIGURE 4. The sh^1-classes in $\mathrm{Div}(\Delta_3)$ and $\mathrm{Div}(\Delta_3^2)$—we recall that the plain arrows represent σ_1, while the dotted arrows represent σ_2; in the case of B_3, being $\equiv_1$-equivalent just means that the quotient is a power of σ_2.

The connection between σ_i-jumps and sh^i-classes easily follows from their definitions:

LEMMA 3.9. *For $\beta < \beta'$, the relation $\beta \equiv_i \beta'$ holds if and only if there is no σ_i-jump between β and β'.*

Here comes the main structure result about sh^i-classes.

PROPOSITION 3.10. *Assume that β is a positive braid, and C is a sh^i-class in $\mathrm{Div}(\beta)$. Let β_* and β^* be the $<$-extremal elements of C. Then β_* is a left divisor of every element of C, and the left translation by β_* defines an isomorphism of $(\mathrm{Div}(\beta_*^{-1}\beta^*), \preccurlyeq, <)$ onto $(C, \preccurlyeq, <)$. In particular, $(C, \preccurlyeq)$ is a lattice.*

PROOF. By Lemma 3.9, C is the $<$-interval determined by β_* and β^*, *i.e.*, we have $C = \{\gamma \in \mathrm{Div}(\beta) \mid \beta_* < \gamma < \beta^*\}$.

First, we claim that C is closed under the gcd operation. Assume $\gamma, \gamma' \in C$. Let γ_0, γ_0' be defined by $\gamma = \gcd(\gamma, \gamma')\gamma_0$ and $\gamma' = \gcd(\gamma, \gamma')\gamma_0'$. The hypothesis that $\gamma^{-1}\gamma'$ belongs to the image of sh^i implies that there exist positive braids γ_1, γ_1' satisfying $\gamma^{-1}\gamma' = \mathrm{sh}^i({\gamma_1}^{-1}\gamma_1')$. By definition of the gcd, there exists a positive braid β_1 satisfying $\mathrm{sh}^i(\gamma_1) = \beta_1\gamma_0$ and $\mathrm{sh}^i(\gamma_1') = \beta_1\gamma_0'$. Because β_1 is positive, this implies that γ_0 belongs to the image of sh^i, and, therefore, that $\gcd(\gamma, \gamma')$ belongs to C.

Then, C is closed under the lcm operation. Indeed, the definition of γ_0 and γ_0' and the compatibility of the lcm operation with multiplication on the left imply

$$\mathrm{lcm}(\gamma, \gamma') = \gcd(\gamma, \gamma') \cdot \mathrm{lcm}(\gamma_0, \gamma_0'). \tag{3.5}$$

As γ_0 and γ_0' lie in the image of sh^i, so does $\mathrm{lcm}(\gamma_0, \gamma_0')$, and we deduce from (3.5) that $\mathrm{lcm}(\gamma, \gamma')$ belongs to C.

As C is finite, it admits a global gcd which, by the above closure result, has to lie in C. By construction, the linear ordering $<$ extends the partial divisibility ordering $\preccurlyeq$. Hence, this global gcd must be the $<$-minimum β_* of C. Symmetrically, C admits a global lcm, which must be the $<$-maximum β^*. So, at this point, we know that β_* is a left divisor of every element in C, and β^* is a right multiple of each such element; *i.e.*, we have

$$C \subseteq \{\gamma \in B_\infty^+ \mid \beta_* \preccurlyeq \gamma \preccurlyeq \beta^*\}. \tag{3.6}$$

Now, on the other hand, $\beta_* \preccurlyeq \gamma \preccurlyeq \beta^*$ implies $\beta_* \leqslant \gamma \leqslant \beta^*$, hence $\gamma \in C$, in all cases, and so the inclusion in (3.6) is an equality.

Now, let f be the left translation by β_* defined on $\mathrm{Div}(\beta_*^{-1}\beta^*)$. As the monoid B_∞^+ is left-cancellative, f is injective. By construction, it preserves the partial divisibility order $\preccurlyeq$, so the image of $\mathrm{Div}(\beta_*^{-1}\beta^*)$, which is the set $\{\gamma \mid 1 \preccurlyeq \gamma \preccurlyeq \beta_*^{-1}\beta^*\}$, is $\{\gamma \in B_\infty^+ \mid \beta_* \preccurlyeq \gamma \preccurlyeq \beta^*\}$, which is C. Moreover, by construction, f preserves the linear order $<$, so it provides the expected isomorphism. □

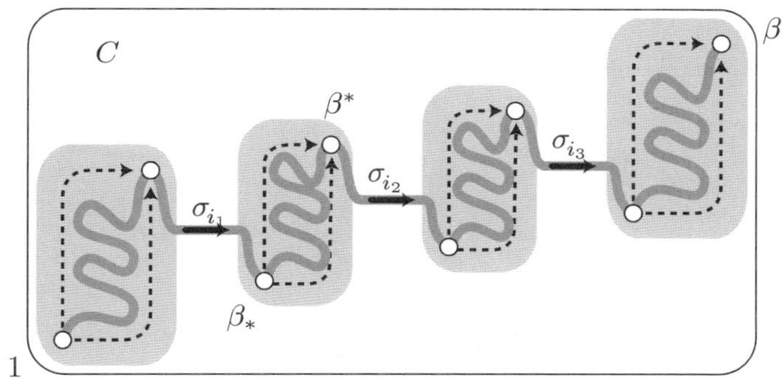

FIGURE 5. Decomposition of $(\mathrm{Div}(\beta), <)$ into sh^i-classes: each class C is a lattice with respect to divisibility, and it is itself a translated copy of some chain $(\mathrm{Div}(\beta'), <)$ with β' of lower index than β; the increasing enumeration of $\mathrm{Div}(\beta)$ exhausts the first class, then jumps to the next one by an σ_i-jump, etc.; the number of classes is $h_i(\beta)$.

So, for every n-strand braid β, the increasing enumeration of $(\mathrm{Div}(\beta), <)$ is obtained by concatenating $h_i(\beta)$ copies of chains of the form $(\mathrm{Div}(\beta'), <)$, where, by construction, β' lies in $\mathrm{sh}^i(B_{n-i}^+)$. In other words, $\mathrm{Div}(\beta)$ is obtained by concatenating $h_i(\beta)$ shifted copies of chains of the form $(\mathrm{Div}(\beta'), <)$ with β' in B_{n-i}^+; see Figure 5. In particular, for $\beta = \Delta_n^d$, we find

PROPOSITION 3.11. *For each n and each i with $i < n$, the sequence $S_{n,d}$ is obtained by concatenating $h_i(\Delta_n^d)$ subsequences, each of which is a shifted and translated copy of some initial fragment of $S_{n-i,d}$.*

PROOF. Let C be a sh^i-class in $\mathrm{Div}(\Delta_n^d)$. By Proposition 3.10, there exist divisors β_* and β^* of Δ_n^d such that $(C, \preccurlyeq, <)$ is isomorphic to $(\mathrm{Div}(\beta'), \preccurlyeq, <)$, with $\beta' = \beta_*^{-1}\beta^*$. By construction, we have $1 \preccurlyeq \beta_* \preccurlyeq \beta^* \preccurlyeq \Delta_n^d$, so β', which is a right divisor of the left divisor β^* of Δ_n^d, is itself a divisor of Δ_n^d. Hence, by Lemma 1.8, β' has degree at most d. On the other hand, β' belongs to the image of sh^i, *i.e.*, there exists a positive braid β'_1 in B_{n-i}^+ such that $\beta' = \mathrm{sh}^i(\beta'_1)$ holds. As the shift endomorphism preserves the normal form, and therefore the degree, β'_1 has degree at most d, so it is a divisor of Δ_{n-i}^d. □

The cases of Δ_3^2 and Δ_4 are illustrated in Figures 6 and 7, respectively. The reader can check that Notation 2.4 exactly reflects Proposition 3.11.

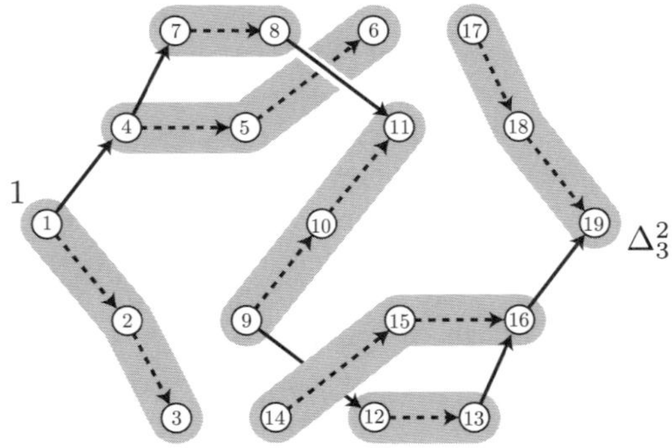

FIGURE 6. The sequence $S_{3,2}$ viewed as a concatenation of sh^1-classes, *i.e.*, of intervals of the form $S_{2,e}$ separated by σ_1-jumps—plain arrows.

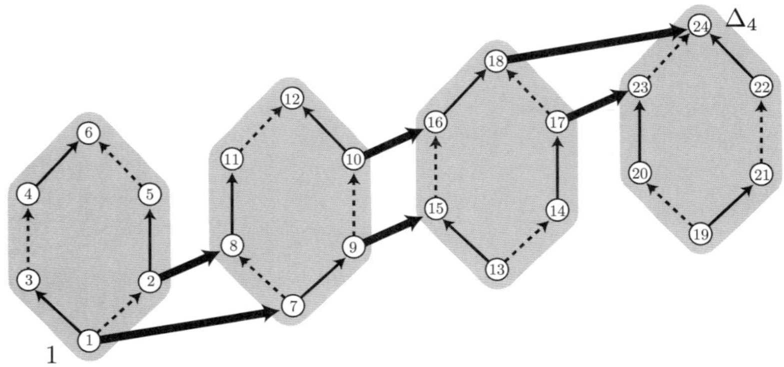

FIGURE 7. The sequence $S_{4,1}$ viewed as a concatenation of sh^1-classes separated by σ_1-jumps; in this case, each sh^1-class is isomorphic to the lattice $(\mathrm{Div}(\Delta_3), <, \preccurlyeq)$, hence its enumeration is a copy of $S_{3,1}$—a general fact in degree one.

3.5. The σ_i-height of Δ_n^d. We shall complete our description of the sequences $S_{n,d}$ by determining the height parameters $h_i(\Delta_n^d)$ explicitly. We recall that, for β a positive braid, the σ_i-height $h_i(\beta)$ of β is the number of σ_i-jumps in the $<$-increasing enumeration of β.

By Lemma 3.9, $h_i(\beta)$ is the number of sh^i-classes in $\mathrm{Div}(\beta)$, so the point is to count sh^i-classes in $\mathrm{Div}(\beta)$. To this end, we shall characterize the maximal elements of each class, and then count such elements using the results of Section 1.

LEMMA 3.12. *Assume $\beta' \preccurlyeq \beta \in B_\infty^+$. Then β' is the maximum of its sh^i-class in $\mathrm{Div}(\beta)$ if and only if the relation $\beta'\sigma_j \preccurlyeq \beta$ fails for $j \geqslant i+1$.*

PROOF. The condition is necessary: if $\beta'\sigma_j$ lies in $\mathrm{Div}(\beta)$ for some $j \geqslant i+1$, then we have $\beta'\sigma_j \equiv_i \beta'$ and $\beta'\sigma_j > \beta'$, so β' cannot be maximal in its sh^i-class. Conversely, assume that β' is not maximal in its sh^i-class. Then there exists β'' satisfying $\beta' < \beta''$ and $\beta'' \equiv_i \beta'$. By Proposition 3.10, the lcm of β' and β'' is also $\equiv_i$-equivalent to β'. Let $\beta_1 = \beta'^{-1}\mathrm{lcm}(\beta', \beta'')$. By construction, β_1 is a nontrivial positive braid that lies in the image of sh^i. Therefore, there exists at least one index $j \geqslant i+1$ such that σ_j is a left divisor of β_1: for that j, we have $\beta' \prec \beta'\sigma_j \preccurlyeq \mathrm{lcm}(\beta', \beta'') \preccurlyeq \beta$, which shows that $\beta'\sigma_j$ lies in $\mathrm{Div}(\Delta_n^d)$. □

Applying the previous criterion to the braids Δ_n^d, we obtain:

LEMMA 3.13. *For β in $\mathrm{Div}(\Delta_n^d)$ and $1 \leqslant i < n$, the following are equivalent:*
(i) *The braid β is $<$-maximal in its sh^i-class;*
(ii) *The dth factor of β is right divisible by $\mathrm{sh}^i(\Delta_{n-i})$;*
(iii) *The $(d+1)$st factor of $\beta\,\mathrm{sh}^i(\Delta_{n-i})$ is $\mathrm{sh}^i(\Delta_{n-i})$.*

PROOF. By Lemma 3.12, (i) is equivalent to
(i)' The braid $\beta\sigma_j$ does not belong to $\mathrm{Div}(\Delta_n^d)$ for any j with $j \geqslant i+1$;
and it suffices to establish the equivalence of (i)' with (ii) and (iii). For $i = n-1$, (i)' is vacuously true, while (ii) and (iii) always hold. So the expected equivalences are true. Henceforth, we assume $i \leqslant n-2$.

Let β belong to $\mathrm{Div}(\Delta_n^d)$, and let β_d be the dth factor in the normal form of β. For $j < n$, saying that $\beta\sigma_j$ does not belong to $\mathrm{Div}(\Delta_n^d)$ means that the normal form of $\beta\sigma_j$ has length $d+1$, hence, equivalently, that the normal form of $\beta_d\sigma_j$ has length 2. This occurs if and only if σ_j is a right divisor of β_d. So (i)' is equivalent to β_d being right divisible by all σ_j's with $j \geqslant i+1$, hence to β_d being right divisible by the (left) lcm of these elements, which is $\mathrm{sh}^i(\Delta_{n-i})$. So (i)' and (ii) are equivalent.

With the same notation, assume that the dth factor β_d in the normal form of β is divisible by $\mathrm{sh}^i(\Delta_{n-i})$ on the right, then $(\beta_d, \mathrm{sh}^i(\Delta_{n-i}))$ is a normal sequence as no σ_j with $j \geqslant i+1$ from $\mathrm{sh}^i(\Delta_{n-i})$ may pass to β_d. Hence $(\beta_1, \dots, \beta_d, \mathrm{sh}^i(\Delta_{n-i}))$ is a normal sequence, necessarily the normal form of $\beta\,\mathrm{sh}^i(\Delta_{n-i})$. So (ii) implies (iii). Conversely, assume that the normal form of $\beta\,\mathrm{sh}^i(\Delta_{n-i})$ is $(\beta_1, \dots, \beta_d, \mathrm{sh}^i(\Delta_{n-i}))$. Then $(\beta_d, \mathrm{sh}^i(\Delta_{n-i}))$ is normal, hence β_d is divisible on the right by each σ_j that is a left divisor of $\mathrm{sh}^i(\Delta_{n-i})$, *i.e.*, by each σ_j with $j \geqslant i+1$. Hence β_d is divisible on the right by the left lcm of these σ_j's, which is $\mathrm{sh}^i(\Delta_{n-i})$. So (iii) implies (ii). □

Appealing to the results of Section 1 about the number of braids whose normal form satisfies given constraints, we can now complete the computation.

PROPOSITION 3.14. *For $1 \leqslant i < n$, we have*

$$h_i(\Delta_n^d) = \sum_{\beta \text{ right divisible by } \Delta_{n-i}} b_{n,d}(\beta) = b_{n,d+1}(\Delta_{n-i}). \tag{3.7}$$

PROOF. By Lemma 3.9, $h_i(\Delta_n^d)$ is the number of sh^i-classes in $\mathrm{Div}(\Delta_n^d)$. Each class contains exactly one maximum element. By Lemma 3.13, these maximal elements are characterized both by their dth factor being right divisible by $\mathrm{sh}^i(\Delta_{n-i})$ and by the $(d+1)$st factor of their product by $\mathrm{sh}^i(\Delta_{n-i})$ being the latter. Owing to the definition of the numbers $b_{n,d}(\beta)$—Notation 1.11—we deduce

$$h_i(\Delta_n^d) = \sum_{\beta \text{ right divisible by } \mathrm{sh}^i(\Delta_{n-i})} b_{n,d}(\beta) = b_{n,d+1}(\mathrm{sh}^i(\Delta_{n-i})).$$

Now, the flip mapping Φ_n is an automorphism of the monoid B_n^+, hence we have $b_{n,d}(\Phi_n(\beta)) = b_{n,d}(\beta)$ for each simple n-strand braid, and we obtain (3.7) by using the equality $\mathrm{sh}^i(\Delta_{n-i}) = \Phi_n(\Delta_{n-i})$, which follows for instance from the fact that both terms represent the lcm of $\sigma_{n-i}, \dots, \sigma_{n-1}$. □

For $i = n-1$, as every simple braid is divisible by 1 on the right, relation (3.7) reduces to $h_{n-1}(\Delta_n^d) = \sum_{\beta \text{ simple}} b_{n,d}(\beta) = b_{n,d+1}(1) = \#\mathrm{Div}(\Delta_n^d)$, a special case of the relation $h_{n-1}(\beta) = \#\mathrm{Div}(\beta)$ of Proposition 3.7—and a trivial equality as, by definition, every jump in $S_{n,d}$ is a σ_{n-1}-jump.

Using Proposition 3.14, it is easy to read the first values of $h_i(\Delta_n^d)$ in Table 1, for instance $h_1(\Delta_3) = b_{3,2}(\Delta_2) = 3$, etc. By solving the recursive characterization of Lemma 1.13, one obtains closed formulas for $n = 3$ and 4:

PROPOSITION 3.15. *For $d \geqslant 1$, we have*

$$h_1(\Delta_3^d) = 2 \cdot 2^d - 1, \quad h_1(\Delta_4^d) = \sum_{\pm} \frac{1}{20}(4 \pm \sqrt{6})(3 \pm \sqrt{6})^d + \frac{8}{5} \cdot 2^d - 1.$$

3.6. The σ_1-content of a braid. As an application of the previous results, we can now precisely determine the value of the parameter called σ_1-content, which was introduced in Section V.2.6 for analyzing handle reduction. We recall that, for β a positive braid, the σ_1-content $c_1(\beta)$ of β is defined to be the maximal number of letters σ_1 in a σ-positive braid word drawn in $\mathrm{Div}(\beta)$. It is not surprising that this number is connected with the σ_1-height of β, *i.e.*, the number of σ_1-jumps in $(\mathrm{Div}(\beta), <)$. However, establishing the connection requires a little care.

PROPOSITION 3.16. *For each positive braid β, the σ_1-content of β is $h_1(\beta) - 1$.*

PROOF. Assume that w is a σ-positive word drawn in $\mathrm{Div}(\beta)$ from β_0. At the expense of adding to w a positive prefix that represents β_0, we may assume $\beta_0 = 1$. Let m be the number of letters σ_1 in w, and let $\beta_0(= 1), \dots, \beta_N(= \beta)$ be the $<$-increasing enumeration of $\mathrm{Div}(\beta)$. By definition, all prefixes of w represent divisors of β, so, letting ℓ be the length of w, there exists a map $f : \{0, \dots, \ell\} \to \{0, \dots, N\}$ such that, for each r, the length r prefix of w represents $\beta_{f(r)}$. By construction, we have $f(0) = 0$ and $f(\ell) = N$. In general, the function f need not be increasing. Now, let $r_1, \dots, r_m$ be the m positions in w where σ_1 occurs, completed with $r_0 = 0$. Then, in the prefix of w of length r_1, *i.e.*, in the subword of w corresponding to positions from r_0+1 to r_1, there is one σ_1, plus letters $\sigma_i^{\pm 1}$ with $i \geqslant 2$ (see Figure 8). This subword is therefore σ-positive, hence we must have $\beta_{f(r_0)} < \beta_{f(r_1)}$, which requires $f(r_0) < f(r_1)$. Moreover, $\beta_{f(r_0)}^{-1}\beta_{f(r_1)}$ is a braid that admits at least one σ_1-positive word representative, so the pair $(\beta_{f(r_0)}, \beta_{f(r_1)})$ is a σ_1-jump. The same is true for $(f(r_1), f(r_2))$, etc. Hence the number of σ_1-jumps in the increasing enumeration of $\mathrm{Div}(\beta)$ is at least m, *i.e.*, we have $h_1(\beta) \geqslant m + 1$.

Conversely, assume that β is a positive n-strand braid. We have to prove that there exists a σ-positive word representing β and containing $h_1(\beta) - 1$ letters σ_1 that is drawn in $\mathrm{Div}(b)$: as was already observed in Section 3.2, this is what Proposition V.3.3 says. Indeed, such a word is obtained by concatenating σ-positive words drawn in $\mathrm{Div}(\beta)$ and witnessing the successive transitions in the $<$-increasing enumeration of $\mathrm{Div}(\beta)$. □

Thus the results of Section 3.5 apply directly, and, in particular, they provide explicit values for $n = 3$ and 4, namely

$$c_1(\Delta_3^d) = 2 \cdot 2^d - 2 \quad \text{and} \quad c_1(\Delta_4^d) = \sum_{\pm} \frac{1}{20}(4 \pm \sqrt{6})(3 \pm \sqrt{6})^d + \frac{8}{5} \cdot 2^d - 2.$$

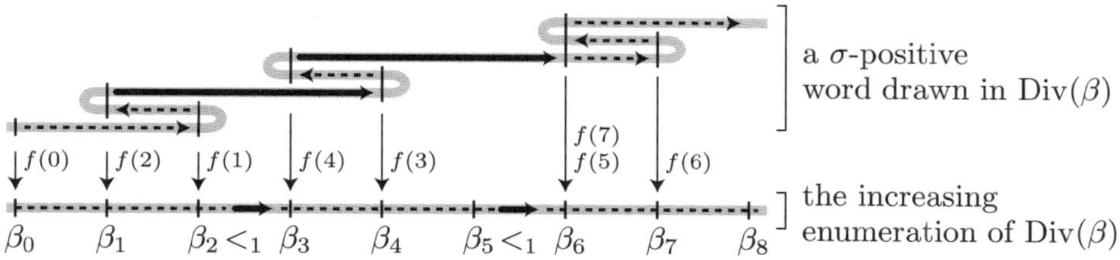

FIGURE 8. Proof of Proposition 3.16: as usual, thick arrows represent σ_1; the function f need not be increasing, but the projection of a thick arrow upstairs must include at least one thick arrow downstairs, *i.e.*, at least one σ_1-jump.

Even for 3-strand braids, $c_1(\Delta_n^d)$ grows exponentially in the degree d. So we cannot expect to obtain a polynomial complexity bound for handle reduction in this way.

REMARK 3.17. In **[61]**, the flipped version $<^{\Phi}$ of the σ-ordering is considered instead of $<$, *i.e.*, one takes into account the generator with highest index rather than that with lowest index. Then some statements take a slightly better form: for instance, $\mathrm{sh}^i(B_\infty^+)$ is replaced with B_i, and $\mathrm{sh}^i(\Delta_{n-i})$ is replaced with Δ_i. Both frameworks are equivalent, as going from one to the other just means applying the flip automorphism Φ_n.

CHAPTER VII

Alternating Decompositions

We have seen in Chapter VI that there exists no simple connection between the σ-ordering of braids and the greedy normal form of braids associated with the Garside structure of B_n^+. The aim of this chapter is to describe another normal form that, contrary to the greedy normal form, provides the expected connection, allowing us in particular to compare braids by a mere inspection of their normal forms. This normal form stems from a new idea, namely decomposing a positive braid into a product of factors alternately lying in two submonoids, as in the case of an amalgamated sum. However, it is very close in spirit to the greedy normal form, and, in particular, it is equally easy to compute.

The current approach is a continuation and an outcome of the combinatorial approach developed by Serge Burckel in his unpublished PhD thesis [**26**] and in the papers [**27, 28, 29**]. Burckel's method consists in encoding positive braid words by finite trees and introducing a well-ordering of the positive braid words using some natural ordering of the associated trees. The main result is that, if β, β' are positive braids, then $\beta <^{\Phi} \beta'$ holds if and only if the least word representing β is smaller than the least word representing β'. This enabled Burckel to prove that $(B_n^+, <)$ is order-isomorphic to the ordinal $\omega^{\omega^{n-2}}$, thus obtaining a constructive version of the well-orderability result previously established by Laver in [**137**] using the methods described in Section 4 of Chapter IV.

Burckel's remarkable work remained confidential because of its strongly syntactical flavour combined with a high combinatorial complexity. Only recently an alternative approach was discovered that bypasses the use of trees, and directly constructs Burckel's normal form of positive braids by a simple algorithm similar to the computation of the greedy normal form [**62**]. This new approach does not replace Burckel's results, but, hopefully, it helps make them more easily understandable. Another benefit of the new construction is that it extends to a wide class of monoids, in particular to the so-called dual braid monoids, as will be explained in Chapter VIII.

As for the various possible constructions of the braid ordering, the current approach leads to a proof of Properties **C** and **S**, but it seems irrelevant to proving Property **A**. On the other hand, as for properties of the braid ordering that can be proved once its existence is known, the main result is the recursive characterization of the ordering of B_n^+ from the ordering of B_{n-1}^+ stated as Proposition II.4.6. This result leads in particular to a simple comparison algorithm with proven quadratic complexity, and to Burckel's above-mentioned result about the ordinal type of the ordering of B_n^+.

The organization of the chapter is as follows. In Section 1, we describe the so-called Φ_n-splitting, a distinguished decomposition for every of B_n^+ in terms of a finite sequence of braids of B_{n-1}^+. In Section 2, we deduce from the splitting procedure both a normal form and a linear ordering of positive braids. In Section 3, we describe Burckel's notion of an irreducible braid word and, building on the results proved by Burckel, we show that the Φ-normal words of Section 2 and Burckel's irreducible words do coincide. Finally, in Section 4, we list the applications of the previous results, in particular the above-mentioned recursive definition of the ordering of B_n^+ from the ordering of B_{n-1}^+.

1. The Φ_n-splitting of a braid in B_n^+

Here we explain how to associate with every braid β of B_n^+ a finite sequence of braids of B_{n-1}^+ that unambigously specifies β. The construction is based on the easy remark that B_{n-1}^+ is a submonoid of B_n^+ that is closed under formation of least common multiples and, therefore, every braid of B_n^+ admits a unique maximal left divisor that lies in B_{n-1}^+.

1.1. Heads and tails. The key point in the construction of the greedy normal form for B_n^+ as described in Section VI.1 is the fact that, for each braid β in B_n^+, there exists a maximal simple braid β_1 that is a left divisor of β—or a right divisor of β if we consider the right greedy form of Remark VI.1.10. We shall see that, in a monoid like B_n^+ where least common multiples exist, not much is needed to guarantee the existence of such an element.

First we recall our notation. A positive braid β' is said to be a *left divisor* of β, denoted $\beta' \preccurlyeq \beta$—or, equivalently, that β is a *right multiple* of β'—if there exists a positive braid γ satisfying $\beta = \beta'\gamma$. We use $\mathrm{Div}(\beta)$ for the set of all left divisors of β. Symmetrically, β' is a *right divisor* of β if $\beta = \gamma\beta'$ holds for some positive γ. The length $\ell(\beta)$ of a positive braid β is the common length of all positive braid words representing β. As $\ell(\beta\beta') = \ell(\beta) + \ell(\beta')$ always holds, β' being a left (or a right) divisor of β implies $\ell(\beta') \leqslant \ell(\beta)$. By Proposition VI.3.10, any two braids β, β' in B_n^+ admit a unique least common right multiple, denoted $\mathrm{lcm}(\beta, \beta')$, and a unique greatest left common divisor, denoted $\mathrm{gcd}(\beta, \beta')$.

We shall start from the following basic observation.

LEMMA 1.1. *Assume that X is a subset of B_n^+ that contains 1 and is closed under right lcm (resp. under left lcm). Then, for each β in B_n^+, there exists a unique maximal left (resp. right) divisor of β that lies in X.*

PROOF. For γ in $\mathrm{Div}(\beta)$ the length of γ belongs to the finite set $\{0, 1, \dots, \ell(\beta)\}$. Hence, the set $\mathrm{Div}(\beta) \cap X$, which is nonempty as it contains at least 1, contains an element β_1 whose length is maximal. Let β' be any element of $\mathrm{Div}(\beta) \cap X$. Then, by hypothesis, $\mathrm{lcm}(\beta', \beta_1)$ lies in X, and we have $\beta_1 \preccurlyeq \mathrm{lcm}(\beta', \beta_1)$, hence $\ell(\beta_1) \leqslant \ell(\mathrm{lcm}(\beta', \beta_1))$. By the choice of β_1, we must have $\ell(\mathrm{lcm}(\beta', \beta_1)) = \ell(\beta_1)$, hence $\mathrm{lcm}(\beta', \beta_1) = \beta_1$, which means that β' is a left divisor of β_1. So β_1 is a maximal left divisor of β lying in X. Moreover, it is unique, as, if β_1' is another such element, we have $\beta_1 \preccurlyeq \beta_1' \preccurlyeq \beta_1$, hence $\beta_1 = \beta_1'$ since 1 is the only invertible element in the monoid B_n^+. □

DEFINITION 1.2. Under the hypotheses of Lemma 1.1, the maximal left divisor (*resp.* right divisor) of β that lies in X is called the *X-head* (*resp.* the *X-tail*) of β.

By definition, the simple n-strand braids of Chapter VI are both the left and the right divisors of Δ_n, so their family is closed both under right lcm and left lcm, and Lemma 1.1 is precisely what is needed to extract the first factor in the left—or in the right—greedy normal form: thus the first factor in the left greedy normal form of a positive n-strand braid is its $\mathrm{Div}(\Delta_n)$-head, while the last (rightmost) factor in its right greedy normal form is its $\mathrm{Div}(\Delta_n)$-tail.

Many other natural subsets of B_n^+ are closed under lcm and, therefore, are potentially eligible for playing the role of X in Lemma 1.1. So are in particular the so-called parabolic submonoids, *i.e.*, the submonoids generated by some subset of the set of standard Artin generators.

NOTATION 1.3. For each nonempty subset I of $\{1, \dots, n-1\}$, we denote by B_I^+ the submonoid of B_n^+ generated by $\{\sigma_i \mid i \in I\}$ and by Δ_I the right lcm of these elements.

For instance, $B^+_{\{1,\dots,n-1\}}$ is B_n^+ itself, $B^+_{\{1,\dots,n-2\}}$ is B^+_{n-1}, viewed as a submonoid of B_n^+, while $B^+_{\{2,\dots,n-1\}}$ is the image of B^+_{n-1} under the shift mapping. We recall from Chapter VI that a positive n-strand braid β is said to have degree d if β is a left divisor of Δ_n^d, but not of Δ_n^{d-1}.

LEMMA 1.4. *Assume that I is a nonempty subset of $\{1, \dots, n-1\}$.*
(i) *Every left divisor and every right divisor of an element of B_I^+ lies in B_I^+.*
(ii) *The submonoid B_I^+ is closed under left and right lcm.*
(iii) *A degree d element β of B_n^+ belongs to B_I^+ if and only if it is a left divisor of Δ_I^d, if and only if it is a right divisor of Δ_I^d.*

The proof uses standard techniques, and we skip it.

Hence Lemma 1.1 applies with $X = B_I^+$, and the notions of the B_I^+-head and the B_I^+-tail of a braid make sense. Examples are postponed until the next section.

1.2. Determining heads and tails. We now explain how the B_I^+-head and the B_I^+-tail of a braid can be recognized and computed in practice. This is an easy task.

PROPOSITION 1.5. *Assume that I is a nonempty subset of $\{1, \dots, n-1\}$. Then, for all positive braids β, β_1 in B_n^+, the following are equivalent:*
(i) *The braid β_1 is the B_I^+-head (resp. B_I^+-tail) of β;*
(ii) *We have $\beta = \beta_1\beta'$ (resp. $\beta = \beta'\beta_1$) for some braid β' that is left (resp. right) divisible by no σ_i with $i \in I$;*
(iii) *The braid β_1 is the left gcd (resp. the right gcd) of β and Δ_I^d, where d is the degree of β;*
(iv) *There exists e such that β_1 is both the left gcd (resp. the right gcd) of β and Δ_I^e, and of β and Δ_I^{e+1}.*

PROOF. Let β_1 denote the B_I^+-head of β, and β' denote the unique positive braid for which $\beta = \beta_1\beta'$ is satisfied.

First, (i) implies (ii). Indeed, if some σ_i with $i \in I$ were a left divisor of β', then $\beta_1\sigma_i$ would be a left divisor of β, contradicting the definition of the B_I^+-head.

We now claim that (ii) implies (i). Indeed, assume $\beta = \gamma_1\gamma'$ with γ' left divisible by no σ_i with $i \in I$. By Lemma 1.1, we have $\gamma_1 \preccurlyeq \beta_1$, hence $\beta_1 = \gamma_1\delta$ for some δ in B_n^+. As we have $\beta_1 \preccurlyeq \beta$, *i.e.*, $\gamma_1\delta \preccurlyeq \gamma_1\gamma'$, we deduce $\delta \preccurlyeq \gamma'$, and, therefore, every σ_i left dividing δ also left divides γ'. Owing to the hypothesis on γ', we deduce that $\sigma_i \preccurlyeq \delta$ holds for no σ_i with $i \in I$. On the other hand, we have $\beta_1 \in B_I^+$ by hypothesis, hence $\delta \in B_I^+$ as δ is a right divisor of β_1. Therefore, $\sigma_i \preccurlyeq \delta$ holds for no σ_i with $i \notin I$. Then the only possibility is $\delta = 1$, *i.e.*, $\gamma_1 = \beta_1$.

Next, we show that (i) is equivalent to (iii). Indeed, put $\beta^{(e)} = \gcd(\beta, \Delta_I^e)$ for each e. Then, we have $\Delta_I^e \in B_I^+$, hence $\beta^{(e)} \in B_I^+$, whence $\beta^{(e)} \preccurlyeq \beta_1$ by definition of β_1. On the other hand, $\beta_1 \preccurlyeq \beta$ implies that the degree of β_1 is at most d, so, by Lemma 1.4(iii), we deduce $\beta_1 \preccurlyeq \Delta_I^d$, hence $\beta_1 \preccurlyeq \beta^{(d)}$, and, finally, $\beta_1 = \beta^{(d)}$.

Then, (iii) clearly implies (iv) with e at most equal to the degree d of β.

Conversely, (iv) implies (iii). Indeed, assume $\beta^{(e)} = \beta^{(e+1)}$. By Lemma 1.4(iii), every left divisor of β lying in B_I^+ with degree $\leqslant e+1$ has degree $\leqslant e$. As B_I^+ is closed under left divisor, the latter implies that, for every $e' \geqslant e$, every left divisor of β lying in B_I^+ with degree at most $\leqslant e'$ has degree $\leqslant e$. So (iv) implies $\beta^{(e)} = \beta^{(e')}$ for each e' large enough. Then, in particular, $\beta^{(e)}$ is the limit value $\beta^{(d)}$ of (iii). □

EXAMPLE 1.6. In order to recognize that a braid β_1 of B_{n-1}^+ is the B_{n-1}^+-tail of a braid β in B_n^+, it suffices to find a decomposition $\beta = \beta'\beta_1$ such that β' is right divisible by no σ_i with $i < n-1$, hence is either 1 or is right divisible by σ_{n-1} only. Consider for instance Δ_n. By definition, we have $\Delta_n = \sigma_1\sigma_2...\sigma_{n-1}\Delta_{n-1}$. The only generator σ_i such that $\sigma_1\sigma_2...\sigma_{n-1}$ is right divisible by σ_i is σ_{n-1}. Hence, by Proposition 1.5(ii), the B_{n-1}^+-tail of Δ_n is Δ_{n-1}. Symmetrically, using the equality $\Delta_n = \Delta_{n-1}\sigma_{n-1}...\sigma_2\sigma_1$—which can be checked by an easy induction—one sees that the B_{n-1}^+-head of Δ_n is also Δ_{n-1}. More generally, the reader can check that for each $p \leqslant n$ both the B_p^+-head and the B_p^+-tail of Δ_n are Δ_p, as might be expected. Perhaps slightly less expected, the formulas

$$\Delta_n^{2d} = (\sigma_{n-1}...\sigma_2\sigma_1 \cdot \sigma_1\sigma_2...\sigma_{n-1})^d \cdot \Delta_{n-1}^{2d},$$
$$\Delta_n^{2d+1} = (\sigma_1\sigma_2...\sigma_{n-1}) \cdot (\sigma_{n-1}...\sigma_2\sigma_1\sigma_1\sigma_2...\sigma_{n-1})^d \cdot \Delta_{n-1}^{2d+1}$$

show that for each d the B_{n-1}^+-tail of Δ_n^d is Δ_{n-1}^d since the terms on the left of the power of Δ_{n-1} are braids that admit only one positive expression, and, therefore, that are right divisible by σ_{n-1} only.

It follows from Proposition 1.5(iv) that the B_I^+-head of the B_I^+-tail of a positive n-strand braid β can be computed in a way that is entirely similar to the computation of the first factor in its left or right greedy normal form: in the latter case, we compute the left or the right gcd of β and Δ_n, while, in the former, we compute the left or the right gcd of β and Δ_I, and simply repeat the operation until no more divisor is found.

COROLLARY 1.7. *For each I, the B_I^+-head and the B_I^+-tail of a braid represented by a positive n-strand braid word of length ℓ can be computed in time $O(\ell^2)$.*

PROOF. There is not even any need to compute gcd's. The only tool we need is a division algorithm $\mathcal{A}_i$ that, running on a positive braid word w, returns a positive word w' representing $\sigma_i^{-1}\overline{w}$ if the latter is a positive braid, and an error message otherwise. In connection with the so-called automatic structure provided by Garside theory [**77**, chapter 9], very simple such procedures exist, and they run in time $O(\ell(w))$. Now, to determine the B_I^+-head of a braid $\overline{w}$, it suffices to iteratively run the procedures $\mathcal{A}_i$ with $i \in I$ from w, until no more division is possible. □

1.3. Alternating decompositions. We now show how to iterate the previous construction and obtain distinguished decompositions at the expense of appealing to two—or more—parabolic submonoids of the ground monoid B_n^+.

From now on in this chapter, we privilege right divisibility, and therefore resort to tails rather than to heads. Of course, results would be similar with heads and left divisibility, but our goal of establishing a connection with the braid ordering dictates the choice of the right side.

When constructing the right greedy normal form of a braid β in B_n^+, we first extract the $\mathrm{Div}(\Delta_n)$-tail of β, thus obtaining a distinguished decomposition

$$\beta = \beta' \cdot \beta_1, \tag{1.1}$$

and then we iterate and decompose β' similarly. Replacing $\mathrm{Div}(\Delta_n)$ with B_I^+ leads to a decomposition similar to (1.1), but iterating makes no sense as, by construction, the B_I^+-tail of β' is 1, and we are stuck. The reason is that B_I^+ is closed under multiplication and, therefore, β_1 exhausts all possibilities of extracting B_I^+-factors.

Now, assume that J is another subset of $\{1, \dots, n-1\}$ and that $J \cup I$ covers $\{1, \dots, n-1\}$. By definition, the unit braid 1 is the only braid in B_n^+ whose B_I^+- and B_J^+-tails both are trivial. Therefore, in the decomposition (1.1) associated with the B_I^+-tail, only two cases are possible: either β' is 1, or it admits a nontrivial B_J^+-head, say β_2, and we obtain a new decomposition

$$\beta = \beta'' \cdot \beta_2 \cdot \beta_1, \tag{1.2}$$

in which β_2 is the B_J^+-tail of $\beta''\beta_2$ and β'' is right divisible by no σ_i with $i \in J$. We can then consider the B_I^+-tail of β'', and iterate. After finitely many steps, we shall obtain a distinguished decomposition in which the factors with odd index lie in B_I^+ while those with even index lie in B_J^+.

PROPOSITION 1.8. *If I, J are nonempty subsets covering $\{1, \dots, n-1\}$, then for every nontrivial braid β in B_n^+ there exists a unique sequence $(\beta_p, \dots, \beta_1)$ satisfying $\beta = \beta_p \dots \beta_1$ with $\beta_p \neq 1$ and such that,*
- *for odd r, we have $\beta_r \in B_I^+$ and no σ_i with $i \in I$ right divides $\beta_p \dots \beta_{r+1}$,*
- *for even r, we have $\beta_r \in B_J^+$ and no σ_i with $i \in J$ right divides $\beta_p \dots \beta_{r+1}$.*

The proof is an easy induction from Lemma 1.1, and we skip it. The termination of the iterative process in finitely many steps is clear because the length of the remainder decreases in each step until the trivial braid 1 is reached.

DEFINITION 1.9. In the framework of Proposition 1.8, the sequence $(\beta_p, \dots, \beta_1)$ is called the (B_J^+, B_I^+)-*decomposition* of the braid β.

EXAMPLE 1.10. Consider B_4^+ with $I = \{1, 2\}$ and $J = \{2, 3\}$. We look for the (J, I)-decomposition of Δ_4^2. The first step is to find the B_I^+-tail of Δ_4^2, which, as was mentioned in Example 1.6, is Δ_3^2. Writing $\mathtt{a}, \mathtt{b}$... for σ_1, σ_2... as usual, we identified the rightmost factor of the decomposition:

$\Delta_4^2 = \mathtt{cbaabc} \cdot \Delta_3^2$.

Then we look at the B_J^+-tail of the current remainder, *i.e.*, of $\mathtt{cbaabc}$: as the latter braid admits a unique representative word, its B_J^+-tail corresponds to the longest suffix that does not contain the letter $\mathtt{a}$, namely $\mathtt{bc}$. We now have the two rightmost factors of the decomposition:

$\Delta_4^2 = \mathtt{cbaa} \cdot \mathtt{bc} \cdot \Delta_3^2$.

For the same reason as above, the B_I^+-tail of the remainder, here $\mathtt{cbaa}$, corresponds to the longest suffix that does not contain the letter $\mathtt{c}$, namely $\mathtt{baa}$, and we have

$\Delta_4^2 = \mathtt{c} \cdot \mathtt{baa} \cdot \mathtt{bc} \cdot \Delta_3^2$.

The last remainder is c, an element of B_J^+, and we are done. We conclude that the $(B_{\{2,3\}}^+, B_{\{1,2\}}^+)$-decomposition of Δ_4^2 is the length 4 sequence $(\mathtt{c}, \mathtt{baa}, \mathtt{bc}, \Delta_3^2)$.

Of course, other choices for J and I give different decompositions. For instance, the reader can check that the $(B_{\{2\}}^+, B_{\{1,3\}}^+)$-decomposition of Δ_4^2 is the length 6 sequence $(\mathtt{b}, \mathtt{ca}, \mathtt{bb}, \mathtt{ca}, \mathtt{b}, \mathtt{ccaa})$.

As for the algorithmic complexity, we easily obtain

PROPOSITION 1.11. *For all I, J covering $\{1, \ldots, n-1\}$, the (B_J^+, B_I^+)-decomposition of a braid represented by a positive n-strand braid word of length ℓ can be computed in time $O(\ell^2)$.*

PROOF. *A priori*, Corollary 1.7 would imply the upper bound $O(\ell^3)$. Actually, the overall computation is quadratic as what we have to do is to repeat ℓ times the operation of determining which generators σ_i right divide the current remainder, which, at step t, is a word of length $\ell - t$. □

1.4. The Φ_n-splitting of a braid. We arrive at the main point and introduce the Φ_n-splitting of a braid, which is a distinguished decomposition of every braid of B_n^+ in terms of a finite sequence of braids of B_{n-1}^+.

We recall that Φ_n denotes the *flip* automorphism of the group B_n and of the monoid B_n^+ that maps σ_i to σ_{n-i}, thus corresponding to a symmetry in the braid diagrams.

In what follows, we shall consider the alternating decompositions associated with the covering of $\{1, \ldots, n-1\}$ by the union of $\{1, \ldots, n-2\}$ and $\{2, \ldots, n-1\}$. By definition, the monoid $B_{\{1,\ldots,n-2\}}^+$ is B_{n-1}^+, while $B_{\{2,\ldots,n-1\}}^+$ is the image of B_{n-1}^+ under the flip automorphism Φ_n, so that every element of $B_{\{2,\ldots,n-1\}}^+$ can be uniquely expressed as $\Phi_n(\beta)$ for some β of B_{n-1}^+. This enables one to specify $(\Phi_n(B_{n-1}^+), B_{n-1}^+)$-decompositions using elements of B_{n-1}^+ only.

DEFINITION 1.12. For β in B_n^+, we define the *Φ_n-splitting* of β to be the sequence $(\beta_p, \ldots, \beta_1)$ in B_{n-1}^+ such that $(\Phi_n^{p-1}(\beta_p), \ldots, \Phi_n(\beta_2), \beta_1)$ is the $(\Phi_n(B_{n-1}^+), B_{n-1}^+)$-decomposition of β. The parameter p is called the *Φ_n-breadth* of β.

An equivalent way of stating Definition 1.12 is to say that $(\beta_p, \ldots, \beta_1)$ is the Φ_n-splitting of β if β_r is the B_{n-1}^+-tail of the braid $\Phi_n^{p-r}(\beta_p) \cdot \ldots \cdot \beta_r$ for each r.

As Φ_n is involutive, $\Phi_n^r(\beta)$ just means $\Phi_n(\beta)$ for odd r, and β for even r. Figure 1 illustrates the definition. Note that for every β in B_{n-1}^+ we have

$$\Phi_n(\beta) = \mathrm{sh}(\Phi_{n-1}(\beta)), \tag{1.3}$$

hence the Φ_n-image of an $(n-1)$-strand braid is the shifted version of its Φ_{n-1}-image.

EXAMPLE 1.13. We have seen in Example 1.10 that the $(B_{\{2,3\}}^+, B_{\{1,2\}}^+)$-decomposition of Δ_4^2, *i.e.*, its $(\Phi_4(B_3^+), B_3^+)$-decomposition, is $(\mathtt{c}, \mathtt{baa}, \mathtt{bc}, \Delta_3^2)$. By applying the flip Φ_4 to every other entry, we obtain the Φ_4-splitting of Δ_4^2, namely $(\mathtt{a}, \mathtt{baa}, \mathtt{ba}, \Delta_3^2)$, or $(\sigma_1, \sigma_2\sigma_1^2, \sigma_2\sigma_1, \Delta_3^2)$.

Rephrasing Proposition 1.8, we immediately obtain

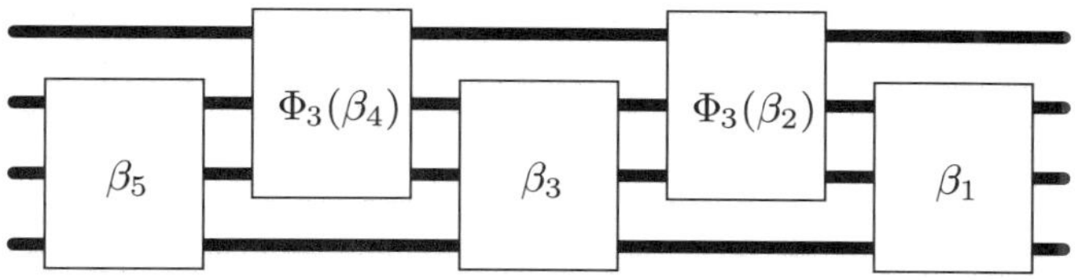

FIGURE 1. The Φ_4-splitting of a 4-strand braid of Φ_4-breadth 5: first we extract the maximal right divisor β_1 that can be expressed without σ_3, then we extract the maximal right divisor of the remainder that can be expressed without σ_1, hence the Φ_4-image of some braid β_2 of B_3^+—*i.e.*, by (1.3), the shifted version of $\Phi_3(\beta_2)$—then we iterate until the remainder is trivial.

PROPOSITION 1.14. *For each nontrivial braid β of B_n^+, the Φ_n-splitting of β is the unique sequence $(\beta_p, \dots, \beta_1)$ in B_{n-1}^+ such that $\beta = \Phi_n^{p-1}(\beta_p) \cdot \dots \cdot \Phi_n(\beta_2) \cdot \beta_1$ holds with $\beta_p \neq 1$ and, for each $r \geqslant 2$, the only σ_i right dividing $\Phi_n^{p-r}(\beta_p) \cdot \dots \cdot \beta_r$ is σ_1.*

As any (J, I)-decomposition, the Φ_n-splitting of a braid represented by a positive n-strand braid word of length ℓ can be computed in time $O(\ell^2)$.

REMARK 1.15. Instead of using the flip automorphism Φ_n, we could also appeal to the shift endomorphism: instead of seeing a braid of $B^+_{\{2,\dots,n-1\}}$ as the Φ_n-image of a braid of B_{n-1}^+, we might as well see it as the sh-image of such a braid, and obtain another distinguished decomposition—corresponding to removing the mappings Φ_3 in Figure 1. Although reasonable, this option would not induce the expected connection with the braid ordering.

2. The Φ-normal form

First, having obtained a distinguished decomposition for every braid of B_n^+ in terms of a finite sequence of braids in B_{n-1}^+, we can naturally iterate the decomposition process and, in this way, obtain for each positive braid β a distinguished word representing β. This word will be called the Φ-normal form of β. Then, we shall observe that the specific construction of Φ-normal words immediately leads to introducing for each n a certain linear ordering on B_n^+. Finally, we show that this ordering coincides with the σ-ordering in the special case $n = 3$, a result that will be subsequently extended to all values of n.

2.1. The Φ-normal form, case of three strands. In the case of 3-strand braids, the Φ_4-splitting procedure directly provides a normal form. Indeed, the monoid B_2^+ is a free monoid generated by σ_1, so every positive 2-strand braid has a unique expression as σ_1^e with e a nonnegative integer. The Φ_3-splitting associates with every positive 3-strand braid β a distinguished sequence $(\beta_p, \dots, \beta_1)$ of elements of B_2^+. By considering the unique expression σ_1^e for each factor β_r in the latter sequence, we obtain a distinguished word representing β.

First we fix some terminology. By definition, a positive 3-strand braid word is a sequence of σ_1's and σ_2's. By grouping together equal letters, we can describe it as a sequence consisting of alternating blocks of σ_1's and of σ_2's, which in turn can be unambiguously specified by the sizes of the successive blocks provided we insist that no intermediate block is empty and the last block is a block of σ_1's:

DEFINITION 2.1. If w is a nonempty positive 3-strand braid word, we define the *exponent sequence* of w to be the unique sequence of nonnegative integers $(e_p, \dots, e_1)$ such that $e_r \geqslant 1$ holds for $r \geqslant 2$ and w is the word $\sigma_{[p]}^{e_p} \dots \sigma_1^{e_3} \sigma_2^{e_2} \sigma_1^{e_1}$, where $[p]$ stands for 2 if p is even, and for 1 if p is odd.

Thus, for instance, the exponent sequence of the word $\sigma_1\sigma_2\sigma_2\sigma_1\sigma_1\sigma_1$ is $(1,2,3)$, while that of $\sigma_2\sigma_1\sigma_2$ is $(1,1,1,0)$, since the rightmost block has to be a block of σ_1's, hence an empty block in the current case.

Then we can easily introduce the normal form, called Φ-normal both to avoid confusion with the greedy normal form and to emphasize the role of the flip automorphism in the construction.

DEFINITION 2.2. Assume $\beta \in B_3^+$, and that $(\sigma_1^{e_p}, \dots, \sigma_1^{e_1})$ is the Φ_3-splitting of β. We define the *code* of β to be the sequence $(e_p, \dots, e_1)$, and its Φ*-normal form* to be the word $\sigma_{[p]}^{e_p} \dots \sigma_1^{e_3} \sigma_2^{e_2} \sigma_1^{e_1}$.

By construction, the Φ-normal form of a positive 3-strand braid β is the word obtained by concatenating the (unique word representing) the entries in its $(\Phi_3(B_2^+), B_2^+)$-decomposition.

EXAMPLE 2.3. As in Example 1.10, one sees that for even d the $(\Phi_3(B_2^+), B_2^+)$-decomposition of Δ_3^d is the length $d+2$ sequence $(\mathtt{b}, \mathtt{a}^2, \mathtt{b}^2, \dots, \mathtt{b}^2, \mathtt{a}^2, \mathtt{b}, \mathtt{a}^d)$. Hence, the Φ-normal form of Δ_3^d is the length $3d$ word $\mathtt{b}\mathtt{a}^2\mathtt{b}^2\mathtt{a}^2 \dots \mathtt{a}^2\mathtt{b}\mathtt{a}^d$, and its code is $(1, 2, \dots, 2, 1, d)$, with 2 repeated $d-1$ times. For odd d, the Φ-normal form is similar, but starting with $\mathtt{b}$, while the formula for the code remains the same.

We now give an intrinsic characterization of Φ-normal words.

PROPOSITION 2.4. *Set $e_1^{\min} = 0$, $e_2^{\min} = 1$, and $e_r^{\min} = 2$ for $r \geqslant 3$. Then a positive 3-strand braid word with exponent sequence $(e_p, \dots, e_1)$ is Φ-normal if and only if the condition $e_r \geqslant e_r^{\min}$ holds for $p > r$.*

PROOF. Assume $w = \sigma_{[p]}^{e_p} \dots \sigma_1^{e_3} \sigma_2^{e_2} \sigma_1^{e_1}$, and, for $r \geqslant 1$, let β_r be the braid represented by $\sigma_{[p-r-1]}^{e_p} \dots \sigma_2^{e_{r+1}} \sigma_1^{e_r}$, *i.e.*, up to a possible flip, the prefix of w that stops at the rth block from the right.

Assume first that $e_r \geqslant 2$ holds for $p > r \geqslant 3$, together with $e_2 \geqslant 1$ if p is at least 3. We claim that $(\sigma_1^{e_p}, \dots, \sigma_1^{e_1})$ is the $(\Phi_3(B_2^+), B_2^+)$-decomposition of the braid $\overline{w}$ represented by w. By Proposition 1.14, it suffices to check that, for $r \geqslant 2$, the only σ_i that right divides β_p is σ_1. Now, $r \geqslant 2$ implies $r+1 \geqslant 3$, hence, by hypothesis, all exponents between e_{p-1} and e_{r+1} are at least 2. Therefore β_p has only one representative word, that finishes with σ_1, so σ_1 is the only σ_i that right divides β_p, and w is the Φ-normal form of β.

Assume now that we have $p \geqslant 3$ and $e_2 = 0$. Then, by definition, $(e_p, \dots, e_1)$ is not an exponent sequence, and, therefore, w is not Φ-normal. Assume now that, for some r with $p > r \geqslant 3$, we have $e_r = 1$. Then $p > r$ implies $e_{r+1} \geqslant 1$, and $r \geqslant 3$ implies $e_{r-1} \geqslant 1$. So, by definition, β_{r-1} has an expression that finishes with $\sigma_1\sigma_2\sigma_1^{e_{r-1}}$. Now, we have $\sigma_1\sigma_2\sigma_1^e = \sigma_2^e\sigma_1\sigma_2$ for each $e \geqslant 1$. Hence β_{r-1} is right divisible by σ_2. By Proposition 1.14, this shows that the word w is not Φ-normal. ☐

In this way, we obtain a result that was alluded to in Section VI.2.5:

COROLLARY 2.5. *Every positive 3-strand braid admits a unique representative word that satisfies the conditions of Proposition* 2.4.

REMARK 2.6. The definition of the Φ-normal form of a braid looks quite different from that of its greedy normal forms. However, in the (very special) case of 3-strand braids, there happens to exist the following rather simple connection, as was first observed by J. Mairesse. Assume that β is a braid of B_3^+ whose right greedy normal form is $(\beta_d, \dots, \beta_1)$. Let e be the number of final factors Δ_3 in $(\beta_d, \dots, \beta_1)$, *i.e.*, the maximal integer e such that Δ_3^e right divides β, and let w be the word obtained by concatenating the words representing $\beta_d, \dots, \beta_{e+1}$, which are unique in this case. Displaying the last block in w (if any), let us write $w = w'\sigma_i^{e'}$. Then the Φ-normal form of β turns out to be

$$\begin{cases} w'(\sigma_2\sigma_1^2\sigma_2)^{\lfloor e/2\rfloor}\sigma_1^{e+e'} & \text{for } i = 1 \text{ and } e \text{ even,}\\ w\,\sigma_1\sigma_2(\sigma_2\sigma_1^2\sigma_2)^{\lfloor e/2\rfloor}\sigma_1^{e} & \text{for } i = 1 \text{ and } e \text{ odd,}\\ w\,(\sigma_2\sigma_1^2\sigma_2)^{\lfloor e/2\rfloor}\sigma_1^{e} & \text{for } i = 2 \text{ and } e \text{ even,}\\ w'\sigma_1\sigma_2(\sigma_2\sigma_1^2\sigma_2)^{\lfloor e/2\rfloor}\sigma_1^{e+e'} & \text{for } i = 2 \text{ and } e \text{ odd.}\end{cases}$$

This shows that the Φ-normal word of a 3-strand braid β can be obtained from its greedy normal form by a simple transformation: one replaces the final e entries Δ_3 by the Φ-normal form of Δ_3^e, namely $(\sigma_2\sigma_1^2\sigma_2)^{\lfloor e/2\rfloor}\sigma_1^e$ or $\sigma_1\sigma_2(\sigma_2\sigma_1^2\sigma_2)^{\lfloor e/2\rfloor}\sigma_1^e$, according to the parity of e, and, in some cases, one pushes through this block the letters σ_1 that lie immediately on the left.

2.2. The Φ-normal form, general case. For β in B_4^+, the Φ_4-splitting provides a distinguished decomposition for β in terms of braids of B_3^+, but not yet a distinguished word expression, as a braid of B_3^+ admits in general several word expressions. Now, we defined in Section 2.1 such a distinguished word, namely the Φ-normal form. We deduce a distinguished word representing β by using the Φ-normal forms of the entries in its Φ_4-splitting. Of course, the construction iteratively extends to every braid index n.

EXAMPLE 2.7. We saw in Example 1.13 that the Φ_4-splitting of Δ_4^2 is the sequence $(\mathtt{a}, \mathtt{baa}, \mathtt{ba}, \Delta_3^2)$, corresponding to the decomposition

$$\Delta_4^2 = \Phi_4(\mathtt{a}) \cdot \mathtt{ba}^2 \cdot \Phi_4(\mathtt{ba}) \cdot \Delta_3^2$$

(see Figure 3). The Φ-normal forms of the four entries in this sequence, respectively, are $\mathtt{a}$, $\mathtt{ba}^2$, $\mathtt{ba}$, and $\mathtt{ba}^2\mathtt{ba}^2$. So we obtained a distinguished word representing Δ_4^2 by concatenating these words after applying the needed flips, *i.e.*, in the current case, $\mathtt{c} \cdot \mathtt{ba}^2 \cdot \mathtt{bc} \cdot \mathtt{ba}^2\mathtt{ba}^2$, or $\sigma_3\sigma_2\sigma_1^2\sigma_2\sigma_3\sigma_2\sigma_1^2\sigma_2\sigma_1^2$. The latter word will naturally be defined to be the Φ-normal form of Δ_4^2.

DEFINITION 2.8. For β in B_n^+, we define the Φ*-normal form* of β to be
• for $n = 2$, the unique word representing β;
• for $n \geqslant 3$, the word $\Phi_n^{p-1}(w_p) \cdot \ldots \cdot \Phi_n(w_2) \cdot w_1$, where $(\beta_p, \dots, \beta_1)$ is the Φ_n-splitting of β and w_r is the Φ-normal form of β_r for each r.

By construction, every braid of B_n^+ has a well-defined, unique Φ-normal form. Its computation is easy:

PROPOSITION 2.9. *For each fixed n and each positive n-strand braid word w of length ℓ, the Φ-normal form of the braid $\overline{w}$ can be computed in time $O(\ell^2)$.*

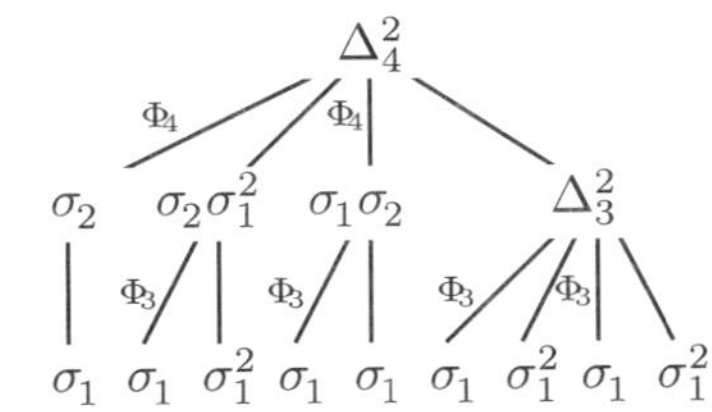

After reintroducing the flips:

$$\sigma_3 \ \sigma_2 \ \sigma_1^2 \ \sigma_2 \ \sigma_3 \ \sigma_2 \ \sigma_1^2 \ \sigma_2 \ \sigma_1^2$$

FIGURE 2. Two-step construction of the Φ-normal form of Δ_4^2: we first split Δ_4^2 into a sequence of 3-strand braids, then split each of them into a sequence of 2-strand braids, *i.e.*, of powers of σ_1's; taking the flips into account, we obtain a distinguished word representing Δ_4^2.

PROOF. We use induction on n. By Proposition 1.11, we can first compute the Φ_n-splitting of $\overline{w}$ in time $O(\ell^2)$, obtaining a sequence of $(n-1)$-strand braid words $(w_p, \dots, w_1)$ satisfying $\ell(w_1) + \dots + \ell(w_p) = \ell$. Then, by induction hypothesis, the Φ-normal forms of $w_p, \dots, w_1$ can be computed in time $O(\ell(w_1)^2 + \dots + \ell(w_p)^2)$, hence in time $O(\ell^2)$. $\square$

The definition of the Φ-normal form given above is explicit and simple, but it appeals to a recursion, and it is natural to ask for a direct definition. Such a definition exists, but, as we shall not use it explicitly, we only give a sketchy description. The idea is that a word w is Φ-normal if and only if each letter in w is the smallest σ_i—in a sense that will be described below—that right divides the braid represented by the prefix that ends at that letter. For instance, in the framework of Example 2.7, σ_1 right divides Δ_4^2, and, therefore, the Φ-normal form of Δ_4^2 will finish with σ_1. After dividing by that σ_1, the remainder can still be divided by σ_1, so the second letter from the right in the Φ-normal form is a σ_1 again. Then, the remainder is not right divisible by σ_1, but it is right divisible by σ_2, so the third letter from the right in the Φ-normal form will be σ_2. If we were to continue in this way, we would obtain the lexicographically minimal word representing the considered braid. Actually, the construction is more subtle. At each step, we indeed look at the smallest σ_i that right divides the current remainder, but smallest here refers to an ordering that need not always be the standard ordering $\sigma_1 < \dots < \sigma_{n-1}$: the latter is the initial ordering, but then the ordering is updated at each step, due to the underlying flips that reverse orientations. One description is as follows:

PROPOSITION 2.10. *A positive n-strand braid word $\sigma_{i_\ell} \dots \sigma_{i_1}$ is Φ-normal if and only if there exist linear orderings $<_0, \dots, <_{\ell-1}$ of $\{1, \dots, n-1\}$ such that $<_0$ is the standard ordering $1 < \dots < n-1$ and, for each r,*

- *i_r is the $<_{r-1}$-smallest i such that σ_i is a right divisor of $\overline{\sigma_{i_\ell} \dots \sigma_{i_r}}$;*
- *$<_r$ is obtained from $<_{r-1}$ as follows: assume that i_r is the kth element in the increasing enumeration of $<_{r-1}$; then the increasing enumeration of $<_r$ consists of i_r, followed by the first $k-1$ elements of $<_{r-1}$ enumerated in standard increasing order if they are larger than i_r (in the standard order), and in standard decreasing order if they are smaller, followed by the $n-1-k$ remaining elements, which keep their position of $<_{r-1}$.*

The previous statement makes sense because the integers that occur before i_r in $<_{r-1}$ must all be smaller, or all larger, than i_r, as an easy induction shows.

EXAMPLE 2.11. Assume $3 <_{r-1} 2 <_{r-1} 1 <_{r-1} 4$ and we found $i_r = 1$; then $<_r$ is 1 first, then $2 < 3$, as 2 and 3 precede 1 in $<_{r-1}$ and are larger than 1, then 4, resulting in $1 <_r 2 <_r 3 <_r 4$. For $2 <_{r-1} 1 <_{r-1} 3 <_{r-1} 4$ and $i_r = 4$, we obtain $4 <_r 3 <_r 2 <_r 1$, as all numbers that precede 4 are smaller than 4.

REMARK 2.12. Like the greedy normal form, the Φ-normal form comes in two versions, left or right, according to whether we privilege the head or the tail of the considered braid. Actually, it comes in four versions because, once one side is chosen, for instance the right—tail—side as above, we still can choose to consider the B_{n-1}^+- or the $\Phi_n(B_{n-1}^+)$-tail first. Above we have chosen the former, thus corresponding to what can be called the *lower right* version. Of course, relations exist: for β in B_n^+, the lower right Φ-normal form of β is the Φ_n-image of the upper right Φ-normal form of $\Phi_n(\beta)$.

2.3. A linear ordering on positive braids. We shall now introduce a linear ordering on the monoid B_n^+ that directly derives from the Φ_n-splitting. We shall see in Section 4 that this ordering coincides with the standard braid ordering $<^{\Phi}$, but, for the moment, we leave the question open.

As the monoid B_2^+ is free, hence isomorphic to the monoid of natural numbers, it admits a canonical ordering, namely the ordering defined by $\sigma_1^e < \sigma_1^{e'}$ if and only if $e < e'$ holds.

When a linear ordering $\prec$ on a set Ω is given, there exists several natural ways to extend it into a linear ordering of the set of all finite sequences from Ω. We shall consider the following one; terminology is borrowed from [**77**].

DEFINITION 2.13. If $\prec$ is a linear ordering on Ω, and $\boldsymbol{x}, \boldsymbol{x}'$ are finite sequences in Ω, we say that $\boldsymbol{x}$ is `ShortLex`-*smaller* than $\boldsymbol{x}'$ if the length of $\boldsymbol{x}$ is strictly smaller than the length of $\boldsymbol{x}'$, or the lengths are equal and $\boldsymbol{x}$ is smaller than $\boldsymbol{x}'$ for the lexicographical extension of $\prec$, *i.e.*, at the first discrepancy between $\boldsymbol{x}$ and $\boldsymbol{x}'$ starting from the left, the $\boldsymbol{x}$-entry is $\prec$-smaller than the $\boldsymbol{x}'$-entry.

The `ShortLex`-extension is a variant of the lexicographical extension in which the length is given an overall priority. Notice that, when comparing lexicographically, we start from the left—and we shall always do so, although we work with splittings that are constructed from the right. What we shall do now should be obvious: starting with the standard ordering of B_2^+, we shall extend it to B_3^+, B_4^+, etc., using the Φ_n-splitting.

DEFINITION 2.14. For $n \geqslant 2$, we recursively define $<_n^+$ on B_n^+ as follows:

• For β, β' in B_2^+, we declare that $\beta <_2^+ \beta'$ holds if we have $\beta = \sigma_1^e$ and $\beta' = \sigma_1^{e'}$ with $e < e'$;

• For β, β' in B_n with $n \geqslant 3$, we declare that $\beta <_n^+ \beta'$ holds if the Φ_n-splitting of β is smaller than the Φ_n-splitting of β' for the `ShortLex`-extension of $<_{n-1}^+$.

PROPOSITION 2.15. (i) *For $n \geqslant 2$, the relation $<_n^+$ is a well-ordering of B_n^+. For each β in B_n^+, the immediate $<_n^+$-successor of β is $\beta\sigma_1$.*

(ii) *For $n \geqslant 3$, the ordering $<_n^+$ extends the ordering $<_{n-1}^+$, and B_{n-1}^+ is the initial segment of B_n^+ determined by σ_{n-1}, i.e., $B_{n-1}^+ = \{\beta \in B_n^+ \mid \beta <_n^+ \sigma_{n-1}\}$ holds.*

PROOF. (i) That $<^+_n$ is a strict linear ordering follows from an immediate induction on $n \geqslant 2$, and so does the stronger property that it is a well-ordering.

Let β be any braid in B^+_n. Let $(\beta_p, \dots, \beta_1)$ be the Φ_n-splitting of β. Then the Φ_n-splitting of $\beta\sigma_1$ is $(\beta_p, \dots, \beta_1\sigma_1)$. So, if, by induction hypothesis, $\beta_1\sigma_1$ is the immediate $<^+_{n-1}$-successor for β_1, then, by definition of the `ShortLex`-extension, $\beta\sigma_1$ is the immediate $<^+_n$ successor of β.

(ii) For β, β' in B^+_{n-1}, the Φ_n-splittings of β and β' are the length 1 sequences (β) and (β'), so, by definition, $\beta <^+_n \beta'$ is equivalent to $\beta <^+_{n-1} \beta'$. On the other hand, the Φ_n-splitting of σ_{n-1} is $(\sigma_1, 1)$, so $\beta <^+_n \sigma_{n-1}$ holds for each β in B^+_{n-1}. Conversely, assume $\beta \in B^+_n$ and $\beta <^+_n \sigma_{n-1}$. By construction, if (β_2, β_1) is a Φ_n-splitting, β_2 is not 1. By (i), σ_1 is the immediate successor of 1 in $(B^+_{n-1}, <^+)$, hence we have $\sigma_1 \leqslant^+_{n-1} \beta_2$. As the Φ_n-splitting of σ_{n-1} is $(\sigma_1, 1)$, we deduce $\sigma_{n-1} \leqslant^+_n \beta$. So $\beta <^+_n \sigma_{n-1}$ implies that the length of the Φ_n-splitting is 1 at most, hence that β belongs to B^+_{n-1}. □

Owing to Proposition 2.15(ii), we shall skip the index n and write $<^+$ for $<^+_n$.

EXAMPLE 2.16. The Φ_4-splittings of σ_1 and σ_2, respectively, are (σ_1) and $(\sigma_1, 1)$. The first sequence is shorter, hence we have $\sigma_1 <^+ \sigma_2$.

On the other hand, the Φ_4-splittings of $\sigma_1\sigma_2\sigma_1^4$ and $\sigma_1\sigma_2^2$ are $(\sigma_1, \sigma_1, \sigma_1^4)$ and $(\sigma_1, \sigma_1^2, 1)$, both of length 3. Hence we compare lexicographically. The first entries coincide, but $\sigma_1 <^+ \sigma_1^2$ holds, hence we have $\sigma_1\sigma_2\sigma_1^4 <^+ \sigma_1\sigma_2^2$.

Proposition 1.11 easily implies

PROPOSITION 2.17. *For each fixed n, comparing with respect to $<^+$ two braids represented by n-strand braid words of length at most ℓ can be done in time $O(\ell^2)$.*

Indeed, finding the Φ_n-splittings can be done in quadratic time, and, by induction hypothesis, so does the lexicographic comparison of their entries, if needed. So, at this point, we have an excellent ordering of positive braids, whose existence and properties so far have been very easy to check. Everything is not so simple.

QUESTION 2.18. *Is the ordering $<^+$ compatible with multiplication on the left?*

We shall see in Section 4 that the answer to Question 2.18 turns out to be positive, but no direct proof is known: the only, rather indirect proof known so far consists in showing that $<^+$ coincides with the standard ordering $<^\Phi$ by using Burckel's results described in Section 3.

2.4. The case of B_3. In the case of 3-strand braids, answering Question 2.18 and proving that the linear orderings $<^+$ and $<^\Phi$ coincide is an easy task, and we shall do it now. The technical result we prove is the following one.

PROPOSITION 2.19. *For β, β' in B^+_3, the relation $\beta <^+ \beta'$ implies that $\beta^{-1}\beta'$ is σ^Φ-positive.*

We recall that a braid word is said to be σ^Φ-positive if the generator σ_i with highest index occurs positively only. Proposition 2.19 follows from a direct computation in which an important role is played by certain special braids that already appeared in Chapter VI and which are closely connected with the Φ_3-splitting of Δ_3^d, and, more generally, with the Φ_n-splitting of Δ_n^d.

DEFINITION 2.20 (See Figure 3). For $n \geqslant 2$ and $d \geqslant -1$, we define $\widehat{\Delta}_{n,d}$ by $\widehat{\Delta}_{n,-1} = 1$, $\widehat{\Delta}_{n,0} = \sigma_{n-1}$ and, for $d \geqslant 1$,

$$\widehat{\Delta}_{n,d} = \begin{cases} (\sigma_{n-1} \ldots \sigma_2 \sigma_1 \cdot \sigma_1 \sigma_2 \ldots \sigma_{n-1})^e & \text{for } d = 2e, \\ (\sigma_1 \sigma_2 \ldots \sigma_{n-1}) \cdot (\sigma_{n-1} \ldots \sigma_2 \sigma_1 \cdot \sigma_1 \sigma_2 \ldots \sigma_{n-1})^e & \text{for } d = 2e+1. \end{cases}$$

FIGURE 3. The braids $\widehat{\Delta}_{3,4}$ (left) and $\widehat{\Delta}_{4,3}$ (right): starting from the right, the upper strand forms d half-twists around all other strands.

An easy induction gives, for all $n \geqslant 2$ and $d \geqslant 1$, the relation

$$\Delta_n^d = \widehat{\Delta}_{n,d} \cdot \Delta_{n-1}^d. \tag{2.1}$$

So $\widehat{\Delta}_{n,d}$ is the remainder of Δ_n^d when its B_{n-1}^+-tail, namely Δ_{n-1}^d, is removed.

The idea now is to consider the increasing sequence

$$1 = \widehat{\Delta}_{3,-1} <^{\Phi} \widehat{\Delta}_{3,0} <^{\Phi} \widehat{\Delta}_{3,1} <^{\Phi} \widehat{\Delta}_{3,2} <^{\Phi} \ldots$$

(that this sequence is $<^{\Phi}$-increasing will be clear from the next result) and, given a positive braid β, to look where β is situated relative to this sequence. The answer is illustrated in Figure 4, and established below.

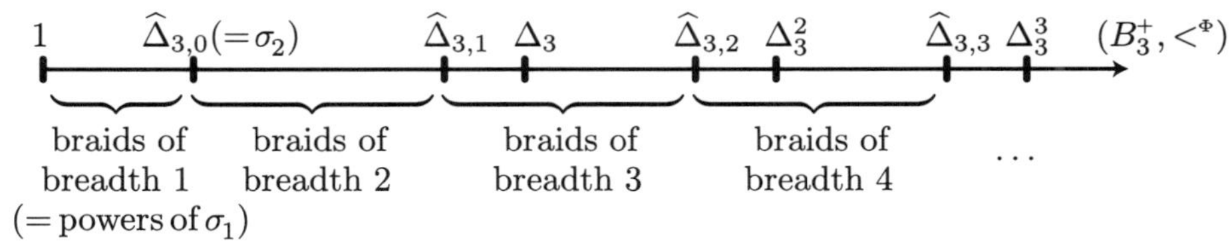

FIGURE 4. The braids $\widehat{\Delta}_{3,d}$ as separators in the ordered line $(B_3^+, <^{\Phi})$: every positive 3-strand braid with Φ_3-breadth p lies between $\widehat{\Delta}_{3,p-2}$ and $\widehat{\Delta}_{3,p-1}$.

LEMMA 2.21. *For every positive 3-strand braid β with Φ_3-breadth p, we have*

$$\widehat{\Delta}_{3,p-2} \leqslant^{\Phi} \beta <^{\Phi} \widehat{\Delta}_{3,p-1}. \tag{2.2}$$

PROOF. Let $\sigma_{[p]}^{e_p} \ldots \sigma_2^{e_2} \sigma_1^{e_1}$ be the Φ-normal form of β. The case $p = 1$ is special: by definition, $\beta <^{\Phi} \sigma_{n-1}$ holds for each β in B_{n-1}^+, and, by Property **S**, the inequality $\sigma_{n-1} \leqslant^{\Phi} \beta$ holds for each braid β of $B_n^+ \setminus B_{n-1}^+$. We assume now $p \geqslant 2$. We have to prove two results, namely that

(i) the braid $\widehat{\Delta}_{3,p-2}^{-1} \beta$ is σ^{Φ}-positive or trivial, and

(ii) the braid $\beta^{-1} \widehat{\Delta}_{3,p-1}$ is σ^{Φ}-positive.

We begin with (ii), which is easier. Using (2.1), we obtain

$$\beta^{-1} \cdot \widehat{\Delta}_{3,p-1} = \sigma_1^{-e_1} \sigma_2^{-e_2} \ldots \sigma_{[p]}^{-e_p} \cdot \Delta_3^{p-1} \cdot \sigma_1^{-p+1}. \tag{2.3}$$

The braid relations imply $\sigma_i \cdot \Delta_3 = \Delta_3 \cdot \Phi_3(\sigma_i)$ for $i = 1, 2$. So we can push the $p-1$ factors Δ_3 of (2.3) to the left, at the expense of applying Φ_3. In this way, we deduce

$$\beta^{-1} \cdot \widehat{\Delta}_{3,p-1} = \sigma_1^{-e_1} \cdot \Delta_3 \cdot \sigma_1^{-e_2} \cdot \Delta_3 \cdot \ldots \cdot \Delta_3 \cdot \sigma_1^{-e_p} \cdot \sigma_1^{-p+1},$$

whence, using $\Delta_3 = \sigma_1\sigma_2\sigma_1$,

$$\beta^{-1} \cdot \widehat{\Delta}_{3,p-1} = \sigma_1^{-e_1+1}\, \sigma_2\, \sigma_1^{-e_2+2}\, \sigma_2 \ldots \sigma_2\, \sigma_1^{-e_r+2}\, \sigma_2\, \sigma_1^{-e_p}\, \sigma_1^{-p+1} :$$

the generator σ_2 occurs $p-1$ times in the latter decomposition, while σ_2^{-1} does not occur, and we have obtained a σ^{Φ}-positive word representing $\beta^{-1}\, \widehat{\Delta}_{3,p-1}$.

We turn to (i). Computing as above, we have

$$\widehat{\Delta}_{3,p-2}^{-1} \cdot \beta = \sigma_1^{p-2} \cdot \Delta_3^{-p-2} \cdot \sigma_{[p]}^{e_p} \ldots \sigma_2^{e_2}\, \sigma_1^{e_1}.$$

Pushing the $p-2$ factors Δ_3^{-1} to the right, and using the relation $\Delta_3^{-1} = \sigma_2^{-1}\sigma_1^{-1}\sigma_2^{-1}$, we deduce

$$\begin{aligned}
\widehat{\Delta}_{3,p-2}^{-1} \cdot \beta &= \sigma_1^{p-2} \cdot \sigma_2^{e_p} \cdot \Delta_3^{-1} \cdot \sigma_2^{e_{p-1}} \cdot \Delta_3^{-1} \cdot \ldots \cdot \Delta_3^{-1} \cdot \sigma_2^{e_2}\, \sigma_1^{e_1} \\
&= \sigma_1^{p-2} \cdot \sigma_2^{e_p-1}\, \sigma_1^{-1}\, \sigma_2^{e_{p-1}-2}\, \sigma_1^{-1} \ldots \sigma_1^{-1}\, \sigma_2^{e_3-2}\, \sigma_1^{-1}\, \sigma_2^{e_2-1}\, \sigma_1^{e_1}.
\end{aligned} \tag{2.4}$$

By Proposition 2.4, the hypothesis that $\sigma_{[p]}^{e_p} \ldots \sigma_2^{e_2}\, \sigma_1^{e_1}$ is Φ-normal implies the inequalities $e_p \geqslant 1$, $e_p \geqslant 2$, ..., $e_3 \geqslant 2$, $e_2 \geqslant 1$, and, therefore, when the expression (2.4) is expanded, no letter σ_2^{-1} occurs. Then three cases are possible. If at least one of the above inequalities is strict, the letter σ_2 occurs in the word deduced from (2.4), and we have obtained a $\sigma^{\Phi}{}_2$-positive word representing $\widehat{\Delta}_{3,p-2}^{-1}\, \beta$. Otherwise, what remains from (2.4) is the equality

$$\widehat{\Delta}_{3,p-2}^{-1}\, \beta = \sigma_1^{p-2} \cdot \sigma_1^{-p+2+e_1} = \sigma_1^{e_1}. \tag{2.5}$$

If e_1 is positive, we have a $\sigma^{\Phi}{}_1$-positive word representing $\widehat{\Delta}_{3,p-2}^{-1}\, \beta$. Finally, $e_1 = 0$ corresponds to $\beta = \widehat{\Delta}_{3,p-2}$. □

Proof of Proposition 2.19. Assume $\beta, \beta' \in B_3^+$ with $\beta <^+ \beta'$. Let p and p' be the respective Φ_3-breadths of β and β'. Two cases are possible. If $p < p'$ holds, Lemma 2.21 shows that $\beta^{-1}\widehat{\Delta}_{3,p+2}$ is $\sigma^{\Phi}{}_2$-positive, and that $\widehat{\Delta}_{3,p+2}^{-1}\beta'$ is σ^{Φ}-positive or trivial. Therefore, $\beta^{-1}\beta'$ is $\sigma^{\Phi}{}_2$-positive.

Assume now $p' = p$. Let $(e_p, \ldots, e_1)$ be the code of β, and $(e'_p, \ldots, e'_1)$ be that of β'. By hypothesis, there exists q satisfying $e'_r = e_r$ for $r > q$, and $e'_q > e_q$. Let

$$\beta_1 = \sigma_{[q-1]}^{e_{q-1}} \cdot \ldots \cdot \sigma_1^{e_1} \quad \text{and} \quad \beta'_1 = \sigma_{[q]}^{e'_q - e_q} \cdot \sigma_{[q-1]}^{e'_{q-1}} \cdot \ldots \cdot \sigma_1^{e'_1}.$$

By hypothesis, we have $\beta = \gamma\beta_1$ and $\beta' = \gamma\beta'_1$ for some braid γ, and, therefore, $\beta^{-1}\beta' = \beta_1^{-1}\beta'_1$. Now, by construction, the code of β_1 is $(e_{q-1}, \ldots, e_1)$, and that of β'_1 is $(e'_q - e_q, e'_{q-1}, \ldots, e'_1)$, as truncating an exponent sequence on the left preserves the normality conditions. Hence β_1 has breadth $q-1$, while β'_1 has breadth q. Then Lemma 2.21 gives a σ^{Φ}-positive word representing $\beta_1^{-1}\beta'_1$, hence for $\beta^{-1}\beta$. □

Proposition 2.19 has many direct consequences, such as a new proof of Property **C** for 3-strand braids. As most results will be generalized in Section 4, we do not develop them here. However, we mention one application that has no direct extension in the case $n \geqslant 4$, namely a direct computation for the rank of a 3-strand braid in the well-ordering $(B_3^+, <^{\Phi})$.

COROLLARY 2.22. *The restriction of $<^{\Phi}$ to B_3^+ is a well-ordering of ordinal type ω^ω. For every β in B_3^+, the rank of β in the $<^{\Phi}$-increasing enumeration of B_3^+ is the ordinal*

$$\omega^{p-1} \cdot e_p + \sum_{p>r\geqslant 1} \omega^{r-1} \cdot (e_r - e_r^{\min}), \tag{2.6}$$

where $(e_p, \dots, e_1)$ is the code of β.

PROOF. Proposition 2.19 shows that, as a set of pairs, the ordering $<^+$ is included in the ordering $<^{\Phi}$. As $<^{\Phi}$ is a linear ordering, this implies that $<^+$ and $<^{\Phi}$ coincide on B_3^+. Hence, by Proposition 2.15(i), $(B_3^+, <^{\Phi})$, which is also $(B_3^+, <^+)$, is a well-ordering.

Assume that β is a braid with code $(e_p, \dots, e_1)$. Then the rank of β in $(B_3^+, <^{\Phi})$ and $(B_3^+, <^+)$ is the ordinal type of the family of all $<^+$-predecessors of β. There exists a one-to-one correspondence between positive 3-strand braids and their codes, and, by construction, the ordering $<^+$ on braids corresponds to the `ShortLex`-extension of the standard ordering of integers. Thus we look for the ordinal type of the set of codes that are `ShortLex`-smaller than $(e_p, \dots, e_1)$. By Proposition 2.4, a sequence $(e'_{p'}, \dots, e'_1)$ is a code if and only if it satisfies the inequalities $e'_r \geqslant e_r^{\min}$ for $r < p'$. Let F be the function defined on nonempty codes by

$$F((e'_{p'}, \dots, e'_1)) = (e'_{p'}, e'_{p'-1} - e_{p'-1}^{\min}, \dots, e'_1 - e_1^{\min}).$$

Let $(d_p, \dots, d_1) = F((e_p, \dots, e_1))$. Then F establishes an isomorphism between the codes below $(e_p, \dots, e_1)$ and the sequences of natural numbers $(d_{p'}, \dots, d'_1)$ satisfying $d_{p'} > 0$ below $(d_p, \dots, d_1)$, both equipped with the `ShortLex`-extension of the standard ordering of integers. It is then standard that the ordinal type of the latter set is the ordinal $\sum_{p\geqslant r} \omega^{r-1} \cdot d_r$.

Finally, the ordinal type of $(B_3^+, <^{\Phi})$ is the supremum of the ranks of its elements, hence it is the ordinal ω^ω. □

For instance, we saw in Example 2.3 that the code of Δ_3^d is the length $d+2$ sequence $(1, 2, \dots, 2, 1, d)$. We deduce that, in the well-ordering $<^{\Phi}$ of B_3^+, the rank of Δ_3^d is the ordinal $\omega^{d+1} + d$. Similarly, the rank of $\widehat{\Delta}_{3,d}$ is ω^{d+1}, as its code is the length $d+2$ sequence $(1, 2, \dots, 2, 1, 0)$. We refer to Figure II.4 for more examples.

3. Burckel's approach

Proposition 2.19 extends to arbitrary positive braids. However, the argument used for the upper bound result in Lemma 2.21 remains valid, but the one for the lower bound becomes problematic, and a direct proof, which is likely to exist, remains to be found. Instead, we shall appeal to the prior approach developed by S. Burckel in **[26]** and describe his results. In order to make the description more easily compatible with the previous sections, we shall not follow Burckel's language exactly, but it would be easy to see that the description given below is equivalent to the one of **[26]**.

3.1. Splitting braid words. Hereafter, we use $\underline{B}_n^+$ for the monoid of all positive n-strand braid words. By definition, $\underline{B}_n^+$ is a free monoid, hence quite different from its quotient the braid monoid B_n^+. However, we observe that most of the notions introduced in Section 1 make sense in the case of the monoid $\underline{B}_n^+$.

DEFINITION 3.1. (i) For w a word in $\underline{B}_n^+$ and $I \subseteq \{1, \ldots, n-1\}$, we define the $\underline{B}_I^+$*-tail* of w to be the maximal suffix of w that only contains letters σ_i with $i \in I$.

(ii) For w a word in $\underline{B}_n^+$, we define the Φ_n*-splitting* of w to be the sequence $(w_p, \ldots, w_1)$ in $\underline{B}_{n-1}^+$ such that, for each r, the word $\Phi_n^{r-1}(w_r)$ is the $\Phi_n^{r-1}(\underline{B}_{n-1}^+)$-tail of $\Phi_n^{p-1}(w_p) \cdot \ldots \cdot \Phi_n^{r-1}(w_r)$; the number p is called the Φ_n*-breadth* of w.

(iii) For $n \geqslant 2$, we recursively define a relation $\sqsubset_n$ on $\underline{B}_n^+$ as follows:

• For w, w' in $\underline{B}_2^+$, we declare that $w \sqsubset_2 w'$ holds if we have $w = \sigma_1^e$ and $w' = \sigma_1^{e'}$ with $e < e'$;

• For w, w' in $\underline{B}_n^+$ with $n \geqslant 3$, we declare that $w \sqsubset_n w'$ holds if the Φ_n-splitting of w is smaller than the Φ_n-splitting of w' for the ShortLex-extension of $\sqsubset_{n-1}$.

Thus, as in Section 1, we consider maximal right divisors, but, as we are in a free monoid, right divisor simply means suffix. The results of Proposition 2.15 remain true in the context of words—and, in this case, they are trivial as one works in a free monoid.

PROPOSITION 3.2. (i) *For $n \geqslant 2$, the relation $\sqsubset_n$ is a strict well-ordering of $\underline{B}_n^+$ of ordinal type $\omega^{\omega^{n-2}}$. For each word w, the immediate $\sqsubset_n$-successor of w is $w\sigma_1$.*

(ii) *For $n \geqslant 3$, the ordering $\sqsubset_n$ extends $\sqsubset_{n-1}$, and $\underline{B}_{n-1}^+$ is the initial segment of $\underline{B}_n^+$ determined by σ_{n-1}, i.e., we have $\underline{B}_{n-1}^+ = \{w \in \underline{B}_n^+ \mid w \sqsubset_n \sigma_{n-1}\}$.*

Owing to (ii), we shall drop the index n in $\sqsubset_n$ on $\underline{B}_n^+$.

EXAMPLE 3.3. Let w be the word cbaabcbaabaa, in which we recognize the Φ-normal word representing Δ_4^2. The longest suffix of w that lies in $\underline{B}_3^+$ is baabaa, and the remaining prefix is cbaabc, *i.e.*, $\Phi_4(w')$ with $w' =$ abccba. The longest suffix of w' that does not contain c is ba, with remaining prefix abcc, *i.e.*, $\Phi_4(w'')$ with $w'' =$ cbaa, etc. Finally, the Φ_4-splitting of w is the sequence of words (a , baa , ba , baabaa).

Consider now the word $\Phi_4(w)$, *i.e.*, abccbabccbcc, which is another word representing Δ_4^2, since $\Phi_4(\Delta_4^2) = \Delta_4^2$ holds. Arguing as above, one sees that the Φ_4-splitting of $\Phi_4(w)$ is the length 5 sequence (a , baa , ba , baabaa , ε). Thus, by definition, we have $w \sqsubset \Phi_4(w)$ in this case.

3.2. Burckel's braid word reduction. Burckel's method consists in considering the words that are $\sqsubset$-minimal in their equivalence class—as usual, two braid words are said to be equivalent if they represent the same braid. We recall that $\overline{w}$ denotes the braid represented by the word w.

DEFINITION 3.4. A positive braid word w is said to be *normal in the sense of Burckel*, or *Burckel normal*, if w is $\sqsubset$-minimal among all words representing $\overline{w}$.

LEMMA 3.5. *Every positive braid is represented by a unique Burckel normal word.*

PROOF. As $\sqsubset$ is a well-ordering on $\underline{B}_n^+$, each nonempty subset of $\underline{B}_n^+$ admits a unique $\sqsubset$-minimal element. So does in particular every equivalence class. □

At this point, how to practically compute the Burckel normal word representing a braid is not clear. What Burckel does in [**27**] is to identify a certain reduction procedure that, starting with a positive braid word w, either says that w is Burckel normal, or returns an equivalent braid word red(w) satisfying red(w) $\sqsubset w$. The definition of reduction is simple in the case of $\underline{B}_3^+$, but very intricate in the general case.

For the convenience of the reader, we shall first concentrate on the case of 3 strands, which is simple and can be explained easily. The indications about the general case are postponed to Section 3.4.

DEFINITION 3.6. Assume that w is a positive 3-strand braid word with exponent sequence $(e_p, \dots, e_1)$. Then w is said to be *reducible* if one has $e_r = 1$ for some r satisfying $p > r \geqslant 3$; otherwise, w is said to be *irreducible*. If w is reducible and r is minimal satisfying $e_r = 1$ with $p > r \geqslant 3$, then $\mathrm{red}(w)$ is defined to be the word with exponent sequence

$$(e_p, \dots, e_{r+2}, e_{r+1} - 1, e_{r-1}, 1, e_{r-2} + 1, e_{r-3}, \dots, e_1) \quad \text{if } e_{r+1} \geqslant 2 \text{ holds,} \tag{3.1}$$

$$(e_p, \dots, e_{r+2} + e_{r-1}, 1, e_{r-2} + 1, e_{r-3}, \dots, e_1) \quad \text{if } e_{r+1} = 1 \text{ holds.} \tag{3.2}$$

By definition, a word w is reducible if and only if it contains the pattern $\sigma_1\sigma_2\sigma_1$, or a pattern of the form $\sigma_2\sigma_1\sigma_2^e\sigma_1$ with $e \geqslant 1$.

EXAMPLE 3.7. Let $w = \sigma_1\sigma_2\sigma_1^2\sigma_2\sigma_1$, one of the words representing Δ_3^2. Its exponent sequence is $(1, \underline{1}, 2, \underline{1}, 1, 0)$, and the two underlined 1's show that w is reducible. Using (3.1) with $r = 3$ here yields the exponent sequence $(1, \underline{1}, \underline{1}, \underline{1}, 1, 1)$, corresponding to $\mathrm{red}(w) = \sigma_1\sigma_2\sigma_1\sigma_2\sigma_1\sigma_2$. As the underlined entries show, the latter word is reducible as well. Using (3.2) with $r = 2$ leads to the code $(1, 2, 1, 2)$, corresponding to $\mathrm{red}^2(u) = \sigma_1\sigma_2^2\sigma_1\sigma_2^2$. The latter word is irreducible, *i.e.*, no reduction may be applied to it.

Then reduction has the expected property, *i.e.*, it maps a 3-strand braid to an equivalent braid word that is smaller in the sense of the word ordering $\sqsubset$.

LEMMA 3.8. *If w is a reducible positive 3-strand braid word, then* $\mathrm{red}(w)$ *is equivalent to w, i.e., it represents the same braid, and* $\mathrm{red}(w) \sqsubset w$ *holds.*

PROOF. Constructing $\mathrm{red}(w)$ from w consists in replacing a subword of the form $\sigma_i^e\sigma_j\sigma_i^{e'}\sigma_j^{e''}$, with $e, e' > 0$ and $\{i, j\} = \{1, 2\}$, by the word $\sigma_i^{e-1}\sigma_j^{e'}\sigma_i\sigma_j^{e''+1}$. It is easy to check that the latter patterns are equivalent.

Now, with the notation of Definition 3.6. In the case $e_{r+1} = 1$, the Φ_3-breadth decreases so $\mathrm{red}(w) \sqsubset w$ holds. Otherwise, the Φ_3-breadths are equal, and the $p - r + 3$ leftmost entries of the exponent sequences are preserved, while e_{r+2} in w is replaced with $e_{r+2} - 1$ in $\mathrm{red}(w)$. So $\mathrm{red}(w) \sqsubset w$ holds as well. □

As the word ordering $\sqsubset$ is a well-ordering, it has no infinite descending chain, and we deduce

LEMMA 3.9. *For each 3-strand braid word w, there exists an integer m such that* $\mathrm{red}^m(w)$ *is irreducible.*

3.3. The key result, case of 3 strands. The point in Burckel's approach consists in establishing the following result.

PROPOSITION 3.10 (Burckel [**27**]). *Assume that w, w' are positive 3-strand braid words, w' is irreducible, and $w \sqsubset w'$ holds. Then the word $w^{-1}w'$ is equivalent to a σ^Φ-positive word.*

Actually, in the case of 3 strands, the result has already been established in Section 2.4: indeed, by definition, irreducible words satisfy the criterion of Proposition 2.4, and, therefore, they are Φ-normal words as defined in Section 1. Moreover, always by definition, $w \sqsubset w'$ is equivalent to $\overline{w} <^+ \overline{w'}$ whenever w and w' are irreducible. Then, Proposition 2.19 applies and guarantees that the braid represented by $w^{-1}w'$ is σ^Φ-positive, *i.e.*, that $w^{-1}w'$ is equivalent to a σ^Φ-positive word.

However, it is interesting to describe Burckel's method even in that simple case, because the argument of Section 2.3 is not known to extend to the general case, while Burckel's does. The main step in Burckel's method consists in using an induction on the ordinal rank of the word w' in the well-ordered set $(\underline{B}_n^+, \sqsubset)$—here with $n = 3$. For ρ an ordinal, let $(\mathcal{S}_\rho)$ denote the statement:

If w' is irreducible and the rank of w' in $(\underline{B}_3^+, \sqsubset)$ is at most ρ, then $w \sqsubset w'$ implies that $w^{-1}w'$ is equivalent to a σ^Φ-positive word.

In our current case $n = 3$, the problem is to prove $(\mathcal{S}_\rho)$ for every ordinal ρ with $\rho < \omega^\omega$. First, $(\mathcal{S}_0)$ is vacuously true, and the point is to prove that, if $(\mathcal{S}_\xi)$ holds for $\xi < \rho$, which will be denoted by $(\mathcal{S}_{<\rho})$, then $(\mathcal{S}_\rho)$ holds as well. When one starts with an arbitrary word w satisfying $w \sqsubset w'$, it seems difficult to prove the result, namely that $w^{-1}w'$ is equivalent to a σ^Φ-positive word, because little is known about w in general. Burckel's idea consists in introducing some special words w below w' for which the result can be proved directly, and to show that this is sufficient to conclude. In the case of 3-strand braid words, the task is easy.

DEFINITION 3.11. Assume $w, w' \in \underline{B}_3^+$, and $e > 0$. Let p' be the Φ_3-breadth of w'. A word w is a called an *e-companion of w'* if w has Φ_3-breadth $p' - 1$ and its exponent sequence begins with e.

By definition, if w is a companion of w', then $w \sqsubset w'$ holds. The converse is almost true, in the sense that every predecessor of w' is a predecessor of some companion of w', or it has a special form.

LEMMA 3.12. *Assume that w, w' belong to $\underline{B}_3^+$ and $w \sqsubset w'$ holds. Then at least one of the following holds:*

(i) *There exist i, w_0, w_0' satisfying $w = \sigma_i w_0$, $w' = \sigma_i w_0'$ and $w_0 \sqsubset w_0'$;*

(ii) *For e sufficiently large, $w \sqsubset w'' \sqsubset w'$ holds for every e-companion w'' of w'.*

The proof is an easy verification. Then comes the key point:

LEMMA 3.13. *Assume that $(\mathcal{S}_{<\rho})$ holds and w' is an irreducible 3-strand positive braid word whose rank in $(\underline{B}_3^+, \sqsubset)$ is ρ. Then, for each e, there exists an irreducible e-companion w of w' such that $w^{-1}w'$ is equivalent to a σ^Φ-positive word.*

SKETCH OF PROOF. Let $(e_p, \dots, e_1)$ be the exponent sequence of w'. Assume $p \geqslant 5$ with, say, p odd. Define w to admit the exponent sequence $(e, 2, 2, \dots, 2, 1, 1)$ with length $p - 1$. Put $w' = \sigma_2 w_0'$, and $w = \sigma_1^e \sigma_2^2 w_0$. Then $\sigma_1 \sigma_2^{e+2} w_0 \sqsubset w_0' \sqsubset w'$ holds. Indeed, either w' begins with at least two σ_2's, and w_0' has breadth p while $\sigma_1 \sigma_2^{e+2} w_0$ has breadth $p-1$, or w' begins with only one σ_2, and both $\sigma_1 \sigma_2^{e+2} w_0$ and w_0' have breadth $p - 1$, but the first entry in the exponent sequence of $\sigma_1 \sigma_2^{e+2} w_0$ is 1, while the first entry in the exponent sequence of w_0', namely e_2, is at least 2 by hypothesis. Now, w_0' is irreducible, hence the induction hypothesis implies that $(\sigma_1 \sigma_2^{e+2} w_0)^{-1} w_0'$ is equivalent to some σ^Φ-positive word w_1, yielding

$$w_1 \equiv (\sigma_1 \sigma_2^{e+2} w_0)^{-1} w_0' \equiv (\sigma_2 \sigma_1 \sigma_2^{e+2} w_0)^{-1} w' \equiv (\sigma_1^e \sigma_2 \sigma_1 \sigma_2^2 w_0)^{-1} w'.$$

Then the word $\sigma_1\sigma_2^2w_0$ is irreducible by construction, and $\sigma_1\sigma_2^2w_0 \sqsubset w'$ holds. Now $\sigma_2w_0 \sqsubset \sigma_1\sigma_2^2w_0$ holds, so the hypothesis $(\mathcal{S}_{<\rho})$ implies that $(\sigma_2w_0)^{-1}(\sigma_1\sigma_2^2w_0)$ is equivalent to some σ_1-positive word w_2. One deduces

$$w^{-1}w' \equiv (\sigma_2^2w_0)^{-1}(\sigma_2\sigma_1\sigma_2^2w_0)(\sigma_1^e\sigma_2\sigma_1\sigma_2^2w_0)^{-1}w' \equiv w_1w_2,$$

hence $w^{-1}w'$ is equivalent to the σ^Φ-positive word w_2w_1. The argument is similar when e is even. Finally, the special cases $p \leqslant 4$ are treated directly using specific—and easier—computations. □

Then the argument is completed as follows.

PROOF OF PROPOSITION 3.10. Assume $(\mathcal{S}_{<\rho})$. Let w' be an irreducible word with rank ρ. Assume $w \sqsubset w'$. By Lemma 3.12 two cases are possible. Assume first $w = \sigma_i w_0$ and $w' = \sigma_i w_0'$ for some i, w_0, w_0'. Then, by definition, w_0' is irreducible, and $w_0' \sqsubset w'$ holds. Hence, by the induction hypothesis, $w_0^{-1}w_0'$ is equivalent to a σ^Φ-positive word, and so is $w^{-1}w'$.

Assume now that $w \sqsubset w''$ holds for every e-companion w'' of w' with e large enough. By Lemma 3.13, there exists such a companion w'' that is irreducible and such that $w''^{-1}w'$ is equivalent to a σ^Φ-positive word. Now, as $w'' \sqsubset w'$ holds, the induction hypothesis implies that $w^{-1}w''$ is also equivalent to a σ^Φ-positive word. Hence $w^{-1}w'$, which is equivalent to $(w^{-1}w'')(w''^{-1}w')$, is equivalent to a σ^Φ-positive word, and $(\mathcal{S}_\rho)$ is true. □

3.4. The general case. In the general case, Burckel uses the same ideas. First, he identifies patterns that cannot occur in a braid word that is $\sqsubset$-minimal in its class, *i.e.*, patterns v so that there exists v' equivalent to v and $\sqsubset$-smaller than v. The problem is to find a limited number of such patterns to keep the definition of reduction tractable. Burckel attaches with every positive n-strand braid word a certain depth $(n-2)$ tree that, in the current framework, corresponds to iterating the operation of Φ_n-splitting: taking the Φ_{n-1}-splitting of each entry of the Φ_n-splitting means constructing a sequence of sequences, hence a depth 2 tree, and this can be iterated until $\underline{B}_2^+$ is reached. Next he introduces a notion of reduction that heavily depends on geometric parameters attached to these trees. Lemmas 3.8 and 3.9 then extend. Note that, at this point, it is clear by definition that a Burckel normal word is irreducible, but it is not clear that, conversely, any irreducible word has to be normal in the sense of Burckel.

Then Burckel's main result is similar to Proposition 3.10:

PROPOSITION 3.14 (Burckel [**27**]). *Assume that w, w' are positive braid words, w' is irreducible, and $w \sqsubset w'$ holds. Then the word $w^{-1}w'$ is equivalent to a σ^Φ-positive word.*

The scheme of the proof is similar to the one explained above in the special case of three strands. Burckel defines a convenient notion of companion. Then, a counterpart of Lemma 3.8 is satisfied, and an induction argument similar to the one described above enables him to establish Proposition 3.14.

COROLLARY 3.15. *A word is normal in the sense of Burckel if and only if it is irreducible.*

PROOF. The condition has already been observed to be necessary. Conversely, assume that w' is irreducible, and w is equivalent to w'. Then, by Property **A**, it is impossible that $w^{-1}w'$ be equivalent to a σ^{Φ}-positive word. By Proposition 3.14, it is therefore impossible that $w \sqsubset w'$ holds, and w' is $\sqsubset$-minimal in its equivalence class. □

4. Applications

Once Proposition 3.14 is proved, many applications can be deduced. As for the construction of the braid ordering, one obtains a new proof of Property **C**—which does not rely on a prior knowledge of Property **A**. As for the properties of the σ^{Φ}-ordering, one obtains a proof that the restriction of the σ^{Φ}-ordering to B_n^+ is a well-ordering, and a determination of the ordinal type of this well-ordering. And, using the connection with the Φ_n-splitting of Section 1, we obtain a simple recursive description of the σ^{Φ}-ordering of B_n^+ from the σ^{Φ}-ordering of B_{n-1}^+.

4.1. A proof of Property C. A direct application of Burckel's Proposition 3.14 is a proof of Property **C**. We recall that every braid is a quotient of positive braids, and, therefore, establishing Property **C** for braids that are quotients of positive braids is not a restriction.

PROPOSITION 4.1 (Property **C**). *If β, β' are distinct positive braids, then the braid $\beta^{-1}\beta'$ is σ-positive or σ-negative.*

PROOF. Assume that β and β' lie in B_n. Let w and w' be the Burckel normal words representing β and β'. As $\sqsubset$ is a linear ordering, one of $\Phi_n(w) \sqsubset \Phi_n(w')$, $\Phi_n(w') \sqsubset \Phi_n(w)$, or $\Phi_n(w) = \Phi_n(w')$ holds. By Proposition 3.14, the first relation implies that $\Phi_n(w^{-1}w')$ is equivalent to a σ^{Φ}-positive word, hence that $w^{-1}w'$ is equivalent to a σ-positive word. The second relation similarly implies that $w'^{-1}w$ is equivalent to a σ-positive word, hence that $w^{-1}w'$ is equivalent to a σ-negative word. The third relation implies that w and w' are equivalent, hence that $w^{-1}w'$ is equivalent to the empty word. □

4.2. Burckel normal *vs.* Φ-normal form. Another application of Proposition 3.14 is that the Φ-normal form of Section 1 coincides with the Burckel normal form of Section 3. This coincidence result is important, for it implies that each normal form inherits the nice properties from the other, namely conceptual simplicity in the case of the Φ-normal form, and connection with the σ^{Φ}-ordering in the case of the Burckel normal form.

Both normal forms arise as the iteration of some splitting operation that associates with every braid of B_n^+ a finite sequence of braids of B_{n-1}^+: this is clear for the Φ-normal form, which is introduced as the iteration of the Φ_n-splitting. This is also true for the Burckel normal form. To explain that, let us introduce the following general notion:

DEFINITION 4.2. Assume $\beta \in B_n^+$. We say that a sequence $(\beta_p, \dots, \beta_1)$ is a *Φ_n-factorization* of β if we have $\beta = \Phi_n^{p-1}(\beta_p) \cdot \dots \cdot \Phi_n(\beta_2) \cdot \beta_1$.

The Φ_n-splitting of β is a particular Φ_n-factorization of β, namely the one obtained by taking the maximal possible right divisor at each step—so it could be adequately called the *right greedy Φ_n-factorization* of β. Now we can consider other distinguished factorizations. In particular, there is the natural notion of a shortest Φ_n-factorization, *i.e.*, the one(s) with the minimal number of entries.

LEMMA 4.3. *For every positive braid β, the Burckel normal form of β comes from iterating the shortest factorization: if w is the Burckel normal form of β, and if $(w_p, \dots, w_1)$ is the Φ_n-splitting of w, then $(\overline{w_p}, \dots, \overline{w_1})$ is a shortest Φ_n-factorization of β.*

PROOF. Assume that $(\beta_{p'}, \dots, \beta_1)$ is a Φ_n-factorization of β satisfying $p' < p$. Then, by concatenating the Burckel normal forms of $\beta_{p'}, \dots, \beta_1$ and inserting the required flips, we obtain a new word w' representing β that, by construction, satisfies $w' \sqsubset w$, so w cannot be Burckel normal. □

Thus the difference between the Φ-normal form and the Burckel normal form is clear: in one case, we maximize the right entries, in the other, we minimize the length. We prove now that the two approaches lead to the same result.

PROPOSITION 4.4. *For each positive braid β, the Φ-normal form of β and its Burckel normal form coincide.*

PROOF. As each positive braid admits a unique Burckel normal representative word and a unique Φ-normal representative word, proving one implication is sufficient. We shall prove that an n-strand braid word that is not Φ-normal cannot be Burckel normal using induction on $n \geqslant 2$ —we could start from $n = 3$ as well, since we already know that the Φ-normal and Burckel normal 3-strand words are characterized by the same exponent constraints. For $n = 2$, the result is obvious, as each positive 2 braid admits only one representative word. Assume now $n \geqslant 3$, let β be a positive n-strand braid, and assume that w is a word representing β that is not Φ-normal. We shall prove that w cannot be the Burckel normal form of β, *i.e.*, that there exists another word w' representing β that satisfies $w' \sqsubset w$.

Let $(w_p, \dots, w_1)$ be the Φ_n-splitting of w. By definition of the Φ-normal form, w being not Φ-normal can have two causes, namely that there exists r such that w_r is not Φ-normal, or that $(\overline{w_p}, \dots, \overline{w_1})$ is not the Φ_n-splitting of β.

Assume first that some word w_r is not Φ-normal. By the induction hypothesis, the word w_r is not Burckel normal. Hence there exists w'_r equivalent to w_r satisfying $w'_r \sqsubset w_r$. Let w' be the word obtained from w by replacing the factor that comes from w_r—that is, w_r or $\Phi_n(w_r)$ according to the parity of r—with w'_r. Then w' is equivalent to w and we have $w' \sqsubset w$, so w is not Burckel normal.

Assume now that $(\overline{w_p}, \dots, \overline{w_1})$ is not the Φ_n-splitting of $\overline{w}$. The hypothesis implies that, for some $r \geqslant 2$, putting $v = \Phi_n^{p-r}(w_p) \cdot \dots \cdot \Phi_n(w_{r+1}) \cdot w_r$, the braid represented by v is right-divisible by some σ_i with $i \geqslant 2$. We shall see that the factor σ_i can be extracted from v and incorporated in the next factor w_{r-1}, giving rise to a new word w' representing β and satisfying $w' \sqsubset w$.

Indeed, by hypothesis, there exists a positive braid β' satisfying $\overline{v} = \beta' \sigma_i$. Let v' be the Burckel normal form of β', and let w' be the word

$$\Phi_n^{r-1}(v') \cdot \Phi_n^{r-2}(\sigma_{n-i} w_{r-1}) \cdot \dots \cdot w_1.$$

By construction, w' is equivalent to w. As we have $i \geqslant 2$, the word $\sigma_{n-i} w_{r-1}$ belongs to $\underline{B}_{n-1}^+$, so the Φ_n-breadth of w' is $r - 1$ plus the Φ_n-breadth of v'. Hence, if we can prove $v' \sqsubset v$, this will imply $w' \sqsubset w$, as expected.

So our aim is to prove $v' \sqsubset v$. As $\sqsubset$ is a linear ordering, it suffices to prove that $v = v'$ and $v \sqsubset v'$ are impossible. Clearly $v = v'$ is impossible, since v is equivalent to $v' \sigma_i^{-1}$, and it only remains to exclude $v \sqsubset v'$. Now, by hypothesis, $v^{-1} v'$ is equivalent to σ_i^{-1}, and, therefore, by Property **A**, it cannot be equivalent to a σ^Φ-positive word. On the other hand, Proposition 3.14 says that, if $v \sqsubset v'$ holds, then $v^{-1}v'$ is equivalent to a σ^Φ-positive word, and we have the expected contradiction. Hence we have $v' \sqsubset v$, whence $w' \sqsubset w$, and w is not Burckel normal. □

REMARK 4.5. The coincidence result of Proposition 4.4 is quite specific. When the parabolic submonoids are changed, the right greedy factorization need not be a shortest factorization: for instance, the $(B^+_{\{2,3\}}, B^+_{\{1,3\}})$-decomposition of $\mathtt{cbabc^2b}$—*i.e.*, the sequence obtained by taking the maximal right divisor at each step—is $(\mathtt{a}, \mathtt{b}^2, \mathtt{a}, \mathtt{cb}, \mathtt{a})$, which has length 5, while $(\mathtt{cb}, \mathtt{a}, \mathtt{bc^2b}, \varepsilon)$ is a another decomposition of β into alternating entries in $B^+_{\{2,3\}}$ and $B^+_{\{1,3\}}$ that has length 4 only.

Proposition 4.4 implies that the Φ-normal form and the Burckel normal form each inherit the properties of the other. In particular, we noted that the Φ-normal form, exactly as the greedy normal form, can be computed in quadratic time. We deduce that the same holds for the Burckel normal form—what was unclear from the construction as the final result of a reduction process.

Even more interesting are the applications involving the σ^Φ-ordering.

COROLLARY 4.6. *The ordering $<^+$ of Section 2.3 coincides with the braid ordering $<^\Phi$. In particular, $<^+$ is compatible with multiplication on the left.*

PROOF. Let β, β' be positive braids satisfying $\beta <^+ \beta'$. Then, by construction, the Φ-normal form of β is $<^+$-smaller than the Φ-normal form of β'. Owing to Proposition 4.4, this means that the Burckel normal form of β is $<^+$-smaller than the Burckel normal form of β'. By Proposition 3.14, the latter implies that $\beta^{-1}\beta'$ is σ^Φ-positive, hence that $\beta <^\Phi \beta'$ holds. □

Thus we established a positive answer to Question 2.18.

By Proposition 2.17, the $<^+$-comparison of two braids has a quadratic complexity. So another consequence of Corollary 4.6 is

COROLLARY 4.7. *For each n, the braid orders $<$ and $<^\Phi$ on B_n can be decided in quadratic time: if w is a (not necessarily positive) n-strand braid word of length ℓ, then whether $\overline{w} > 1$ holds can be decided in time $O(\ell^2)$.*

It also directly follows from Corollary 4.6 that the recursive definition of the ordering $<^+$ becomes a recursive characterization of the ordering $<^\Phi$—as announced in Proposition II.4.6:

PROPOSITION 4.8. *Assume $n \geqslant 3$ and $\beta, \beta' \in B_n^+$. Then $\beta <^\Phi \beta'$ holds if and only if the Φ_n-splitting of β is smaller than the Φ_n-splitting of β' with respect to the* ShortLex-*extension of the σ^Φ-ordering of B_{n-1}^+.*

4.3. The well-order property and the ordinal type. Still another application of Corollary 4.6, hence of Proposition 3.14, is that the σ^Φ-ordering of positive braids is a well-ordering.

PROPOSITION 4.9. *For each $n \geqslant 2$, the restriction of the braid ordering $<^\Phi$ to B_n^+ is a well-ordering of ordinal type $\omega^{\omega^{n-2}}$.*

PROOF. According to Proposition 2.15, the restriction of the braid ordering $<^+$ to B_n^+ is a well-ordering, hence, by Corollary 4.6, the same holds for $<^\Phi$. As for the ordinal type, we first observe that the ShortLex-extension of a well-ordering of ordinal type λ is a well-ordering of ordinal type ω^λ. As $(B_2^+, <_2^+)$, *i.e.*, $(\mathbb{N}, <)$, is a well-ordering of ordinal type ω, *i.e.*, ω^{ω^0}, this inductively implies that $(B_n^+, <^\Phi)$ has ordinal type at most $\omega^{\omega^{n-2}}$. *A priori*, we have an inequality only, because an arbitrary sequence of braids of B_{n-1}^+ need not be the Φ_n-splitting of a braid of B_n^+. In order to obtain an equality, the point is to show that there exist enough braids with a Φ_n-splitting of a given form. A solution consists in considering braids that admit only one representative word—for instance braids represented by words that involve only σ_i^2's and, in addition, are such that, after σ_i^2, only σ_{i-1}^2, σ_i^2, or σ_{i+1}^2 are possible. A one-to-one correspondence between arbitrary positive n-strand braids and those of the form above can easily be described, and this is enough to show inductively that the ordinal type of $(B_n^+, <^\Phi)$ is not less than $\omega^{\omega^{n-2}}$. □

As the automorphism Φ_n of B_n^+ exchanges the original and flipped versions of the braid ordering, *i.e.*, the orders $<$ and $<^\Phi$, we deduce

COROLLARY 4.10. *For each $n \geqslant 2$, the restriction of the braid ordering $<$ to B_n^+ is a well-ordering of ordinal type $\omega^{\omega^{n-2}}$.*

On the other hand, we observed that the well-ordered set $(B_{n-1}^+, <^\Phi)$ is an initial segment of the well-ordered set $(B_n^+, <^\Phi)$. It follows that the union of these well-ordered sets is a well-ordered set, whose ordinal type is the supremum of the order types of its initial intervals:

COROLLARY 4.11. *The restriction of the braid ordering $<^\Phi$ to B_∞^+ is a well-ordering of ordinal type ω^{ω^ω}.*

The result that $(B_n^+, <^\Phi)$ is a well-ordering implies that each braid of B_n^+ can be attributed an ordinal rank that completely specifies its position in the well-ordering. In the case of B_3^+, we saw in Corollary 2.22 that this rank is determined by a simple formula. Things become more complicated for $n \geqslant 4$, and no such closed formula is known. We refer to [**28**] where an iterative method is described.

REMARK 4.12. We observed in Chapter II that there is no global automorphism of B_∞^+ exchanging $<$ and $<^\Phi$, and that $(B_\infty^+, <)$ and $(B_\infty^+, <^\Phi)$ are not isomorphic. In particular, $(B_\infty^+, <)$ is not a well-ordering, as $\sigma_1, \sigma_2, \ldots$ is an infinite descending sequence for $<$.

4.4. The subword property. In Chapter II, we explained how to deduce the result that the restriction of $<$ to B_n^+ is a well-ordering from Property **S**, which asserts that $<$ extends the left-divisibility ordering. The latter property can be proved using Burckel's techniques as well.

LEMMA 4.13. *Property* **S** *is equivalent to:*

(4.1) *For every positive braid β and every i, we have $\sigma_i\beta > \beta$.*

PROOF. Property **S** says that $\sigma_i\beta > \beta$ holds for every braid β and every i. So relation (4.1) is the particular case where β is assumed to be positive.

Conversely, let β be an arbitrary braid in B_n. For d large enough, $\Delta_n^{2d}\beta$ is a positive braid, as was shown in the proof of Proposition I.4.6. If (4.1) is true, we obtain $\sigma_i\Delta_n^{2d}\beta > \Delta_n^{2d}\beta$ for $i < n$, hence $\Delta_n^{2d}\sigma_i\beta > \Delta_n^{2d}\beta$, as Δ_n^2 commutes with σ_i, and finally $\sigma_i\beta > \beta$ by multiplying by Δ_n^{-2d} on the left. □

PROPOSITION 4.14 (Property **S**). *For each positive braid β and each i, we have $\beta < \sigma_i\beta$ and $\beta <^{\Phi} \sigma_i\beta$.*

PROOF (SKETCH). It is enough to prove one of the inequalities, and we consider the case of $<^{\Phi}$. The point is that, if w is a Φ-normal n-strand braid word, then, for each $i \leqslant n-1$, the word $w^{-1}\sigma_i w$ is equivalent to a σ^{Φ}-positive word. In the case $n = 3$, a direct verification is easy, as we can explicitly determine the Φ-normal form of $\sigma_i w$ from that of w. In the general case, no simple direct argument is known. Here again, Burckel's approach proves relevant: the result is easy when $\sigma_i w$ is irreducible; otherwise, by inspecting the various possible reduction cases, Burckel is able to show the above-mentioned implication using induction on the rank of w in the well-ordered set $(\underline{B}_n^+, \sqsubset)$. $\square$

CHAPTER VIII

Dual Braid Monoids

In Chapter VII, we have seen how to exploit the fact that, for $n \geqslant 3$, every braid of the monoid B_n^+ can be decomposed into a product of factors that alternately lie in the submonoid B_{n-1}^+ and its flipped image $\Phi_n(B_{n-1}^+)$. In this chapter, we consider another monoid of which the group B_n is a group of fractions, namely the so-called Birman–Ko–Lee, or dual, monoid B_n^{+*} of [**15, 10**]—the name "dual" has become standard although, so far, no actual duality is involved here, only numerical coincidences in which some parameters attached with B_n^+ and B_n^{+*} are exchanged. Like the monoid B_n^+, the monoid B_n^{+*} admits a Garside structure, but, in this case, the associated automorphism ϕ_n is not an involution, but an order n morphism similar to a rotation. One is naturally led to considering for each element of B_n^{+*} a decomposition into a product of factors that cyclically lie in the submonoid B_{n-1}^{+*} and its successive images under ϕ_n, ϕ_n^2, ..., ϕ_n^{n-1}. One thus obtains for B_n^{+*} a new normal form that compares with the greedy normal form as the Φ-normal form of Chapter VII compares with the standard greedy normal form of B_n^+.

Using this approach, Jean Fromentin analyses in [**90, 91**] the restriction of the σ^Φ-ordering to B_n^{+*}. The results are similar to those of Chapter VII. As for the construction of the braid ordering, one obtains one more proof of Property **C**. As for its properties, one establishes that $(B_n^{+*}, <^\Phi)$ is a well-ordering of ordinal type $\omega^{\omega^{n-2}}$, a result that properly extends those of Chapter VII since B_n^{+*} is a proper extension of B_n^+.

One of the most interesting features in this approach is that, for most results, we obtain proofs that are cleaner and simpler by using the dual monoid B_n^{+*} than by using the monoid B_n^+. In particular, we shall prove by straight computation that, if β, β' are two braids of B_n^{+*} and $\beta <^* \beta'$ holds in a sense similar to Definition VII.2.14, then the quotient $\beta^{-1}\beta'$ is σ^Φ-positive. We recall that the proof of the analog result in B_n^+ given in Section VII.4 was much more delicate, relying on Burckel's sophisticated methods involving a transfinite induction. So, although many questions remain open at the time of writing, this line of research seems quite promising.

The chapter is organized as follows. Section 1 contains a quick introduction to dual braid monoids and their Garside structure. In Section 2, we describe the ϕ_n-splitting of every braid in B_n^{+*} and deduce both a new normal form and a natural linear ordering of B_n^{+*}. In Section 3, we state Fromentin's coincidence result that connects the latter ordering with the σ^Φ-ordering, and we explain both its applications and the main steps in the proof.

1. Dual braid monoids

As shown by Birman, Ko, and Lee in [**15**], there exists for each n a submonoid B_n^{+*} of B_n that properly includes the monoid B_n^+ for $n \geqslant 3$, but still admits what is called a Garside structure, *i.e.*, a good divisibility theory with lcm's and gcd's. Here we briefly describe the construction of the monoid B_n^{+*} and its main algebraic properties.

1.1. Birman–Ko–Lee generators. The braid group B_n admits a standard generating family, namely the Artin generators σ_i. Clearly, $\{\sigma_1, \ldots, \sigma_{n-1}\}$ is minimal in that each proper subfamily generates a proper subgroup of B_n—but B_n admits generating families of smaller cardinality, for instance $\{\sigma_1, \sigma_1\sigma_2 \ldots \sigma_{n-1}\}$.

If we add more braids to the family of Artin generators, we obtain a new, redundant family of generators for the group B_n. Hereafter, we consider one such redundant family, obtained by adding certain conjugates of the Artin generators σ_i.

Definition 1.1. (See Figure 1.) For $1 \leqslant i < j$, we put

$$a_{i,j} = \sigma_i \ldots \sigma_{j-2}\, \sigma_{j-1}\, \sigma_{j-2}^{-1} \ldots \sigma_i^{-1}. \tag{1.1}$$

Figure 1. From left to right: $a_{1,4}$, $a_{1,3}$, and $a_{2,4}$; geometrically, $a_{i,j}$ corresponds to a braid diagram in which the strands at position j crosses over the strand at position i, both passing behind all intermediate strands.

The family of all braids $a_{i,j}$ enjoys nice invariance properties with respect to cyclic permutations of the indices, which are better visualized when $a_{i,j}$ is represented on a cylinder (see Figure 2). Then, it is natural to associate with $a_{i,j}$ the chord connecting the vertices i and j in a circle with n marked vertices.

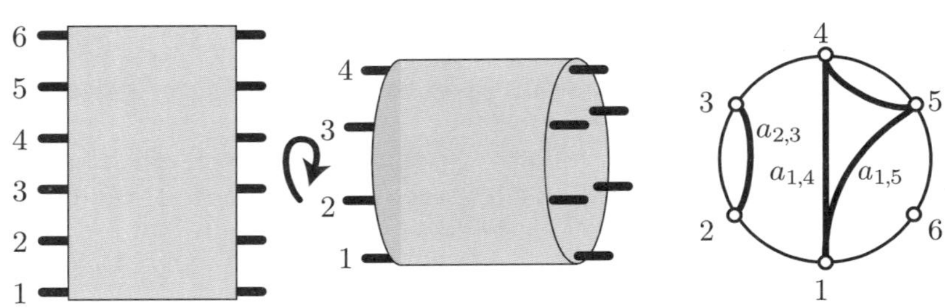

Figure 2. Drawing a braid diagram on a cylinder: folding the usual diagram around a cylinder helps with vizualizing the cyclic symmetries of the family $\{a_{i,j} \mid i < j\}$; viewing the cylinder from the side, we associate to $a_{i,j}$ the chord that connects the ith and the jth point in a circle with n distinguished points.

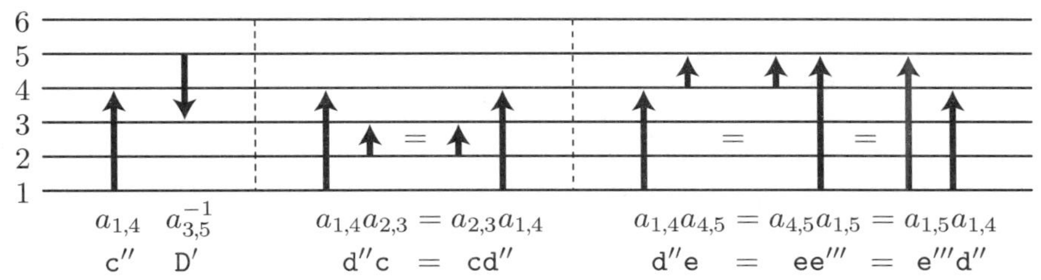

FIGURE 3. Arrow representation of the braids $a_{i,j}$ and their relations: $a_{i,j}$ corresponds to an upward arrow from line i to line j, while $a_{i,j}^{-1}$ is represented with a downward arrow; below are the counterparts using the simplified notational system described in Convention 1.2.

It will also be helpful to adopt a more economical representation in which $a_{i,j}$ is associated with a vertical arrow from the ith line to the jth line on an n-line stave (Figure 3). Then $a_{i,j}^{-1}$ is naturally represented by a similar descending arrow.

CONVENTION 1.2. In concrete examples, we shall extend the previous convention of using a, b,... for σ_1, σ_2,...—hence also for $a_{1,2}, a_{2,3}, \ldots$. We shall use b′, c′,... for $a_{1,3}, a_{2,4}$,... as they are conjugates of b, c,... by one σ_i, then, similarly, c″, d″,... for $a_{1,4}, a_{2,5}$,... which are conjugates of c, d,... by two σ_i's, etc. Thus, the letters refer to the end line in the arrow representation, while the dashes refer to the number of conjugating σ_i's.

In terms of the generators $a_{i,j}$, the group B_n can be presented by the usual braid relations of (I.1.1) in which we substitute $a_{i,i+1}$ to σ_i, plus the defining relations (1.1). A more symmetric and more interesting presentation is as follows. Hereafter, we write $[i, j]$ for the interval $\{i, i+1, \ldots, j\}$ of $\mathbb{N}$, and we say that $[i, j]$ is *nested* in $[i', j']$ if we have $i' < i < j < j'$.

LEMMA 1.3. *In terms of the $a_{i,j}$, the group B_n is presented by the relations*

$$a_{i,j}a_{i',j'} = a_{i',j'}a_{i,j} \quad \text{for } [i,j] \text{ and } [i',j'] \text{ disjoint or nested,} \tag{1.2}$$

$$a_{i,j}a_{j,k} = a_{j,k}a_{i,k} = a_{i,k}a_{i,j} \quad \text{for } 1 \leqslant i < j < k \leqslant n. \tag{1.3}$$

PROOF. It is easy to check from the definition of the $a_{i,j}$'s in terms of the σ_i's that the relations (1.2) and (1.3) are satisfied in B_n. Conversely, it is equally easy to deduce from the previous relations that the generators $a_{i,i+1}$ satisfy the braid relations and that each $a_{i,j}$ can be expressed in terms of the sole elements $a_{i,i+1}$ by relations imitating (1.1). □

In the chord representation, relations of type (1.2) correspond to the fact that, for each chord triangle, the product of two adjacent edges taken clockwise does not depend on the edges: for instance, in Figure 2, the triangle $(1, 4, 5)$ gives $a_{1,4}a_{4,5} = a_{4,5}a_{1,5} = a_{1,5}a_{1,4}$. Relations of type (1.3) say that the generators associated with nonintersecting chords commute: for instance, in Figure 2, we see that $a_{3,4}$ and $a_{2,6}$ commute—but, for instance, we claim nothing about $a_{1,4}$ and $a_{2,6}$. The corresponding arrow representations are given in Figure 3.

REMARK 1.4. In [**15**], $a_{i,j}$ is defined to be $\sigma_i^{-1}\ldots\sigma_{j-2}^{-1}\sigma_{j-1}\sigma_{j-2}\ldots\sigma_i$, *i.e.*, it corresponds to the strands at positions i and j passing in front of all intermediate strands, not behind. Both options lead to isomorphic monoids, but our choice is the only one that naturally leads to the suitable embedding of B_{n-1}^{+*} into B_n^{+*}.

1.2. The dual braid monoid B_n^{+*} and its Garside structure. By definition, the monoid B_n^+ is the submonoid of B_n generated by $\sigma_1, \dots, \sigma_{n-1}$. When we consider the family of all braids $a_{i,j}$ with $i < j \leqslant n$, it also generates a submonoid of B_n, but, because, for $n \geqslant 3$, the family contains elements that do not lie in B_n^+, this submonoid properly includes B_n^+. This monoid is the object we shall study in this chapter.

DEFINITION 1.5. For $n \geqslant 2$, we define the *dual braid monoid* B_n^{+*} to be the submonoid of B_n generated by the braids $a_{i,j}$ with $1 \leqslant i < j \leqslant n$. The elements of B_n^{+*} will be called *dual-positive* braids.

By definition, we have $\sigma_i = a_{i,i+1}$ for each i, and, therefore, the monoid B_n^+ is included in the monoid B_n^{+*}: every positive braid is dual-positive. The inclusion is proper for $n \geqslant 3$ as $a_{1,3}$ does not lie in B_3^+: the dual-positive braid $a_{1,3}$ is not positive.

We saw in Chapter I that B_n^+ turns out to admit, as a monoid, the same presentation as B_n, *i.e.*, it is the monoid presented by the braid relations (I.1.1): this is one of the many consequences of the fact that the monoid presented by the above braid relations is what is now called a Garside monoid. A similar result holds for the monoid B_n^{+*} with respect to the relations of Lemma 1.3.

PROPOSITION 1.6. *For each n, the monoid B_n^{+*} is generated by elements $a_{i,j}$ with $1 \leqslant i < j \leqslant n$ subject to the relations of Lemma* 1.3.

The method for proving this result consists in introducing the abstract monoid M presented by generators $a_{i,j}$ subject to the relations of Lemma 1.3, and in proving that M is isomorphic to B_n^{+*}. Exactly as in the study of the monoid B_n^+ by F.A. Garside in [**95**], the major step consists in showing that M admits a Garside structure: it is cancellative, any two elements admit unique left and right lcm's and gcd's, and $a_{1,2}a_{2,3}...a_{n-2,n-1}$ is a Garside element, *i.e.*, its left and right divisors coincide, they are finite in number, and they generate the monoid. Two types of proofs can be used, namely combinatorial arguments directly inspired from [**95**]—see [**15**] or [**53**]—or more conceptual arguments based on an interpretation of the elements of B_n^{+*} in terms of noncrossing partitions [**10**].

The existence of a Garside structure on B_n^{+*} implies that the properties of divisibility in B_n^{+*} are similar to those of the monoid B_n^+, the role of Δ_n being now played by the braid δ_n of Definition I.4.3, *i.e.*,

$$\delta_n = a_{1,2}\, a_{2,3} \dots a_{n-1,n}, \tag{1.4}$$

in terms of the generators $a_{i,j}$. In particular, every element of B_n^{+*} admits a distinguished decomposition similar to the greedy normal form of Chapter VI. This decomposition involves the elements of B_n^{+*} that are divisors of δ_n—in the sense of B_n^{+*}, *i.e.*, the quotient lies in B_n^{+*}, not necessarily in B_n^+. It turns out that these elements, which can naturally be called *dual-simple*, are in one-to-one correspondence with those permutations of $\{1, \dots, n\}$ that can be realized by parallel descending cycles, or, in another language, with noncrossing partitions of $\{1, \dots, n\}$. This implies that the number of dual-simple n-strand braids is the Catalan number $\frac{1}{n+1}\binom{2n}{n}$, which is strictly smaller than the number $n!$ of simple n-strand braids.

1.3. The cycling automorphism. In order to study the monoid B_n^+ in Chapter VII, we appealed to the flip automorphism Φ_n, which is the inner automorphism associated with Δ_n. In this chapter, we shall similarly use the inner automorphism ϕ_n associated with the braid δ_n, which we recall is defined for β in B_n by

$$\phi_n(\beta) = \delta_n\, \beta\, \delta_n^{-1}.$$

This is quite natural: we mentioned that both monoids B_n^+ and B_n^{+*} are equipped with a Garside structure, with minimal Garside elements Δ_n and δ_n, respectively, and it is well known that, in every Garside monoid, the inner automorphism associated with the minimal Garside element plays an important role [**53**]. However, the situations in B_n^+ and B_n^{+*} are different: while conjugating by Δ_n gives an involutory automorphism that corresponds to a symmetry in braid diagrams, conjugation by δ_n gives an order n automorphism that should be viewed as a rotation.

For our current purpose, the important point is that—contrary to the flip automorphism Φ_n—the automorphism ϕ_n leaves the monoid B_n^{+*} invariant:

LEMMA 1.7. (See Figure 4.) *For all i, j with $i < j \leqslant n$, we have*

$$\phi_n(a_{i,j}) = \begin{cases} a_{i+1,j+1} & \text{for } j \leqslant n-1, \\ a_{1,i+1} & \text{for } j = n. \end{cases} \tag{1.5}$$

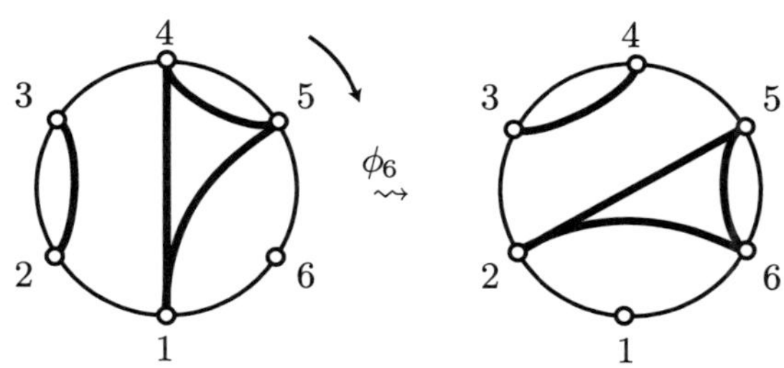

FIGURE 4. The automorphism ϕ_6 viewed as a clockwise rotation by $2\pi/6$: we read $\phi_6(a_{2,3}) = a_{3,4}$—*i.e.*, with our notational conventions, $\phi_6(\mathsf{b}') = \mathsf{c}'$—or $\phi_6(a_{1,5}) = a_{2,6}$, *i.e.*, $\phi_6(\mathsf{d}'') = \mathsf{e}''$; note that $\phi_6(a_{1,6})$ is $a_{1,2}$.

The proof is an easy verification from the formulas of Lemma I.4.4. Note that the relation $\phi_n(a_{i,j}) = a_{i+1,j+1}$ always holds provided indices are taken mod n, *i.e.*, $a_{i+1,n+1}$ means $a_{1,i+1}$.

1.4. The σ^Φ-ordering on B_n^{+*}. As B_n^{+*} is a subset of B_n, it makes sense to investigate the restriction of the σ-ordering $<$ to B_n^{+*}. As in Chapter VII, it will be convenient—actually necessary—to consider the flipped version $<^\Phi$ of the ordering, *i.e.*, to take into account the letters σ_i with highest index occurring in a braid word, and not the one with lowest index.

EXAMPLE 1.8. An easy direct verification shows that $a_{i,j} <^\Phi a_{i',j'}$ holds if and only if we have either $j < j'$, or $j = j'$ and $i > i'$. Thus, under the ordering $<^\Phi$, the generators $a_{i,j}$ form a well-ordered sequence that starts with

$$1 <^\Phi a_{1,2} <^\Phi a_{2,3} <^\Phi a_{1,3} <^\Phi a_{3,4} <^\Phi a_{2,4} <^\Phi a_{1,4} <^\Phi a_{4,5} <^\Phi a_{3,5} <^\Phi a_{2,5} <^\Phi a_{1,5} <^\Phi \dots,$$

i.e., using Convention 1.2 to improve readability,

(1.6) $1 <^\Phi \mathtt{a} <^\Phi \mathtt{b} <^\Phi \mathtt{b}' <^\Phi \mathtt{c} <^\Phi \mathtt{c}' <^\Phi \mathtt{c}'' <^\Phi \mathtt{d} <^\Phi \mathtt{d}' <^\Phi \mathtt{d}'' <^\Phi \mathtt{d}''' <^\Phi \dots .$

For information, when we consider the σ-ordering rather than the σ^Φ-ordering, the sequence goes in the other direction, but with a much more complicated pattern:

$$\mathtt{a} > \mathtt{b}' > \mathtt{c}'' > \mathtt{d}''' > \dots > \mathtt{b} > \mathtt{c}' > \mathtt{d}'' > \dots > \mathtt{c} > \mathtt{d}' > \dots > \mathtt{d} > \dots > 1.$$

Our aim in what follows will be to understand the σ^Φ-ordering of B_n^{+*} as completely as possible. To do that, we shall have to work with words, and we first fix some notation. Two alphabets are involved here, namely that of the Artin generators σ_i and that of Birman–Ko–Lee generators $a_{i,j}$.

Definition 1.9. (i) A word on the alphabet $\{a_{i,j}^{\pm 1} \mid 1 \leqslant i < j \leqslant n\}$ is called a *dual n-strand braid word*. It is said to be *positive* if it contains no letter $a_{i,j}^{-1}$.

(ii) Let $\widetilde{a}_{i,j}$ be the braid word specified by the right term in (1.1). For w a dual braid word, we denote by $\widetilde{w}$ the word in the letters $\sigma_i^{\pm 1}$ obtained by replacing each letter $a_{i,j}^{\pm 1}$ by the corresponding word $\widetilde{a}_{i,j}^{\pm 1}$.

So, for instance, if w is the dual braid word $a_{2,4}a_{1,3}^{-1}$—*i.e.*, $\mathtt{c}'\mathtt{B}'$—then $\widetilde{w}$ is the braid word $\sigma_2\sigma_3\sigma_2^{-1}\sigma_1\sigma_2^{-1}\sigma_1^{-1}$—*i.e.*, $\mathtt{bcBaBA}$.

In Chapter II, a braid word (in the letters $\sigma_i^{\pm 1}$) has been called σ^Φ-positive if the letter σ_i with highest index i occurs positively only. Here it will be useful to introduce the notion of a σ_i^Φ-positive word that takes into account the specific index i, and is therefore the counterpart of the notion of a σ_i-positive word.

Definition 1.10. (i) A braid word is said to be *σ_i^Φ-positive* if it contains at least one letter σ_i, but no σ_i^{-1} and no $\sigma_j^{\pm 1}$ with $j > i$. Similarly, it is said to be *σ_i^Φ-negative* if it contains at least one σ_i^{-1}, but no σ_i and no $\sigma_j^{\pm 1}$ with $j > i$.

(ii) A braid β is said to be *σ^Φ-positive* (*resp.* *σ^Φ-negative*) if it admits at least one σ^Φ-positive (*resp.* σ^Φ-negative) representative braid word.

By definition, the word $\widetilde{a}_{i,j}$ is σ_{j-1}^Φ-positive for each i and j. So, the first obvious observation is

Lemma 1.11. *Assume that w is a dual braid word. Then the braid word $\widetilde{w}$ is σ_{j-1}^Φ-positive if and only if w contains at least one letter $a_{i,j}$, no letter $a_{i',j}^{-1}$, and no letter $a_{i',j'}^{\pm 1}$ with $j' > j$.*

Proposition 1.12. *Every braid β in B_n^{+*} satisfies $\beta \geqslant^\Phi 1$ and $\beta \geqslant 1$.*

Proof. As for the σ^Φ-ordering, Lemma 1.11 shows that each braid $a_{i,j}$ is σ^Φ-positive, and an obvious induction on the length then shows that, for every nonempty positive dual braid word w, the word $\widetilde{w}$ is σ^Φ-positive.

As for the σ-ordering, we cannot conclude directly, as, for $j \geqslant i+2$, the word $\widetilde{a}_{i,j}$ is neither σ-positive nor σ-negative. However, it is immediate to check that $\sigma_{j-1}^{-1} \dots \sigma_{i+1}^{-1}\,\sigma_i\,\sigma_{i+1} \dots \sigma_{j-1}$ is another expression of $a_{i,j}$, which is σ-positive. Then, we deduce as above that every nontrivial braid in B_n^{+*} is σ-positive. □

As the automorphism ϕ_n corresponds to a rotation, it need not preserve the braid ordering: for instance, for $n \geqslant 3$, we have $\phi_n(a_{1,2}) = a_{2,3}$ and $\phi_n(a_{1,n}) = a_{1,2}$; hence, according to (1.6), we get $a_{1,2} <^\Phi a_{1,n}$ but $\phi_n(a_{1,2}) >^\Phi \phi_n(a_{1,n})$. However, such counterexamples may occur only when generators of the form $a_{i,n}$ are involved, and we have the following simple criterion.

LEMMA 1.13. *Assume that β lies in B_n and is σ_i^Φ-positive for some $i \leqslant n-2$. Then $\phi_n(\beta)$ is σ_{i+1}^Φ-positive.*

PROOF. Let w be a σ_i^Φ-positive expression of β. By hypothesis, w contains no letter $\sigma_{n-1}^{\pm 1}$. Hence applying ϕ_n amounts to shifting all indices by 1, and the resulting braid word is therefore σ_{i+1}^Φ-positive. □

2. The ϕ-normal form on B_n^{+*}

As was the case for the σ^Φ-ordering and the Garside structure of B_n^+, the σ^Φ-ordering and the Garside structure of B_n^{+*} are not easily connected. In particular, there is no simple way of comparing dual-positive braids by inspecting their (left or right) greedy normal forms. Following the scheme that was successful in Chapter VII, we shall now describe a new normal form on B_n^{+*} that will turn out to provide a very simple connection with the σ^Φ-ordering.

2.1. The ϕ_n-splitting. In Chapter VII, the initial observation is that each braid in the monoid B_n^+ admits a unique, well-defined, maximal right divisor that lies in the submonoid B_{n-1}^+. The same phenomenon occurs in the dual monoid B_n^{+*}.

LEMMA 2.1. *Assume $n \geqslant 3$. Then every braid β in B_n^{+*} admits a unique maximal right divisor β_1 that lies in B_{n-1}^{+*}.*

PROOF. The condition for applying Lemma VII.1.1 is that the considered submonoid, here B_{n-1}^{+*}, is closed under right divisor and left lcm, two conditions that can be easily checked in the current case. □

The unique braid β_1 provided by Lemma 2.1 is naturally called the *B_{n-1}^{+*}-tail* of β. It admits the following characterization:

LEMMA 2.2. *For every braid β in B_n^{+*}, $n \geqslant 3$, the following are equivalent:*
(i) *The braid β_1 is the B_{n-1}^{+*}-tail of β;*
(ii) *We have $\beta = \beta'\beta_1$ for some β' that is right divisible by no $a_{i,j}$ with $j < n$;*
(iii) *The braid β_1 is the right gcd of β and δ_{n-1}^d, where d is the degree of β.*

The proof is similar to that of Proposition VII.1.5.

Here comes the difference between B_n^+ and B_n^{+*}. In the case of B_n^+, the flip automorphism Φ_n has order 2, so we obtained a distinguished B_{n-1}^+-decomposition for every braid in B_n^+ by using the tail construction for B_{n-1}^+ and $\Phi_n(B_{n-1}^+)$ alternately. In our current case, the automorphism ϕ_n has order n, and we shall therefore consider the n submonoids B_{n-1}^{+*}, $\phi_n(B_{n-1}^{+*})$, ..., $\phi_n^{n-1}(B_{n-1}^{+*})$ cyclically. In this way, we find, for every braid β in B_n^{+*}, a distinguished decomposition

$$\beta = \phi_n^{p-1}(\beta_p) \cdot \ldots \cdot \phi_n(\beta_2) \cdot \beta_1, \tag{2.1}$$

where all braids β_r belong to B_{n-1}^{+*}.

DEFINITION 2.3. (See Figure 5.) Assume $n \geqslant 3$ and β belongs to B_n^{+*}. The *ϕ_n-splitting* of β is the unique sequence $(\beta_p, \ldots, \beta_1)$ in B_{n-1}^{+*} such that (2.1) holds and, for each r, the braid β_r is the B_{n-1}^{+*}-tail of $\phi_n^{p-r}(\beta_p) \cdot \ldots \cdot \phi_n(\beta_{r+1}) \cdot \beta_r$. The parameter p is called the *ϕ_n-breadth* of β.

As in the case of the Φ_n-splitting in B_n^+, the ϕ_n-splitting admits a simple characterization that follows from Lemma 2.2.

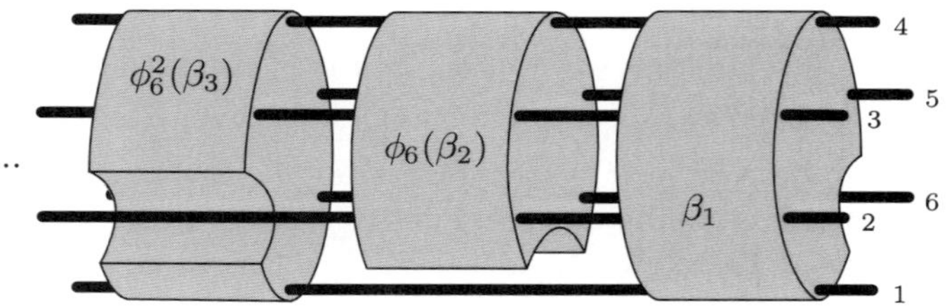

FIGURE 5. The ϕ_n-splitting of a braid in B_n^{+*}—here with $n = 6$: starting from the right, we extract the B_5^{+*}-tail, *i.e.*, the maximal fragment that leaves the last strand, here the 6th one, unbraided, then extract from the remainder the maximal fragment that leaves the first strand unbraided, then extract from the remainder the maximal fragment that leaves the second strand unbraided, etc.

PROPOSITION 2.4. *For each nontrivial braid β in B_n^{+*}, the ϕ_n-splitting of β is the unique sequence $(\beta_p, \dots, \beta_1)$ in B_{n-1}^{+*} such that $\beta = \phi_n^{p-1}(\beta_p) \cdot \dots \cdot \phi_n(\beta_2) \cdot \beta_1$ holds with $\beta_p \neq 1$ and, for each $r \geqslant 2$, the braid $\phi_n^{p-r+1}(\beta_p) \cdot \dots \cdot \phi_n(\beta_r)$ is right-divisible by no $a_{i,j}$ with $j < n$.*

As the notion of ϕ_n-splitting is crucial in what follows, we describe several examples.

EXAMPLE 2.5. We start with the ϕ_3-splitting of the three standard generators of B_3^{+*}, namely $a_{1,2}, a_{2,3}, a_{1,3}$, or $\mathtt{a}, \mathtt{b}$, and $\mathtt{b}'$ using Convention 1.2. As $\mathtt{a}$ belongs to B_2^{+*}, its ϕ_3-splitting is the length 1 sequence $(\mathtt{a})$. Then $\mathtt{b}$ does not lie in B_2^{+*}, but its image under ϕ_3^{-1}, namely $\mathtt{a}$, does. So the ϕ_3-splitting of $\mathtt{b}$ is $(\mathtt{a}, 1)$, corresponding to the decomposition

$$\mathtt{b} = \phi_3(\mathtt{a}) \cdot 1.$$

As for $\mathtt{b}'$, neither $\mathtt{b}'$ nor $\phi_3^{-1}(\mathtt{b}')$, which is $\mathtt{b}$, lies in B_2^{+*}. But $\phi_3^{-2}(\mathtt{b}')$ equals $\mathtt{a}$, so the ϕ_3-splitting of $\mathtt{b}'$ is $(\mathtt{a}, 1, 1)$, corresponding to the decomposition

$$\mathtt{b}' = \phi_3^2(\mathtt{a}) \cdot \phi_3(1) \cdot 1.$$

The reader can similarly check that, for each $n \geqslant 3$, the ϕ_n-splitting of $a_{i,j}$ is

$$\begin{cases} (a_{i,j}) & \text{for } i < j \leqslant n-1, \\ (a_{i-1,n-1}, 1) & \text{for } 2 \leqslant i \text{ and } j = n, \\ (a_{n-2,n-1}, 1, 1) & \text{for } i = 1 \text{ and } j = n. \end{cases} \tag{2.2}$$

EXAMPLE 2.6. We consider now a few more cases. Lemma 2.2(iii) provides a systematic algorithm for computing the B_{n-1}^{+*}-tail of a braid in B_n^{+*} from an implementation of the gcd operation. However, in simple cases, it is in general sufficient to play with the braid relations to recognize the expected tail using the fact that β_1 is the B_{n-1}^{+*}-tail of a braid β of B_n^{+*} if it is a right-divisor of β and the quotient $\beta\beta_1^{-1}$ is right-divisible by no $a_{i,j}$ with $j < n$.

Let us compute the ϕ_3-splitting of the braid δ_3^2—*i.e.*, $\mathtt{abab}$. The first step is to determine the B_2^{+*}-tail of δ_3^2. Using braid relations, we obtain $\delta_3^2 = \mathtt{ab'aa}$, and it is easy to check that $\mathtt{a}$ is not a right-divisor of $\mathtt{ab'}$. So $\mathtt{a}^2$ is the expected tail. Applying ϕ_3^{-1} to the remainder $\mathtt{ab'}$, we find the first step in the decomposition:

$$\delta_3^2 = \phi_3(\mathtt{b'b}) \cdot \mathtt{a}^2.$$

Then we look for the B_2^{+*}-tail of $\mathtt{b}'\mathtt{b}$. It turns out that $\mathtt{a}$ is not a right-divisor of $\mathtt{b}'\mathtt{b}$, so the B_2^{+*}-tail of $\mathtt{b}'\mathtt{b}$ is 1, and, applying ϕ_3^{-1} again, we find

$$\delta_3^2 = \phi_3^2(\mathtt{ba}) \cdot \phi_3(1) \cdot \mathtt{a}^2.$$

We now look for the B_2^{+*}-tail of $\mathtt{ba}$, which is $\mathtt{a}$, and we obtain

$$\delta_3^2 = \phi_3^3(\mathtt{a}) \cdot \phi_3^2(\mathtt{a}) \cdot \phi_3(1) \cdot \mathtt{a}^2.$$

We look for the B_2^{+*}-tail of $\mathtt{a}$, which is $\mathtt{a}$ itself, and we are done as the remainder is 1. So the ϕ_3-splitting of the braid δ_3^2 is the length 4 sequence $(\mathtt{a}, \mathtt{a}, 1, \mathtt{a}^2)$, *i.e.*, $(a_{1,2}, a_{1,2}, 1, a_{1,2}^2)$. Applying successive rotations, this corresponds to the factorization $\delta_3^2 = \mathtt{a} \cdot \mathtt{b}' \cdot 1 \cdot \mathtt{a}^2$.

The reader is invited to check similarly that, for each $d \geqslant 1$, the ϕ_3-splitting of δ_3^d is the length $d+2$ sequence $(\mathtt{a}, \mathtt{a}, \dots, \mathtt{a}, 1, \mathtt{a}^d)$, and that the ϕ_4-splitting of δ_4^d is the length $d+2$ sequence $(\mathtt{b}, \dots, \mathtt{b}, 1, \delta_3^2)$, corresponding for $d = 2$ to the factorization $\delta_4^2 = \mathtt{a} \cdot \mathtt{c}'' \cdot 1 \cdot \delta_3^2$. Note that the formula for δ_4^2 is quite similar to that for δ_3^2; in particular, it is not more complicated, a general phenomenon that illustrates the technical advantages of B_n^{+*} as compared with B_n^+.

2.2. The ϕ-normal form, case of 3 strands. As B_2^{+*} is a free monoid generated by $a_{1,2}$—*i.e.*, by σ_1—every element of B_2^{+*} has a unique expression as $a_{1,2}^e$ with $e \geqslant 0$. With the ϕ_3-splitting, we have a distinguished decomposition of every braid β in B_3^{+*} as a product of ϕ_3-images of braids in B_2^{+*}. From there, we deduce a distinguished expresssion of β.

DEFINITION 2.7. Assume $\beta \in B_3^{+*}$. Let $(a_{1,2}^{e_p}, \dots, a_{1,2}^{e_1})$ be the ϕ_3-splitting of β. We define the *code* of β to be the sequence $(e_p, \dots, e_1)$, and its *ϕ-normal form* to be the word $a_{[p,p+1]}^{e_p} \dots a_{1,3}^{e_3}\, a_{2,3}^{e_2}\, a_{1,2}^{e_1}$, where $a_{[p,p+1]}$ denotes $a_{1,2}$ (*resp.* $a_{2,3}$, $a_{1,3}$) for $p = 1$ (*resp.* 2, 3) mod 3.

By construction, the ϕ-normal form of a braid β of B_3^{+*} is the word obtained by concatenating the (unique words representing the) entries in its ϕ_3-splitting after applying the required rotations ϕ_3.

EXAMPLE 2.8. We saw in Example 2.6 that the ϕ_3-splitting of δ_3^2 is the sequence $(\mathtt{a}, \mathtt{a}, 1, \mathtt{a}^2)$ corresponding to the decomposition $\delta_3^2 = \phi_3^3(\mathtt{a}) \cdot \phi_3^2(\mathtt{a}) \cdot \phi_3(1) \cdot \mathtt{a}^2$, *i.e.*, $\delta_3^2 = \mathtt{a} \cdot \mathtt{b}' \cdot 1 \cdot \mathtt{a}^2$. Hence the ϕ-normal form of δ_3^2 is the word $\mathtt{ab'aa}$.

Like Φ-normal words in Chapter VII, ϕ-normal words can easily be characterized in terms of their exponents, *i.e.*, the sizes of the blocks of successive letters. We say that a word w on the alphabet $\{a_{1,2}, a_{2,3}, a_{1,3}\}$ admits the exponent sequence $(e_p, \dots, e_1)$ if, with the notational convention of Definition 2.7, we have $w = a_{[p,p+1]}^{e_p} \dots a_{1,3}^{e_3}\, a_{2,3}^{e_2}\, a_{1,2}^{e_1}$ and p is minimal, *i.e.*, we have $e_p \geqslant 1$ and $e_{r+1} = e_r = 0$ may occur only for $r = 1$.

PROPOSITION 2.9. *A word on the alphabet $\{a_{1,2}, a_{2,3}, a_{1,3}\}$ with the exponent sequence $(e_p, \dots, e_1)$ is ϕ-normal if and only if $e_r \geqslant 1$ holds for $r \geqslant 3$.*

2.3. The ϕ-normal form, general case. For β in B_4^{+*}, the ϕ_4-splitting provides a distinguished decomposition for β in terms of braids in B_3^{+*}, so, as in Section VII.2.2, we can obtain a unique normal expression for a braid of B_4^{+*} by using the ϕ-normal form of the successive entries in its ϕ_4-splitting, and so on iteratively.

EXAMPLE 2.10. (See Figure 6.) We saw in Example 2.6 that the ϕ_4-splitting of δ_4^2 is the sequence $(\mathtt{b}, \mathtt{b}, 1, \delta_3^2)$, corresponding to the decomposition

$$\delta_4^2 = \phi_4^3(\mathtt{b}) \cdot \phi_4^2(\mathtt{b}) \cdot \phi_4(1) \cdot \delta_3^2. \tag{2.3}$$

Then we saw in Example 2.8 that the ϕ-normal form of δ_3^2 is $\mathtt{ab'aa}$. Similarly, according to Example 2.5, the ϕ_3-splitting of $\mathtt{b}$ is $(\mathtt{a}, 1)$, leading to the ϕ-normal form $\mathtt{b}$. When we insert these expressions in (2.3) and apply the rotations, we obtain a distinguished representing word for δ_4^2, namely $\mathtt{a} \cdot \mathtt{c''} \cdot \varepsilon \cdot \mathtt{ab'aa}$, *i.e.*, $\mathtt{ac''ab'aa}$. The latter word will naturally be defined to be the ϕ-normal form of δ_4^2. It corresponds to the iterated splitting illustrated in Figure 6.

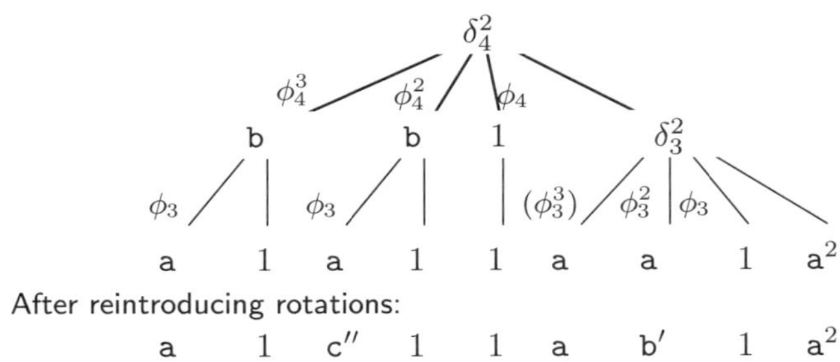

FIGURE 6. Two-step construction of the ϕ-normal form of δ_4^2: we first split δ_4^2 into a sequence of braids in B_3^{+*}, then split each of them into a sequence of braids in B_2^{+*}, *i.e.*, of powers of a; taking rotations into account, we obtain a distinguished word representing δ_4^2, here $\mathtt{ac''ab'a^2}$.

DEFINITION 2.11. For β in B_n^{+*}, we define the *ϕ-normal form* of β to be

- for $n = 2$, the unique word representing β;
- for $n \geqslant 3$, the word $\phi_n^{p-1}(w_p) \cdot \ldots \cdot \phi_n(w_2) \cdot w_1$, where $(\beta_p, \ldots, \beta_1)$ is the ϕ_n-splitting of β and w_r is the ϕ-normal form of β_r for each r.

The construction is exactly similar to that of Definition VII.2.8. As for computational complexity, all procedures mentioned here for B_n^{+*} have the same characteristics as their B_n^+-counterparts. In particular, the ϕ-normal form of an n-strand braid specified by a braid word of length ℓ on the alphabet of the letters $a_{i,j}$ can be computed in time $O(\ell^2)$.

2.4. A linear ordering of B_n^{+*}. Using the ϕ_n-splitting, we recursively introduce a linear ordering on B_n^{+*}. As in Chapter VII, we shall eventually see in Section 3 that it coincides with the σ^Φ-ordering. The construction is entirely similar to that of Section VII.2.3, and we shall only insist on the specificities of the dual case.

DEFINITION 2.12. For $n \geqslant 2$, we recursively define $<_n^*$ on B_n^{+*} as follows:

- For β, β' in B_2^{+*}, we declare $\beta <_2^* \beta'$ if we have $\beta = a_{1,2}^e$ and $\beta' = a_{1,2}^{e'}$ with $e < e'$;
- For β, β' in B_n^{+*} with $n \geqslant 3$, we declare $\beta <_n^* \beta'$ if the ϕ_n-splitting of β is smaller than the ϕ_n-splitting of β' for the ShortLex-extension of $<_{n-1}^*$.

EXAMPLE 2.13. We saw in Example 2.6 that the ϕ_3-splittings of $\mathtt{a}, \mathtt{b}$, and $\mathtt{b}'$, respectively are $(\mathtt{a})$, $(\mathtt{a}, 1)$, and $(\mathtt{a}, 1, 1)$, of lengths $1, 2$, and 3. Hence, by definition, we have $\mathtt{a} <^*_3 \mathtt{b} <^*_3 \mathtt{b}'$. Similarly, using (2.2), an easy induction on $n \geqslant 3$ shows that, for $1 \leqslant i < j \leqslant n$ and $1 \leqslant i' < j' \leqslant n$, the relation $a_{i,j} <^*_n a_{i',j'}$ holds if and only if we have either $j < j'$, or $j = j'$ and $i > i'$. So, in particular, the $<^*_n$-smallest $a_{i,j}$ in B^{+*}_n that does not lie in B^{+*}_{n-1} is $a_{n-1,n}$.

PROPOSITION 2.14. (i) *For $n \geqslant 2$, the relation $<^*_n$ is a well-ordering of B^{+*}_n. For each braid β in B^{+*}_n, the immediate $<^*_n$-successor of β is $\beta\, a_{1,2}$.*

(ii) *For $n \geqslant 3$, the ordering $<^*_n$ extends the ordering $<^*_{n-1}$, and B^{+*}_{n-1} is the initial segment of B^{+*}_n determined by $a_{n-1,n}$, i.e., $B^{+*}_{n-1} = \{\beta \in B^{+*}_n \mid \beta <^*_n a_{n-1,n}\}$ holds.*

PROOF. Everything is similar to Proposition VII.2.15, except the last point. For the latter, we observe that the braids of B^{+*}_n whose ϕ_n-breadth is 1 are the elements of B^{+*}_{n-1}.

On the other hand, the braid $a_{n-1,n}$ does not lie in B^{+*}_{n-1}, and its ϕ_n-splitting is $(a_{n-2,n-1}, 1)$. Let β be a braid in B^{+*}_{n-1} that satisfies $\beta <^*_{n-1} a_{n-2,n-1}$. By the induction hypothesis, β lies in B^{+*}_{n-2}—with the convention $B^{+*}_1 = \{1\}$—hence $\phi_n(\beta)$ lies in B^{+*}_{n-1}. It follows that $(\beta, 1)$ cannot be the ϕ_n-splitting of a braid in B^{+*}_n. Thus $(a_{n-2,n-1}, 1)$ is the smallest length 2 splitting in B^{+*}_n. Hence, the associated braid, namely $a_{n-1,n}$, is the $<^*_n$-minimal braid that has a length 2 splitting, *i.e.*, that does not lie in B^{+*}_{n-1}. □

Owing to Proposition 2.14(ii), we skip the index n and write $<^*$ for $<^*_n$. According to Example 2.13, the $<^*$-increasing enumeration of the generators $a_{i,j}$ is

$$\mathtt{a} <^* \mathtt{b} <^* \mathtt{b}' <^* \mathtt{c} <^* \mathtt{c}' <^* \mathtt{c}'' <^* \mathtt{d} <^* \mathtt{d}' <^* \mathtt{d}'' <^* \mathtt{d}''' <^* \ldots.$$

We can observe that the above sequence coincides with the one of (1.6), *i.e.*, that the linear orders $<^\Phi$ and $<^*$ coincide on the set $\{a_{i,j} \mid i < j\}$. We can naturally ask whether the coincidence extends to all braids in B^{+*}_∞. The positive answer will be given below.

3. Connection between orders

At this point, we have two *a priori* unrelated linear orderings of the dual braid monoid B^{+*}_n, namely the one induced by the σ^Φ-ordering, and the ordering $<^*$ constructed in Section 2.4 by a recursive definition based on the ϕ_n-splitting. We shall see now that these orders actually coincide, a recent result by Fromentin [**90**]. This beautiful result gives a new proof for Property **C**, as well as a complete and simple description for the σ^Φ-ordering of B^{+*}_n.

3.1. The main result and its applications. As in Chapter VII with the monoid B^+_n and the ordering $<^+$, the method for proving that the ordering $<^*$ deduced from the ϕ_n-splitting coincides with the σ^Φ-ordering consists in explicitly constructing σ^Φ-positive expressions. The key statement is the following result:

PROPOSITION 3.1 (Fromentin [**90**]). *Assume that β, β' are braids in B^{+*}_n that satisfy $\beta <^* \beta'$. Then $\beta^{-1}\beta'$ is σ^Φ-positive.*

Before sketching the main steps of the proof, we mention applications. First, exactly as in Chapter VII, we obtain a new proof for Property **C**:

COROLLARY 3.2 (Property **C**). *If β, β' are distinct braids in B_∞^{+*}—hence in particular in B_∞^+—then $\beta^{-1}\beta'$ is σ-positive or σ-negative.*

PROOF. The relation $<^*$ is a linear ordering of B_∞^{+*}, hence, for all β, β' in B_∞^{+*}, one of $\beta <^* \beta'$, $\beta' <^* \beta$, $\beta' = \beta$ holds, and we apply Proposition 3.1. □

Next, we deduce the expected coincidence between the orders $<^\Phi$ and $<^*$.

COROLLARY 3.3. *The ordering $<^*$ of Section 2.4 coincides with the σ^Φ-ordering. In particular, $<^*$ is compatible with multiplication on the left.*

PROOF. Proposition 3.1 asserts that, for β, β' in B_n^{+*}, the relation $\beta <^* \beta'$ implies $\beta <^\Phi \beta'$. Conversely, as $<^*$ is a linear ordering, $\beta \not<^* \beta'$ implies $\beta' \leqslant^* \beta$, whence $\beta' \leqslant^\Phi \beta$, and, using Property **A**, we deduce $\beta \not<^\Phi \beta'$. □

From the previous coincidence result, we deduce that the recursive definition of $<^*$ is a definition for the restriction of the σ^Φ-ordering to B_n^{+*}:

PROPOSITION 3.4. *Assume $n \geqslant 3$ and $\beta, \beta' \in B_n^{+*}$. Then $\beta <^\Phi \beta'$ holds if and only if the ϕ_n-splitting of β is smaller than the ϕ_n-splitting of β' with respect to the* `ShortLex`*-extension of the restriction of $<^\Phi$ to B_{n-1}^{+*}.*

Another application of Corollary 3.3, hence of Proposition 3.1, is that the restriction of the σ^Φ-ordering to the dual braid monoid B_n^{+*} is a well-ordering.

PROPOSITION 3.5. *For each $n \geqslant 2$, the restriction of the σ^Φ-ordering to B_n^{+*} is a well-ordering of ordinal type $\omega^{\omega^{n-2}}$.*

PROOF. The argument is the same as for Proposition VII.4.9: as the `ShortLex`-extension of a well-ordering of ordinal type λ is a well-ordering of ordinal type λ^ω, we inductively obtain that $(B_n^{+*}, <^\Phi)$ is a well-ordering of ordinal type at most $\omega^{\omega^{n-2}}$. In order to prove that the ordinal type is exactly $\omega^{\omega^{n-2}}$, it is enough to exhibit a subset X of B_n^{+*} such that the ordinal type of $(X, <^\Phi)$ is at least $\omega^{\omega^{n-2}}$: we saw in Chapter VII that B_n^+ is such a subset. □

REMARK 3.6. The current result that the restriction of the σ^Φ-ordering to B_n^{+*} is a well-ordering gives a new proof of the previous result that its restriction to B_n^+ is a well-ordering: indeed, B_n^+ is included in B_n^{+*}, and any restriction of a well-ordering is a well-ordering.

Finally, as the well-ordered set $(B_n^{+*}, <^\Phi)$ is an initial segment of the well-ordered set $(B_{n+1}^{+*}, <^\Phi)$, the union B_∞^{+*} of these well-ordered sets is a well-ordered set, whose ordinal type is the supremum of the order types of its initial intervals:

COROLLARY 3.7. *The σ^Φ-ordering of B_∞^{+*} is a well-ordering of ordinal type ω^{ω^ω}.*

REMARK 3.8. No result involving the σ-ordering of B_n^{+*} directly follows from the previous results, because, contrary to B_n^+, the monoid B_n^{+*} is not invariant under the flip morphism Φ_n.

3.2. Splitting the problem. We shall now describe the main steps in Fromentin's proof of Proposition 3.1. Although relatively delicate, the argument is nevertheless very natural: it simply consists in directly computing the quotient of two braids β, β' satisfying $\beta <^* \beta'$ and using the relations of the monoid B_n^{+*} to obtain a σ^Φ-positive expression of this quotient. The proof uses inductions on the braid index, but, for each value of n, it is essentially a computational verification—but a tricky one: as the Birman–Ko–Lee generators $a_{i,j}$ form a redundant family of generators, the quotient-braid $\beta^{-1}\beta'$ involved in Proposition 3.1 generally admits a huge number of expressions in terms of the generators $a_{i,j}$, and extracting one that witnesses σ^{Φ}-positivity is not so easy.

As in Chapter VII, the first step of the proof consists in replacing the initial problem, which involves two braids β, β' with two easier questions involving only one braid at a time. To this end, one introduces landmark braids playing the role of a separator with respect to the ϕ_n-breadth.

DEFINITION 3.9. For $n \geqslant 2$, we put $\widehat{\delta}_{n,-1} = 1$, $\widehat{\delta}_{n,0} = a_{n-1,n}$ and, for $d \geqslant 1$,

$$\widehat{\delta}_{n,d} = \phi_n^{d+1}(a_{n-2,n-1}) \cdot \ldots \cdot \phi_n^2(a_{n-2,n-1}). \tag{3.1}$$

For $d \geqslant 1$, the braid $\widehat{\delta}_{n,d}$ is the braid in B_n^{+*} whose ϕ_n-splitting is the length $d+2$ sequence $(a_{n-2,n-1}, \ldots, a_{n-2,n-1}, 1, 1)$. For instance, we find $\widehat{\delta}_{3,1} = \mathtt{b}'$, $\widehat{\delta}_{3,2} = \mathtt{ab}'$, $\widehat{\delta}_{3,3} = \mathtt{bab}'$, $\widehat{\delta}_{3,4} = \mathtt{b'bab}'$, and so on with the letters $\mathtt{a}, \mathtt{b}, \mathtt{b}'$ cyclically repeated from right to left. Similarly, we find $\widehat{\delta}_{4,1} = \mathtt{c}''$, $\widehat{\delta}_{4,2} = \mathtt{ac}''$, $\widehat{\delta}_{4,3} = \mathtt{bac}''$, $\widehat{\delta}_{4,4} = \mathtt{cbac}''$, now with period 4. The connection between the braid $\widehat{\delta}_{n,d}$ and the powers of the Garside element δ_n of B_n^{+*} is similar to the connection between the braid $\widehat{\Delta}_{n,d}$ and the powers of Δ_n in Chapter VII.

LEMMA 3.10. *For $d \geqslant 1$, we have $\delta_n^d = \widehat{\delta}_{n,d} \cdot \delta_{n-1}^d$.*

PROOF. We use induction on $d \geqslant 1$. Consider $d = 1$. By definition, we have $a_{1,n} = \delta_{n-1} a_{n-1,n} \delta_{n-1}^{-1}$, hence $\delta_{n-1} a_{n-1,n} = a_{1,n} \delta_{n-1}$. Using the relation $\phi_n(\beta) \cdot \delta_n = \delta_n \cdot \beta$ that follows from the definition of ϕ_n, we deduce

$$\delta_n = \delta_{n-1} \cdot a_{n-1,n} = a_{1,n} \cdot \delta_{n-1} = \phi_n(a_{n-1,n}) \cdot \delta_{n-1} = \phi_n^2(a_{n-2,n-1}) \cdot \delta_{n-1} = \widehat{\delta}_{n,1} \cdot \delta_{n-1}.$$

Assume now $d \geqslant 2$. Using the induction hypothesis, we find

$$\begin{aligned} \delta_n^d = \delta_n^{d-1} \cdot \delta_n &= \delta_n^{d-1} \cdot \phi_n^2(a_{n-2,n-1}) \cdot \delta_{n-1} \\ &= \phi_n^{d+1}(a_{n-2,n-1}) \cdot \delta_n^{d-1} \cdot \delta_{n-1} \\ &= \phi_n^{d+1}(a_{n-2,n-1}) \cdot \widehat{\delta}_{n,d-1} \cdot \delta_{n-1}^{d-1} \cdot \delta_{n-1} = \widehat{\delta}_{n,d} \cdot \delta_{n-1}^d, \end{aligned}$$

which is the expected formula. □

It follows that $\widehat{\delta}_{n,d}$ is the remainder of δ_n^d when its B_{n-1}^{+*}-tail, which is δ_{n-1}^d, is removed.

Our aim will be to establish for B_n^{+*} and the braids $\widehat{\delta}_{n,d}$ a result similar to Lemma VII.2.21, namely that every braid in B_n^{+*} that has ϕ_n-breadth p lies between $\widehat{\delta}_{n,p-2}$ and $\widehat{\delta}_{n,p-1}$, according to the image in Figure 7.

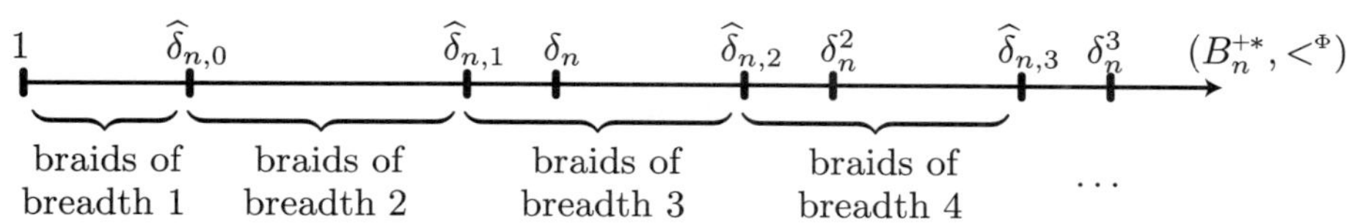

FIGURE 7. The braids $\widehat{\delta}_{n,d}$ as separators in the ordered line $(B_n^{+*}, <^{\Phi})$.

Establishing the upper bound result is easy.

PROPOSITION 3.11. *If β is a braid in B_n^{+*} whose ϕ_n-breadth is p, then $\beta^{-1}\,\widehat{\delta}_{n,p-1}$ is σ_{n-1}^{Φ}-positive.*

PROOF. For $p=1$, the result is clear, as the hypothesis means that β^{-1} belongs to B_{n-1}^{+*}, while $\widehat{\delta}_{n,0}$, which is σ_{n-1} by definition, is σ_{n-1}^{Φ}-positive. So assume $p \geqslant 2$. The computation is similar to that used for Lemma VII.2.21. Let $(\beta_p, \dots, \beta_1)$ be the ϕ_n-splitting of β. Using Lemma 3.10 to let δ_n^{p-1} appear, and then pushing the factors δ_n to the left appealing to the equality $\phi_n^r(\gamma) \cdot \delta_n = \delta_n \cdot \phi_n^{r-1}(\gamma)$, we obtain

$$\begin{aligned}\beta^{-1}\widehat{\delta}_{n,p-1} &= \beta^{-1} \cdot \delta_n^{p-1} \cdot \delta_{n-1}^{-p+1}\\ &= \beta_1^{-1} \cdot \phi_n(\beta_2)^{-1} \cdot \dots \cdot \phi_n^{p-1}(\beta_p)^{-1} \cdot \delta_n^{p-1} \cdot \delta_{n-1}^{-p+1}\\ &= \beta_1^{-1} \cdot \delta_n \cdot \beta_2^{-1} \cdot \delta_n \cdot \dots \cdot \delta_n \cdot \beta_p^{-1} \cdot \delta_{n-1}^{-p+1}.\end{aligned}$$

Choosing for each braid β_r^{-1} and for δ_{n-1} a representative that contains no σ_{n-1}, and choosing for δ_n the representative $\sigma_1 \dots \sigma_{n-1}$, we obtain from the above decomposition an expression for $\beta^{-1}\widehat{\delta}_{n,p-1}$ that contains $p-1$ letters σ_{n-1} and no letter σ_{n-1}^{-1}. □

Our aim will be to prove the following counterpart of Proposition 3.11:

PROPOSITION 3.12. *If β is a braid in B_n^{+*} whose ϕ_n-breadth is p, then $\widehat{\delta}_{n,p-2}^{-1}\,\beta$ is trivial or σ^{Φ}-positive.*

Gathering Propositions 3.11 and 3.12 immediately gives the result illustrated in Figure 7:

COROLLARY 3.13. *If β is a braid in B_n^{+*} whose ϕ_n-breadth is p, then we have*

$$\widehat{\delta}_{n,p-2} \leqslant^{\Phi} \beta <^{\Phi} \widehat{\delta}_{n,p-1}. \tag{3.2}$$

When Proposition 3.12 is proved, one can easily deduce that, if β, β' belong to B_n^{+*} and the ϕ_n-breadth of β is strictly less than that of β'—and that the latter is at least 3—then the quotient $\beta^{-1}\beta'$ is σ^{Φ}-positive—so, we essentially obtain the first half of Proposition 3.1, namely the `Short` part in the `ShortLex`-comparison. Handling the `Lex` part requires an induction on n and more precise statements, that will be described below.

3.3. Dangerous elements. Proving Proposition 3.12 turns out to be rather difficult. To explain the problem, we shall consider a weaker statement that is more suitable for subsequent developments.

DEFINITION 3.14. A braid β is said to be *σ_i^{Φ}-nonnegative* if it is σ_i^{Φ}-positive, or it belongs to B_i.

In other words, a braid is σ_i^{Φ}-nonnegative if it admits an expression by a braid word that contains no σ_i^{-1} and no $\sigma_j^{\pm 1}$ with $j > i$.

Let us consider for a while the following statement:

$$\begin{aligned}&\text{For every } \beta \text{ in } B_n^{+*} \text{ with } \phi_n\text{-breadth } p,\\ &\qquad\qquad \text{the quotient } \widehat{\delta}_{n,p-2}^{-1}\,\beta \text{ is } \sigma_{n-1}^{\Phi}\text{-nonnegative.}\end{aligned} \tag{3.3}$$

The trivial braid and all σ^{Φ}-positive n-strand braids are σ_{n-1}^{Φ}-nonnegative, so (3.3) is a weakening of Proposition 3.12. Let us try to prove it using the method of Lemma VII.2.21, *i.e.*, by reversing the argument used for Proposition 3.11. Our aim is to prove that $\widehat{\delta}_{n,p-2}^{-1} \cdot \beta$ is σ_{n-1}^{Φ}-nonnegative. As $\widehat{\delta}_{n,p-2}^{-1}$ equals $\delta_{n-1}^{p-2} \cdot \delta_n^{-p+2}$ and δ_{n-1}^{p-2} is σ_{n-1}^{Φ}-nonnegative, we can forget about that factor, and we are left with proving that, if $(\beta_p, ..., \beta_1)$ is the ϕ_n-splitting of β, then

$$\delta_n^{-p+2} \cdot \phi_n^{p-1}(\beta_p) \cdot \phi_n^{p-2}(\beta_{p-1}) \cdot ... \cdot \beta_1 \tag{3.4}$$

is σ_{n-1}^{Φ}-nonnegative. As in Proposition 3.11, we can push the negative factors δ_n^{-1} to the right with the hope that each of them will be absorbed by the neighbouring factor $\phi_n^{r-1}(\beta_r)$ and result in a σ_{n-1}^{Φ}-nonnegative factor. Unfortunately, this naive approach need *not* work.

EXAMPLE 3.15. Let $\beta = \mathtt{c'b'bc'}$. The ϕ_4-splitting of β is $(\mathtt{b'}, \mathtt{b'}, \mathtt{ab'}, 1)$, so β has ϕ_4-breadth 4, and we wish to compare it with $\widehat{\delta}_{4,2}$. Our claim is that $\widehat{\delta}_{4,2}^{-1}\beta$ is σ_3^{Φ}-nonnegative. Following the scheme above, we express $\widehat{\delta}_{4,2}^{-1}$ as $\delta_3^2 \cdot \delta_4^{-2}$ and push the δ_4^{-1} factors to the right in the B_3^{+*}-decomposition of β, thus obtaining

$$\begin{aligned}\widehat{\delta}_{4,2}^{-1} \cdot \beta = \delta_3^2 \cdot \delta_4^{-2} \cdot \beta &= \delta_3^2 \cdot \delta_4^{-2} \cdot \phi_4^3(\mathtt{b'}) \cdot \phi_4^2(\mathtt{b'}) \cdot \phi_4(\mathtt{ab'}) \\ &= \delta_3^2 \cdot \underline{\phi_4(\mathtt{b'}) \cdot \delta_4^{-1}} \cdot \underline{\phi_4(\mathtt{b'}) \cdot \delta_4^{-1}} \cdot \phi_4(\mathtt{ab'}).\end{aligned}$$

We would like to show that each positive factor neutralizes the factor δ_4^{-1} that lies on its right. For both the first and the second underlined factors, we find $\phi_4(\mathtt{b'}) \cdot \delta_4^{-1} = \mathtt{CA}$, a σ_3^{Φ}-negative braid. Hence, the expression of $\widehat{\delta}_{4,2}^{-1}\beta$, obtained by simply concatenating expressions of the above factors, is not σ_3^{Φ}-nonnegative—and *a fortiori* not σ_3^{Φ}-positive.

So we have to be more subtle. It is not hard to identify the fragments that are responsible for the problem described above: these are the fragments that can, when multiplied by δ_n^{-1}, potentially lead to a σ_{n-1}^{Φ}-negative factor. Such fragments will be called *dangerous*.

DEFINITION 3.16. (i) For $i \leqslant n-1$, a braid of B_{n-1}^{+*} is called *$a_{i,n}$-dangerous* if it admits a decomposition of the form

$$\delta_{j_1,n-1}^{-1} \cdots \delta_{j_\ell,n-1}^{-1},$$

where $j_1 \geqslant ... \geqslant j_\ell = i$ holds and $\delta_{j,n-1}$ stands for $\sigma_j\sigma_{j+1}...\sigma_{n-2}$.

(ii) A braid of B_n is called *$\sigma_{i,n-1}^{\Phi}$-positive* if it admits a decomposition of the form

$$\beta \cdot \sigma_{n-1} \cdot \beta',$$

where β is σ_{n-1}^{Φ}-nonnegative and β' is $a_{i,n}$-dangerous.

By definition, an $a_{i,n}$-dangerous braid lies in B_{n-1}, so every $\sigma_{i,n-1}^{\Phi}$-positive braid is σ_{n-1}^{Φ}-positive.

EXAMPLE 3.17. Write $(\beta_4, ..., \beta_1)$ for the ϕ_4-splitting of the braid β of Example 3.15. We found $\beta_4 = \beta_3 = \mathtt{b'}$. Now, $\mathtt{b'}$, *i.e.*, $a_{1,3}$, can be decomposed as $\mathtt{b'} = \mathtt{abA} = \sigma_1 \cdot \sigma_2 \cdot \delta_{1,2}^{-1}$, where σ_1 is σ_3-nonnegative and $\delta_{1,2}^{-1}$ is $a_{1,3}$-dangerous: so β_3 and β_4 are $\sigma_{1,3}^{\Phi}$-positive.

3.4. The main induction. The technical statement we shall use to establish the main result, namely Proposition 3.1, is a refinement of (3.3) that takes dangerous fragments into account and somehow keeps them under control. In order to be able to use an induction on the breadth, and, to this end, to maintain the needed induction hypothesis, we shall consider not only the braid of (3.4), but also, more generally, every braid of the form

$$\widehat{\delta}_{n,p-2}^{-1} \cdot \phi_n^{p-1}(\gamma) \cdot \phi_n^{p-2}(\beta_{p-1}) \cdot \ldots \cdot \beta_1, \tag{3.5}$$

where γ resembles β_p enough. Hereafter, for β in B_n^{+*}, the last letter in the ϕ-normal form of β is simply called the *last letter of* β.

PROPOSITION 3.18. *Assume that β is a braid belonging to $B_n^{+*} \setminus B_{n-1}^{+*}$. Let $(\beta_p, \ldots, \beta_1)$ be the ϕ_n-splitting of β, and assume $\beta_r \neq a_{n-2,n-1}$ for $r \geqslant 3$. Let $a_{i-1,n-1}$ be the last letter of β_p, and $a_{j,n}$ be the last letter of $\beta\beta_1^{-1}$. Then, for every $\sigma_{i-1,n-2}^{\Phi}$-positive braid γ, the braid*

$$\widehat{\delta}_{n,p-2}^{-1} \cdot \phi_n^{p-1}(\gamma) \cdot \phi_n^{p-2}(\beta_{p-1}) \cdot \ldots \cdot \beta_1 \tag{3.6}$$

is $\sigma_{j,n-1}^{\Phi}$-positive.

The hypotheses of Proposition 3.18 guarantee that β_p itself is $\sigma_{i-1,n-2}^{\Phi}$-positive, so, provided the hypothesis on the ϕ-normal form of β is satisfied, Proposition 3.18 applies to $\widehat{\delta}_{n,p-2}^{-1} \cdot \beta$ itself. The excluded cases, namely when some nonfinal entry in the ϕ_n-splitting of β is $a_{n-2,n-1}$, turn out to occur only when β is close to the braid $\widehat{\delta}_{n,p-2}$ with respect to the $<^*$-ordering. They require a specific treatment, but the principle remains the same, and the conclusion is again that the quotient is σ_{n-1}^{Φ}-nonnegative.

When Proposition 3.18 is established, it is not very difficult to conclude.

PROOF OF PROPOSITION 3.1 FROM PROPOSITION 3.18 (SKETCH). We prove using induction on $n \geqslant 2$ the following strengthening of Proposition 3.1:

> For β, β' in B_n^{+*} satisfying $\beta <^* \beta'$, the braid $\beta^{-1}\beta'$ is σ^{Φ}-positive; moreover, if the ϕ_n-breadths of β and β' are at least 2, then $\beta^{-1}\beta'$ is $\sigma_{i,n-1}^{\Phi}$-positive, where $a_{i,n}$ is the last letter of $\beta'\beta_1'^{-1}$ and β_1' is the B_{n-1}^{+*}-tail of β'.

We use induction on the braid index n. Everything is obvious for $n = 2$. Assume $n \geqslant 3$. Let β, β' be braids in B_n^{+*} satisfying $\beta <^* \beta'$, and let p, p' be their respective breadths. We consider the two cases in the definition of a `ShortLex`-extension.

`Short` case: $p < p'$. Then, we use $\widehat{\delta}_{n,p-1}$ to separate β and β'. By Proposition 3.11, $\beta^{-1}\widehat{\delta}_{n,p-1}$ is σ_{n-1}^{Φ}-positive. Provided we are not in one of the special cases excluded from Proposition 3.18, the latter implies that $\widehat{\delta}_{n,p-1}^{-1}\beta'$ is σ_{n-1}^{Φ}-nonnegative. Hence $\beta^{-1}\beta'$ is a σ_{n-1}^{Φ}-positive. Moreover, if $p \geqslant 2$ holds and the last letter of $\beta'_{p'}$ is not $a_{n-2,n-1}$, Proposition 3.18 implies that $\widehat{\delta}_{n,p-1}^{-1}\beta'$ is $\sigma_{i,n-1}^{\Phi}$-positive. The same holds for $\beta^{-1}\beta'$, and multiplying on the left by the σ_{n-1}^{Φ}-nonnegative braid $\beta^{-1}\widehat{\delta}_{n,p-1}$ does not change the result. The remaining particular cases are treated directly.

`Lex` case: $p = p'$. For $p = 1$, the braids β and β' lie in B_{n-1}^{+*}, and the result directly follows the induction hypothesis. Assume $p \geqslant 2$. At the expense of left dividing by the common left factors, we may assume $\beta_p <^* \beta'_p$. The induction hypothesis guarantees that $\beta_p^{-1}\beta'_p$ is σ^{Φ}_{n-2}-positive. Thus we are left with

$$\beta_1^{-1} \cdot \ldots \cdot \phi_n^{p-2}(\beta_{p-1}^{-1}) \cdot \phi_n^{p-1}(\beta_p^{-1}\beta'_p) \cdot \phi_n^{p-2}(\beta'_{p-1}) \cdot \ldots \cdot \beta'_1.$$

By Proposition 3.11, $\beta_1^{-1} \cdot \ldots \cdot \phi_n^{p-2}(\beta_{p-1}^{-1}) \cdot \widehat{\delta}_{n,p-2}$ is σ^{Φ}_{n-1}-positive, so, up to neglecting initial σ^{Φ}_{n-1}-nonnegative factors, we are left with proving the σ^{Φ}_{n-1}-positivity of

$$\widehat{\delta}_{n,p-2}^{-1} \cdot \phi_n^{p-1}(\beta_p^{-1}\beta'_p) \cdot \phi_n^{p-2}(\beta'_{p-1}) \cdot \ldots \cdot \beta'_1.$$

By the induction hypothesis, and because the tail of β'_p must be trivial, the factor $\beta_p^{-1}\beta'_p$ is $\sigma^{\Phi}_{i-1,n-2}$-positive, where $a_{i-1,n-1}$ is the last letter of β'_p. Then applying Proposition 3.18 with $\gamma = \beta_p^{-1}\beta'_p$ and β'_r in place of β_r for $r = p-1, \ldots, 1$ gives the result. Once again, some variants of Proposition 3.18 are used to treat the remaining special cases. □

3.5. Stairs. Thus we are left with proving Proposition 3.18. The main problem is to control the dangerous fragments. We shall see that, for each $a_{i,n}$-dangerous fragment possibly occurring in the evaluation of (3.6), there exists in the sequence $(\beta_p, \ldots, \beta_1)$ some factors that neutralize the σ_{n-1}^{-1} factors possibly occurring in that dangerous fragment. Such protecting factors will be called *$a_{i,n}$-stairs*.

DEFINITION 3.19. (See Figure 8.) A word w in the letters $a_{i,j}$ with $1 \leqslant i < j \leqslant n-1$ is called an *$a_{i,n}$-stair lent on $a_{j,n-1}$* if there exists a decomposition

$$w = w_0\, s_1\, w_1 \ldots w_{\ell-1}\, s_\ell\, w_\ell, \tag{3.7}$$

and a sequence $i = k_1 < k_2 < \ldots < k_{\ell+1} = n-1$ such that

- for each $r \leqslant \ell$, the letter s_r is $a_{k,k_{r+1}}$ for some k satisfying $k < k_r$,
- for each $r < \ell$, the word w_r contains no $a_{i,j}$ with $i < k_{r+1} < j$,
- the last letter of $s_\ell w_\ell$ is $a_{j,n-1}$.

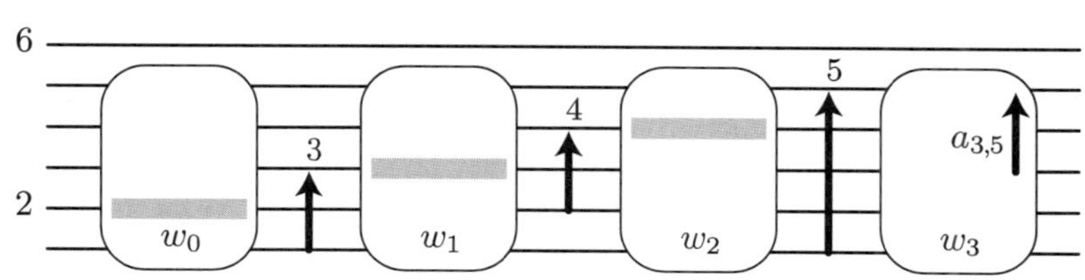

FIGURE 8. An $a_{2,6}$-stair lent on $a_{3,5}$: a word that contains letters $a_{i,j}$ such that the upper indices grow until $n-1$—here 5—and some overlapping condition is guaranteed.

Now, everything works and we can complete the argument. Firstly, explicit computations ensure that the ϕ-normal form of the braids we are interested in is a stair.

LEMMA 3.20. *Assume that $(\beta_p, \ldots, \beta_1)$ is a ϕ_n-splitting. For $1 \leqslant r \leqslant p$, let w_r be the ϕ-normal form of β_r, and let s_r be the last letter in w_r. Then, provided $r \geqslant 3$ and $s_r \neq a_{n-2,n-1}$ hold, the word w_{r-1} is an $\phi_n(s_r)$-stair lent on s_{r-1}.*

Then, the fact that stairs achieve the expected protection against dangerous braids is expressed in the following result, whose proof consists of lengthy but easy computations from the dual braid relations of Lemma 1.3:

LEMMA 3.21. *Assume that β' is $a_{i,n}$-dangerous and β can be represented by an $a_{i,n}$-stair lent on $a_{j,n-1}$. Then $\beta'\,\beta$ is $\sigma^{\Phi}_{j,n-1}$-positive.*

From there, proving Proposition 3.18 is an easy induction.

PROOF OF PROPOSITION 3.18. Our aim is to prove that the expression of (3.6), namely

$$\widehat{\delta}_{n,p-2}^{-1} \cdot \phi_n^{p-1}(\gamma) \cdot \phi_n^{p-2}(\beta_{p-1}) \cdot \ldots \cdot \beta_1, \tag{3.8}$$

is $\sigma^{\Phi}_{j-1,n-1}$-positive, where $a_{j,n}$ is the last letter of $\beta\beta_1^{-1}$—*i.e.*, of β in which we forget the B_{n-1}^{+*}-tail. Write $\gamma = \gamma'\,\sigma_{n-2}\,\gamma''$, where γ' is σ^{Φ}_{n-2}-nonnegative and γ'' is $a_{i-1,n-1}$-dangerous. We use induction on $p \geqslant 2$.

For $p = 2$, (3.8) reduces to $\phi_n(\gamma) \cdot \beta_1$, and we obtain

$$\phi_n(\gamma) \cdot \beta_1 = \phi_n(\gamma') \cdot \sigma_{n-1} \cdot \phi_n(\gamma'') \cdot \beta_1.$$

By Lemma 1.13, the braid $\phi_n(\gamma')$ is σ^{Φ}_{n-1}-nonnegative, and $\phi_n(\gamma'')$ is $a_{i,n}$-dangerous. So is $\phi_n(\gamma'')\cdot\beta_1$ too, hence the above decomposition shows that $\phi_n(\gamma)\cdot\beta_1$ is $\sigma^{\Phi}_{j,n-1}$-positive.

Assume $p \geqslant 3$. First, we compute

$$\begin{aligned}\widehat{\delta}_{n,p-2}^{-1} \cdot \phi_n^{p-1}(\gamma' \cdot \sigma_{n-2}) &= \delta_{n-1}^{p-2} \cdot \delta_n^{-p+2} \cdot \phi_n^{p-1}(\gamma' \cdot \sigma_{n-2}) \\ &= \delta_{n-1}^{p-2} \cdot \phi_n(\gamma' \cdot \sigma_{n-2}) \cdot \delta_n^{-1} \cdot \delta_n^{-p+3} \\ &= \delta_{n-1}^{p-2} \cdot \phi_n(\gamma') \cdot \delta_{n-1}^{-1} \cdot \delta_n^{-p+3}.\end{aligned} \tag{3.9}$$

So, up to an initial σ^{Φ}_{n-1}-nonnegative factor, we are now left with

$$\delta_n^{-p+3} \cdot \underline{\phi_n^{p-1}(\gamma'') \cdot \phi_n^{p-2}(\beta_{p-1})} \cdot \ldots \cdot \beta_1.$$

Put $\widetilde{\gamma} = \phi_n(\gamma'') \cdot \beta_{p-1}$. Then the underlined factor is $\phi_n^{p-2}(\widetilde{\gamma})$. By Lemma 1.13, $\phi_n(\gamma'')$ is $a_{i,n}$-dangerous. By Lemma 3.20, the ϕ-normal form of β_{p-1} is an $a_{i,n}$-stair lent on the last letter of β_{p-1}, say $a_{j-1,n-1}$. Then, Lemma 3.21 says that $\widetilde{\gamma}$ is $\sigma^{\Phi}_{j-1,n-1}$-positive. Thus, up to left multiplying by the σ^{Φ}_{n-1}-nonnegative factor δ_{n-1}^{p-3}, our expression becomes

$$\widehat{\delta}_{n,p-3}^{-1} \cdot \phi_n^{p-2}(\widetilde{\gamma}) \cdot \phi_n^{p-3}(\beta_{p-2}) \cdot \ldots \cdot \beta_1,$$

i.e., we obtained exactly the same form as in (3.8) with $p-1$ replacing p. We conclude using the induction hypothesis. □

So the proof of Proposition 3.18, and, therefore, of Proposition 3.1, is complete. We can see that, in the previous argument, the crucial point is the very simple fact that, in the basic computation of (3.9), the factor $\phi_n(\sigma_{n-2})$, *i.e.*, σ_{n-1}, neutralizes the σ_{n-1}^{-1} factor present in δ_n^{-1}, just leaving the σ^{Φ}_{n-1}-nonnegative—hence safe—factor δ_{n-1}^{-1}. This point is the core of the proof. It relies on the fact that the minimal Garside elements δ_n and δ_{n-1} of B_n^{+*} and B_{n-1}^{+*} are connected by the relation $\delta_n = \delta_{n-1}\cdot\sigma_{n-1}$; whereas the connection between the minimal Garside elements Δ_n and Δ_{n-1} of B_n^+ and B_{n-1}^+ is expressed in the slightly more complicated formula $\Delta_n = \Delta_{n-1} \cdot \sigma_{n-1} \ldots \sigma_2\sigma_1$. This seemingly microscopic difference might explain why the results of this chapter can have easier proofs than their counterparts in Chapter VII, although the latter involve a smaller monoid, and therefore are weaker.

EXAMPLE 3.22. To conclude, let us come back to the case of Example 3.15, *i.e.*, $\widehat{\delta}_{4,2}^{-1} \cdot \beta$ with $\beta = \mathtt{c'b'bc'}$. We wish to prove that $\widehat{\delta}_{4,2}^{-1} \cdot \beta$ is σ_3^{Φ}-positive, but the previous attempt failed. Indeed, we tried the decomposition

$$\widehat{\delta}_{4,2}^{-1} \cdot \beta = \delta_3^2 \cdot \phi_4(\mathtt{b'}) \cdot \delta_4^{-1} \cdot \phi_4(\mathtt{b'}) \cdot \delta_4^{-1} \cdot \phi_4(\mathtt{ab'}),$$

but $\phi_4(\mathtt{b'}) \cdot \delta_4^{-1}$ turns out to be σ_3^{Φ}-negative, and the method does not lead to a σ_3^{Φ}-positive expression.

Now we shall use a stair—here a very simple one, as it consists of only one step—and apply the strategy of Proposition 3.12. So we start from

$$\widehat{\delta}_{4,2}^{-1} \cdot \beta = \delta_3^2 \cdot \underline{\delta_4^{-2} \cdot \phi_4^3(\mathtt{b'})} \cdot \phi_4^2(\mathtt{b'}) \cdot \phi_4(\mathtt{ab'}).$$

We decompose $\mathtt{b'}$ into $\mathtt{abA}$, and the underlined term becomes $\delta_4^{-2}\phi_4^3(\mathtt{a})\cdot\phi_4^3(\mathtt{bA})$, *i.e.*, $\phi_4(\mathtt{a}) \cdot \underline{\delta_4^{-2} \cdot \phi_4^3(\mathtt{bA})}$. The first factor is σ_3^{Φ}-nonnegative, hence it can be forgotten: it will not endanger σ_3^{Φ}-positivity. The underlined term corresponds to the factor $\delta_n^{-p+2} \cdot \phi_n^{p-1}(a_{n-2,n-1}\beta')$ in (3.6), here with $\beta' = \mathtt{A}$, which is $a_{1,3}$-dangerous.

Then we are in situation of applying Lemma 3.21. In this case, we can check directly the conclusion of the lemma: we have $\phi_4(\mathtt{A}) \cdot \mathtt{b'} = \mathtt{a} \cdot \mathtt{b} \cdot \mathtt{A}^2$: here $\mathtt{a}$ is σ_2^{Φ}-nonnegative, and $\mathtt{A}^2$ is $a_{1,3}$-dangerous. So, at this point, we obtain

$$\begin{aligned}
\widehat{\delta}_{1,2}^{-1} \cdot \beta &= \delta_3^2 \cdot \phi_4(\mathtt{a}) \cdot \delta_4^{-2} \cdot \phi_4^3(\mathtt{bA}) \cdot \phi_4^2(\mathtt{b'}) \cdot \phi_4(\mathtt{ab'}) \\
&= \delta_3^2 \cdot \phi_4(\mathtt{a}) \cdot \delta_4^{-2} \cdot \phi_4^3(\mathtt{b}) \cdot \phi_4^2(\phi_4(\mathtt{A}) \cdot \mathtt{b'}) \cdot \phi_4(\mathtt{ab'}) \\
&= \delta_3^2 \cdot \phi_4(\mathtt{a}) \cdot \delta_4^{-2} \cdot \phi_4^3(\mathtt{b}) \cdot \phi_4^2(\mathtt{a} \cdot \mathtt{b} \cdot \mathtt{A}^2) \cdot \phi_4(\mathtt{ab'}).
\end{aligned}$$

Moving the factor δ_4^{-2} to the right, we obtain

$$\begin{aligned}
\widehat{\delta}_{4,2}^{-1} \cdot \beta &= \delta_3^2 \cdot \phi_4(\mathtt{a}) \cdot \phi_4(\mathtt{b}) \cdot \delta_4^{-2} \cdot \phi_4^2(\mathtt{a} \cdot \mathtt{b} \cdot \mathtt{A}^2) \cdot \phi_4(\mathtt{ab'}) \\
&= \delta_3^2 \cdot \phi_4(\mathtt{a}) \cdot \phi_4(\mathtt{b}) \cdot \delta_4^{-1} \cdot \phi_4(\mathtt{a}) \cdot \delta_4^{-1} \cdot \phi_4^2(\mathtt{b} \cdot \mathtt{A}^2) \cdot \phi_4(\mathtt{ab'}) \\
&= \delta_3^2 \cdot \phi_4(\mathtt{a}) \cdot \delta_3^{-1} \cdot \phi_4(\mathtt{a}) \cdot \underline{\delta_4^{-1} \cdot \phi_4^2(\mathtt{b} \cdot \mathtt{A}^2) \cdot \phi_4(\mathtt{ab'})},
\end{aligned}$$

which consists of an initial σ_3^{Φ}-nonnegative term followed by the underlined term, which again has the form involved in Proposition 3.12, now with $p = 1$.

We repeat the process. We have to compute $\phi_4(\mathtt{A}^2)\cdot\mathtt{ab'}$. The normal form of $\mathtt{ab'}$ is an $a_{2,4}$-stair, and Lemma 3.21 here takes the form $\phi_4(\mathtt{A}^2) \cdot \mathtt{ab'} = \mathtt{b'b} \cdot \mathtt{b} \cdot \mathtt{A}^2$, again with an initial σ_2^{Φ}-nonnegative fragment $\mathtt{b'b} \cdot \mathtt{b}$, and a final dangerous fragment $\mathtt{A}^2$. As $\mathtt{b}$ is σ_2^{Φ}-positive, and $\mathtt{A}^2$ is σ_2^{Φ}-nonnegative, the latter expression is σ_2^{Φ}-positive. Then the computation ends, as we find

$$\begin{aligned}
\widehat{\delta}_{4,2}^{-1} \cdot \beta &= \delta_3^2 \cdot \phi_4(\mathtt{a}) \cdot \delta_3^{-1} \cdot \phi_4(\mathtt{a}) \cdot \delta_4^{-1} \cdot \phi_4^2(\mathtt{b} \cdot \mathtt{A}^2) \cdot \phi_4(\mathtt{ab'}) \\
&= \delta_3^2 \cdot \phi_4(\mathtt{a}) \cdot \delta_3^{-1} \cdot \phi_4(\mathtt{a}) \cdot \phi_4(\mathtt{b}) \cdot \delta_4^{-1} \cdot \phi_4(\phi_4(\mathtt{A}^2) \cdot \mathtt{ab'}) \\
&= \delta_3^2 \cdot \phi_4(\mathtt{a}) \cdot \delta_3^{-1} \cdot \phi_4(\mathtt{a}) \cdot \phi_4(\mathtt{b}) \cdot \delta_4^{-1} \cdot \phi_4(\mathtt{b'b} \cdot \mathtt{b} \cdot \mathtt{A}^2) \\
&= \underline{\delta_3^2} \cdot \underline{\phi_4(\mathtt{a}) \cdot \delta_3^{-1}} \cdot \underline{\phi_4(\mathtt{a}) \cdot \delta_3^{-1}} \cdot \underline{\phi_4(\mathtt{b'b} \cdot \mathtt{b} \cdot \mathtt{A}^2)} :
\end{aligned}$$

the first three underlined factors are σ_3^{Φ}-nonnegative; as for the last underlined fragment, we noted above that $\mathtt{b'b} \cdot \mathtt{b} \cdot \mathtt{A}^2$ is σ_2^{Φ}-positive, and, therefore, its image under ϕ_4 is σ_3^{Φ}-positive, as expected. When we expand, we find $\mathtt{abab} \cdot \mathtt{bBA} \cdot \mathtt{b} \cdot \mathtt{BA} \cdot \mathtt{bcBc}^2\mathtt{B}^2$, a word that contains three σ_3 and no σ_3^{-1}.

This completes the verification that the braid $\widehat{\delta}_{4,2}^{-1} \cdot \beta$ is σ_3^{Φ}-positive following the general method used to establish Proposition 3.18.

As the example shows, the above method is effective, and it can be turned into an actual algorithm that, running on a word of the form $w^{-1}w'$ where w and w' are dual braid words and $\overline{w} <^{\Phi} \overline{w'}$ holds, returns a σ^{Φ}-positive equivalent dual braid word—see Section 1.3 of Chapter XVI.

CHAPTER IX

Automorphisms of a Free Group

This short chapter is a transition between the combinatorial and the topological approaches. It is centered on Artin's representation of braid groups inside the group of automorphisms of a free group, which can fruitfully be addressed from either of these viewpoints. The connection with the σ-ordering of braids was mainly developed by David Larue in [**132, 131**]. The approach leads to a very short proof of Property **A**, and to a (not so simple) proof of Property **C**—together with a new very simple proof of the faithfulness of Artin's representation.

Except in Section 1, most of the arguments explained in this chapter will simply be sketched as they can be more naturally viewed from a topological perspective, as will be done in Chapters X and XIII. In order to learn more about the point of view developed in this chapter, we refer the reader to the recent reference [**7**].

The chapter is organized as follows. In Section 1 we define the action of B_n on the free group F_n, study some of its properties, characterize the σ-ordering of B_n in terms of this action, and use it to give a proof of Property **A**. In Section 2 we show how to explicitly reconstruct a braid from its action on the generators of F_n, and we give a proof of Property **C**. In Section 3 we present a different perspective on these ideas: we show how the σ-ordering (and certain other orderings) of B_n can be obtained by ordering the free group F_n, and by pulling this ordering back to obtain an ordering of B_n.

CONVENTION. In this chapter, as well as in all chapters using topological methods (Chapters X, XI, XII, and XIII), braid groups will act on the left. The reason for this convention is that in all these chapters we want to obtain a left-invariant ordering of braid groups using the following principle. We start with some totally ordered set Ω—e.g., Ω is the free group in Chapter IX, it is a set of curve diagrams in Chapter X, a set of triangulations in Chapter XII, and the real line in Chapter XIII. Then we consider an order-preserving action of B_n on Ω, and an element x of Ω with trivial stabilizer. We obtain an ordering of B_n by pulling back the ordering on the orbit of x under the B_n-action: we define a braid β' to be larger than β if the image of x under the action of β' is larger in Ω than the image of x under the β-action. If the braid group acts on the *left* on Ω, then the ordering thus obtained is invariant under left multiplication, whereas a *right* B_n-action leads to a right-invariant ordering of B_n.

1. Artin representation of σ-positive braids

It has been known since Artin [**4**] that braid groups embed in groups of automorphisms of free groups, and it is natural to try to identify the σ-ordering of braids in terms of the associated automorphisms. Here we shall see that the Artin representation of σ-positive braids have specific properties that make them easily recognizable.

1.1. Embedding B_n into $\mathrm{Aut}(F_n)$. For $n \geqslant 1$, we denote by F_n the rank n free group based on $\{x_1, \dots, x_n\}$, and we denote by F_∞ the free group based on $\{x_i \mid 1 \leqslant i < \infty\}$.

DEFINITION 1.1. For $1 \leqslant i < n$, we denote by $\widehat{\sigma_i}$ the automorphism of F_n defined by

$$\widehat{\sigma_i}(x_k) = \begin{cases} x_i x_{i+1} x_i^{-1} & \text{for } k = i, \\ x_i & \text{for } k = i+1, \\ x_k & \text{for } k \neq i, i+1. \end{cases} \tag{1.1}$$

LEMMA 1.2. *For $1 \leqslant n \leqslant \infty$, the mapping $\sigma_i \mapsto \widehat{\sigma_i}$ extends to a homomorphism of B_n into $\mathrm{Aut}(F_n)$.*

PROOF. The automorphisms $\widehat{\sigma_i}$ satisfy the braid relations. Alternatively, we can observe that the action of $\widehat{\sigma_i}$ on $(F_n)^n$ is the right action of conjugacy considered in Section IV.1 in the context of braid diagram colourings, and its compatibility with braid relations follows from conjugacy being a left self-distributive operation. □

The above morphism is called the *Artin representation* of B_n. For each braid β, we shall denote by $\widehat{\beta}$ the associated automorphism. For w a braid word, we denote by $\widehat{w}$ the automorphism associated with the braid represented by w. For further reference, let us note that the action of σ_i^{-1} is given by

$$\widehat{\sigma_i}^{-1}(x_k) = \begin{cases} x_{i+1} & \text{for } k = i, \\ x_{i+1}^{-1} x_i x_{i+1} & \text{for } k = i+1, \\ x_k & \text{for } k \neq i, i+1. \end{cases} \tag{1.2}$$

The embedding of B_n into B_{n+1} induced by identity on σ_i's is compatible with the embedding of $\mathrm{Aut}(F_n)$ into $\mathrm{Aut}(F_{n+1})$ induced by the identity on x_i's, so there is no need to specify n here; equivalently, we may consider that we work with B_∞ and $\mathrm{Aut}(F_\infty)$.

1.2. Images of σ-positive braid words. We shall prove that, if the braid β admits at least one σ-positive representative braid word, then the automorphism $\widehat{\beta}$ has some specific properties that can be read on the words $\widehat{\beta}(x_i)$.

Hereafter, we identify F_∞ with the set of all freely reduced words on $x_1^{\pm 1}$, $x_2^{\pm 1}$, ..., where we say that u is *freely reduced* if it contains no pattern of the form xx^{-1} or $x^{-1}x$. For u an arbitrary word on $x_1^{\pm 1}$, $x_2^{\pm 1}$, ..., we denote by $\mathrm{red}(u)$ the unique reduced word obtained from u by iteratively deleting all patterns xx^{-1} and $x^{-1}x$.

NOTATION 1.3. (i) For x a letter x_i or x_i^{-1}, we denote by $S(x)$ the subset of F_∞ consisting of all freely reduced words that end with x.

(ii) We denote by sh the (shift) endomorphism of F_∞ that maps x_k to x_{k+1} for every k.

(iii) For f in $\mathrm{Aut}(F_\infty)$, we denote by $\mathrm{sh}(f)$ the automorphism of F_∞ defined by $\mathrm{sh}(f)(x_1) = x_1$, and $\mathrm{sh}(f)(x_{k+1}) = \mathrm{sh}(f(x_k))$.

We shall investigate the image of the set $S(x_1^{-1})$ under the automorphism $\widehat{\sigma_i}^{\pm 1}$.

LEMMA 1.4. *Every automorphism $\mathrm{sh}(f)$ maps $S(x_1^{-1})$ into itself.*

PROOF. Consider an arbitrary element of $S(x_1^{-1})$, say ux_1^{-1} with $u \notin S(x_1)$. By construction, we have $\mathrm{sh}(f)(ux_1^{-1}) = \mathrm{red}(\mathrm{sh}(f)(u)\, x_1^{-1})$. Assume that $\mathrm{sh}(f)(ux_1^{-1})$ does not belong to $S(x_1^{-1})$. Then the final letter x_1^{-1} in $\mathrm{sh}(f)(u)x_1^{-1}$ is cancelled by some letter x_1 occurring in $\mathrm{sh}(f)(u)$. Such a letter x_1 in $\mathrm{sh}(f)(u)$ must come from a letter x_1 in u. So there exists a decomposition $u = u_1x_1u_2$ satisfying $\mathrm{sh}(f)(u_2) = 1$. As $\mathrm{sh}(f)$ is injective, the latter condition implies $u_2 = 1$, hence $u \in S(x_1)$, which contradicts the hypothesis. □

LEMMA 1.5. *The automorphism $\widehat{\sigma_i}$ maps both $S(x_i)$ and $S(x_i^{-1})$ into $S(x_i^{-1})$.*

PROOF. Let us consider an arbitrary element of $S(x_i) \cup S(x_i^{-1})$, say ux_i^e with $e = \pm 1$ and $u \notin S(x_i^{-e})$. Then we have $\widehat{\sigma_i}(ux_i^e) = \mathrm{red}(\widehat{\sigma_i}(u)x_ix_{i+1}^e x_i^{-1})$. Assume $\widehat{\sigma_i}(ux_i^e) \notin S(x_i^{-1})$. This means that the final x_i^{-1} in $\widehat{\sigma_i}(ux_i^e)$ is cancelled by some letter x_i in $\widehat{\sigma_i}(u)$. This letter comes either from some x_{i+1} or from some $x_i^{e'}$ in u.

In the first case, we display the letter x_{i+1} involved in the cancellation by writing $u = u_1x_{i+1}u_2$, where u_2 is a reduced word. We find

$$\widehat{\sigma_i}(ux_i^e) = \mathrm{red}(\widehat{\sigma_i}(u_1)x_i\widehat{\sigma_i}(u_2)x_ix_{i+1}^e x_i^{-1}). \tag{1.3}$$

By hypothesis, we have $\mathrm{red}(\widehat{\sigma_i}(u_2)x_ix_{i+1}^e) = \varepsilon$, where we recall ε denotes the empty word. Hence (1.3) implies $\widehat{\sigma_i}(u_2) = x_{i+1}^{-e}x_i^{-1} = \widehat{\sigma_i}(x_{i+1}^{-1}x_i^{-e})$. We deduce $u_2 = x_{i+1}^{-1}x_i^{-e}$, which contradicts $u \notin S(x_i^{-e})$.

In the second case, we write similarly $u = u_1x_i^{e'}u_2$ with $e' = \pm 1$. So we have

$$\widehat{\sigma_i}(ux_i^e) = \mathrm{red}(\widehat{\sigma_i}(u_1)x_ix_{i+1}^{e'}x_i^{-1}\widehat{\sigma_i}(u_2)x_ix_{i+1}^e x_i^{-1}),$$

and the hypothesis is $\mathrm{red}(x_{i+1}^{e'}x_i^{-1}\widehat{\sigma_i}(u_2)x_ix_{i+1}^e) = \varepsilon$. This implies $\mathrm{red}(\widehat{\sigma_i}(u_2)) = x_ix_{i+1}^{-e-e'}x_i^{-1} = \widehat{\sigma_i}(x_i^{-e-e'})$, hence $u_2 = x_i^{-e-e'}$. For $e = +1$, we obtain either $u_2 = x_i^{-2}$ (for $e' = +1$) or $u_2 = \varepsilon$ (for $e' = -1$), and, in both cases, $u \in S(x_i^{-e})$, a contradiction. Similarly, for $e = -1$, we obtain either $u_2 = \varepsilon$ (for $e' = +1$) or $u_2 = x_i^2$ (for $e' = -1$), and, in both cases, $u \in S(x_i^{-e})$, again a contradiction. □

We deduce the following implication:

PROPOSITION 1.6. *Assume that the braid β is σ_1-positive. Then the word $\widehat{\beta}(x_1)$ ends with x_1^{-1}.*

PROOF. Our hypothesis implies that the automorphism $\widehat{\beta}$ admits a decomposition of the form

$$\widehat{\beta} = \mathrm{sh}(f_0) \circ \widehat{\sigma_1} \circ \mathrm{sh}(f_1) \circ \ldots \circ \widehat{\sigma_1} \circ \mathrm{sh}(f_p).$$

Then we have $\mathrm{sh}(f_p)(x_1) = x_1$, and $\widehat{\sigma_1}(x_1) = x_1x_2x_1^{-1}$, an element of $S(x_1^{-1})$. By Lemmas 1.4 and 1.5, every subsequent factor $\mathrm{sh}(f_k)$ and $\widehat{\sigma_1}$ maps $S(x_1^{-1})$ into $S(x_1^{-1})$. □

COROLLARY 1.7 (Property **A**). *A σ_1-positive braid is not trivial.*

PROOF. If β is σ_1-positive, then, by Proposition 1.6, the word $\widehat{\beta}(x_1)$ is not equal to x_1, so $\widehat{\beta}$ is not the identity, and, therefore, β cannot be trivial. □

As mentioned in Chapter III, another application for the previous result is a new proof for the injectivity of the Artin representation, once we know that Property **C**, or, at least, Property $\mathbf{C}_\infty$ is true—see the introduction to Section 2.1 below.

COROLLARY 1.8. *The Artin representation of B_n to* $\mathrm{Aut}(F_n)$ *is an embedding.*

PROOF. Assume that β is a nontrivial braid. We claim that $\widehat{\beta}$ is not the identity. By Proposition 1.6, this is the case when β admits a σ_1-positive representative, and, more generally, when β admits a σ-positive representative (by injectivity of the shift mapping). By applying the result to β^{-1}, we obtain similarly that $\widehat{\beta}$ is not the identity when β admits a σ-negative representative. Using Property $\mathbf{C}_\infty$, we conclude that $\beta = 1$ is the only case that has not been considered. □

REMARK 1.9. We mentioned in Proposition III.2.2 other representations of B_n into $\mathrm{Aut}(F_n)$ for which a counterpart of Proposition 1.6 applies. Each such result gives a new proof of Property **A**. The argument leading to Proposition III.2.2(iii) gives an especially short proof of Property **A**—probably the shortest proof known so far.

1.3. Characterization of the braid ordering. We proved above an implication, namely that β being σ_1-positive results in $\widehat{\beta}(x_1)$ ending with x_1^{-1}. We shall see now that the converse implication is also true—provided Property $\mathbf{C}_\infty$ is known.

LEMMA 1.10. *For $k \neq i, i+1$ and $e = \pm 1$, the automorphism $\widehat{\sigma_i}$ maps $S(x_k^e)$ into itself.*

The easy proof is analogous to that of Proposition 1.6.

LEMMA 1.11. *The automorphism $\widehat{\sigma_i}$ maps $S(x_{i+1})$ into $S(x_i) \cup S(x_{i+1}) \cup S(x_{i+1}^{-1})$, and $S(x_{i+1}^{-1})$ into $S(x_i^{-1})$.*

PROOF. Consider an arbitrary element of $S(x_{i+1})$, say ux_{i+1} with $u \notin S(x_{i+1}^{-1})$. Then we have $\widehat{\sigma_i}(ux_{i+1}) = \mathrm{red}\big(\widehat{\sigma_i}(u)x_i\big)$. Assume $\widehat{\sigma_i}(ux_{i+1}) \notin S(x_i)$. This means that the final letter x_i is cancelled by some letter x_i^{-1} in $\widehat{\sigma_i}(u)$. This letter comes either from some x_{i+1}^{-1} or from some $x_i^{\pm 1}$ in $\widehat{\sigma_i}(u)$. As previously, we consider the possible cases.

The first case is $u = u_1 x_{i+1}^{-1} u_2$. We have now

$$\widehat{\sigma_i}(ux_{i+1}) = \mathrm{red}\big(\widehat{\sigma_i}(u_1)x_i^{-1}\widehat{\sigma_i}(u_2)x_i\big),$$

and the hypothesis is $\mathrm{red}(\widehat{\sigma_i}(u_2)) = \varepsilon$. This implies $u_2 = \varepsilon$, and, therefore, $u \in S(x_{i+1}^{-1})$, a contradiction.

The second case is $u = u_1 x_i^e u_2$ with $e = \pm 1$ and $u_1 \notin S(x_i^{-e})$. We find

$$\widehat{\sigma_i}(ux_{i+1}) = \mathrm{red}\big(\widehat{\sigma_i}(u_1)x_i x_{i+1}^e x_i^{-1}\widehat{\sigma_i}(u_2)x_i\big),$$

and the hypothesis is $\mathrm{red}(\widehat{\sigma_i}(u_2)) = \varepsilon$. This implies $u_2 = \varepsilon$, hence $u = u_1 x_i^e$. Thus we have

$$\widehat{\sigma_i}(ux_{i+1}) = \mathrm{red}\big(\widehat{\sigma_i}(u_1)x_i x_{i+1}^e\big).$$

Assume that the final letter x_{i+1}^e vanishes in the reduction. Then x_{i+1}^e cancels with some letter x_{i+1}^{-e} that necessarily comes from some letter x_i^{-e} in $\widehat{\sigma_i}(u_1)$. So there must exist a decomposition $u_1 = u_1' x_i^{-e} u_1''$, giving

$$\widehat{\sigma_i}(ux_{i+1}) = \mathrm{red}\big(\widehat{\sigma_i}(u_1')x_i x_{i+1}^{-e} x_i^{-1}\widehat{\sigma_i}(u_1'')x_i\big)$$

with $\mathrm{red}\big(x_i^{-1}\widehat{\sigma_i}(u_1'')x_i\big) = \varepsilon$. As above, this implies $u_1'' = \varepsilon$, and, therefore, $u_1 \in S(x_i^{-e})$, a contradiction. So $\widehat{\sigma_i}(u) \in S(x_{i+1}^e)$ is the only possibility.

The argument for the image of $S(x_{i+1}^{-1})$ is similar. □

Using the fact that the sets $S(x_i^{\pm 1})$ form a partition of $F_\infty \setminus \{1\}$ and applying the previous lemmas, we see that the only possibilities for the images under the inverse automorphisms $\widehat{\sigma_i}^{-1}$ are as follows:

LEMMA 1.12. *The automorphism $\widehat{\sigma_i}^{-1}$ maps $S(x_k^e)$ into itself for $k \neq i,\ i+1$ and $e = \pm 1$; it maps $S(x_i)$ to $S(x_{i+1})$, $S(x_i^{-1})$ to $S(x_i) \cup S(x_i^{-1}) \cup S(x_{i+1}^{-1})$, and both $S(x_{i+1})$ and $S(x_{i+1}^{-1})$ to $S(x_{i+1})$.*

Then gathering the results, we obtain

PROPOSITION 1.13. *Let β be an arbitrary braid.*

(i) *If β is σ_1-positive, the word $\widehat{\beta}(x_1)$—a freely reduced word by definition—ends with x_1^{-1};*

(ii) *If β is σ_1-free, the word $\widehat{\beta}(x_1)$ is x_1;*

(iii) *If β is σ_1-negative, the word $\widehat{\beta}(x_1)$ ends with $x_k^{\pm 1}$ for some k with $k \geqslant 2$.*

PROOF. By construction, $\widehat{\beta}(x_1) = x_1$ is true if β admits a σ_1-free representative. If β admits a σ_1-positive representative, we have seen in Proposition 1.6 that $\widehat{\beta}(x_1)$ lies in $S(x_1^{-1})$. Assume finally that β admits a σ_1-negative representative. Thus β can be expressed as $\beta_1 \sigma_1^{-1} \mathrm{sh}(\beta_2)$, where β_1 admits a representative which may contain the letter σ_1^{-1}, but not the letter σ_1. Then $\widehat{\mathrm{sh}(\beta_2)}$ maps x_1 to itself, hence $(\sigma_1^{-1}\mathrm{sh}(\beta_2))^\frown$ maps x_1 to x_2, an element of $S(x_2)$, hence of $\bigcup_{k \geqslant 2} S(x_k^{\pm 1})$. Then $\widehat{\sigma_1}^{-1}$ and all $\widehat{\sigma_k}^{\pm 1}$ with $k \geqslant 2$ map $\bigcup_{k \geqslant 2} S(x_k^{\pm 1})$ into itself: indeed, $\widehat{\sigma_1}$ is the only automorphism in the considered family that possibly maps an element of $S(x_k^{\pm 1})$ with $k \geqslant 2$ into $S(x_1^{\pm 1})$. □

Applying the shift operation, we obtain similarly the following more general result:

PROPOSITION 1.14. *Let β be an arbitrary braid.*

(i) *If β is σ-positive, then there exists i such that $\widehat{\beta}(x_j) = x_j$ holds for $j < i$, and $\widehat{\beta}(x_i)$ ends with x_i^{-1};*

(ii) *If β is σ-negative, then there exists i such that $\widehat{\beta}(x_j) = x_j$ holds for $j < i$, and $\widehat{\beta}(x_i)$ ends with $x_k^{\pm 1}$ for some k satisfying $k \geqslant i+1$.*

Assuming Property $\mathbf{C}_\infty$, every braid is σ_1-positive, σ_1-negative, or σ_1-free, and we deduce that the previous implications are actually equivalences.

COROLLARY 1.15. *Let β be an arbitrary braid.*

(i) *The braid β is σ_1-positive if and only if $\widehat{\beta}(x_1)$ ends with x_1^{-1};*

(ii) *The braid β is σ_1-free if and only if we have $\widehat{\beta}(x_1) = x_1$;*

(iii) *The braid β is σ_1-negative if and only if $\widehat{\beta}(x_1)$ ends with $x_k^{\pm 1}$ for some k satisfying $k \geqslant 2$;*

(iv) *The braid β is σ-positive if and only if there exists i such that $\widehat{\beta}(x_j) = x_j$ holds for $j < i$, and $\widehat{\beta}(x_i)$ ends with x_i^{-1};*

(v) *The braid β is σ-negative if and only if there exists i such that $\widehat{\beta}(x_j) = x_j$ holds for $j < i$, and $\widehat{\beta}(x_i)$ ends with $x_k^{\pm 1}$ for some k satisfying $k \geqslant i+1$.*

In this way we obtain the sixth characterization of the σ-ordering mentioned in the Introduction:

COROLLARY 1.16. *Let β, β' be any braids. Then $\beta < \beta'$ is true if and only if, for some i, the automorphism associated with $\beta^{-1}\beta$ maps x_j to x_j for $j < i$, and it maps x_i to a word that ends with x_i^{-1}.*

We also obtain a new braid comparison algorithm with exponential complexity. In order to decide whether $\beta > 1$ is true, we compute the reduced word $\widehat{\beta}(x_1)$: if it ends with x_1^{-1}, we deduce $1 < \beta$; if it ends with $x_k^{\pm}$ for some $k \geqslant 2$, we deduce $1 > \beta$. Otherwise, we know that $\widehat{\beta}(x_1)$ must be x_1. In this case, we compute $\widehat{\beta}(x_2)$: if the latter word ends with x_2^{-1}, we deduce $1 < \beta$; if it ends with $x_k^{\pm}$ for some $k \geqslant 3$, we deduce $1 > \beta$. Otherwise, $\widehat{\beta}(x_2)$ must be x_2, and we continue similarly with $\widehat{\beta}(x_3)$. By construction, if β is specified using some braid word of length ℓ, the lengths of the words $\widehat{\beta}(x_k)$ are bounded by 3^ℓ, and they can be computed iteratively in a number of steps of the same order.

1.4. Property $\mathbf{A}_i$. Property **A** specifically involves the generator σ_1. In [**132**], Larue extends the argument given above to prove a similar statement involving any generator σ_i: this is Property $\mathbf{A}_i$, already introduced in Remark IV.2.15, where an independent argument was given.

PROPOSITION 1.17. *Let i be any fixed positive integer. Then a braid word that contains at least one letter σ_i and no letter σ_i^{-1} is not trivial, i.e., it does not represent the unit braid.*

Larue's proof of this result—which we skip—uses a symmetric variant of the Artin representation in which the action of σ_i moves x_{i-1}, x_i, and x_{i+1}.

REMARK 1.18. Property $\mathbf{A}_i$ can be used to define a new relation on B_n: say that $\beta <_i \beta'$ holds if $\beta^{-1}\beta'$ admits at least one word representative in which σ_i occurs, but σ_i^{-1} does not. Then, for instance, we have $\sigma_2 >_2 \sigma_1$ and $\sigma_2 >_2 \sigma_3$. Property $\mathbf{A}_i$ implies that the relation so defined is a partial ordering in B_n, which is invariant under left multiplication by construction. As it stands, the ordering $<_i$ is not a linear ordering, since we did not decide anything for the braids that admit a σ_i-free representative. But, whatever the extension is, there is no natural way to obtain a linear ordering, except in the cases $i = 1$ and $i = n - 1$, where one recovers the σ-ordering and its flipped version $<^{\Phi}$. The reason for this is that, for $i \neq 1, n-1$, there are braids of B_n *all* of whose representative words contain the letter σ_i and also its inverse σ_i^{-1}, like $\sigma_2\sigma_1\sigma_3^{-1}\sigma_2^{-1}$ with respect to σ_2. More details can be found in H. Sibert's PhD thesis [**186**].

2. From an automorphism back to a braid

Since Artin's representation of B_n in $\mathrm{Aut}(F_n)$ is faithful, we can go the other way around and consider the question of recovering a braid β from its image $\widehat{\beta}$. A solution to the problem is likely to give not only the braid β, but rather a certain braid word representing β. What we shall see here is that, by choosing a convenient strategy, we can obtain a word that is σ-positive, σ-negative, or trivial.

2.1. Untangling automorphisms of free groups. In his thesis [**131**], Larue gave the first proof of Property **C** in its full strength—not just Property $\mathbf{C}_\infty$—taking $\mathbf{C}_\infty$ as a hypothesis. At the time of Larue's work (1994), the only known approach to Property **C** was the one described in Chapter IV, and, as explained in Section IV.1, the latter naturally leads to a proof of $\mathbf{C}_\infty$.

We recall that Property $\mathbf{C}_\infty$ in the assertion that every nontrivial braid is σ-positive or σ-negative: it differs from Property $\mathbf{C}$ in that we do not demand that a nontrivial n-strand braid be σ-positive or σ-negative inside B_n, *i.e.*, we accept that the σ-positive or σ-negative representative braid word can involve more than the minimal number of strands.

Larue's proof of 1994 turns out to be similar to the topological argument given a few years later in the paper [**83**], whose authors were unaware of Larue's unpublished work. His argument, which takes place entirely in the group-theoretic setting and uses almost no plane topology, is quite intricate; we shall only sketch it here, and refer the reader to Chapter X for a somewhat more detailed account of the argument using curve diagrams—which, in addition, does not require taking Property $\mathbf{C}_\infty$ as an hypothesis. The core of the problem is to be able to recover a braid from its image in $\mathrm{Aut}(F_n)$. The main result is the following:

PROPOSITION 2.1. *Assume that w is an n-strand braid word of length ℓ such that $\widehat{w}(x_1)$ ends with x_1^{-1}. Then w is equivalent to a σ_1-positive n-strand braid word w' of length at most $(3^\ell+1)n^2/4+\ell$.*

In order to prove Proposition 2.1, we shall use a different generating set for the free group. First, we introduce a dummy generator x_0 on which B_n acts trivially. Thus our free group F_n is included in a free group on $n+1$ generators. Then we define generators y_i for i in $\{0,\dots,n\}$ by $y_i = x_i^{-1}\dots x_1^{-1}x_0^{-1}$. We have the following relation between a reduced word u with letters $x_i^{\pm1}$ $(1 \leqslant i \leqslant n)$ and a reduced word u' with letters $y_i^{\pm1}$ $(0 \leqslant i \leqslant n)$ representing the same element of F_n: u ends with x_i or with x_{i+1}^{-1} if and only if u' ends with y_i^{-1} for $i = 0,\dots,n$, and in particular u ends with x_1^{-1} if and only if u' ends with y_0^{-1}. Moreover, u equals x_1 if and only if u' equals $y_0y_1^{-1}$. Finally, the action by the braid σ_i sends the generator y_i to $y_{i-1}y_i^{-1}y_{i+1}$, and leaves all other generators fixed.

Geometrically, we can identify the free group on $n+1$ generators $x_0, x_1, \dots, x_n$ with the fundamental group of the $n+1$ times punctured disk D_{n+1} included in $\mathbb{C}$ whose base point is the point -1 of $\mathbb{C}$, and whose punctures, labelled $0, 1, \dots, n$, are contained in the real line—see Section 3 of Chapter I.

Under this identification, the generators y_i $(i = 0,\dots,n)$ can be represented by $n+1$ simple loops in D_{n+1} which are disjoint except in the basepoint, and where the curve corresponding to y_i winds once around the punctures labelled $0,\dots,i$ in an anticlockwise sense, as shown in Figure I.3. Moreover, if an element y of $\pi_1(D_{n+1})$ is given by a reduced word of length ℓ in the letters $y_0,\dots,y_n$, then y can be represented by a path in D_{n+1} whose interior has exactly ℓ intersections with the horizontal axis $\mathbb{R}$, not counting the start and endpoint of the path in the basepoint, and this is the minimum intersection number among all paths representing y.

It is easy to prove inductively that for any braid word w of length ℓ, the element $\widehat{w}(x_1) = \widehat{w}(y_0y_1^{-1})$ of F_n is given by a word of length at most $3^\ell+1$ in the generators $y_0,\dots,y_n$.

Our aim is to write the braid represented by w^{-1} as a σ_1-negative braid word, and more precisely as a product of braid words of a particular form. For $1 \leqslant r < s < t \leqslant n$, we define

$$w_{r,s,t} = (\sigma_s\sigma_{s-1}\cdots\sigma_{r+1})(\sigma_{s+1}\cdots\sigma_{r+2})\ \cdots\ (\sigma_{t-1}\cdots\sigma_{t-s+r}),$$

and, for $0 \leqslant r < s < t \leqslant n$, we define

$$w'_{r,s,t} = (\sigma_s^{-1}\sigma_{s-1}^{-1} \dots \sigma_{r+1}^{-1})(\sigma_{s+1}^{-1} \dots \sigma_{r+2}^{-1}) \ \dots \ (\sigma_{t-1}^{-1} \dots \sigma_{t-s+r}^{-1}).$$

We observe that the braid words $w_{r,s,t}$ and $w'_{r,s,t}$ are σ_1-free for $r \geqslant 1$, and that $w'_{r,s,t}$ is σ_1-negative for $r = 0$. Note that the length of $w_{r,s,t}$ and of $w'_{r,s,t}$ is at most $n^2/4$. We intend to find a product of such words equivalent to w^{-1}.

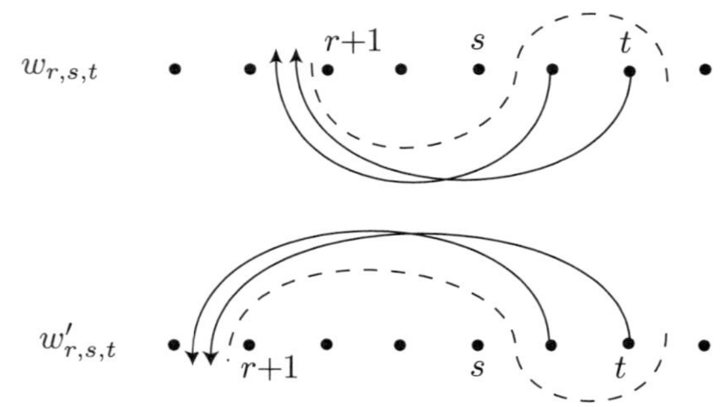

FIGURE 1. Our aim is to find a representative of any braid by a concatenation of certain braid words $w_{r,s,t}$ and $w'_{r,s,t}$. The braids represented by these words are shown above: strands $s+1, \dots, t$ are moved in front of (respectively behind) the other stands, and are reinserted between strands number r and $r+1$.

In Figure 1 the braids represented by $w_{r,s,t}$ and $w'_{r,s,t}$ are sketched. The meaning of the dashed arcs will be explained shortly. The key lemma is now

LEMMA 2.2. *If w is a braid word such that $\widehat{w}(x_1)$, written as a reduced word in the generators $y_0^{\pm 1}, \dots, y_n^{\pm 1}$, ends with y_0^{-1}, then there exists a braid word v equal to some $w_{r,s,t}$ or $w'_{r,s,t}$ as above such that*

(i) *the length of $\widehat{v}(\widehat{w}(x_1))$ is at least 2 less than that of $\widehat{w}(x_1)$,*

(ii) *we have either $\widehat{v}(\widehat{w}(x_1)) = x_1 = y_0 y_1^{-1}$, or the reduced form of $\widehat{v}(\widehat{w}(x_1))$ still ends with y_0^{-1}.*

PROOF (SKETCH). We define an $(r \downarrow s \uparrow t)$-arc—with $1 \leqslant r < s < t \leqslant n$—and an $(r \uparrow s \downarrow t)$-arc—with $0 \leqslant r < s < t \leqslant n$—of $\widehat{w}(x_1)$ to be a (nonoriented) subarc of the loop $\widehat{w}(x_1)$ whose extremal points lie between the rth and the $(r+1)$st, and between the tth and $(t+1)$st puncture, respectively, in the horizontal axis, and which intersect the axis precisely once more, namely between the sth and $(s+1)$st puncture, as indicated in Figure 1.

One can prove that if $\widehat{w}(x_1)$ ends with y_0^{-1}, then $\widehat{w}(x_1)$ contains an $(r \downarrow s \uparrow t)$-arc or an $(r \uparrow s \downarrow t)$-arc, for some r, s, t in the legal range, as a subarc.

Now it is not too hard to see—and should be intuitively clear from Figure 1—that applying the braid (represented by) $w_{r,s,t}$ or $w'_{r,s,t}$, respectively, to such a loop $\widehat{w}(x_1)$ reduces the number of intersections of the loop with the horizontal line by at least two. In other words, with $v = w_{r,s,t}$ or $v = w'_{r,s,t}$ we have

$$\ell(\widehat{v}(\widehat{w}(y_1))) \leqslant \ell(\widehat{w}(y_1)) - 2,$$

where we recall that $\ell(u)$ denotes the length of u.

Moreover, one can show that the resulting element $\widehat{v}(\widehat{w}(y_1))$ of F_n still ends with y_0^{-1}, unless $\widehat{v}\widehat{w}$ is a σ_1-free braid, in which case we have $\widehat{v}(\widehat{w}(y_1)) = y_1$. □

Now Proposition 2.1 can be deduced from Lemma 2.2, by the following induction argument.

Proof of Proposition 2.1 (Sketch). For w a braid word of length ℓ, the element $\widehat{w}(x_1)$ of F_n has length at most $3^\ell+1$ when written in the letters $y_0^{\pm 1}, \dots, y_n^{\pm 1}$. Now we can apply a sequence of braids, each with at most $n^2/4$ crossings, to this element, reducing its length by at least two in each step. This process yields eventually a σ_1-negative braid word w_1 of length at most $(3^\ell + 1)n^2/8$ such that the braid represented by $w_1 w$ acts trivially on the generator x_1; notice that $w_1 w$ is of length at most $(3^\ell + 1)n^2/8 + \ell$. It is then possible, but nontrivial, to prove that there exists another braid word w_2, say, equivalent to $w_1 w$, whose length is also bounded by $(3^\ell+1)n^2/8+\ell$, but which does not contain any letter $\sigma_1^{\pm 1}$. Now w and $w_1^{-1} w_2$ represent the same braid, and the latter word is σ_1-positive and of length at most $(3^\ell + 1)n^2/4 + \ell$. □

2.2. A proof of Property C. With Proposition 2.1 at hand, we can now easily deduce Property **C** from Property $\mathbf{C}_\infty$, as in [**132**].

Proposition 2.3. *Property $\mathbf{C}_\infty$ implies Property* **C**, *i.e., assuming that every n-strand braid word w is equivalent to some braid word w' that is σ_1-positive, σ_1-negative, or σ_1-free, we can require in addition that w' is an n-strand braid word.*

Proof. Suppose that w is an n-strand braid word. By Property $\mathbf{C}_\infty$, there exists some braid word w' with letters $\sigma_1^{\pm 1}, \dots, \sigma_{m-1}^{\pm 1}$, possibly with $m > n$, which is equivalent to w and which is σ_1-positive, σ_1-negative, or σ_1-free. In the first case, Proposition 1.6 implies that the automorphism $\widehat{\beta}$ maps x_1 to some word that ends with x_1^{-1}. Thus the original word w satisfies the hypotheses of Proposition 2.1, so there exists an n-strand braid word w'' equivalent to w and w' that is σ_1-positive. Applying the same argument to w^{-1} gives the result when w' is σ_1-negative. Finally, if w' is σ_1-free, then we can obtain a braid word w'' equivalent to w' with letters $\sigma_2^{\pm 1}, \dots, \sigma_{n-1}^{\pm 1}$ (*i.e.*, both σ_1-free *and* involving only n strands), as follows: we consider the braid represented by the braid word w', remove from it the strands numbered $n+1, \dots, m$, and write down the braid word w'' corresponding to the resulting n-strand braid. In order to see that w'' is indeed equivalent to w, it suffices to remove strands number $n+1, \dots, m$ from an isotopy connecting the m-strand braid represented by w to the one represented by w'. □

The previous result is a little frustrating, as we would like to obtain a self-contained proof of Property **C**, one that does not take Property $\mathbf{C}_\infty$ as an hypothesis. Using the geometrical interpretation of the action of B_n on F_n, we can argue as follows.

Corollary 2.4 (Property **C**). *Every braid of B_n is σ_1-positive, σ_1-negative, or σ_1-free inside B_n.*

Proof. Let β be a braid of B_n, and let w be an n-strand braid word representing β. Let us look at $\widehat{w}(x_1)$, considered as an element of $\pi_1(D_n)$. Thus, we have a disk with n punctures labelled $1, \dots, n$, and we do *not* have $n+1$ punctures as in the proof of Proposition 2.1. Now there are three possibilities.

If $\widehat{w}(x_1)$ ends with x_1^{-1}, then, by Proposition 2.1, w is equivalent to a σ_1-positive n-strand braid word. A second possibility is $\widehat{w}(x_1) = x_1$; in that case we can deduce that $\widehat{w}$ has a σ_1-free representative, as in the proof of Proposition 2.1. If neither of these two possibilities is satisfied, then the image of $\widehat{w}(x_1)$, under the automorphisms $\pi_1(D_n) \to \pi_1(D_n)$ given by reflection of D_n in the horizontal axis, ends with x_1^{-1}. This means that the image of the braid β under the reflection–automorphism of B_n, which sends every generator of B_n to its inverse, has a σ_1-positive representative. This implies that β itself has a σ_1-negative representative. □

3. Pulling back orderings of free groups

In this section we explain a different point of view on the material explained so far in this chapter. One can equip the free group with a linear ordering which is not invariant under the action of F_n on itself by left multiplication, but which *is* invariant under the action of B_n on F_n. Thus any element of F_n on which B_n acts freely gives rise to a linear left-invariant ordering of B_n, by pulling back the ordering on the orbit. In this way we obtain many different orderings of B_n. This approach, which is due to Jonathon Funk [**93**], is by and large equivalent to the Nielsen–Thurston type approach described in Chapter XIII, and some of the constructions in this section, which at first sight may seem contrived, take their motivation from the Nielsen–Thurston approach. On the positive side, the constructions in this chapter are very explicit and, and since they avoid all geometrical tools, they generalize neatly to braid groups with a countable infinity of strands. This approach brings into focus the connections between braid orderings and notions from topos theory.

3.1. From a path to a word. Let us consider the straight line segments $e_1, \dots, e_n$ between punctures of the n times punctured disk D_n as shown in Figure 2(a). Also, we suppose that D_n is equipped with a basepoint $*$ at its left extremity, also shown in the figure. To each element of the fundamental group $\pi_1(D_n)$ we attribute a sequence of letters $y_i^{\pm 1}$ in the following way. We choose a path Γ in D_n representing this element. We go along Γ, starting from the basepoint $*$, and each time we cross e_i, we write y_i if we are going in the upward direction, and y_i^{-1} if we are going in the downward direction. Then we perform free reductions on the resulting word by eliminating subwords of the form $y_i y_i^{-1}$ and $y_i^{-1} y_i$—geometrically, this corresponds to homotopies of the path reducing the number of intersections with the horizontal axis.

We remark that conversely, to every freely reduced word in the letters $y_i^{\pm 1}$ we can associate a loop D_n. So the preceding procedure amounted to a very explicit construction of an isomorphism from $\pi_1(D_n)$ to the free group generated by $y_1, \dots, y_n$.

Similarly, even to an infinite path in D_n without backtracking—avoiding successive intersections with one of the segments e_i in opposite directions—we can associate an *infinite word* with letters $y_i^{\pm 1}$ in the same manner, by writing down its cutting sequence with the segments e_i.

For instance, the paths shown in Figures 2(b), 2(c), and 2(d) are attributed the words $y_1^{-1} y_2^{-1} y_1 y_3^{-1} y_2 y_4^{-1} y_3$, $y_2^{-1} y_3^{-1} y_1$ and $y_2^{-1} y_3 y_4^{-1} y_1^{-1} y_4 y_3^{-1} y_2 y_3^{-1} \dots$, respectively.

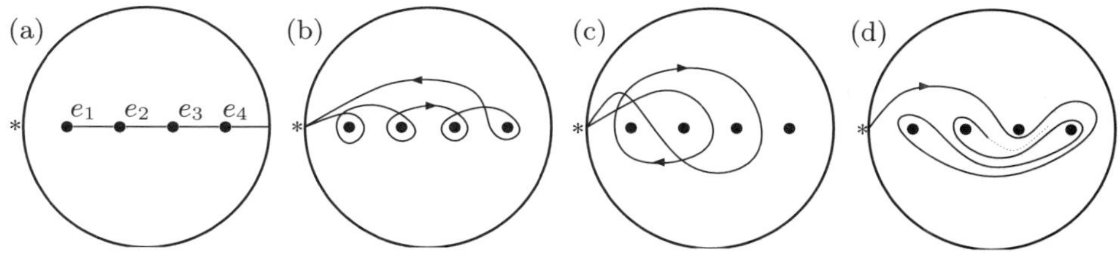

FIGURE 2. One can code elements of the group $\pi_1(D_4)$, which is isomorphic to the free group F_4, by their intersections with e_1, e_2, e_3, e_4. Figures (b) and (c) give examples of elements of this group, whereas the infinite path in Figure (d) would be coded by an infinite word.

Even more generally, we can consider paths in a disk with a countable infinity of punctures which are lined up from left to right on the horizontal line, with an accumulation point at the right extremity of the disk; homotopy classes of such paths can be identified with words in the free group with infinitely many generators $y_1, y_2, \ldots$.

3.2. Ordering free groups and braid groups. Let us denote by $\widehat{F_\infty}$ the set of all freely reduced nonempty words in the letters $y_1^{\pm 1}, y_2^{\pm 1}, \ldots$ that are finite or infinite to the right. An example of an element of $\widehat{F_\infty}$ which will turn out to be essential is the infinite word $y_1^{-1} y_2^{-1} y_1 y_3^{-1} y_2 y_4^{-1} y_3 \ldots$.

Our next aim is to define an ordering $\lhd$ of $\widehat{F_\infty}$. The details of this construction may look complicated, but the reader should keep in mind that we are simply trying to capture the idea of one path going "more to the left" than another one.

Consider the following *circular* list L involving the letters $y_i^{\pm 1}$ plus one additional symbol y_∞:

$$L: \; y_\infty \twoheadrightarrow \ldots \twoheadrightarrow y_2 \twoheadrightarrow y_1 \twoheadrightarrow y_1^{-1} \twoheadrightarrow y_2^{-1} \twoheadrightarrow \ldots \twoheadrightarrow y_\infty .$$

Here $\twoheadrightarrow$ is meant to indicate that our circular list L has a distinguished direction.

DEFINITION 3.1. For three pairwise distinct elements a, b, c of L, we say that a, b, c *go in the right order* if b is met before c when one goes along the list, starting from a, in the preferred direction, *i.e.*, along the arrows $\twoheadrightarrow$.

If we denote y_i^{-1} as y_{-i}, then this definition is equivalent to saying that y_i, y_j, y_k are in the right order when we have

$$\left(\frac{1}{i} - \frac{1}{j}\right)\left(\frac{1}{j} - \frac{1}{k}\right)\left(\frac{1}{i} - \frac{1}{k}\right) > 0.$$

DEFINITION 3.2. Let w_1, w_2 be distinct elements of $\widehat{F_\infty}$. Let w denote the longest common prefix (*i.e.*, left subword) of w_1 and w_2. Let y_i^e be the last (rightmost) letter of w, let z_0 be the inverse letter y_i^{-e}, and let z_1, z_2 be the letters right adjacent to w in w_1 and w_2, respectively. If w is empty, we set $z_0 = y_\infty$; if w equals w_i, we set $z_i = y_\infty$. Then we declare that $w_1 \lhd w_2$ is true if z_0, z_1, z_2 are in the right order.

This definition always makes sense as, by construction, the letters z_0, z_1, and z_2 are pairwise distinct. The proof of the following statement is straightforward.

LEMMA 3.3. *The relation $\lhd$ is a linear ordering on $\widehat{F_\infty}$.*

Note that we do not claim that the ordering $\triangleleft$ is invariant under left or right multiplication.

Having defined an ordering of $\widehat{F_\infty}$, we shall now construct an action of B_∞ on $\widehat{F_\infty}$. First, we set

$$\sigma_i \bullet y_k = \begin{cases} y_1^{-1} y_2 & \text{for } i = k = 1, \\ y_{i-1} y_i^{-1} y_{i+1} & \text{for } i = k > 1, \\ y_k & \text{for } i \neq k, \end{cases} \quad \sigma_i \bullet y_k^{-1} = \begin{cases} y_2^{-1} y_1 & \text{for } i = k = 1, \\ y_{i+1}^{-1} y_i y_{i-1}^{-1} & \text{for } i = k > 1, \\ y_k^{-1} & \text{for } i \neq k, \end{cases}$$

and define the action of σ_i on a (finite or infinite) word w to be the result of freely reducing the concatenation of the images of the successive letters of w. In this way, the action of σ_i induces a bijection of $\widehat{F_\infty}$, so we can define the action of σ_i^{-1} to be the inverse bijection, and, finally, define the action of an arbitrary braid word on $\widehat{F_\infty}$.

For finite words, the action coincides with the action of B_∞ on $F_\infty \setminus \{1\}$ considered in Section 2, so it is not hard to check that the above formulas provide an action of B_∞ on $\widehat{F_\infty}$.

PROPOSITION 3.4. *The action of B_∞ on $\widehat{F_\infty}$ equipped with $\triangleleft$ is order-preserving.*

Since the linear ordering $\triangleleft$ of $\widehat{F_\infty}$ is described explicitly, Proposition 3.4 can be established by a direct verification of cases similar to those of Section 1. This has been done in [**93**] in a slightly different setting.

Proposition 3.4 implies that each element x of $\widehat{F_\infty}$ defines a partial ordering of B_∞ by setting $\beta <_x \beta'$ for $\beta(x) \triangleleft \beta'(x)$. This ordering $<_x$ is linear if and only if the stabilizer of x in B_∞ is trivial.

Let

$$z = y_1^{-1} y_2^{-1} y_1 y_3^{-1} y_2 y_4^{-1} y_3 \cdots,$$

and let z_n denote the truncated version: $z_n = y_1^{-1} y_2^{-1} y_1 \ldots y_n^{-1} y_{n-1}$. See Figure 2(b) for a picture of this path in the $n = 4$. A detailed study of the orderings induced by these words will be undertaken in Chapter XIII—see in particular Section XIII.1.3. Either by using the results of that study, or by a more elementary argument involving the combinatorics of words (see [**93**]), one can prove the following results:

PROPOSITION 3.5. *For each n, the relation $<_{z_n}$ is a linear ordering on B_n for each n, and the relation $<_z$ is a linear ordering on B_∞.*

The counterpart of Proposition XIII.1.8 is then

PROPOSITION 3.6. *The linear ordering $<_{z_n}$ on B_n coincides with the σ-ordering for every n, and, therefore, so does the linear ordering $<_z$ on B_∞.*

In this way, we obtain the seventh of the many definitions of the σ-ordering mentioned in the Introduction:

COROLLARY 3.7. *For β, β' in B_n, the relation $\beta < \beta'$ is true if and only if we have $\beta(z_n) \triangleleft \beta'(z_n)$ in $\widehat{F_\infty}$—actually, in $F_\infty \setminus \{1\}$, since z_n is a finite word.*

We refer the reader to [**93**] for a further connection of the previous interpretation with the theory of toposes.

CHAPTER X

Curve Diagrams

We turn to a very different construction of the braid ordering based on a topological approach. A crucial fact concerning the braid group B_n which will be used throughout this chapter is that B_n is isomorphic to the mapping class group $\mathcal{MCG}(D_n)$, *i.e.*, the group of isotopy classes of self-homeomorphisms of a disk with n punctures (Proposition I.3.3). Thus the task of ordering braid groups is equivalent to that of ordering mapping class groups of punctured disks: given two self-homeomorphisms φ, φ' of D_n representing two elements of $\mathcal{MCG}(D_n)$, we want to define which of the two is the larger.

The strategy in this chapter for doing so is as follows: if E denotes the main (horizontal) diameter of D_n, we are going to define the relative order of the elements represented by φ and φ' in terms of the relative position of the arcs $\varphi(E)$ and $\varphi'(E)$.

We will see that the linear ordering defined in this way coincides with the σ-ordering. This construction works not only for the braid groups, but more generally for mapping class groups of compact surfaces with nonempty boundary.

The approach described in this chapter was developed in [**83**] and [**182**], and, like the self-distributivity approach, it is complete in the sense that it leads to proofs of all of Properties **A**, **C**, and **S**.

The principle of the proof of Property **C** explained in Section 2.2 is very powerful and well known: it is the idea of relaxation of curve diagrams. This idea will be pursued further in Chapter XI.

The structure of this chapter is as follows. In Section 1 we define the curve diagram associated to an element of the braid group, and we show how curve diagrams can be used to define an ordering of the braid group. We also point out generalizations of this construction for mapping class groups of surfaces with nonempty boundary. In Section 2 we give new proofs of Properties **A**, **C**, and **S**, and along the way we prove that our ordering based on curve diagrams coincides with the σ-ordering of braids.

1. A braid ordering using curve diagrams

In this section we define curve diagrams of braids and explain how any two curve diagrams can be put into a preferred position relative to each other. Then we show how this yields an ordering of braids: we can define a relative order of any two given elements of the braid group by inspection of the relative position of their associated curve diagrams. Finally, we point out that these ideas generalize readily to the framework of mapping class groups and surface braid groups of surfaces with nonempty boundary.

1.1. Mapping class groups and curve diagrams. Throughout this chapter, we shall denote by D^2 the unit disk in $\mathbb{C}$ with centre 0, and by D_n the same disk with n uniformly spaced points in the real axis $\mathbb{R} \cap D^2$ marked as distinguished points. These points will be called the *punctures*, and denoted $P_1, \dots, P_n$, from left to right. We also introduce notation for some diagrams in D_n: we write $e_0, \dots, e_n$ for the $n+1$ horizontal open line segments $(-1, P_1)$, (P_1, P_2), ..., (P_{n-1}, P_n), $(P_n, +1)$, respectively, and E for their union; see Figure 1(a). All homeomorphisms of D_n, and indeed any surface, are supposed to permute the punctures and to fix the boundary pointwise.

Our first aim is to formalize the following idea, which was hinted at in the Introduction. In order to specify the isotopy class of a homeomorphism φ of D_n to itself which permutes the punctures and fixes the boundary pointwise, it suffices to specify the image under φ of the horizontal line E in D_n. This reduces the problem of understanding isotopy classes of homeomorphisms of a surface to the more intuitively accessible one of understanding isotopy classes of curves on the surface.

DEFINITION 1.1. The *curve diagram* of a homeomorphism $\varphi\colon D_n \to D_n$ is the image of E under φ. Two curve diagrams are *isotopic* if there exists an isotopy of D_n which deforms one diagram into the other and leaves $P_1, \dots, P_n$ and the boundary of D^2 fixed during the isotopy.

Obviously, isotopic homeomorphisms give rise to isotopic curve diagrams. This means that to every element of the mapping class group $\mathcal{MCG}(D_n)$, or equivalently to every element of the braid group B_n, we can associate a curve diagram which is uniquely defined up to isotopy. Some examples of curve diagrams are given in Figure 1. For one more example, see Figure 3: the leftmost curve diagram in this figure is the curve diagram of the braid $\sigma_2\sigma_1\sigma_2^{-1}\sigma_1\sigma_2$, which is drawn at the bottom of the figure.

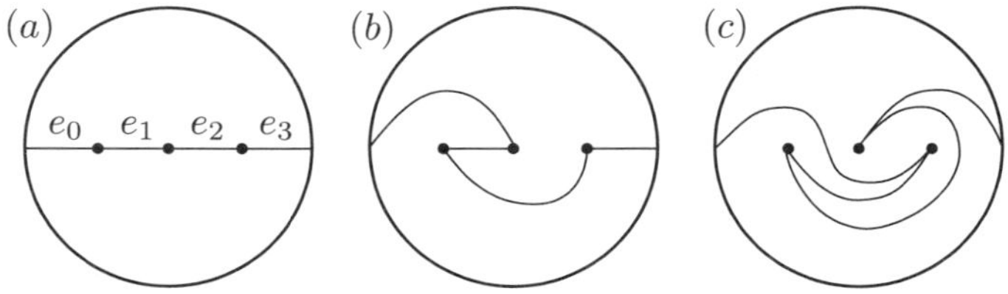

FIGURE 1. Examples of curve diagrams: the curve diagrams of the trivial braid, of σ_1, and of $\sigma_2^{-1}\sigma_1$. Note that the braid group acts on the left, so the latter curve diagram is obtained from the trivial curve diagram (a) by the action of σ_1, followed by the action of σ_2^{-1}.

Next we claim that there is in fact a natural one-to-one correspondence between elements of $\mathcal{MCG}(D_n)$—which in turn is naturally isomorphic to the braid group B_n—and isotopy classes of curve diagrams in D_n. In other words, braids are uniquely determined by their associated curve diagrams:

PROPOSITION 1.2. *Two homeomorphisms $\varphi, \varphi'\colon D_n \to D_n$ are isotopic if and only if their associated curve diagrams are isotopic—isotopies in both instances are to be fixed on the punctures and the boundary.*

PROOF. A formal proof can be found in [**83**]. We shall only give an informal explanation of how geometric braids give rise to curve diagrams and, conversely, given the curve diagram of a braid β in B_n, how to recover a geometric braid representing β. A particular example of these two procedures is shown in Figure 3.

Let β be a geometric n-strand braid, sitting in the cylinder $[0,1] \times D^2$, whose n strands are starting at the puncture points of $\{0\} \times D^2$ and ending at the puncture points of $\{1\} \times D^2$. A good way to imagine the curve diagram of a homeomorphism φ associated with β is as follows. We think of the diagram E in D_n as consisting of $n+1$ segments of rubber band in the disk D_n, and we imagine that this disk D_n is itself embedded in the face $\{1\} \times D^2$ of the cylinder. If we now slide the diagram E from the 1-level in the cylinder back to the 0-level, then the braid β corresponds to a dance of the n puncture points in the disk D^2. During the dance, the rubber bands get stretched and deformed, and the resulting picture in $\{0\} \times D^2$ is the curve diagram of the homeomorphism φ.

Conversely, suppose that we are given the curve diagram of some braid β in B_n, *i.e.*, in $\mathcal{MCG}(D_n)$, but we are not given β itself, and we want to find a geometric n-strand braid representing β. Our procedure for doing this will play a key role in what follows. We can place the curve diagram in the 0-level of the cylinder. If we now authorize the puncture points to move, then due to the elastic force of the rubber bands the diagram will untangle and become the straight horizontal line in D_n. While this is happening, we can slide the disk along the cylinder into the 1-level. Then the punctures will trace out the geometric braid β, which we have thus recovered from the curve diagram. □

1.2. Definition of the ordering. Our aim now is to construct an ordering $<_{\mathrm{CD}}$ of B_n using curve diagrams. We shall see later that this ordering coincides with the σ-ordering which is the main subject of this book.

Given two braids, or equivalently two elements of $\mathcal{MCG}(D_n)$ represented by two homeomorphisms φ and ψ, we want to define which one is larger. The first step is to superimpose the curve diagrams of φ and ψ in D_n. A priori, these diagrams may intersect each other unnecessarily; for instance, φ-curves and ψ-curves may have points of tangency, or may intersect in infinitely many isolated points, or may simply enclose digons; see Figure 2.

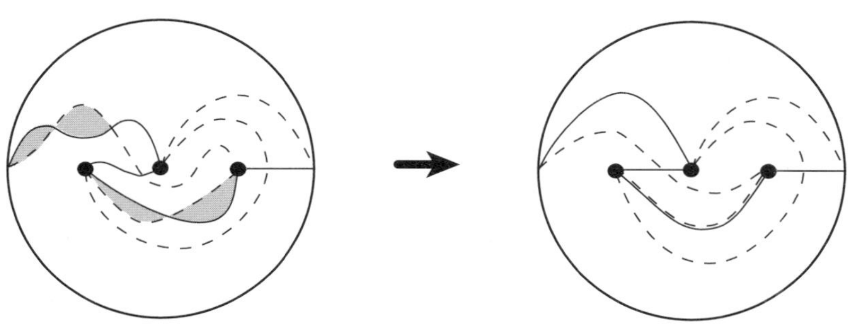

FIGURE 2. Removing four digons in the curve diagrams for σ_1 and $\sigma_2^{-1}\sigma_1$. The diagrams in the resulting picture are tight with respect to each other. In the left figure, the intersection number is 5; in the right one it is 1.

So the second step is to pull the curve diagrams tight. In order to explain what that means, we introduce the intersection number. To this end, it is necessary to have a precise notion of transverse curve diagrams.

DEFINITION 1.3. Two curve diagrams C_1, C_2 are said to be *transverse* if every arc of C_1 either is transverse (*i.e.*, nontangent) to every arc of C_2 or coincides with one of them.

Hereafter, all curve diagrams we mention are assumed to be pairwise transverse.

DEFINITION 1.4. The *intersection number* of two curve diagrams C_1, C_2 is defined to be the number of transverse intersection points in the interiors of the arcs, minus the number of coincident arcs of the two curve diagrams; *i.e.*, if one of the $n+1$ φ-curves coincides exactly with one of the ψ-curves, we count this as -1 intersection point.

Then we can define the notion of tight position as follows.

DEFINITION 1.5. We say that two curve diagrams C_1, C_2 are *in tight position* if, for all curve diagrams C_1', C_2' such that C_1' is isotopic to C_1 and C_2' is isotopic to C_2 (not necessarily by the same isotopy), the intersection number of C_1' and C_2' is greater than or equal to the intersection numbers of C_1 and C_2.

Now, *pulling two curve diagrams* C_1, C_2 *tight* means that we apply independent isotopies to them until we obtain diagrams that are in tight position. One possible way to find such a tight positioning of the curves is to start with two diagrams with finite intersection number and successively decrease the intersection number by removing digons until no more digon is left. It is a nonobvious fact [**83**] that pairs of curve diagrams without digons are always tight—a similar but more general statement will be proved in Section 3 of Chapter XII. This opens up a theoretically neat (although not practical) method for tightening curve diagrams: one equips D_n with a hyperbolic metric, in which the n puncture points are cusps, and lets the $2(n+1)$ curves of the diagrams flow into geodesics. Since geodesics never form digons, this yields a tight pair of diagrams. This observation is the basis of the techniques which will be introduced in Chapter XIII. The crucial fact for our purposes is that such a tight positioning of two curve diagrams is essentially unique:

PROPOSITION 1.6. *Suppose that C_0 and C_1 are two curve diagrams, which are isotopic, and are both in tight position with respect to another curve diagram C_2. Then there exists an isotopy of the disk which fixes the diagram C_2 setwise, fixes the punctures and the boundary of the disk, and transforms C_0 into C_1.*

A simple and elementary proof is given in [**83**]. Very similar results with similar proofs were, however, well known before this paper, and we refer the reader to Section XII.3 for an argument for this result, but in a slightly different setting.

As a third step, we want to define a relation between curve diagrams. Suppose that C and C' are two nonisotopic curve diagrams. Let us imagine that we sit down at the point -1 of D_n and walk along the curves of C. For some initial period of time the curves of C' coincide with those of C. At some moment, however, either at -1 or at one of the puncture points, the diagrams C and C' will diverge (otherwise we would have $C = C'$). At this divergence point, C' will either set out into the upper component of $D_n \setminus C$ (the one containing the point $\sqrt{-1}$) or into the lower component (which contains $-\sqrt{-1}$). In the first case we say C' *goes more to the left than* C, and in the second that C' *goes more to the right than* C. Note that C' goes more to the right than C if and only if C goes more to the left than C'.

DEFINITION 1.7. A relation $<_{\text{CD}}$ on B_n is defined as follows. If β and β' are two distinct braids, then we superimpose their curve diagrams, C and C' say, pull them tight, and define that $\beta >_{\text{CD}} \beta'$ is true if C goes more to the left than C' and $\beta <_{\text{CD}} \beta'$ is true if C' goes more to the left than C.

LEMMA 1.8. *The relation* $<_{\text{CD}}$ *is a linear ordering on* B_n *and it is left-invariant.*

PROOF. In order to see that $<_{\text{CD}}$ is transitive, we only need to know that any three curve diagrams can be drawn in D_n in such a way that they are mutually tight. In [**83**], this is proved using the so-called triple reduction lemma. Alternatively, we can use hyperbolic geometry: if all the curves of all three curve diagrams consist of geodesics of the same hyperbolic structure on D_n, then the three diagrams will be mutually tight.

Seeing that the ordering $<_{\text{CD}}$ is left-invariant is easy. Suppose that $\varphi_1, \varphi_2, \varphi$ are homeomorphisms of D_n representing three braids β_1, β_2, β, and that the φ_1-diagram goes more to the left than the φ_2-diagram. If we apply φ to the curve diagrams of φ_1 and φ_2, we obtain curve diagrams for $\varphi\varphi_1$ and $\varphi\varphi_2$, respectively. Since we have applied the same homeomorphism φ to the two diagrams, their relative position is unchanged: the diagram of $\varphi\varphi_1$ will still be to the left of the diagram of $\varphi\varphi_2$. This proves that $\beta_1 >_{\text{CD}} \beta_2$ implies $\beta\beta_1 >_{\text{CD}} \beta\beta_2$. □

2. Proof of Properties A, C, and S

In the previous section we defined a linear, left-invariant ordering $<_{\text{CD}}$ of the braid group B_n. In this section we prove that this ordering coincides with the σ-ordering, and we give new proofs of Properties **A**, **C**, and **S**; these proofs only use arguments about planar diagrams.

2.1. A proof of Property A. Let us first prove that the positive cone of the ordering $<_{\text{CD}}$ contains the positive cone of the σ-ordering.

PROPOSITION 2.1. *Every* σ*-positive braid* β *satisfies* $\beta >_{\text{CD}} 1$. *Every* σ*-negative braid* β *satisfies* $\beta <_{\text{CD}} 1$.

PROOF. Let w be a σ-positive braid word representing β, with curve diagram C. For simplicity, let us first assume that w is in fact σ_1-positive; then w is of the form $w_0\sigma_1 w_1 \sigma_1 w_2 ... \sigma_1 w_k$, where the words w_i contain no letter $\sigma_1^{\pm 1}$. We shall now consider the curve diagrams of various braids, with particular attention to their first curves, *i.e.*, those starting at -1. Our aim is to prove that an initial segment of the first curve of C lies in the upper half of D_n, which implies $\beta >_{\text{CD}} 1$.

The first curve of the curve diagram of the braid represented by w_k coincides with that of the trivial braid: it is just a horizontal line segment, because the first strand does not cross any other strands. Acting on this curve diagram by σ_1 yields the curve diagram of $\sigma_1 w_k$. Its first curve, *i.e.*, the image of the arc e_0 under $\sigma_1 w_k$, is an arc in the upper half of D_n connecting -1 to the second leftmost puncture P_2. In particular, this first curve has the following property: its first intersection with the vertical line through the leftmost puncture P_1 lies in the upper half of D_n. Now we observe that successively applying any sequence of braids $\sigma_2^{\pm 1}, \dots, \sigma_{n-1}^{\pm 1}$ and σ_1 (but not σ_1^{-1}) cannot change this property. In particular, the curve diagram C of the braid word w has this property, which implies that an initial segment of C lies in the upper half of D_n.

This completes the proof for σ_1-positive braids. For braids that admit a representative braid word containing at least one σ_i, and no σ_i^{-1} and no $\sigma_j^{\pm 1}$ with $j < i$, the argument is similar, guaranteeing that the first $i-1$ curves of the curve diagram are horizontal line segments, and the ith sets off into the upper half of D_n.

Finally, the argument for σ-negative braids is symmetric. □

This technical result has two important consequences. First, we have proved that the curve diagram of a σ-positive word has to diverge (to the left, in fact) from the trivial curve diagram, so we deduce

COROLLARY 2.2 (Property **A**). *A braid that admits at least one σ_1-positive representative braid word is not trivial.*

Second, Proposition 2.1 tells us that the σ-ordering of Chapter II is included in the ordering $<_{\mathrm{CD}}$. Since both are linear (total) orderings, we deduce that the ordering $<_{\mathrm{CD}}$ and the σ-ordering coincide. In particular, we have established the eighth equivalence mentioned in the Introduction:

PROPOSITION 2.3. *For all braids β, β', the relation $\beta < \beta'$ is true if and only if we have $\beta <_{\mathrm{CD}} \beta'$, i.e., the standardized curve diagram associated with β' first diverges from that associated with β towards the left.*

2.2. A proof of Property C. The aim of this section is to use curve diagram techniques in order to prove the comparison property (Property **C**).

PROPOSITION 2.4 (Property **C**). *Every nontrivial braid of B_n can be represented by an n-strand braid word that is σ-positive or σ-negative.*

PROOF (SKETCH). Here we shall only outline the method of proof, and refer to [**83**] for details. This method is also illustrated in Figure 3. The strategy is to prove that any braid β with $\beta >_{\mathrm{CD}} 1$, *i.e.*, whose curve diagram first diverges from the trivial curve diagram into the upper half of D_n, has a σ-positive representative.

For simplicity we will assume the curve diagram of β diverges *immediately* to the left, and conclude that β admits a σ_1-positive representative. The other possibility is that the the first curve of the diagram coincides with e_0 and the diagram only diverges later; in this case β admits σ_1-free representative, and a similar argument (which we leave to the reader) will show that at least one representative of β is σ_i-positive with $i > 1$.

We recall the technique which was introduced in the proof of Proposition 1.2: we can reconstruct a braid from its curve diagram by placing the curve diagram in the 0-level of a cylinder $[0,1]\times D^2$, untangling the diagram while sliding the disk into the 1-level, and observing the trace of the puncture points under this movement. We shall apply this technique, but very carefully avoid using the letter σ_1^{-1}.

So let C be a curve diagram which is tight with respect to the trivial curve diagram (the straight horizontal line), and whose first curve sets out immediately into the upper half of D_n. Our aim is to transform C into the trivial diagram by using an isotopy. At every moment in time we can imagine a vertical line through the leftmost puncture. What we have to avoid during the isotopy is any puncture ever hitting this line *below* the leftmost puncture, in order to turn itself into the new leftmost puncture. Punctures are only allowed to travel into the leftmost position by hitting the *upper* half of the vertical line through the current leftmost puncture. This condition is equivalent to the requirement that the resulting geometric braid is described by a σ_1-positive word.

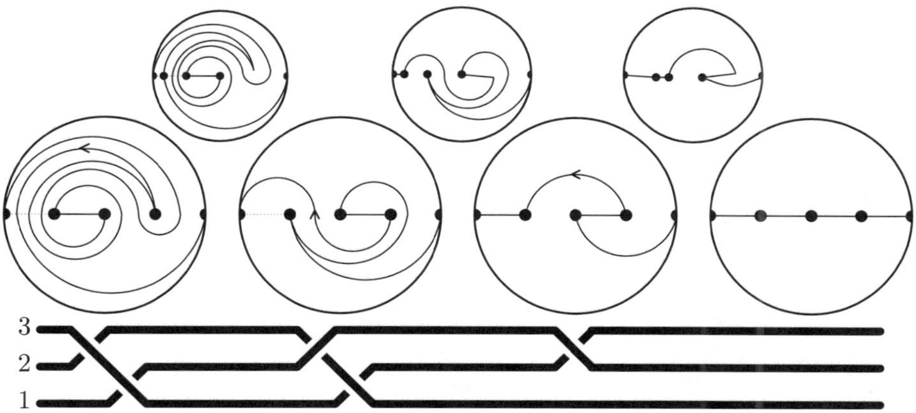

FIGURE 3. How to find a σ-positive representative of a braid word, given the curve diagram. To untangle it, we have to alternately slide punctures along useful arcs and tighten the diagram. The useful arcs are drawn in bold lines.

The idea for actually finding such an untangling movement is as follows. We recall that e_0 denotes the horizontal line segment from -1 to the leftmost puncture. The curve diagram C intersects e_0 in a finite number of points (one of which is the point -1). We define a *useful arc* to be a subarc of one of the curves of C which starts at one of the intersection points with e_0, has an initial segment which lies in the upper half of D_n, does not intersect e_0 again, and terminates at any puncture (P say) other than the leftmost one. A simple geometric argument (which we leave as an exercise) guarantees that, in any curve diagram whose first curve sets off into the upper half of D_n, such a useful arc can indeed be found.

Now we are ready to start untangling the diagram C. We slide the puncture P back along the useful arc, all the way to its starting point in e_0. Note that during this movement the puncture P will slide exactly once (near the end of its voyage) *over* the current leftmost puncture to achieve itself the new leftmost position, but it never slides *under* the leftmost puncture. This completes the first stage of our untangling process. The curve diagram thus deformed may well fail to be tight with respect to the trivial curve diagram, so we pull it tight, and obtain a new curve diagram C'.

Now we claim that the first curve of C' still sets off into the upper half of D_n, or it might possibly coincide with e_0. This can be proved by another easy geometric argument—and it is actually very plausible if one thinks of the curve diagram as being realised by rubber bands. In the first case, we can iterate our construction, *i.e.*, find a useful arc in C', etc. In the second case we can simply untangle the curve diagram C' without any special care.

This process has to terminate with the trivial curve diagram—indeed, with respect to a suitably defined notion of complexity of curve diagrams, C' is simpler than C, and the complexity decreases further in each successive iteration. □

We remark that the algorithm introduced in this section for finding a σ-positive representative of a braid β that satisfies $\beta >_{\rm CD} 1$ by first constructing its curve diagram and then undoing this diagram again is very inefficient: the length of the resulting σ-positive braid word depends, in general, exponentially on the length of the input braid word. For instance, the reader may check that, for the 4-strand braid represented by the word $(\sigma_3\sigma_2^{-1})^p\sigma_1$, the algorithm presented here outputs a σ_1-positive word whose length grows exponentially with p. This is particularly dissatisfying, given that the input braid word was itself σ_1-positive, so the algorithm did not really need to do anything!

However, this disadvantage is not inherent in the basic method of this algorithm. As we shall see in Chapter XI, the idea of relaxing, or untangling, curve diagrams is in fact a very powerful one. For more comments on the algorithmic aspect, also see Chapter XVI.

2.3. A proof of Property S. It is possible to establish Property **S** using the machinery of this chapter [**194**]. We shall outline such a proof here, using the notion of a Dehn half-twist. A detailed proof along only slightly different lines can be found in [**194**].

Definition 2.5. Let e be a simple arc connecting two distinct points, say P and Q, in the plane—or, more generally, in an arbitrary oriented surface. A *Dehn half-twist* around e is a homeomoprhism φ of the plane such that φ is the identity outside a small neighbourhood of e, it flips e, and it screws clockwise a small neighbourhood U of e as shown in Figure 4.

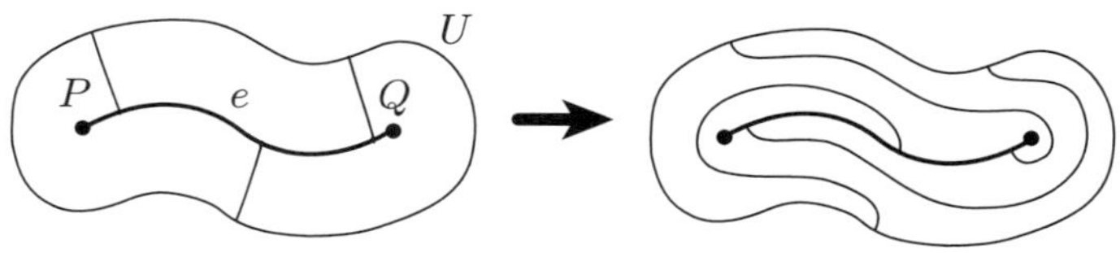

Figure 4. Dehn half-twist around the arc e.

For instance, the element of $\mathcal{MCG}(D_n)$ defined by a Dehn half-twist around the segment e_i is the image of the braid σ_i under the isomorphism of B_n to $\mathcal{MCG}(D_n)$.

Let φ be a Dehn half-twist around an arc e, and let Γ be a curve intersecting e transversely at some point, say R. The image curve $\varphi(\Gamma)$ behaves as follows. First, $\varphi(\Gamma)$ walks along Γ until coming close enough to R, then it *turns to the left* and goes to one endpoint of e, keeping close to e, then it turns clockwise around that point, goes to the second endpoint of e, always nearby e and crossing e at $\varphi(R)$ in the opposite direction, then it turns counterclockwise around the second endpoint and returns along e into a small neighbourhood of R. Finally, it continues moving along Γ.

With this notion at hand, we can prove

Proposition 2.6. *For every braid β and every i, we have $1 <_{\rm CD} \beta\sigma_i\beta^{-1}$.*

Proof. Let ψ be the homeomorphism of D_n corresponding to β; let e be the arc $\psi(e_i)$, and let φ be the Dehn half-twist around e. Then φ represents the braid $\beta\sigma_i\beta^{-1}$. So, by definition of $<_{\rm CD}$, the inequality $1 <_{\rm CD} \beta\sigma_i\beta^{-1}$ follows from the above observation that $\varphi(E)$ first diverges from E to the left, provided we make sure that φ can be chosen so that $\varphi(E)$ is tight with respect to E.

Now, we can assume without loss of generality that e is tight with respect to the main diameter E. Then we claim that the Dehn half-twist φ can be chosen so that there is no digon bounded by E and $\varphi(E)$. Indeed, in addition to parts of E and $\varphi(E)$ coinciding, there are only unavoidable crossing points of E and $\varphi(E)$ that appear near crossing points of E and e. Between two such points the curve $\varphi(E)$ goes parallel to e. Now it is not hard to see that the existence of a digon bounded by E and $\varphi(E)$ would imply the existence of a digon bounded by E and $\varphi(e)$. Now one can show that only a small perturbation is needed for $\varphi(E)$ to become tight with respect to E. So the proof is complete. □

By Proposition 2.3, the σ-ordering and the $<_{\mathrm{CD}}$-ordering coincide, so we deduce

COROLLARY 2.7 (Property **S**). *Every braid of the form $\beta\sigma_i\beta^{-1}$ is σ-positive.*

More general results in the above direction can be proved using the refinements of the curve diagram technique introduced in Section 3 of Chapter XIII.

CHAPTER XI

Relaxation Algorithms

Suppose we are given a braid word representing some element of the braid group B_n. It is a very common situation that we have to find a canonical representative braid word of that element which has some desirable property. Typically in the context of the present book, we might be interested in finding a σ-positive or σ-negative word representing the given braid. In this chapter we shall discuss one class of methods for obtaining canonical representative braid words, namely the method of untangling, or relaxing, curve diagrams.

This general method works as follows. Given a braid β in B_n, in a first step we construct its curve diagram. In a second step, we successively untangle this diagram—according to certain rules and with certain restrictions, we search for a braid word whose action simplifies the diagram . We repeat this procedure until the trivial curve diagram is reached. If we concatenate the braid words chosen in the successive steps, we obtain a braid word representing β.

We have already encountered this method twice in this book. Once, in disguise, in Chapter IX, in Larue's proof of Property **C** (Proposition IX.2.1). Our second encounter with this principle was in Chapter X, in the diagrammatic proof of Property **C** (Proposition X.2.4). One more use of this principle will occur in Chapter XII: Mosher's automatic normal form for braids—and more general mapping classes—has this idea at its base, even though it uses it in a modified, very subtle form.

In the current chapter we shall present two algorithms using this principle in order to find σ-positive or σ-negative representative braid words of a given element of B_n. In this, they are like the algorithm in the proof of Property **C** in Proposition X.2.4. However, both algorithms and normal forms presented here have some useful and interesting additional properties.

The first normal form of braids that we shall present, due to Bressaud [**19**], is remarkable in that it can be found by two completely different algorithms: either by a relaxation method of curve diagrams, as explained above, called the delooping algorithm, or by a rewriting system operating on braid words, called the Tetris algorithm. This surprising double perspective makes it reasonable to hope that this normal form may serve as a powerful theoretical tool.

The second normal form of braids, which we shall call the transmission-relaxation normal form, was introduced in [**74**]. Roughly speaking, it has two remarkable properties. First, the normal form of a σ-positive braid is a σ-positive braid word—and our proof of this fact leads to a new proof of Property **C**. Second, it tends to write braids as a product of relatively few factors, where each factor is, on the other hand, a relatively high power of a single generator. Among the geometrical consequences of this second property is that the transmission-relaxation normal form of a braid can serve to estimate the complexity of the curve diagram of the braid. As shall be mentioned in Chapter XVI (see [**74, 178**] for a full explanation), it can also serve to estimate distances in the Teichmüller space.

We make one general remark concerning algorithmic complexity; see also Section XVI.1.3. In general, relaxation-type algorithms do not output σ-positive or σ-negative braid words, and they can fail to be efficient; see for instance the algorithm used in the proof of Property **C** in Chapter X. Nevertheless, it appears that they tend to be very robust: most reasonable ways of filling in details in the above vague description of a general relaxation algorithm seem to yield algorithms that are very efficient in practice, but whose efficiency is difficult to prove. For instance, the algorithms of Larue and Bressaud, as well as the transmission-relaxation algorithm, are all believed to have output of linearly bounded length and to have polynomial running time—in fact, the latter is known to hold in the case of the transmission-relaxation algorithm. Moreover, it appears that even quite naive relaxation algorithms tend to yield uniformly quasi-geodesic representatives of elements of mapping class groups. One obstacle to explaining this fascinating phenomenon is that we are currently far from having a complete understanding of quasi-geodesics in mapping class groups, despite some very deep results which explore the relation between the geometry of the mapping class group and the geometry of the train track complex, the curve complex, the pants complex and the Teichmüller space; see for instance [**151, 178, 104, 24**].

The structure of this chapter is as follows. In Section 1 we present Bressaud's normal form of braids and the two different algorithms to compute it. In Section 2 we explain an algorithm for counting orbits of strip decompositions due to Agol, Hass and Thurston, and how this algorithm leads to the transmission-relaxation method.

1. Bressaud's regular language of relaxation braids

In this section we shall present a normal form of braids which is due to Xavier Bressaud [**19**]—also see [**35**] for a good exposition of Bressaud's work. This normal form arises by untangling curve diagrams, and it yields σ-positive representatives for elements of B_n that are positive in the linear ordering—but the analogue statement with positive replaced by negative is false. On the other hand, the language of normal forms has the advantage of being asynchronously automatic [**77**], and moreover it is conjectured to be uniformly quasi-geodesic. Thus we have two points of view on this normal form—one curve diagrammatic and one language combinatorial—which give rise to two different algorithms, and it is rather surprising that they output the same normal form braid words. Like the handle reduction algorithm, these algorithms are conjectured to be quadratic time, but the only known upper bound on their efficiency is exponential.

Bressaud introduced this normal form while studying random walks on the braid group and its Poisson boundary. For more information on this connection, we refer the reader to Section XVI.2.8 and to the original paper [**19**].

1.1. Normal words. Throughout this section we will use a new family of generators of the n-strand braid group B_n.

DEFINITION 1.1. For $1 \leqslant p < n$, and for $p \leqslant i \leqslant n$ and $j \in \{p, n\}$ with $j \neq i$, we put

$$\sigma_{i,j,p} = \begin{cases} \sigma_i^{-1}\sigma_{i+1}^{-1}\ldots\sigma_{j-1}^{-1} & \text{for } j = n, \\ \sigma_{i-1}\sigma_{i-2}\ldots\sigma_j & \text{for } j = p. \end{cases} \tag{1.1}$$

For instance, the braid diagram associated with $\sigma_{4,1,1}\sigma_{3,1,1}\sigma_{2,4,1}\sigma_{3,1,1}\sigma_{4,2,2}\sigma_{4,3,3}$ is presented in Figure 1.

To explain it in words, braids of the form $\sigma_{*,*,p}$ are trivial on strands 1 through $p-1$ (and possibly some other strands). The generator $\sigma_{i,j,p}$ moves the ith strand in front of the other strands into the jth position, and the strands between these two positions get pushed one notch to the left or right. Note that for instance $\sigma_{2,4,1}$ and $\sigma_{2,4,2}$ represent the same braid, but still we wish to distinguish them.

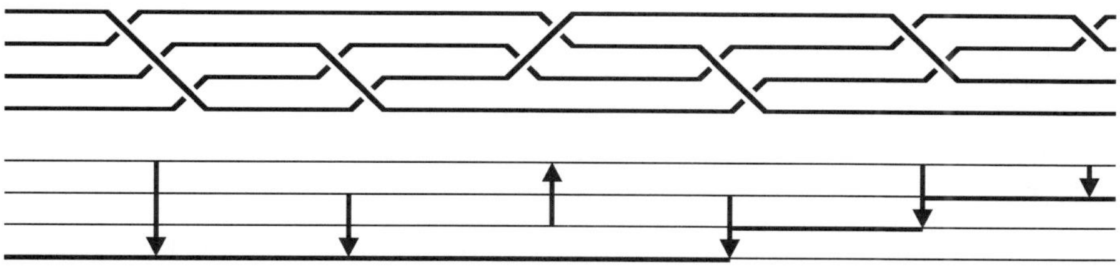

FIGURE 1. The braid represented by $\sigma_{4,1,1}\sigma_{3,1,1}\sigma_{2,4,1}\sigma_{3,1,1}\sigma_{4,2,2}\sigma_{4,3,3}$, and the corresponding Tetris diagram—the meaning is *not* the same as for similar diagrams used in Chapter VIII.

A useful way of representing braid words in this generating set are Bressaud's Tetris diagrams:

DEFINITION 1.2. The *Tetris diagram* representing the generator $\sigma_{i,j,p}$ consists of n horizontal lines and a vertical arrow from the ith to the jth line. In order to specify p, we may draw the pth horizontal line in bold. For any word in the generators $\sigma_{i,j,p}$, the Tetris diagram representing the word is obtained by stacking the diagrams corresponding to each of its letters.

For instance, the Tetris diagram representing the word $\sigma_{4,1,1}\sigma_{3,1,1}\sigma_{2,4,1}\sigma_{3,1,1}$ $\sigma_{4,2,2}\sigma_{4,3,3}$ can be seen in Figure 1.

DEFINITION 1.3. A word w in the generators $\sigma_{i,j,p}$ is *normal in the sense of Bressaud* if, for every letter $\sigma_{i,j,p}$ of w, one of the following conditions holds:

- we have $i = p$ or $j = p$, and this is the last letter of the word,
- we have $i = p$ or $j = p$, and, in the following letter $\sigma_{i',j',p'}$, we have $p' > p$,
- in the following letter $\sigma_{i',j',p'}$, we have $p' = p$ and $i' \in [i, j)$ or $i' \in (j, i]$, depending on whether $i < j$ or $i > j$.

So a normal word can be decomposed as $w = w^{(1)}w^{(2)}\ldots w^{(n-1)}$, where some of the factors may be empty, and where $w^{(p)}$ consists of letters of the form $\sigma_{*,*,p}$. Roughly speaking, the second condition tells us that at the transition from one subword to the next, the last letter of the first subword could not be integrated into the second subword. The third condition says that for two subsequent letters within the same subword, i' must lie between i and j, but we insist that $i' \neq j$, because in case $i' = j$ the two generators could be merged into one.

1.2. The language of normal words. We shall see that the set of all normal words is a regular language, *i.e.*, it is recognized by a finite state automaton. First, we recall the needed terminology and refer to [**77**] for further development.

Let A be a finite set, which will be referred to as an *alphabet* and its elements as *letters.* The set of all words on A will be denoted by A^*.

DEFINITION 1.4. A *finite state automaton* over the alphabet A is the collection of the following data:

• a finite oriented graph, whose vertices are called *states* and whose edges are called *arrows*;

• for each state s, a bijection between A and the set of arrows coming from s, which are said to be *marked* with the corresponding letter (an arrow marked with a letter a will also be called an *a-arrow* for short);

• a subset of states, whose elements are called *accept states* and all the other states are called *failure states*;

• a distinguished state s_*, which is called the *start state.*

Clearly, for any word w on the alphabet A, say $w = a_1 \dots a_k$, and any state s of a finite state automaton M, there is a unique sequence $\eta_1, \dots, \eta_k$ of arrows of M such that η_1 starts at s, η_{i+1} starts at the end of η_i for $1 \leqslant i \leqslant k-1$, and the arrow η_i is marked with a_i for $1 \leqslant i \leqslant k$. If η_k points to s', we say that the word w *reads* from the state s to s'.

DEFINITION 1.5. Let M be an automaton over A. A word w in A^* is said to be *accepted* by M if it reads from the start state s_* to an accept state. A subset L of A^* is said to be a *regular language* if there exists a finite state automaton over A that accepts a word w of A^* if and only if w lies in L.

Coming back to words that are normal in the sense of Definition 1.3, we have

PROPOSITION 1.6. *The set of all words that are normal in the sense of Bressaud is a regular language.*

PROOF. As in the case of the greedy normal form of Chapter VI, being normal in the sense of Bressaud is a *local* condition: there exists a (finite) list of letters T—the legal terminal letters—and for each letter s, there exists a set of letters $F(s)$—the locally forbidden letters—such that a word w is normal if and only if the following two conditions hold. First, for each pair (s, s') of consecutive letters in w, we have $s' \notin F(s)$, and, second, the last letter of w belongs to T.

To construct an automaton recognizing the language of normal forms, one defines the set of states to be the alphabet plus one initial state and one failure state, and, from each state s, one draws an s'-arrow to the state s' for each s' in the complement of $F(s)$, and an s'-arrow to the failure state for each s' in $F(s)$. Moreover, for every state s one defines s to be an accept state if $s \in T$ holds, and a failure state otherwise. □

Figure 2 shows the local condition in the case of the 4-strand braid group.

1.3. An untangling procedure for curve diagrams. The main result is that normal words in the sense of Bressaud provide a unique normal form for braids.

PROPOSITION 1.7. *Every element of B_n has exactly one representative that is normal in the sense of Bressaud.*

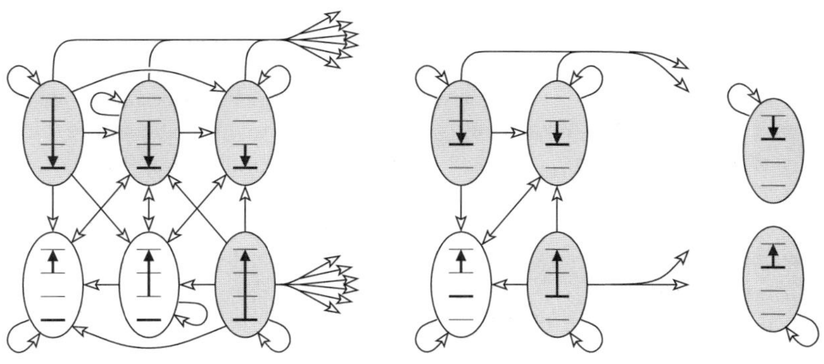

FIGURE 2. The local condition, in the sense of the proof of Proposition 1.6, on words to be normal in the sense of Bressaud. Locally legal sequences of letters are indicated by arrows, and the shaded vertices indicate legal terminal letters.

Our aim now is to prove Proposition 1.7. For the rest of this section we shall consider as the trivial curve diagram on the n-punctured disk D_n the diagram E shown in solid lines in Figure 3(i). Also, whenever we speak of the intersection number of two arcs, it shall be understood that the arcs have been pulled tight with respect to each other, in the sense of Definition X.1.5.

DEFINITION 1.8. The *complexity* of a curve diagram $\beta(E)$ is defined as follows. On the boundary ∂D_n the starting points of the n arcs are lined up from left to right, and some of them may already be trivial (just vertical line segments). Suppose the pth is the leftmost nontrivial segment. Then the complexity is the number of intersections of this line segment with the interiors of the dotted arcs shown in Figure 3.

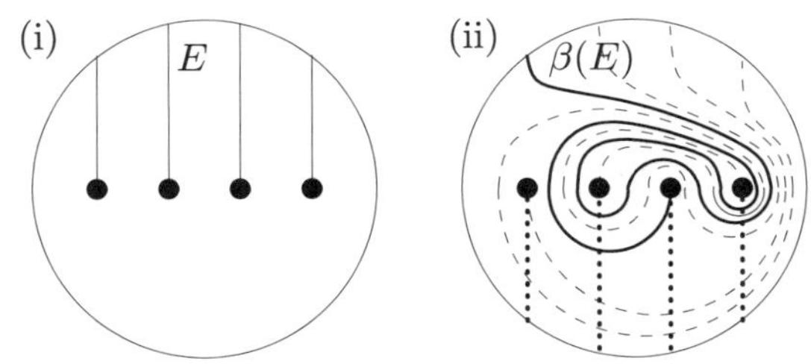

FIGURE 3. On the left, the trivial curve diagram E is drawn in solid lines; on the right, the curve diagram of the braid $\beta = \sigma_3\sigma_2^2\sigma_1^2\sigma_3^{-1}\sigma_2\sigma_1\sigma_3\sigma_2$. In this diagram, the leftmost nontrivial arc is drawn in solid lines, the others in dashed lines. The solid arc has four intersections with the dotted lines, so by definition the complexity is 4.

Our untangling procedure is defined as follows. In order to find the normal form of a braid β of B_n, we consider the curve diagram $\beta(E)$. Then we apply repeatedly the following elementary step. Suppose the pth arc from the left is the first nontrivial one.

• Case 1. The first intersection of the pth arc with the horizontal axis of D_n is between the ith and $i+1$st puncture, and it turns left after this intersection. Then we write down the generator $\sigma_{i+1,n,p}$, and also act on the curve diagram by the inverse of this generator.

• Case 2. The first intersection of the pth arc with the horizontal axis of D_n is between the ith and $i+1$st puncture and it turns right after this intersection. Then we write down the generator $\sigma_{i,p,p}$, and also act on the curve diagram by the inverse of this generator.

• Case 3. The complexity of the curve diagram equals zero, *i.e.*, the pth arc runs straight into the ith puncture ($i > p$). Then we write down the generator $\sigma_{i,p,p}$, and also act on the curve diagram by the inverse of this generator, yielding a diagram where the pth arc is also trivial.

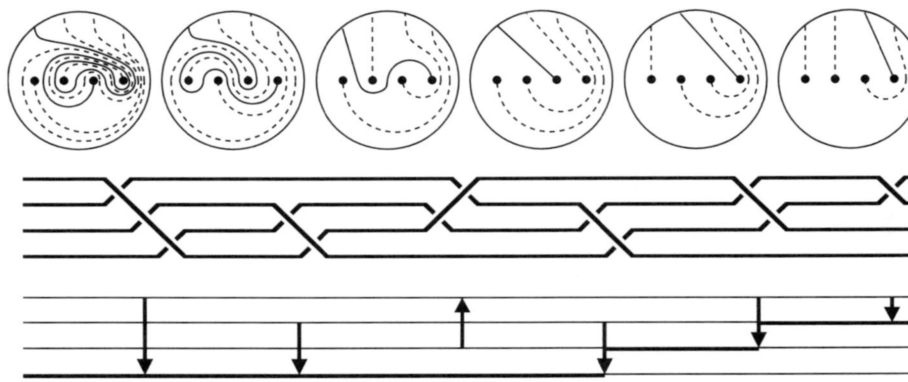

FIGURE 4. The relaxation procedure for the braid $\beta = \sigma_3\sigma_2^2\sigma_1^2\sigma_3^{-1}\sigma_2\sigma_1\sigma_3\sigma_2$. The curve diagram of β is in the top left. The calculation shown here proves that the normal form of β is $\sigma_{4,1,1}\sigma_{3,1,1}\sigma_{2,4,1}\sigma_{3,1,1}\sigma_{4,2,2}\sigma_{4,3,3}$. In each step, the leftmost nontrivial arc is drawn in solid, the others in dashed lines.

The proof of the existence part of Proposition 1.7 now follows immediately from the following lemma whose proof is left as an exercise.

LEMMA 1.9. (i) *Relaxation steps corresponding to Cases* 1 *and* 2 *above decrease the complexity of the curve diagram. Steps corresponding to Case* 3 *do not, but they increase the index* p.

(ii) *A braid word obtained by the relaxation procedure is normal.*

We now turn to the uniqueness part of Proposition 1.7. Suppose that some braid β in B_n is represented by a normal word w. The first letter of w is either of the form $\sigma_{i,p,p}$ for some i and p with $p < i \leqslant n$ or of the form $\sigma_{i+1,n,p}$ where $p-1 \leqslant i < n-1$. In order to prove uniqueness, it suffices to show that, given the curve diagram $\beta(E)$, we can reconstruct this first letter. This is achieved by the following lemma.

LEMMA 1.10. *If the first letter of w is of the form $\sigma_{*,*,p}$, then the leftmost nontrivial arc of the diagram $\beta(E)$ is the pth one. Moreover, if the first letter is $\sigma_{i,p,p}$ with $p < i \leqslant n$, then in the curve diagram $\beta(E)$ the pth arc first intersects* *the horizontal axis between the* i*th and the* $(i+1)$*st puncture and then turns right. Similarly, if the first letter is* $\sigma_{i+1,n,p}$*, then in the curve diagram the* p*th arc first cuts between the* i*th and* $(i+1)$*st puncture and then turns left.*

PROOF. This is an easy induction on the length of the word w. □

This completes the proof of Proposition 1.7. Our next aim is to give a new proof of Property **C**.

DEFINITION 1.11. The curve diagram $\beta(E)$ is said to be *k-positive* if the leftmost nontrivial arc is the kth one and it deviates from its initial position towards left, *i.e.*, its first intersection with the horizontal axis occurs on the right of the kth puncture.

DEFINITION 1.12. A word w in the alphabet $\{\sigma_{i,j,p}\}$ is said to be σ_k-positive if it becomes a σ_k-positive braid word in standard generators after unpacking (1.1).

PROPOSITION 1.13. *If the curve diagram* $\beta(E)$ *is* k*-positive, then the normal form of* β *is* σ_k*-positive.*

We remark that the converse of Proposition 1.13 is also true—this is essentially equivalent to Proposition 2.1 in Chapter X.

PROOF OF PROPOSITION 1.13. Assume that $\beta(E)$ is k-positive. Then the generators $\sigma_{i,j,p}$ with $p<k$ do not appear in the normal form. This implies that, after transition to the standard generators, the letters $\sigma_i^{\pm 1}$ with $i<k$ are absent. The only generator $\sigma_{i,j,p}$ that may give rise to σ_k^{-1} is therefore $\sigma_{k,n,k}$.

Since the diagram $\beta(E)$ is k-positive, the normal form of β starts with a letter $\sigma_{i,k,n}$ with $i>k$, which contains σ_k after unpacking. According to our rules, the generator $\sigma_{k,n,k}$ cannot then appear anywhere in the normal form. □

COROLLARY 1.14 (Property **C**). *Every nontrivial braid in* B_n *is* σ*-positive or* σ*-negative.*

PROOF. For a curve diagram Γ, we denote by $\Gamma^{\updownarrow}$ the image of Γ under the reflection in the horizontal axis. Also, by $\beta \mapsto \beta^{\updownarrow}$ we denote the involution of B_n that sends each generator σ_i to its inverse. Obviously, we have $\beta^{\updownarrow}(E^{\updownarrow}) = \beta(E)^{\updownarrow}$.

The union $E \cup E^{\updownarrow}$ consists of vertical chords of D_n passing through the punctures. Suppose that β is some nontrivial braid and that the kth vertical chord is the leftmost one on which β acts nontrivially. Then one can easily see that at least one of the two arcs making up this chord—one belonging to $\beta(E)$ and the other to $\beta(E^{\updownarrow})$—has its first intersection with the horizontal axis *not* in the segment between punctures number $k-1$ and k, but further to the right. From this, we deduce that either $\beta(E)$ or $\beta(E^{\updownarrow})^{\updownarrow}$ is k-positive for some k. □

1.4. A rewriting procedure for braid words. Bressaud also constructs a purely algebraic algorithm for transforming any braid word in the generators $\sigma_{i,j,p}$ into a normal word representing the same element of B_n. The proof that this algorithm stops in finite time is quite technical, and we shall only present the general idea here.

Suppose inductively that we already have a normal word u, and a generator s of the form $\sigma_{i,j,p}$; our aim is to find the normal form of the product su.

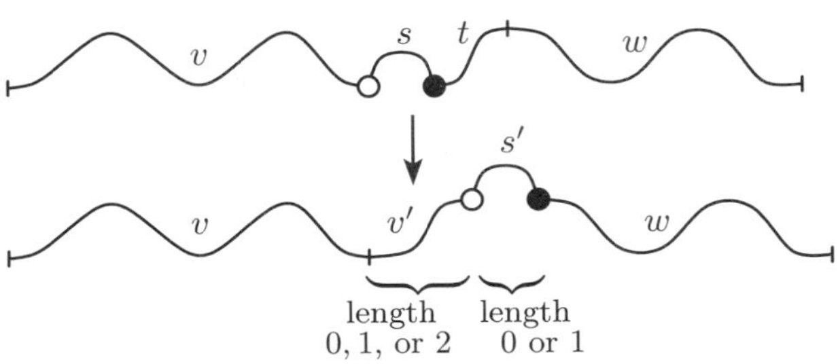

FIGURE 5. A symbolic representation of one step of the Tetris algorithm. The symbol ∘ indicates that two subsequent letters are either disjoint or they form a legal sequence. The symbol • indicates that a sequence of two letters may be illegal. Curve segments that fit together smoothly represent normal words. For instance, in the upper picture, the the word vtw is normal.

This is done again by an iterative procedure; see Figure 5. Suppose that we have already found a way of writing the element represented by su as a word $vstw$, where s and t are single letters, and v and w are words, satisfying the following inductive hypothesis. First, removing the letter s yields a normal word (in this case vtw). Second, the transition from the last letter of v to s is either disjoint or legal—the meaning of these words will be defined momentarily. Then one elementary step of the Tetris algorithm replaces the word st by an equivalent word $v's'$, in such a way that the word $vv's'w$ again satisfies the induction hypothesis. We have made progress in the sense that the disturbing letter s has been transformed into a letter s', which is one step closer to the end of the word, and the price we have paid is that we may have increased the length of the word by 1.

DEFINITION 1.15. A sequence of two letters $\sigma_{i,j,p}\sigma_{i',j',p'}$ is called *disjoint* if the closed intervals $[i,j]$ (or $[j,i]$) and $[i',j']$ (or $[j',i']$) are disjoint. It is called *legal* if i' lies between i and j, but is different from j.

A nearly complete list of rules for the elementary transformations can be found in Figure 6, and the reader is encouraged to check that these rules do indeed satisfy the requirements of the inductive step. An explicit example is worked out in Figure 7.

We see from Figure 5 that the normal representative of su is an asynchronous fellow traveller of u. This means that the traces of the two paths in the Cayley graph are at universally bounded Hausdorff distance from each other—in this case the Hausdorff distance is at most 2; also see [**77**, Section 7.2]. Note that we do not claim that the two paths are necessarily close to each other when viewed as parametrized paths, *i.e.*, that they are synchronous fellow travellers—prefixes of the same length of the two words may represent braids which are quite distant in the Cayley graph. Bressaud gave an example which proved that, for s a generator, the braids β and $s\beta$ may have normal forms whose lengths may be arbitrarily large, but different by a factor 2. Nevertheless, computer experiments lend strong support to the following conjecture. Transforming a word in the generators $\sigma_{i,j,p}$

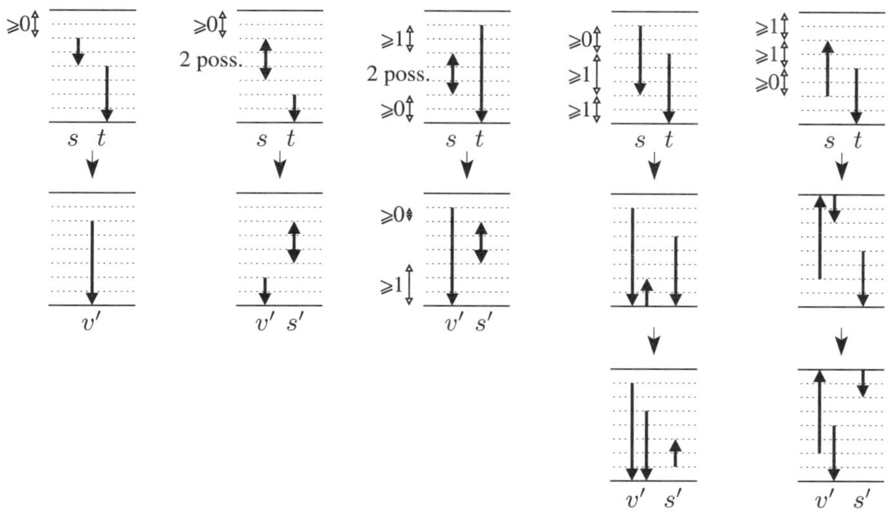

FIGURE 6. The rules for moving the disturbing letter s' to the right through the braid word. Not listed are obvious cancellations. The list is only nearly complete for two reasons. First, obvious cancellations like $\sigma_{3,4,1}\sigma_{4,1,1} \mapsto \sigma_{3,1,1}$ are not listed. Second, immediately following the transformations listed here, we sometimes modify the resulting letter s' by tranforming $\sigma_{i,j,p}$ to $\sigma_{i,j,p'}$ with $p' > p$, depending on the subsequent letters. For examples of this, see the second and fifth steps of the procedure in Figure 7.

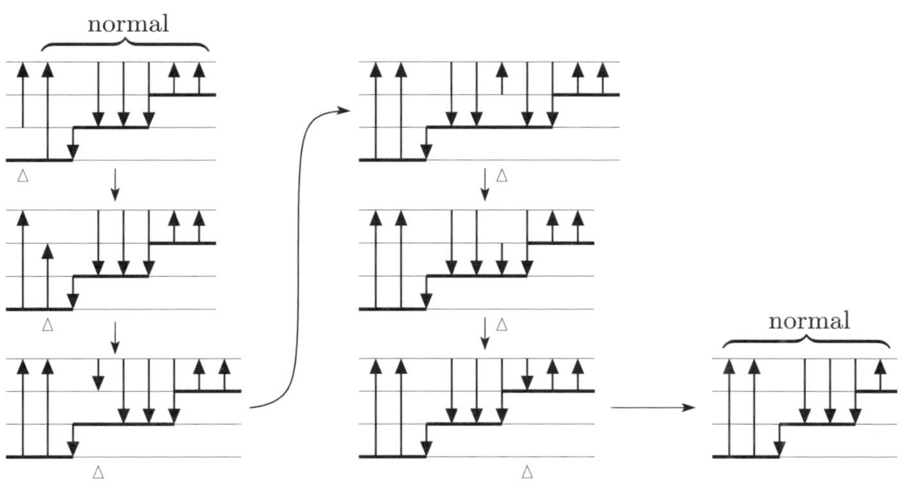

FIGURE 7. An example of the Tetris algorithm. The symbol $\triangle$ marks the position of the current letter s.

into an equivalent normal word may only increase its length by a linear factor, with a calculation time depending quadratically on the length.

2. The transmission-relaxation normal form of braids

As mentioned in the introduction to this chapter, any reasonable way of untangling gives an algorithm that is quite efficient in practice for constructing a sort of a braid normal form. However, in most cases no proof of the observed efficiency of the algorithm is known.

In this section we describe a curve diagram untangling method developed in [**74**]. The main purpose of this approach was to find an algorithm for recovering a braid from its curve diagram for which a nice upper bound for the complexity can be *proved*. The effort to make the algorithm as efficient as possible resulted, quite surprisingly, in yet another proof of Property **C** and a procedure that produces a σ-definite braid word representing the braid whose curve diagram is given as the input.

Probably the main difficulty with proving the efficiency of numerous untangling algorithms is the fact that the braid length defined as the length of a shortest word representative (in some set of generators) is a measure of complexity that is quite different from the one suggested by curve diagrams. To illustrate the difficulty, we consider the much simpler group $\mathrm{SL}(2,\mathbb{Z})$ of integral unimodular $2{\times}2$ matrices. Let us look at two different notions of complexity of such matrices. Define the *norm* $\|X\|$ of a matrix $X = \begin{pmatrix} a & b \\ c & d \end{pmatrix}$ in $\mathrm{SL}(2,\mathbb{Z})$ as $\|X\| = \max(|a|,|b|,|c|,|d|)$, and the *length* $\ell(X)$ as the length of a shortest word representing X in the alphabet $\{u_1^{\pm1}, u_2^{\pm1}\}$ with

$$u_1 = \begin{pmatrix} 1 & 1 \\ 0 & 1 \end{pmatrix}, \quad u_2 = \begin{pmatrix} 1 & 0 \\ 1 & 1 \end{pmatrix}.$$

One can immediately see that these two measures of complexity are quite different by looking at the example

$$u_1^{30}u_2^{30} = \begin{pmatrix} 901 & 30 \\ 30 & 1 \end{pmatrix}, \quad (u_1u_2)^{30} = \begin{pmatrix} 2504730781961 & 1548008755920 \\ 1548008755920 & 956722026041 \end{pmatrix},$$

where the two words are the shortest word representatives of the corresponding matrices.

There are two algorithms for finding a word representing a matrix X in $\mathrm{SL}(2,\mathbb{Z})$ that we call the *slow Euclidean algorithm* and the *fast Euclidean algorithm*. Given X in $\mathrm{SL}(2,\mathbb{Z})$, both produce a word representing X by simplifying X step by step. If $\|X\| > 1$ holds, then the slow Euclidean algorithm spells out one letter a in $\{u_1, u_1^{-1}, u_2, u_2^{-1}\}$ guaranteeing $\|a^{-1}X\| < \|X\|$—such a letter is always unique—and proceeds as before with the simpler matrix $a^{-1}X$. In the same situation, the fast Euclidean algorithm spells out a pair (a,k), where a is the same as above and k is an integer that minimizes $\|a^{-k}X\|$—there may be at most three choices for k—and then proceeds with $a^{-k}X$. Such a k is found by rounding off the ratio of the greatest element of X to the other element in the same column. If $\|X\| = 1$ holds, both algorithms spell out a shortest word representing X, which can be preprogrammed as there are just 16 such matrices.

For example, if $X = u_1^{1000000}$ is given as the input, then the slow Euclidean algorithm will spell out the word $u_1u_1 \dots u_1$ with 1000000 letters having performed 1000000 steps, whereas the fast one will output $(u_1, 1000000)$ in just one step. The running time for the fast Euclidean algorithm is bounded by a polynomial in $\log(\|X\|)$, which can be established by observing that the value $\log(\|X\|)$ is comparable to the following modified length function of X:

$$\ell'(X) = \min_{u_1^{k_1} u_2^{k_2} u_1^{k_3} u_2^{k_4} \ldots = X} \sum_j \log_2(|k_j| + 1).$$

One deduces from this and from analyzing the general step of the algorithm that the fast Euclidean algorithm produces a word representing X whose ℓ'-length is bounded by a linear factor of $\ell'(X)$.

The slow Euclidean algorithm is not polynomial time in $\log(\|X\|)$, but it *is* polynomial time in $\ell(X)$ and it also produces a word representative of length $O(\ell(X))$. However, the value $\ell(X)$ is not something that can be easily extracted from X—without implementing the Euclidean algorithm—so, to establish the result, one needs to analyze the group structure of $\mathrm{SL}(2, \mathbb{Z})$.

A similar thing happens to the braid groups. The reader is invited to find and compare curve diagrams of braids $(\sigma_1\sigma_2)^{10}$ and $(\sigma_1\sigma_2^{-1})^{10}$ to see the analogy with the case of $\mathrm{SL}(2, \mathbb{Z})$.

Roughly speaking, the principle of most of the untangling algorithms for curve diagrams of braids is close to the one of the slow Euclidean algorithm. Since the group structure of B_n is in a sense more complicated than that of $\mathrm{SL}(2, \mathbb{Z})$, it is not easy to establish nice bounds for the running time of such an algorithm and the length of the output. The algorithm of [**74**] described here is close in nature to the fast Euclidean algorithm, but all the complexity bounds are therefore given in terms of a nonstandard notion of braid length, similar to ℓ' above, which we now define.

2.1. Length of braids and complexity of curve diagrams. We denote by $\Delta_{i,j}$ with $1 \leqslant i < j \leqslant n$ the following element of B_n:

(2.1) $$\Delta_{i,j} = (\sigma_i \ldots \sigma_{j-1})(\sigma_i \ldots \sigma_{j-2}) \ldots \sigma_i.$$

Geometrically, this corresponds to the half-twist involving all strands between strands number i and j, inclusive.

For each i, the generator σ_i is equal to $\Delta_{i,i+1}$, and, therefore, every n-strand braid word w admits decompositions of the form

(2.2) $$w = \Delta_{i_1,j_1}^{d_1} \ldots \Delta_{i_m,j_m}^{d_m}.$$

DEFINITION 2.1. For w a word in the generators $\Delta_{i,j}$, the *Δ-length* of w is defined to be

$$\ell_\Delta(w) = \sum_{r=1}^{m} \log_2(|d_r| + 1),$$

where $\Delta_{i_1,j_1}^{d_1} \ldots \Delta_{i_m,j_m}^{d_m}$ is the shortest decomposition of w of the form (2.2). For β in B_n, we define the *Δ-length* of β, denoted $\ell_\Delta(\beta)$, to be the minimal value of $\ell_\Delta(w)$ for w representing β.

DEFINITION 2.2. A word w in the generators $\Delta_{i,j}$ is said to be *σ-positive* or *σ-negative* if the word in standard generators σ_i obtained from w by the expansion (2.1) is.

The main result of Section 2, which contains Property **C**, is the following.

PROPOSITION 2.3 (Property **C** with length bounds). *There is an algorithm that, given a nontrivial word w in the generators $\Delta_{i,j}$ with $1 \leqslant i < j \leqslant n$, constructs* *an equivalent word w' that is either σ-positive or σ-negative, and the following inequality holds:*

$$\ell_\Delta(w') \leqslant \mathrm{const} \cdot n \cdot \ell_\Delta(w). \tag{2.3}$$

The word w' provided by Proposition 2.3 depends only on the braid represented by w. It will be called the *normal form* of the braid $\overline{w}$ hereafter.

REMARK 2.4. One can even show that the running time of the algorithm from Proposition 2.3 is bounded from above by $O(n^2 \cdot \ell_\Delta(w)^2)$ provided the algorithm is implemented on a random access machine, but *not* on a Turing machine, for instance.

The proof of this result, which we shall outline below, really shows something which is of considerable theoretical interest, beyond questions of orderability, namely that the complexity of the curve diagram of a braid β and the Δ-length $\ell_\Delta(\beta)$ are comparable.

More precisely, let us use as a basic curve diagram in D_n the diagram E shown in Figure 12, and let $\beta(E)$ be the image of E under the action of a braid β.

DEFINITION 2.5. We define the *geometrical complexity* of the braid of β to be $c(\beta) = \log_2 \|\beta(E)\| - \log_2 \|E\|$, where the symbol $\|\cdot\|$ denotes the number of intersections with the horizontal diameter of D_n provided that the diagram is tight with respect to this diameter.

Then the following can be proved using the techniques of this section.

PROPOSITION 2.6. *For each n, the geometric complexity and the Δ-length are comparable, i.e., there exist constants C_1 and C_2 such that, for every braid β in B_n, we have $C_1 \cdot c(\beta) \leqslant \ell_\Delta(\beta) \leqslant C_2 \cdot c(\beta)$.*

In fact, one can choose $C_1 = 1/\log_2 3$ (independently of n) and $C_2 = 9n$.

In what follows, we shall only discuss Proposition 2.3 whose proof consists of the following steps. First, we need to endow the curve diagram $\Gamma_0 = \beta(E)$ of the braid β with an additional structure, which we call here a strip decomposition, and define a special complexity function called AHT-complexity depending on this structure. This AHT-complexity is bounded from above by $\mathrm{const} \cdot n \cdot \ell_\Delta(w)$—for braids that are long enough, see below.

Then we describe the simplification step that, given the curve diagram Γ_0 with the previously constructed strip decomposition, produces a word w_1 representing a braid β_1, the curve diagram Γ_1 of $\beta_1^{-1}\beta$ and a strip decomposition of Γ_1 such that the AHT-complexity of Γ_1 drops by some $a_1 \geqslant 1$ compared to that of Γ_0, and the following holds: $\ell_\Delta(w_1) \leqslant \mathrm{const} \cdot a_1$. The simplification is then applied recursively to the new curve diagram and its strip decomposition, producing braids $\beta_2, \beta_3, \dots$. The procedure stops when the AHT-complexity becomes zero, which means by construction that the curve diagram is untangled and, therefore, we have $\beta_1\beta_2... = \beta$. By analyzing the simplification step we see that either we get a σ-positive braid word $w_1w_2...$ representing β or a symmetric procedure produces a σ-negative word representing β. Then we need to estimate the number of operations required for implementing this procedure.

Intermediate results needed for the proof require a lot of routine checking that we will skip in this presentation, referring the reader to the paper [**74**]. A simple method for computing the curve diagram of a braid by using a special coordinate system will be described in Chapter XII.

In this section we only describe the geometric background of the untangling process and provide details when a delicate choice of the construction should be made. We also demonstrate how the untangling procedure works for a concrete example.

Of course, in order to get an actual algorithm, one should translate our geometrical description into combinatorial language. This is not hard, and for exposition purposes we will stick with the geometrical language.

2.2. Agol–Hass–Thurston orbit counting algorithm. We are going to forget about braids for a while and consider the following question that may appear irrelevant at first sight. Suppose we have a union Γ of pairwise disjoint simple closed curves consisting of semicircular arcs having the endpoints on a distinguished horizontal line—which will be referred to as the *axis*—and intersecting it transversely. An example is shown in Figure 8. The puzzle is: How many connected components does Γ have? It is assumed that Γ is presented in combinatorial terms that we now describe.

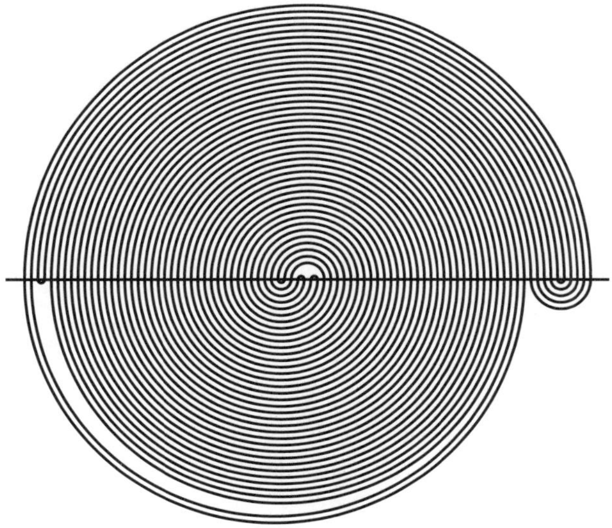

FIGURE 8. How can one count the number of curves?

DEFINITION 2.7. A *simple strip* in Γ is a collection of parallel (concentric) semicircular arcs lying in the same half-plane such that their left ends as well as their right ends form a family of consecutive intersection points of Γ with the axis. These two families of consecutive intersection points are called *bases* of the simple strip. The number of arcs in the strip is called the *width* of the simple strip.

In order to describe the combinatorial structure of Γ it suffices to specify a collection of simple strips that cover Γ without overlapping. This means that one should specify relative positions of the bases and the widths of the strips. For

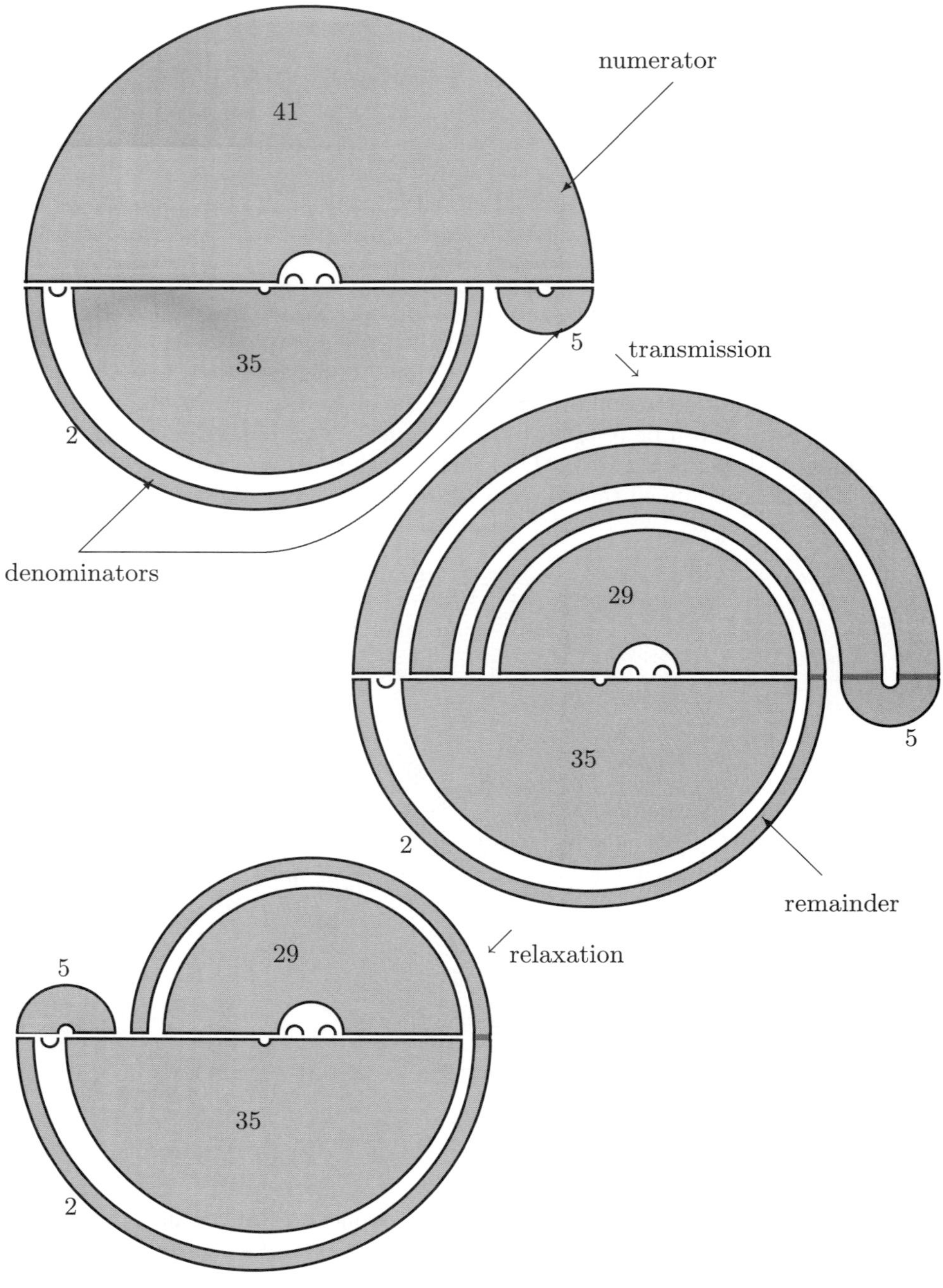

FIGURE 9. Transmission and relaxation of a collection of curves endowed with a strip decomposition.

example the Γ shown in Figure 8 can be covered by 7 simple strips of widths $41, 1, 1, 2, 1, 35$, and 5 as indicated in Figure 9.

So, we suppose Γ is given in such a combinatorial form. What we describe below is a particular case of the Agol–Hass–Thurston algorithm—AHT-algorithm for short [**2**]—which allows us to compute the number of connected components of Γ in time polynomial in the number of strips and the logarithm of the width of the widest strip. The general AHT-algorithm was designed for counting the number of connected components of a normal surface provided its normal coordinates are given. The simple geometrical interpretation we use here does not work so well in the general case.

We need more definitions in order to describe the simplifying step.

DEFINITION 2.8. (i) A *strip* in Γ is a union $s = s_1 \cup \cdots \cup s_p$ of pairwise non-overlapping simple strips of equal width such that, for each $i = 1, \dots, p-1$, the strips s_i and s_{i+1} share a common base that must be the right base of s_1 or s_p if $i = 1$ or $i = p-1$, respectively. The left bases of s_1 and s_p are then called the *bases* of the strip s. The number p is called the *length* of s.

(ii) By a *strip decomposition* of Γ we mean any collection $\{s_1, \dots, s_r\}$ of strips that cover Γ without overlapping and satisfy the following condition: there exists a point K on the axis such that all bases of the strips s_i are on the left of K and all the intersections of interiors of strips are on the right of K.

For a large number of examples of what such a strip decomposition may look like, the reader can consult Figures 14 through 17, where the point K corresponds to the right endpoint of the horizontal line segment drawn in each of the figures.

The idea of the algorithm is to replace the pair $(\Gamma, \{s_1, \dots, s_r\})$ by a simpler one $(\Gamma', \{s'_1, \dots, s'_{r'}\})$, but such that Γ' has the same number of connected components as Γ, and then to proceed recursively until a collection of circles is obtained, which are easy to count. The simplification step consists of two operations that we call *transmission* and *relaxation*. The former changes only the strip decomposition and the latter changes the whole picture keeping the position of the strip bases fixed.

We proceed with the definition of a transmission—for an example of this operation, see Figure 9.

DEFINITION 2.9. (i) Let some strip decomposition of Γ be fixed. Let N be the rightmost point on the axis covered by a base of a strip—there must be exactly two bases covering this point. The strip that is the owner of the larger one will be called the *numerator* of the transmission. If the two bases covering N are of the same size and belong to different strips, we choose any of the strips to be the numerator. The strips whose right bases are covered completely by the right base of the numerator are called the *denominators* of the transmission.

(ii) A *transmission* consists of splitting the numerator into a collection of parallel strips such that all their right bases, except at most one, match the bases of the denominators (those that are covered by the right base of the numerator). Then one glues up the matched bases of denominators and of the new strips so as to get longer strips. The unmatched strip, if nontrivial, is called the *remainder* of the transmission.

There are three levels of luckiness that we may have. We are most lucky if the numerator forms a circular ring, thus being the denominator at the same time. In this case, it covers circular connected components of Γ, and the number of these

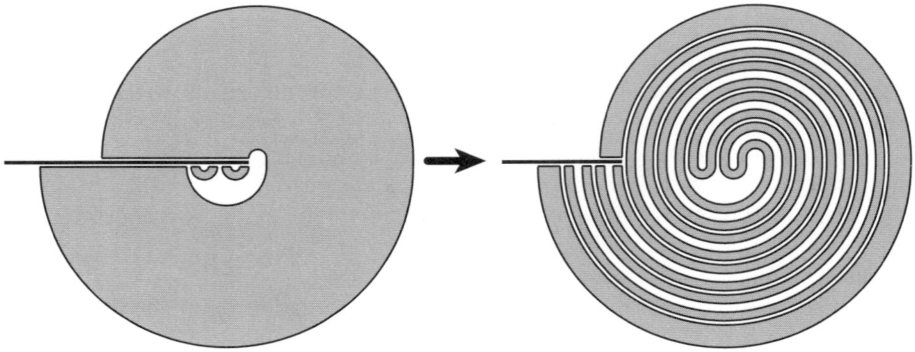

FIGURE 10. A twice spiralling transmission.

circles equals the width of the numerator. We advance the connected component counter (which is initially set to zero) by this number and just erase the revealed circles.

We are least lucky if the two bases of the numerator have a large overlap. More precisely, let μ be the width of the numerator and $\mu_1, \ldots, \mu_j$ the widths of the denominators, where the width of a denominator is listed twice if both its bases are covered by the right base of the numerator. If the bases of the numerator overlap and we have $\mu/(\mu_1 + \cdots + \mu_j) \gg 1$, then the width of the remainder will be comparable to that of the numerator, and the remainder will become the numerator of the next transmission. We will need to apply at least $d = [\mu/(\mu_1 + \cdots + \mu_j)]$ transmissions—with $[x]$ denoting the integral part of x—which can be a very large number. However, if $d > 1$ holds, we can easily predict the final result of these d transmissions.

DEFINITION 2.10. We call the operation just described consisting of d transmissions, a *d-times spiralling transmission.*

See Figure 10 for an example.

This is where the similarity to the *fast* Euclidean algorithm occurs: roughly speaking, instead of d subtractions we implement one division with remainder.

In other cases, when we are moderately lucky, the remainder is either empty or becomes a denominator of the next transmission.

DEFINITION 2.11. By *relaxation* we mean here just a replacement of strips of lengths at least 3 by shorter ones having the same bases.

Note that for orbit-counting purposes we do not need to know the geometry of strips. We only need to know for each strip its bases and whether they are coupled with the same or opposite orientation. Thus, in the actual implementation of this algorithm, the relaxation operation disappears.

EXAMPLE 2.12. If we start from the collection of curves shown in Figure 8, just six transmission-relaxation steps will be needed to get a union of three circles. One of the transmissions will be 3-times spiralling. The whole process is shown in Figure 11.

FIGURE 11. An example of running the AHT-algorithm for counting the number of connected components. The third transmission is 3-times spiralling.

DEFINITION 2.13. Let Γ be as before, $\{s_1, \dots, s_r\}$ be some strip decomposition of Γ, and $\mu_1, \dots, \mu_r$ be the widths of the strips $s_1, \dots, s_r$, respectively. Let $\widetilde{\mu}$ be the width of the denominator of the forthcoming transmission whose right base is the rightmost among all bases of the denominators. We define the *AHT-complexity* $c_{\text{AHT}}(\Gamma, \{s_1, \dots, s_r\})$ to be

$$c_{\text{AHT}}(\Gamma, \{s_1, \dots, s_r\}) = r + \sum_{i=1}^{r} \log_2 \mu_i - \frac{\log_2 \widetilde{\mu}}{2}.$$

The last term in this formula looks quite strange and has no natural explanation, but it is very useful for getting good estimates for the complexity of the algorithm.

The following lemma is one of the key ingredients in the proof of Proposition 2.3.

LEMMA 2.14. *A nonspiralling transmission drops the AHT-complexity by at least* 1. *A d-times spiralling transmission drops the AHT-complexity by at least* $\log_2(d+1)$.

The proof, which is easy, can be found in [**74**].

2.3. Transmission-relaxation algorithm for braids. We now recall that our purpose is not to count the number of connected components of a strange looking curve, but to recover a braid from a curve diagram. However, we will do exactly the same thing as before with some more restrictions and with paying attention to the punctures.

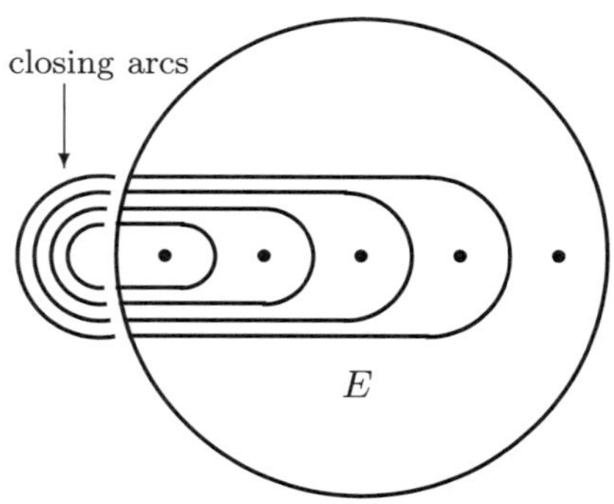

FIGURE 12. The base curve diagram for the transmission-relaxation algorithm.

Recall that here for the base curve diagram E we take the collection of $n-1$ arcs shown in Figure 12. For technical reasons we also *close up* every curve diagram by adding $n-1$ arcs outside the disk D_n (on the left) so as to get $n-1$ closed curves. The union of these curves will still be referred to as a curve diagram. The curve diagrams are assumed to be tight with respect to the horizontal axis at the beginning of the untangling process. However, during the process, the curve diagrams that appear may not be tight.

In the pictures we do not draw the border of the disk D_n, whose position is obvious. We also view all pictures up to a *horizontal rescaling* meaning a homeomorphism of the plane of the form $(x, y) \mapsto (\varphi(x, y), y)$, where $\varphi(\cdot, y)$ is a monotonically

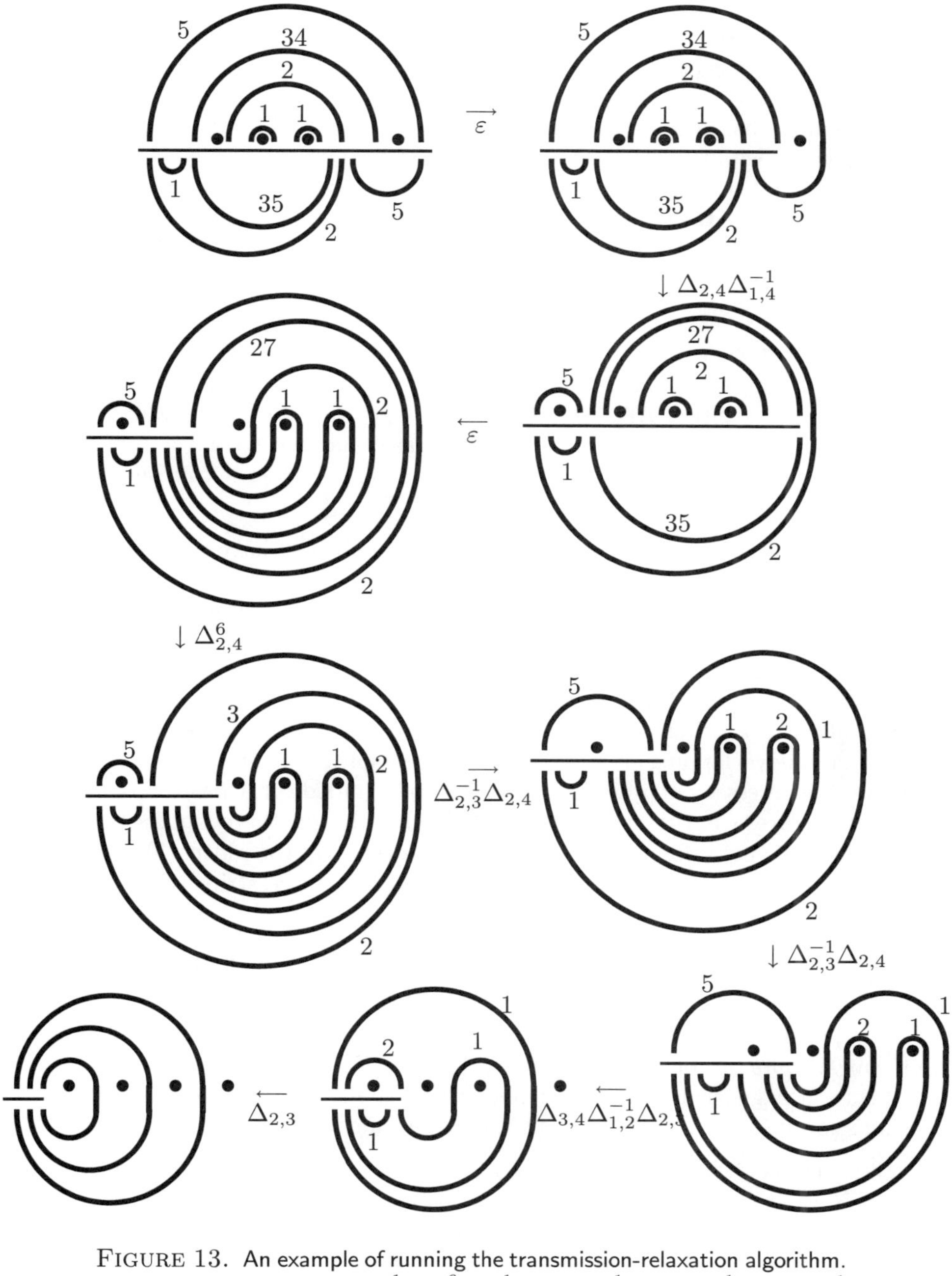

FIGURE 13. An example of running the transmission-relaxation algorithm. The output word is $\Delta_{1,4}\ \Delta_{2,4}^{-1}\ \Delta_{2,4}^{-6}\ \Delta_{2,4}^{-1}\ \Delta_{2,3}\ \Delta_{2,4}^{-1}\ \Delta_{2,3}\ \Delta_{2,3}^{-1}\ \Delta_{1,2}\ \Delta_{3,4}^{-1}\ \Delta_{2,3}^{-1}$, which is σ-positive.

increasing function for every y. By such a homeomorphism, any curve transversal to the horizontal axis can be transformed to one consisting of semicircular arcs with endpoints on the axis.

We now describe the transmission-relaxation method for recovering a braid from its curve diagram. The general principles are as follows. We start with a strip decomposition of the given curve diagram, and we apply alternatingly a transmission and, if possible, a move consisting of a *puncture slide*, followed by a relaxation, until the trivial curve diagram is reached. Any relaxation should replace strips by isotopic ones in the punctured plane; this is why between each transmission and the subsequent relaxation we need to perform puncture slides that allow some strips to relax.

DEFINITION 2.15. A base of a strip will be called an *A-base* if the strip approaches it from above, and a *B-base* otherwise. To each strip, we associate its *type*, which can be either AA, AB, BA, or BB depending on the types of the bases: the first letter indicates the type of the left base, and the second letter indicates the type of the right base. If the bases of the strip coincide, it can be thought of as an AB- or BA-strip, this does not matter.

We arrange the transmission-relaxation process so as to have the following rules hold at every step:

- No puncture is located inside an A-base of a strip (meaning in between two points in this base) or inside the interior of a strip;
- After each relaxation, all BB-strips are of length at most three, and all the other strips are of length at most two.

Initially this is achieved by covering the curve diagram by the smallest possible number of nonoverlapping simple strips and then splitting some strips in the upper half-plane as necessary in order for punctures to be outside of A-bases. For example, shown in Figure 8 is the curve diagram of a braid, and the strip decomposition from Figure 9 is not good as the bases of the widest strip (of width 41) contain two punctures. This is resolved by cutting this strip into three strips of widths 5, 34, and 2; see Figure 13.

LEMMA 2.16. *The initial strip decomposition of the curve diagram of a braid β has AHT-complexity bounded from above by* $\mathrm{const}\cdot n\cdot \ell_\Delta(\beta) + \mathrm{const}'\cdot n\cdot \log_2 n$.

The proof consists in showing that $\|\beta(E)\|$ is at most $\log_2 3\cdot \ell_\Delta(\beta) + \log_2 n$, and that the number of strips is at most $3n$; see [**74**].

There is a general rule that defines which punctures should be slid after a transmission in order to comply with the above rules.

DEFINITION 2.17. At every stage of the untangling procedure when a transmission has just been performed, we call a semicircular arc α of the current curve diagram *essential* if it lies in the lower half-plane and the left endpoint of α is located in the right base of the numerator of the transmission but not in the right base of the remainder. Any puncture between the endpoints of an essential arc is called *obstructing*.

The relaxation following a transmission consists in pushing all essential arcs across the axis to the upper half-plane. In order to do so we must first slide all the obstructing punctures to somewhere. An example of running the transmission-relaxation algorithm is shown in Figure 13. Note that the sequence of transmissions is different from the one in Figure 11 since the initial strip decomposition is different, though the initial curve system is exactly the same. Note also that each strip is shown schematically, as a single curve with a number specifying the width. Therefore, the relative position of punctures and the endpoints of the B-bases cannot always be seen from the picture. However, the relative position of the punctures and the A-bases is reflected in the pictures, which is possible as no puncture is allowed to sit inside an A-base.

We now briefly review which puncture slides and relaxations are needed after each transmission. We shall distinguish several cases, depending on the type of the numerator of the transmission.

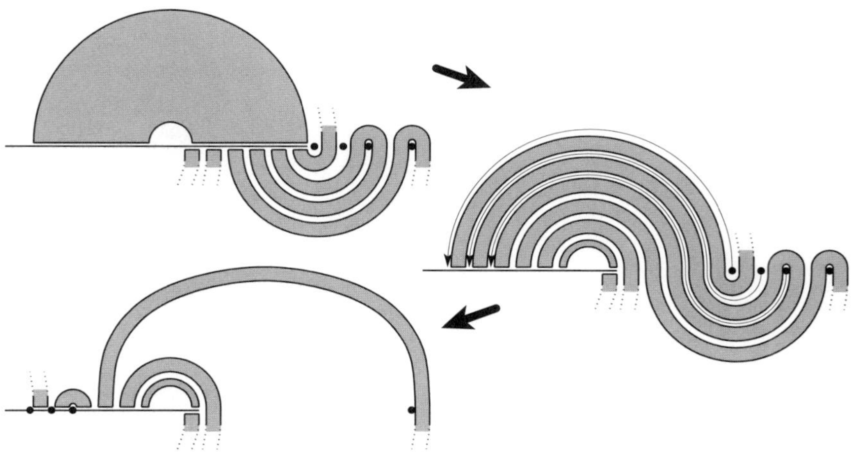

FIGURE 14. Case of a type AA numerator.

• Case AA (Figure 14). In this case BB-denominators of length 1 with both bases participating in the transmission cannot occur. Indeed, between those bases there must be a puncture, which contradicts the requirement that all A-bases are free of punctures. Thus, any length one BB-denominator has one of its bases further to the left. Such a denominator gives rise to a length two AB- or BA-strip, which does not need to be simplified.

All the other denominators are of type AB and length two, or of type BB and length three. The obstructing punctures should be slid along arcs parallel to the denominators toward the right base of the numerator, and then along the numerator toward the left base; see the figure. The braid corresponding to this sliding (if there is at least one obstructing puncture) has the form $\Delta_{k+i-j+1,k}\Delta_{i,k}^{-1}\Delta_{j,k}$ with $i < j \leqslant k$—so the word spelled out is $\Delta_{j,k}^{-1}\Delta_{i,k}\Delta_{k+i-j+1,k}^{-1}$—where we put $\Delta_{j,k} = 1$ if $j = k$.

We make, however, an exception to this rule if the left endpoint of the left base of the numerator is outside of the disk D_n, and some punctures would also be slid outside if we followed the rule. In this case, we slide each of those punctures once to the right along an arc in the upper half-plane (there will be just one option for that), each time spelling out a braid of the form $\Delta_{i+1,j}\Delta_{i,j}^{-1}$, and after that apply the usual rule to the remaining obstructing punctures. One can show that each puncture can participate in an exceptional sliding at most once during the whole process, because such a sliding delivers the puncture to the final destination. So, these slidings will contribute at most $2n$ to the Δ-length of the output, which is dominated by $2 \cdot n \cdot \ell_\Delta(w)$. For an example of an exceptional slide, see the last two transformations in Figure 13.

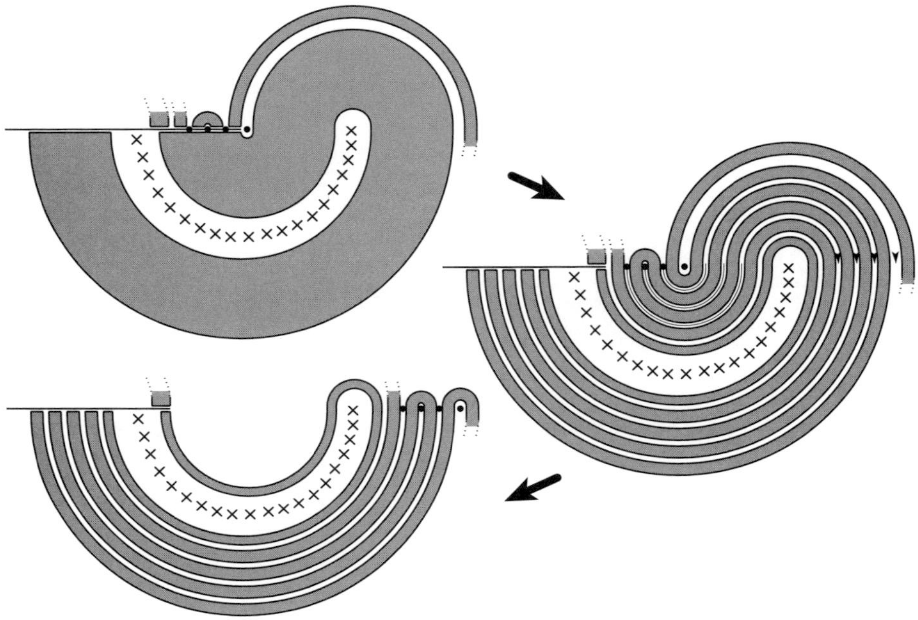

FIGURE 15. Case of a type BB numerator of length 3. The numerator is the fat band visible in the top picture. Saying that it is of length 3 means that cutting it along its intersections with the axis decomposes it into three connected components.

• Case BB, length 1. No relaxation is needed at this point, since every strip that is created during the transmission is of type AB and length two, or of type BB and length 3.

• Case BB, length 3 (Figure 15). The obstructing punctures may be inside the right base of the numerator and on the immediate right of that base. They are slid twice along the numerator to the right. If the set of obstructing punctures is not empty, the corresponding braid has the form $\Delta_{i,k}\Delta_{j+1,k}^{-1}\Delta_{i,j}^{-1}$ with $i \leqslant j < k$.

• Case AB, nonspiralling (Figure 16). The obstructing punctures, which are inside and on the immediate right of the B-base of the numerator, are slid twice along the numerator. The braid has the form $\Delta_{k+i-j+1,k}\Delta_{i,k}^{-1}\Delta_{j,k}^{-1}$ with $i < j \leqslant k$.

• Case BA, nonspiralling (Figure 17). All the denominators are of type BB. Those of length one must have the other base further to the left. After the transmission, they give rise to BB-strips of length three, which do not need to be relaxed for the moment. The denominators of length three give rise to strips of length five or seven. The obstructing punctures are first slid along arcs parallel to essential ones,

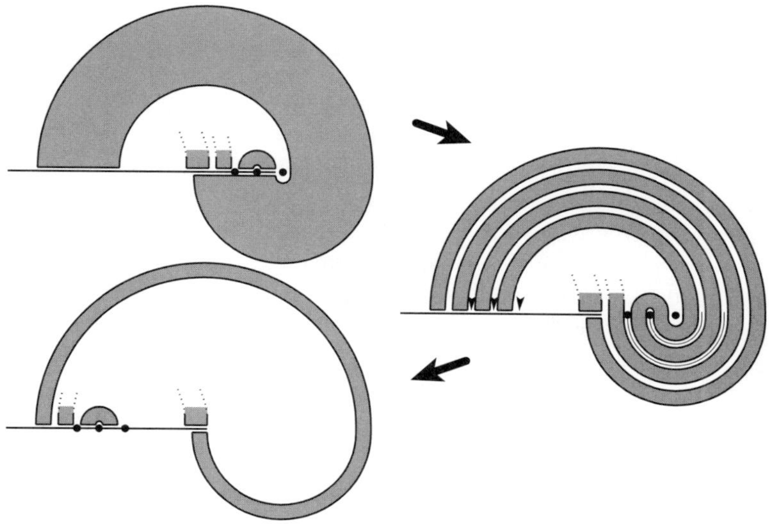

FIGURE 16. Case of a type AB numerator, nonspiralling.

and then once along the numerator. The corresponding braid has the form $\Delta_{i,j}^{2}$ with $i \leqslant j$.

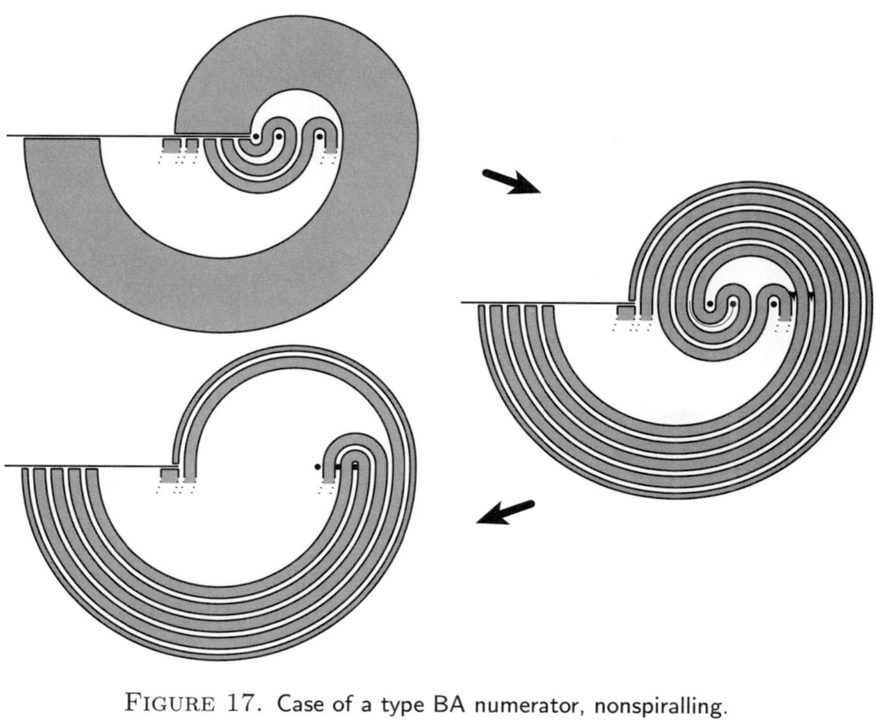

FIGURE 17. Case of a type BA numerator, nonspiralling.

• Cases AB and BA, d-spiralling. We simply apply $\Delta_{i,j}^{2d}$ in the BA-case and $\Delta_{i,j}^{-2d}$ in the AB-case, where the half-twist $\Delta_{i,j}$ involves the punctures inside the spiral. Note that in the BA-spiralling case all the denominators are of type BB and of length three. After the relaxation, the strips they give rise to are also of length three.

It is immediately seen that, with one exception, the word spelled out at each step has Δ-length bounded above by a linear factor of the amount by which the AHT-complexity decreases and that the exceptions contribute to the Δ-length of the output word w' an amount that can be taken care of by choosing an appropriate constant in (2.3). Combining this with Lemma 2.16 we get an upper bound for $\ell_\Delta(w')$ of the following form

$$\mathrm{const}_1 \cdot n \cdot \ell_\Delta(\beta) + \mathrm{const}_2 \cdot n \cdot \log_2 n,$$

which has the extra term $\mathrm{const}_2 \cdot n \cdot \log_2 n$ compared to the right-hand side of (2.3). However, this term is obviously dominated by the other one for long enough braids. For short braids, namely, with Δ-length smaller than $\log_2 n$, one can show that $\ell_\Delta(w')$ is bounded from above by

$$\mathrm{const} \cdot \log_2 n \cdot (\ell_\Delta(w) + \log_2 n), \tag{2.4}$$

which is also dominated by $\mathrm{const} \cdot n \cdot \ell_\Delta(\beta)$ under an appropriate choice of the constant. One does so by observing that such a short braid, being presented geometrically, consists of at most $2\log_2 n + 1$ groups of strands in each of which the strands do not tangle, but go in parallel and may get twisted all together. Leaving just two strands (if there are at least two present) in each group gives a braid β' for which the untangling procedure will go in parallel with that for β. In particular, the Δ-length of the output word will be the same. But β' has at most $4\log_2 n + 2$ strands, which implies (2.4). Thus, we can omit the $n \log_2 n$ term in all the cases.

It still remains to establish the connection to the σ-ordering.

The algorithm described above has the same connection to σ-ordering as Bressaud's Tetris algorithm does, *i.e.*, the transmission-relaxation algorithm produces σ-positive representatives for σ-positive braids, but not necessarily σ-negative representatives for σ-negative braids.

DEFINITION 2.18. For a braid β, the *transmission-relaxation normal form* of β, denoted $\mathrm{NF}^+_{\mathrm{t.r.}}(\beta)$, is defined to be the word spelled out by the original algorithm given the curve diagram of β as the input. Symmetrically, we denote by $\mathrm{NF}^-_{\mathrm{t.r.}}(\beta)$ the word spelled out by the algorithm that is defined in the same way with the roles of the upper and lower half-planes exchanged, *i.e.*, $(\mathrm{NF}^+_{\mathrm{t.r.}}(\beta^\updownarrow))^\updownarrow$.

PROPOSITION 2.19. *For every braid β, either $\mathrm{NF}^+_{\mathrm{t.r.}}(\beta)$ is a σ-positive word or $\mathrm{NF}^-_{\mathrm{t.r.}}(\beta)$ is a σ-negative word.*

PROOF. We mention here only the idea of the proof. Let Γ be the curve diagram of β, and let P_k be the first puncture such that the number of arcs in Γ passing over P_i is not equal to the number of arcs passing under P_i. If the number of arcs over P_k is larger, then we claim that the diagram Γ is σ_i-positive, and otherwise σ_i-negative. The sliding and relaxation rules have been designed so as not to turn a σ_i-positive curve diagram into a σ_i-negative one. So, once a σ_i-positive diagram is given, it remains σ_i-positive during the whole untangling process. Then, revising all cases of the algorithm step, one finds out that the words spelled out are always

σ_i-positive. So if the initial diagram is σ_i-positive, then the word $\text{NF}^+_{\text{t.r.}}(\beta)$ is σ_i-positive. Otherwise, the same argument about the symmetric algorithm shows that $\text{NF}^-_{\text{t.r.}}(\beta)$ is σ_i-negative. □

This also concludes the proof of Proposition 2.3.

CHAPTER XII

Triangulations

In this chapter we appeal to a technique that is frequently employed for studying mapping class groups and various geometric structures on surfaces—hyperbolic metrics, foliations, etc.—namely using triangulations. A triangulation of a surface plays a role similar to that of a basis in a vector space. In particular, there is a natural way to associate with every triangulation a coordinate system on the set of topological objects of a certain type. For instance, the isotopy class of a simple closed curve disjoint from the vertices of a triangulation can be uniquely determined from the knowledge of its intersection numbers with the edges of the triangulation, provided that the curve is tight with respect to the triangulation.

There is a natural operation on triangulations, called a flip, which is an analogue of an elementary transformation of a matrix in linear algebra. Any two triangulations having the same set of vertices can be obtained from each other by finitely many flips. This allows us to express geometrical ideas from the previous two chapters in purely combinatorial terms and to construct algorithms for detecting the order in braid groups and, more generally, in mapping class groups.

We describe here two approaches, both using triangulations. The first one is based on the notion of an integral lamination, which is a finite collection of closed curves satisfying certain conditions. It was originally developed by one of the authors (I.D.) **[73]** to detect braid triviality efficiently. Stepan Orevkov suggested that the action of braids on laminations should give a simple method for comparing braids with respect to the σ-ordering. These ideas result in a very efficient comparison algorithm, and they provide a very short proof of Property **A** that can be given without any reference to the geometric origin of the approach. A nice feature is that the running time of the algorithm deduced from this approach does not depend on the number of strands when the algorithm is implemented on a random access machine.

The second approach is based on the Mosher normal form of a braid. It relies on letting braids—viewed as homeomorphisms—act on a triangulation and choosing a distinguished sequence of flips that connect a fixed base triangulation to its image. It leads to an automatic ordering of the braid group, *i.e.*, a description by means of a finite state automaton.

The chapter is organized as follows. The lamination approach leads to computational formulas that are easy both to state and to use independently of the underlying theory. For the convenience of the reader, these formulas and some of their applications and corollaries are stated in Section 1. In Section 2, we introduce the framework of singular triangulations and integral laminations and use it to motivate and to prove easily the complicated-looking formulas of Section 1. Finally, in Section 3, we describe the Mosher normal form of braids, which requires more delicate arguments.

1. The coordinates of a braid

The approach described below leads to attributing to every n-strand braid a sequence of $2n$ integers that can be viewed as a sequence of coordinates. In particular, this sequence specifies the braid unambiguously. In this section, we describe this system of coordinates and deduce various applications, in particular in terms of Property **A** and the braid ordering. The explanation for the formulas, which may look strange at first, will be given subsequently, in Section 2.

1.1. Curious formulas. Our coordinate system stems from defining an action of the braid group B_n on the set $\mathbb{Z}^{2n}$. A curious feature of this system is that it is reminiscent of tropical algebra, involving the operations $+$ and max, rather than the usual $\times$ and $+$ of most algebraic systems. This property, which will become quite natural in Section 2, is important in practice, as it ensures a low space complexity.

In what follows, for x in $\mathbb{Z}$, we write x^+ for $\max(0, x)$, and x^- for $\min(x, 0)$.

Definition 1.1. First, we introduce two functions F^+, F^- of $\mathbb{Z}^4$ to $\mathbb{Z}^4$ by $F^+ = (F_1^+, \dots, F_4^+)$, $F^- = (F_1^-, \dots, F_4^-)$ with

$$\begin{aligned}
F_1^+(x_1, y_1, x_2, y_2) &= x_1 + y_1^+ + (y_2^+ - z_1)^+, \\
F_2^+(x_1, y_1, x_2, y_2) &= y_2 - z_1^+, \\
F_3^+(x_1, y_1, x_2, y_2) &= x_2 + y_2^- + (y_1^- + z_1)^-, \\
F_4^+(x_1, y_1, x_2, y_2) &= y_1 + z_1^+, \\
F_1^-(x_1, y_1, x_2, y_2) &= x_1 - y_1^+ - (y_2^+ + z_2)^+, \\
F_2^-(x_1, y_1, x_2, y_2) &= y_2 + z_2^-, \\
F_3^-(x_1, y_1, x_2, y_2) &= x_2 - y_2^- - (y_1^- - z_2)^-, \\
F_4^-(x_1, y_1, x_2, y_2) &= y_1 - z_2^-,
\end{aligned} \tag{1.1}$$

where we put $z_1 = x_1 - y_1^- - x_2 + y_2^+$ and $z_2 = x_1 + y_1^- - x_2 - y_2^+$.

Then, we define a left action of n-strand braid words on $\mathbb{Z}^{2n}$ by

$$\sigma_i^e \bullet (a_1, b_1, \dots, a_n, b_n) = (a_1', b_1', \dots, a_n', b_n') \tag{1.2}$$

with $a_k' = a_k$ and $b_k' = b_k$ for $k \neq i, i+1$, and

$$(a_i', b_i', a_{i+1}', b_{i+1}') = \begin{cases} F^+(a_i, b_i, a_{i+1}, b_{i+1}) & \text{for } e = +1, \\ F^-(a_i, b_i, a_{i+1}, b_{i+1}) & \text{for } e = -1. \end{cases}$$

Finally, we define the *coordinates* of an n-strand braid word w to be the sequence $w \bullet (0, 1, 0, 1, \dots, 0, 1)$.

So the coordinates of w are computed by the recursive rule

$$w \bullet (a_1, b_1, \dots, a_n, b_n) = \begin{cases} (a_1, b_1, \dots, a_n, b_n) & \text{for } w = \varepsilon, \\ \sigma_i^e \bullet (w' \bullet (a_1, b_1, \dots, a_n, b_n)) & \text{for } w = \sigma_i^e w'. \end{cases}$$

Example 1.2. The reader may check that the coordinates of the 3-strand braid words σ_1, $\sigma_2^{-1}\sigma_1$, and $\sigma_1\sigma_2\sigma_1$ are the sequences $(1, 0, 0, 2, 0, 1)$, $(1, 0, -2, 0, 0, 3)$, and $(2, 0, 1, 0, 0, 3)$, respectively.

Of course, we are interested in getting coordinates for braids and not only for braid words. So the first result is quite natural.

PROPOSITION 1.3. *The coordinates of an n-strand braid word w only depend on the braid it represents: if w, w' are equivalent n-strand braid words, then they have the same coordinates.*

PROOF. It will be seen in Section 2 that the coordinates of a braid word w have a simple interpretation in terms of the braid represented by w—actually of the homeomorphism of a punctured disk associated with the latter—which will make the result clear.

A direct elementary verification is also possible. Owing to the presentation of B_n, it suffices to show, for every sequence $(a_1, b_1, \dots, a_n, b_n)$ in $\mathbb{Z}^{2n}$, the equalities

$$\sigma_i\sigma_j \bullet (a_1, b_1, \dots, a_n, b_n) = \sigma_j\sigma_i \bullet (a_1, b_1, \dots, a_n, b_n) \qquad \text{for } |i-j| \geqslant 2, \tag{1.3}$$

$$\sigma_i\sigma_j\sigma_i \bullet (a_1, b_1, \dots, a_n, b_n) = \sigma_j\sigma_i\sigma_j \bullet (a_1, b_1, \dots, a_n, b_n) \qquad \text{for } |i-j| = 1, \tag{1.4}$$

$$\sigma_i\sigma_i^{-1} \bullet (a_1, b_1, \dots, a_n, b_n) = \sigma_i^{-1}\sigma_i \bullet (a_1, b_1, \dots, a_n, b_n) = (a_1, b_1, \dots, a_n, b_n). \tag{1.5}$$

Relation (1.3) is trivial as, for $|i-j| \geqslant 2$, the actions of σ_i and σ_j involve disjoint blocks of coordinates. As for (1.4), it is enough to check it for B_3 with $i = 1$ and $j = 2$, *i.e.*, to prove

$$\sigma_1\sigma_2\sigma_1 \bullet (a_1, b_1, a_2, b_2, a_3, b_3) = \sigma_2\sigma_1\sigma_2 \bullet (a_1, b_1, a_2, b_2, a_3, b_3), \tag{1.6}$$

and similarly, for (1.5), it is enough to check the equalities

$$\sigma_1\sigma_1^{-1} \bullet (a_1, b_1, a_2, b_2) = \sigma_1^{-1}\sigma_1 \bullet (a_1, b_1, a_2, b_2) = (a_1, b_1, a_2, b_2). \tag{1.7}$$

Relations (1.6) and (1.7) can be checked by a direct computation. However, since computing in $(\mathbb{N}, +, -, \max, 0)$ is not so usual, we outline an alternative method for checking these equalities. We observe that the structures $(\mathbb{N}, +, -, \max, 0)$ and $(\mathbb{Q}, \times, /, +, 1)$ share many algebraic properties. So we can translate (1.6) and (1.7) in the language of $(\mathbb{Q}, \times, /, +, 1)$, *i.e.*, we replace $+$ by $\times$, max by $+$, etc. For instance, x^+ becomes $1 + x$, while x^- becomes $x/(1 + x)$. Then it suffices to establish the counterparts of (1.6) and (1.7) in $(\mathbb{Q}, \times, /, +, 1)$, and to verify that all computation rules we use remain valid in both structures—essentially, that we do not use the cancellativity of $+$ in $\mathbb{Q}$. The details are tedious but easy. □

By Proposition 1.3, we have obtained a well-defined action of B_n on $\mathbb{Z}^{2n}$ for each n, and it is natural to introduce the notion of coordinates for a braid.

DEFINITION 1.4. For each n-strand braid β, the *coordinates* of β are defined to be the coordinates of any n-strand braid word that represents β.

REMARK 1.5. It is worth noting that the relation (1.6) means that the function F^+ satisfies the set-theoretical version of the so-called Yang–Baxter equation:

$$(F^+ \times \mathrm{id}_{\mathbb{Z}^2})(\mathrm{id}_{\mathbb{Z}^2} \times F^+)(F^+ \times \mathrm{id}_{\mathbb{Z}^2}) = (\mathrm{id}_{\mathbb{Z}^2} \times F^+)(F^+ \times \mathrm{id}_{\mathbb{Z}^2})(\mathrm{id}_{\mathbb{Z}^2} \times F^+).$$

REMARK 1.6. It can also be noted that the action of σ_i^{-1} can be deduced from that of σ_i directly. Indeed, for each sequence $\boldsymbol{x}$ in $\mathbb{Z}^{2n}$, we have

$$\sigma_i^{-1} \bullet \boldsymbol{x} = (\sigma_i \bullet \boldsymbol{x}^{\#})^{\#}, \tag{1.8}$$

where, for $\boldsymbol{x} = (a_1, b_1, \dots, a_n, b_n)$, we define $\boldsymbol{x}^{\#} = (-a_1, b_1, \dots, -a_n, b_n)$. So in order to compute the coordinates of a braid, we can use (1.8) instead of the function F^- of Definition 1.1.

1.2. A proof of Property A. One of the main interests of the braid coordinates defined above is that they lead to an extremely easy proof of Property **A**—*i.e.*, of the fact that a σ-positive braid is never trivial. We shall also deduce a new characterization of braids bigger than 1 with respect to the σ-ordering.

PROPOSITION 1.7. *Assume that β is a σ_i-positive braid. Let $(a_1, b_1, \dots, a_n, b_n)$ be the coordinates of β. Then we have $a_i > 0$, and $a_j = 0$ for $j < i$.*

PROOF. It is clear from formulas (1.2) that applying a σ_i-free braid (*i.e.*, we recall, one that can be expressed by a braid word that contains no letter $\sigma_j^{\pm 1}$ with $j \leqslant i$) leaves the coordinates a_j and b_j unchanged for $j < i$. Once σ_i is applied, the coordinate a_i becomes positive, since it is replaced with an expression of the form

$$a_i + (b_i + c^+)^+ = 0 + (1 + c^+)^+ \geqslant 1.$$

It is also clear from (1.2) that a_i cannot decrease if we apply a further σ_i-positive or σ_i-free braid word. □

COROLLARY 1.8 (Property **A**). *A σ-positive braid is not trivial.*

We thus obtain one more equivalent definition of the σ-ordering of braids.

PROPOSITION 1.9. *For β, β' in B_n, the relation $\beta < \beta'$ is true if and only if the first nonzero coordinate of $\beta^{-1}\beta'$ of odd index is positive.*

For instance, we saw that the coordinates of $\sigma_2^{-1}\sigma_1$ are $(1, 0, -2, 0, 0, 3)$: the first nonzero coordinate of odd index is the first entry, which is 1, so $\sigma_2^{-1}\sigma_1$ is larger than 1—as could be expected!

On the other hand, provided we take Property **C** for granted, we obtain a solution to the word problem of the braid group—*i.e.*, to the braid isotopy problem.

PROPOSITION 1.10. *A braid is unambiguously determined by its coordinates.*

PROOF. Assume that β, β' are distinct braids. Then, by Property **C**, the braid $\beta^{-1}\beta'$ is σ-positive or σ-negative. In either case, Proposition 1.7 implies that its coordinates are not those of the trivial braid. This means that the action of β^{-1} on the coordinates of β' does not yield the coordinates of the trivial braid, hence that the coordinates of β are not the coordinates of β'. □

1.3. Complexity issues. By Proposition 1.10, in order to recognize whether a braid word represents 1 in the braid group, or even whether it represents a braid larger than 1 in the σ-ordering, it suffices to compute its coordinates. The remarkable point is that, because braid coordinates involve the semiring $(\mathbb{Z}, \max, +, 0)$ only, they give rise to a very efficient algorithm—much more efficient than a counterpart involving the ring $(\mathbb{Q}, +, \times, 1)$. In particular, in terms of space complexity, we have the following obvious upper bound.

PROPOSITION 1.11. *For w a braid word of length ℓ, the size—number of digits in the binary expansion—of each coordinate of the braid $\overline{w}$ represented by w is at most $4\ell + 1$.*

PROOF. When σ_i acts on a sequence of integers, the size of each entry increases by at most 4, because it involves at most 4 additions, each of which increases the number of digits in the binary expansion by at most 1, and the max operation, which does not increase the size. □

A better estimation can be given in terms of a variation of the Δ-length introduced in Section XI.2. Recall that by Δ_{ij} we denote Garside-like half-twist braids involving strands numbered i through j; see Section XI.2.1.

DEFINITION 1.12. For each braid word w, we put

$$\ell'_\Delta(w) = \sum_{r=1}^{m}(\log_2 |d_r| + \log_2(j_r - i_r) + 1), \tag{1.9}$$

where $w = \Delta_{i_1,j_1}^{d_1} \dots \Delta_{i_m,j_m}^{d_m}$ is the shortest decomposition of w as a product of $\Delta_{i,j}$ factors. For a braid β, we define $\ell'_\Delta(\beta)$ to be the minimal value of $\ell'_\Delta(w)$ for w representing β.

The parameter $\ell'_\Delta(w)$ is closely connected with the Δ-length $\ell_\Delta(w)$ as defined in Definition XI.2.1. The only change is that here we take into account the width of the Δ-factors in addition to their exponents. Writing $\ell(w)$ for the length of w, *i.e.*, the number of letters $\sigma_i^{\pm 1}$, we clearly have $\ell'_\Delta(w) \leqslant \ell(w)$ for every braid word w. But, as in the case of $\ell_\Delta(w)$, there exist words w for which $\ell'_\Delta(w)$ is much smaller than $\ell(w)$.

For a sequence $\boldsymbol{x}$ in $\mathbb{Z}^{2n}$, we denote $\max_j |x_j|$ by $\|\boldsymbol{x}\|$.

PROPOSITION 1.13. *For every braid β in B_n, we have*

$$\log_2 \|\beta \bullet (0, 1, \dots, 0, 1)\| \leqslant 2\ell'_\Delta(\beta). \tag{1.10}$$

Moreover, if the braid β is given as a braid word w, then $\beta \bullet (0, 1, \dots, 0, 1)$ can be computed algorithmically using $C \cdot \ell'_\Delta(w) \cdot \ell(w)$ operations on a random access machine, where C is a constant that does not depend on the number of strands n. In particular, the relative order of β and 1 can be detected in $O(\ell'_\Delta(w) \cdot \ell(w))$ time.

PROOF. Anticipating the methods of Section 2, it is not difficult to establish (1.10) by considering the geometrical picture of applying a Garside-like element to a lamination. The other two assertions follow immediately. □

REMARK 1.14. The asymptotic estimation given in Proposition 1.13 for the running time of the algorithm computing $\beta \bullet (0, 1, \dots, 0, 1)$ is sharp in the sense that, for certain class of braid words (for instance those representing the powers of a pseudo-Anosov braid), the number of arithmetical operations needed to compute $w \bullet (0, 1, \dots, 0, 1)$ by using formulas (1.2) will be bounded from below by $C' \cdot \ell(w)^2$, where $C' > 0$ is some constant. For instance, in the case of a pseudo-Anosov braid β, it is known that the coordinates of $\beta^k \bullet (0, 1, \dots, 0, 1)$ grow exponentially in k. An example of such a braid is $\sigma_1\sigma_2^{-1}$.

2. Triangulations and laminations

At first, the formulas (1.1) come out of the blue and seem quite mysterious. Actually, there is no miracle here, but only a tricky use of the simple formula

$$x + x' = \max(x_1 + x_3, x_2 + x_4) \tag{2.1}$$

that compares the number of intersections of a normal curve system with two triangulations obtained one from the other by switching one diagonal in a quadrilateral. This is what we shall explain in this section. The framework consists of considering an n-strand braid as the isotopy class of a homeomorphism of a disk with n punctures. Then we let braids act on a particular collection of closed curves called an integral lamination, and count the intersections with a fixed triangulation. Applying (2.1) repeatedly leads to the curious formulas (1.1).

It can be observed that the geometrical ideas we use here essentially are the same as those used in Chapters X and XI for detecting the order of braids: in both cases, we look at the action of braids on particular curves in a punctured disk. The starting curve system that we use here is almost exactly the same as the one we use in Section XI.2: just one component, which is not changed by the action, is added for symmetry. Let us mention that a method similar to the one we use here was also developed by A. Malyutin in [**146**].

2.1. Singular triangulations. A standard tool for studying a surface $\mathcal{S}$ consists in using triangulations, *i.e.*, in decomposing $\mathcal{S}$ as the union of finitely many triangles, called faces, whose pairwise intersections consist of one common edge, or one common vertex, or is empty. For instance, the sphere S^2 can be triangulated by 4 triangles, reflecting the fact that S^2 is homeomorphic to the boundary of a tetrahedron.

In what follows, the only surface we consider is the sphere S^2. This sphere is supposed to carry a piecewise linear structure and an orientation. All self-homeomorphisms of S^2 considered in this chapter are assumed to preserve the orientation and to be piecewise linear; however, in the figures, we will draw various curves as if they were smooth. We often consider a finite set $\mathcal{P}$ included in S^2, and then call the elements of $\mathcal{P}$ *punctures*.

We make use of triangulations of an extended type, in which some triangles may be degenerate in that two vertices or two edges may coincide.

DEFINITION 2.1. A *singular triangulation* of the sphere S^2 with vertex set $\mathcal{P}$ is a set T of simple proper arcs, called *edges*, such that
- the endpoints of every edge in T belong to $\mathcal{P}$,
- the edges in T do not intersect each other except at the ends,
- the edges in T cut the sphere into triangles.

The latter condition means that each connected component of $S^2 \setminus \bigcup_{e \in T} e$ can be represented as the homeomorphic image of an open two-dimensional simplex Σ under a mapping that can be continuously extended to the boundary $\partial\Sigma$ and, after that, sends each side of the simplex onto an edge in T. (We allow different edges of Σ to map to the same edge in T.) These triangles will be referred to as *faces* of the singular triangulation.

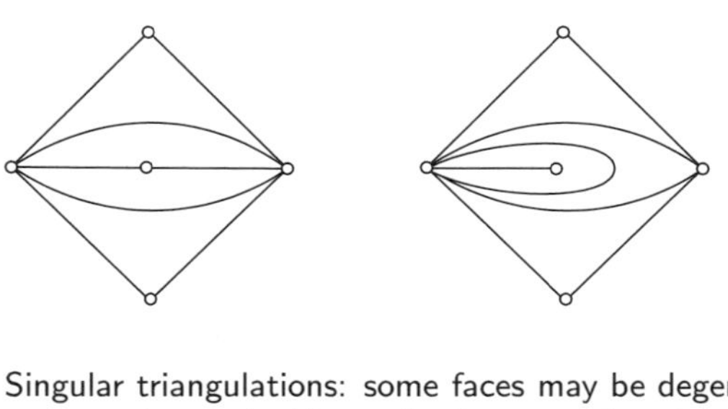

FIGURE 1. Singular triangulations: some faces may be degenerate triangles in which two vertices coincide, as in the examples above; the latter cannot appear in an ordinary triangulation.

Note that the set of edges of any triangulation of S^2 with vertex set $\mathcal{P}$ forms a singular triangulation, but not all singular triangulations are of this type: singular triangulations strictly generalize ordinary triangulations (see Figure 1).

For the subsequent investigation of the n-strand braid group B_n, which corresponds to the homeomorphisms of a disk S^2 with n punctures, it will be useful to choose a realization of the sphere with $n+3$ punctures, and to fix some notation.

NOTATION 2.2 (Figure 2). We identify the sphere S^2 with the real plane $\mathbb{R}^2$ completed with a point at infinity P_∞. Then, for $n \geqslant 2$, we define S^2_{n+3} to be S^2 with $n+3$ punctures, namely the points $P_i = (i, 0)$ for $0 \leqslant i \leqslant n+1$, plus the point at infinity P_∞. Finally, we denote by T_* the distinguished triangulation of S^2 with vertex set $\{P_i\}_{i=0,\dots,n+1,\infty}$ displayed in the figure.

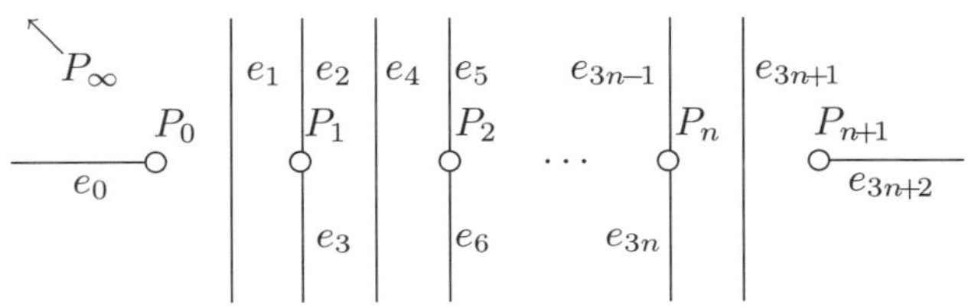

FIGURE 2. Realization of the punctured sphere S^2_{n+3}, and of its distinguished triangulation T_*. It consists of $3n+3$ edges, namely $n+1$ vertical lines, $2n$ vertical half-lines, and 2 horizontal half-lines. Observe that T_* is singular as, for instance, its leftmost face consists of a degenerate triangle where P_∞ is a double vertex with e_0 as a double-edge, similar to the innermost face in the right example in Figure 1.

Hereafter, we refer to a singular triangulation of S^2 with vertex set $\{P_0, P_1, \dots, P_{n+1}, P_\infty\}$ simply as a (singular) triangulation of S^2_{n+3}.

As the Euler characteristic of the sphere S^2 is two, there are $3p-6$ edges in a singular triangulation T of S^2 with vertex set of size p. Hence, with our current notation, there are $3n+3$ edges in a singular triangulation of S^2_{n+3}.

REMARK 2.3. In the literature, the most frequently used name for singular triangulations is *ideal* triangulations, the term emanating from hyperbolic geometry where the edges of a triangulation are geodesic paths.

Also, the triangulation technique considered in this chapter can be equally well developed for an arbitrary compact surface with or without boundary and not necessarily oriented.

2.2. Flips. When two triangulations T, T' admit the same vertex set, one can transform T into T' using a chain of elementary transformations called flips.

Assume that T is a singular triangulation and e is an edge of T that separates two faces F_1, F_2 of T. Then the union of F_1 and F_2 can be cut into two triangles in a different way as shown in Figure 3. By replacing the edge e with the edge e', we obtain another singular triangulation T'.

DEFINITION 2.4. In the situation above, we shall say that the singular triangulation T' is obtained from T by *flipping* the edge e.

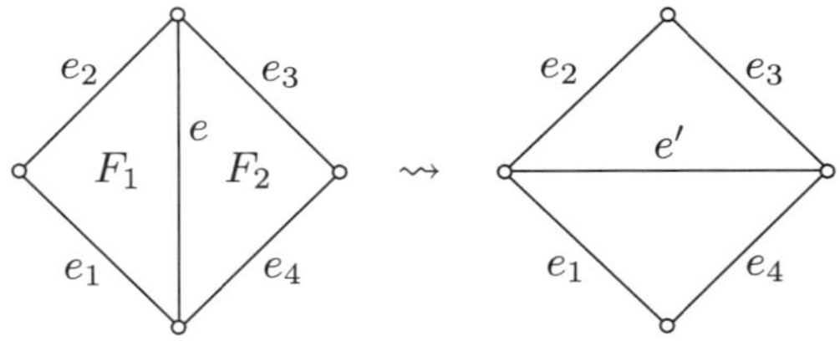

FIGURE 3. Flipping the edge e in a (singular) triangulation.

The four vertices involved in a flip, *i.e.*, the vertices of the faces F_1 and F_2 in the definition above may or may not be distinct. For instance, the two singular triangulations shown in Figure 1 can be obtained from each other by a flip.

The importance of flips is due to the following well known property.

PROPOSITION 2.5. *Any two singular triangulations T and T' with the same vertex set can be obtained from each other by finitely many flips.*

We shall prove this result in Section 3 by providing an algorithm that constructs a distinguished sequence of flips, called the combing sequence, transforming one triangulation into another. The finiteness of the length of the combing sequence will be established in Proposition 3.13, immediately implying Proposition 2.5.

Actually, Proposition 2.5 is not needed to establish the results of Section 1: what will be used is an explicit flip decomposition for a certain transformation between two specific triangulations, which corresponds to checking Proposition 2.5 in a (very) special case. However, although not directly necessary, Proposition 2.5 gives both the motivation and the explanation for what follows.

2.3. Normal curves. The second ingredient in the current approach is the notion of a curve normal with respect to a (singular) triangulation. The general type of curves we are considering is as follows.

DEFINITION 2.6. For $\mathcal{S}$ a two-dimensional surface, a *reduced curve system* on $\mathcal{S}$ is defined to be a compact one-dimensional submanifold of $\mathcal{S}$ without boundary of which no connected component is isotopic to zero, *i.e.*, bounds a disk.

Notice that the curves that we shall use are *not* assumed to be oriented and that they may surround a single puncture. Since a curve system as above may have several connected components, it might better be seen as a finite family of closed curves drawn on the considered surface, here the punctured sphere S^2_{n+3}. When some singular triangulation T is fixed, we shall have to make sure that the curves behave properly with respect to the triangulation.

DEFINITION 2.7. Assume that T is a singular triangulation of the sphere S^2 with vertex set $\mathcal{P}$. We say that a reduced curve system Γ included in $S^2 \setminus \mathcal{P}$ is *normal* with respect to T if

• Γ intersects any edge of T transversely, and

• among the connected components of $S^2 \setminus (\Gamma \cup \bigcup_{e \in T} e)$, there is no disk whose boundary consists of two arcs, one of which is a part of Γ and the other is a part of an edge of T.

By using the pulling-tight process for reduced curves described for curve diagrams in Chapter X, one can prove the following two results.

PROPOSITION 2.8. *Let T be a singular triangulation of S^2 with vertex set $\mathcal{P}$.*

(i) *For every reduced curve system Γ included in $S^2 \setminus \mathcal{P}$, there exists a curve system Γ' isotopic to Γ and which is normal with respect to T.*

(ii) *If Γ_1 and Γ_2 are two reduced curve systems on $S^2 \setminus \mathcal{P}$ that are normal with respect to a singular triangulation T, then Γ_1 and Γ_2 are isotopic if and only if, for every edge e of T, we have $\#(\Gamma_1 \cap e) = \#(\Gamma_2 \cap e)$.*

In other words, the isotopy class of a reduced curve system Γ drawn on $S^2 \setminus \mathcal{P}$ is uniquely determined by the family formed by the number of intersections of Γ with the edges of any fixed singular triangulation with vertex set $\mathcal{P}$. The main idea hereafter will be to use such intersection numbers as coordinates for the curve system Γ.

Then the key point is to be able to control the way these coordinates change when the reference triangulation is changed—but not the vertex set. Owing to Proposition 2.5, it is enough to consider the case of a flip. As only one edge is changed, the problem is to compute the number of intersections with the edge, and this is what the next result does.

LEMMA 2.9. *Assume that T is a singular triangulation of S^2 with vertex set $\mathcal{P}$ and Γ is a curve system on $S^2 \setminus \mathcal{P}$ that is normal with respect to T. Assume that e is an edge of T that separates two triangles (e_1, e_2, e) and (e, e_3, e_4) as shown in Figure 3. Then one can always flip the edge e so as to obtain a new triangulation T' such that the curve system Γ remains normal with respect to T'. Moreover, if e' is the new edge in T', we have*

$$x + x' = \max(x_1 + x_3, x_2 + x_4), \tag{2.2}$$

where we put $x = \#(\Gamma \cap e)$, $x' = \#(\Gamma \cap e')$, and $x_k = \#(\Gamma \cap e_k)$ for $k = 1, 2, 3, 4$.

The idea of the proof of formula (2.2) is as follows. Both sides of the equation count the number of arcs in the intersection of Γ with the square whose edges are e_1, e_2, e_3, e_4, except that arcs that connect opposite sides of the square are counted twice. The details are left to the reader. An example is shown in Figure 4.

2.4. Laminations and their coordinates. We are ready to introduce our last ingredient, which is the notion of an integral lamination of a punctured sphere. The simple idea is that a lamination is just a reduced curve system drawn on the punctured sphere. Actually, one must always be able to deform the curve systems so as to make them normal with respect to a given triangulation and, therefore, we are led to define a lamination not as a curve system but, rather, as the isotopy class of such a system. We recall from Notation 2.2 that S^2_{n+3} denotes the sphere S^2 with the $n+3$ punctures $P_0, \ldots, P_{n+1}, P_\infty$ of Figure 2.

DEFINITION 2.10. A connected component of a reduced curve system on S^2_{n+3} is said to be *trivial* if it bounds a disk on S^2_{n+3} with exactly one puncture inside. An *integral lamination* on the punctured sphere S^2_{n+3} is defined to be the isotopy class of a reduced curve system on S^2_{n+3} having no trivial component; each element of the isotopy class is then called a *representative* of the lamination. We denote by $\mathcal{L}_n$ the set of all integral laminations on the punctured sphere S^2_{n+3}.

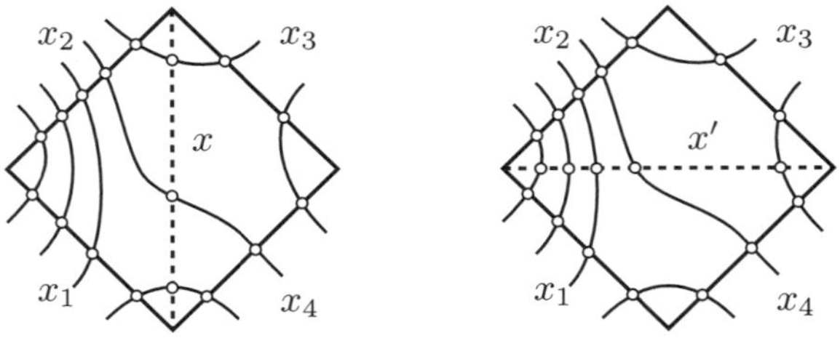

FIGURE 4. Action of a flip on intersęctions of a triangulation with a normal curve system. We count how many curves intersect each edge, and compare the numbers when the diagonal is flipped. The connection between the new numbers and the old numbers is given by (2.1). In the current case, we have $x_1 = 4$, $x_2 = 5$, $x_3 = 2$, $x_4 = 4$, $x = 3$, $x' = 5$, and (2.1) corresponds to the equality $3 + 5 = \max(4 + 2, 5 + 3)$. The hypothesis that the curves are normal means that they are tangent to no edge of the triangulation, and they form no digon with them.

For instance, a typical lamination in $\mathcal{L}_n$ is the lamination L_* consisting of n nested ellipses, as represented in Figure 5.

Assume that L is a lamination in $\mathcal{L}_n$ and T is a triangulation of S^2_{n+3}. By Proposition 2.8(i), we can find a curve system representing L that is normal with respect to T. Then we can count the intersections of a representative of L with the edges of T. By Proposition 2.8(ii), the numbers so obtained do not depend on the choice of the curve system representing L.

DEFINITION 2.11. Assume that T is a triangulation of S^2_{n+3}. Let $(e_0, \dots, e_{3n+2})$ be a fixed enumeration of the edges of T. For each lamination L in $\mathcal{L}_n$ and for each k in $\{0, \dots, 3n+2\}$, the number $\#(L \cap e_k)$ is called the *kth T-coordinate* of L.

EXAMPLE 2.12. Let T_* be the triangulation of S^2_{n+3} introduced in Notation 2.2, and let L_* be the lamination drawn in Figure 5. Then, relative to the enumeration of edges displayed in Figure 2, the T_*-coordinates of the lamination L_* are

$$(n, 2n, n, n, 2n-2, n-1, n-1, 2n-4, \dots, 2, 1, 1, 0, 0). \tag{2.3}$$

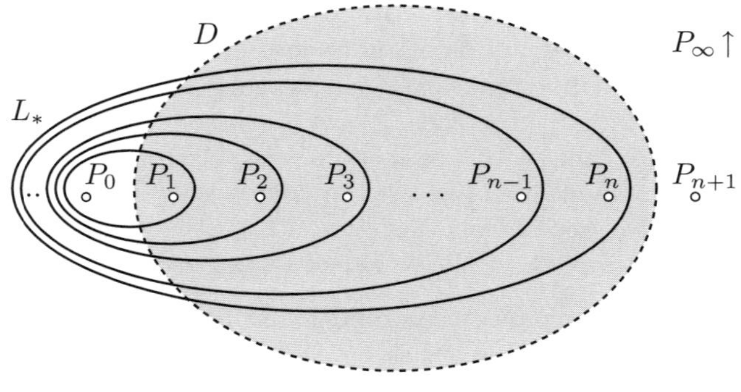

FIGURE 5. The lamination L_* of $\mathcal{L}_n$ and the disk D in our standard realization of the punctured sphere S^2_{n+3}.

Then, Proposition 2.8 implies

PROPOSITION 2.13. *Let T be any fixed triangulation of S^2_{n+3}. Then every lamination in $\mathcal{L}_n$ is determined by its T-coordinates.*

We shall not prove the proposition here, as it is not needed for what follows. Only a special case will be useful, and it will actually be an immediate consequence of the subsequent results.

2.5. Action of braids on integral laminations. Homeomorphisms of the sphere S^2 act on curves drawn on S^2, and, therefore, on curve systems and on triangulations as well. In this section, we explain how to deduce an action of n-strand braids on the space of laminations $\mathcal{L}_n$.

Assume that φ is a homeomorphism of S^2_{n+3} onto itself that globally preserves the punctures. Then φ induces a bijection of $\mathcal{L}_n$ onto itself: if Γ is a curve system representing L, then the isotopy class of $\varphi(\Gamma)$ depends on the isotopy class of Γ only, *i.e.*, on L, and we can define the image of L under φ to be this class. Thus, we get an action of the mapping class group of S^2_{n+3} on $\mathcal{L}_n$.

By Proposition I.3.3, there exists an isomorphism of the braid group B_n with the mapping class group of the disk with n punctures. We fix one such isomorphism as follows. Keeping the framework of Notation 2.2, we choose D to be a disk in the plane $\mathbb{R}^2$ covering the points $P_1, \dots, P_n$, but neither P_0 nor P_{n+1}; see Figure 5.

Each homeomorphism of the disk D that leaves ∂D pointwise fixed can be extended to a homeomorphism of the the whole plane by using the identity map on the complement to the disk D. In this way, we obtain a homomorphism

$$\iota : B_n \to \mathcal{MCG}(S^2_{n+3}). \tag{2.4}$$

This homomorphism is, in fact, an injection: this will be proved later. In this way, we obtain a well-defined action of braids on laminations of S^2_{n+3}.

DEFINITION 2.14 (Figure 6). For each n-strand braid β and each lamination L in $\mathcal{L}_n$, we define $\beta(L)$ to be the lamination $\iota(\beta)(L)$.

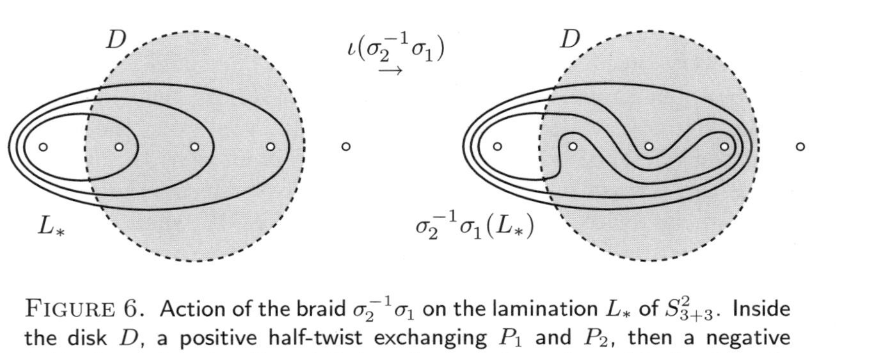

FIGURE 6. Action of the braid $\sigma_2^{-1}\sigma_1$ on the lamination L_* of S^2_{3+3}. Inside the disk D, a positive half-twist exchanging P_1 and P_2, then a negative half-twist exchanging P_2 and P_3 have been performed; outside the disk, the action is trivial.

In Section 2.4, we attributed coordinates to every lamination that specify it unambiguously. More precisely, for each fixed triangulation T of S^2_{n+3}, we associated to every lamination L of $\mathcal{L}_n$ a sequence of T-coordinates of L belonging to $\mathbb{N}^{3n+3}$. It is therefore natural to introduce

DEFINITION 2.15 (Figure 7). For each n-strand braid β, the sequence of *unreduced coordinates* of β is defined to be the sequence of T_*-coordinates of the lamination $\beta(L_*)$ in S^2_{n+3}.

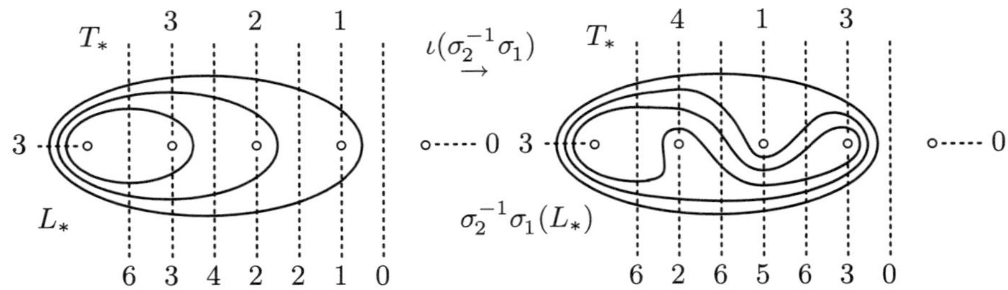

FIGURE 7. Unreduced coordinates of the braid $\sigma_2^{-1}\sigma_1$. We let $\sigma_2^{-1}\sigma_1$ act on the base lamination L_* and take the T_*-coordinates of the lamination $\sigma_2^{-1}\sigma_1(L_*)$ so obtained. Here, we read the length 12 sequence $(3, 6, 4, 2, 6, 0, 4, 6, 3, 3, 0, 0)$—we follow the edge enumeration order of Figure 2.

The name "unreduced coordinates" is chosen because we shall eventually derive another shorter sequence of coordinates. Also, we use the term "coordinates" but, so far, we did not yet prove that these numbers determine the braid unambiguously: for this to be true, it would be sufficient that Proposition 2.13 be true and the morphism ι of (2.4) be injective, but these two results have not yet been established.

We shall return to this below, but for the moment our first task will be to compute the coordinates of a braid. This amounts to solving

QUESTION 2.16. *Let T be a fixed singular triangulation of S^2_{n+3}. Assume that β is an n-strand braid and L is a lamination in $\mathcal{L}_n$. Can we compute the T-coordinates of $\beta(L)$ in terms of β and of the T-coordinates of L?*

We shall see that the answer is beautifully simple.

First, as every braid is a product of σ_i's and σ_i^{-1}'s, it is clear that, for an induction, it is enough to answer Question 2.16 when β is an elementary braid $\sigma_i^{\pm 1}$.

Then the key observation is that applying the braid β to the lamination L and keeping the reference triangulation T fixed is equivalent to keeping L fixed and applying the inverse braid β^{-1} to T. By this we mean that we apply the homeomorphism associated with β^{-1} to the edges of T. We shall naturally denote by $\beta^{-1}(T)$ the new triangulation so obtained—and, more generally, denote by $\varphi(T)$ the image of T under any homeomorphism φ of S^2_{n+3}.

LEMMA 2.17. *Let T be a triangulation of S^2_{n+3} and L be a lamination of $\mathcal{L}_n$. Then, for each homeomorphism φ of S^2 that preserves the punctures of S^2_{n+3}, the T-coordinates of $\varphi(L)$ coincide with the $\varphi^{-1}(T)$-coordinates of L.*

The result is clear: for each edge e of T and for each curve system Γ representing L and normal with respect to T, we have $\#(\varphi(\Gamma) \cap \varphi(e)) = \#(\Gamma \cap e)$.

In particular, using the embedding of B_n into $\mathcal{MCG}(S^2_{n+3})$, we obtain that, for each β in B_n and under the same hypotheses, the T_*-coordinates of $\sigma_i^{\pm 1}(L)$

are the $\sigma_i^{\mp 1}(T_*)$-coordinates of L. Thus, in order to answer Question 2.16, it is enough to be able to compute the $\sigma_i^{\pm 1}(T_*)$-coordinates of a lamination L from the T_*-coordinates of L—*i.e.*, we need a formula for a change of basis of the previous specific type.

Now, by Lemma 2.9, we know how coordinates change under one flip transformation. So the strategy is clear: we shall decompose the transformation of T_* to $\sigma_i^{\mp 1}(T_*)$ into a sequence of flips—we know already from Proposition 2.5 that this is possible, but in fact in the current case this is especially easy. Then, we use the formulas of Lemma 2.9 repeatedly so as to express the $\sigma_i^{\mp 1}(T_*)$-coordinates in terms of the T_*-coordinates.

LEMMA 2.18. *The action of the braid σ_i^{-1} on the triangulation T_* of Figure 2 decomposes into the sequence of four flips shown in Figure 8.*

The graphical argument is entirely contained in the picture.

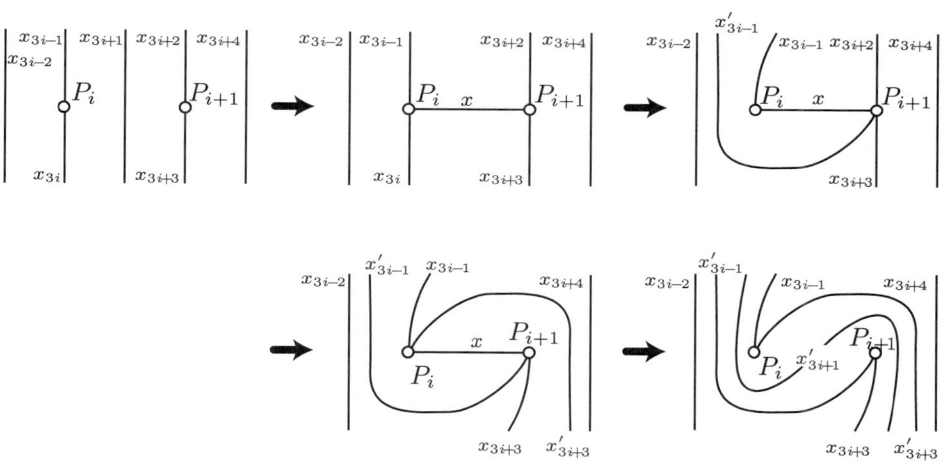

FIGURE 8. Decomposition of the action of the braid σ_i^{-1} on the triangulation T_* into a sequence of four flips. All flips occur in a degenerate quadrilateral with vertices P_i, P_{i+1} and P_∞ repeated twice; first, one flips the edge $P_\infty P_\infty$, then $P_i P_\infty$, then $P_{i+1}P_\infty$, and, finally, $P_i P_{i+1}$.

Using the above scheme, we deduce

LEMMA 2.19. *Assume that $(x_0, \dots, x_{3n+2})$ are the T_*-coordinates of a lamination L in $\mathcal{L}_n$. Then the T_*-coordinates $(x'_0, \dots, x'_{3n+2})$ of $\sigma_i(L)$ are determined by $x'_k = x_k$ for $k \leqslant 3i-2$, and $k \geqslant 3i+4$, and*

$$
(2.5)\qquad \begin{cases}
x'_{3i} = x_{3i+3},\\
x'_{3i+2} = x_{3i-1},\\
x'_{3i-1} = \max(x_{3i-2} - x_{3i+1} + x, x_{3i-1} + x_{3i+3}) - x_{3i},\\
x'_{3i+3} = \max(x_{3i+4} - x_{3i+1} + x, x_{3i-1} + x_{3i+3}) - x_{3i+2},\\
x'_{3i+1} = \max(x_{3i-1} + x_{3i+3}, x'_{3i-1} + x'_{3i+3}) + x_{3i+1} - x,
\end{cases}
$$

with $x = \max(x_{3i-1} + x_{3i+3}, x_{3i} + x_{3i+2}) - x_{3i+1}$.

PROOF. The formulas of (2.5) are what one obtains by using (2.2) for the four flips of Lemma 2.18. Hence the numbers x'_k are the $\sigma_i^{-1}(T_*)$-coordinates of L. By Lemma 2.17, they are also the T_*-coordinates of $\sigma_i(L)$. □

Of course, similar formulas exist for σ_i^{-1} replacing σ_i, which amounts to performing a horizontal symmetry in all pictures of Figure 8. We skip them here. So, we proved

PROPOSITION 2.20. *Assume that β is an n-strand braid represented by a braid word w. Then the unreduced coordinates of β are obtained from the sequence of* (2.3) *by applying the formulas of Lemma* 2.19*—and their counterpart for σ_i^{-1}—to the successive letters in w read from right to left.*

2.6. Reduced coordinates. At this point, we have answered Question 2.16, but one last (easy) step is still needed to obtain the mysterious formulas of Section 1.

It turns out that, in the case of a braid, *i.e.*, when we look at homeomorphisms of S^2_{n+3} that come from a braid, the T_*-coordinates, which live in $\mathbb{N}^{3n+3}$, are redundant. We shall extract a shorter sequence from them which belongs to $\mathbb{Z}^{2n}$. The latter will turn out to be our final coordinates, as introduced in Section 1.

DEFINITION 2.21. For β in B_n with unreduced coordinates $(x_0, \dots, x_{3n+2})$, the *reduced coordinates* of β are defined to be $(a_1, b_1, \dots, a_n, b_n)$ with

$$a_i = \frac{x_{3i-1} - x_{3i}}{2}, \quad b_i = \frac{x_{3i-2} - x_{3i+1}}{2}, \quad \text{for } i = 1, \dots, n. \tag{2.6}$$

Reduction consists in taking into account the differences between associated numbers of intersections rather than these numbers themselves. So the number a_i counts the (half)-difference between the numbers of curves passing above and below the ith puncture—we shall see in a while that the differences are even numbers—while the number b_i counts the (half)-difference between the numbers of curves passing on the right and on the left of the ith puncture.

EXAMPLE 2.22. By definition, the unreduced coordinates of the unit 3-strand braid are $(3, 6, 3, 3, 4, 2, 2, 2, 1, 1, 0, 0)$; see (2.3). When we compute the half-differences as prescribed in Definition 2.21, we obtain the reduced coordinates $(0, 1, 0, 1, 0, 1)$. Similarly, we read in Figure 7 that the sequence of unreduced coordinates of the braid $\sigma_1\sigma_2^{-1}$ is $(3, 6, 4, 2, 6, 0, 4, 6, 3, 3, 0, 0)$; we deduce that its reduced coordinates are $(1, 0, -2, 0, 0, 3)$.

Comparing with Example 1.2, we observe that, in the two cases considered, the reduced coordinates coincide with the coordinates of Section 1. This is a general result.

PROPOSITION 2.23. *For each braid β, the reduced coordinates of β coincide with the coordinates of β as introduced in Definition* 1.4.

PROOF. We consider n-strand braids. The result is true for the unit braid, since in both cases the coordinates are $(0, 1, \dots, 0, 1)$. For an induction, it is enough to prove that, if $(a_1, b_1, \dots, a_n, b_n)$ and $(a'_1, b'_1, \dots, a'_n, b'_n)$ are the reduced coordinates of β and $\sigma_i^{\pm 1}\beta$, they are connected by the formulas of Definition 1.1. This is a direct consequence of Proposition 2.19—and its counterpart for σ_i^{-1}. Indeed, it is easy to check that, if two sequences $(x_0, \dots, x_{3n+2})$ and $(x'_0, \dots, x'_{3n+2})$ are connected by the formulas of (2.5), then the sequences $(a_1, b_1, \dots, a_n, b_n)$ and $(a'_1, b'_1, \dots, a'_n, b'_n)$ deduced using (2.6) are connected by the formulas of (1.1). □

We thus found an explanation for the mysterious formulas of Section 1.

Several consequences follow from Proposition 2.23 directly. First, we conclude that the reduced coordinates of a braid must be integers, *i.e.*, the differences involved in (2.6) must be even numbers—this can actually be proved by an easy direct induction.

More interestingly, we deduce that reduced coordinates, hence *a fortiori* unreduced coordinates, characterize braids. In the latter case, this could be seen by appealing to general results about laminations (Proposition 2.13), but, here, we obtain a more direct and elementary proof—provided we take Property **C** for granted.

COROLLARY 2.24. (i) *Reduced coordinates characterize braids—and so do unreduced coordinates.*

(ii) *The morphism* ι *of* (2.4) *is an embedding of* B_n *into* $\mathcal{MCG}(S^2_{n+3})$.

PROOF. Point (i) follows from Proposition 1.10, which asserts that coordinates characterize braids.

As for (ii), if two braids of B_n have the same image in $\mathcal{MCG}(S^2_{n+3})$, then they must admit the same unreduced coordinates, and therefore the same reduced coordinates, so they must be equal. □

REMARK 2.25. An arbitrary sequence of $3n+3$ natural numbers need not be the sequence of T_*-coordinates of a lamination in $\mathcal{L}_n$, because intersection numbers have to satisfy certain constraints such as triangle inequalities. So, *a priori*, the braid action of Definition 2.14 induces a *partial* action on $\mathbb{N}^{3n+3}$ only: if a sequence $\boldsymbol{x}$ of natural numbers happens to be the T_*-coordinates of some lamination L, then we can define $\beta \bullet \boldsymbol{x}$ to be the T_*-coordinates of $\beta(L)$, but it is not obvious that this partial action extends to an everywhere-defined action. This is true, however, and can be established by a direct verification similar to the proof of Proposition 1.3. Another way to see this is to introduce more general objects called *decorated* laminations, which are in one-to-one correspondence with points in $\mathbb{R}^{3n+3}$—the action extends even to real points.

As for the reduced coordinates, they already define a one-to-one correspondence between integral laminations and sequences of $2n$ integers. Indeed, given a sequence $(a_1, b_1, ..., a_n, b_n)$ of integers we set $x_1 = 2N$, $x_{3n+2} = N$, $x_{3i} = N$ for $i = 0, 1, ..., n$, and then compute the other entries in the sequence $(x_0, x_1, ..., x_{3n+2})$ in order that (2.6) holds. One can show that for large enough N the obtained sequence $(x_0, x_1, ..., x_{3n+2})$ *is* a sequence of T_*-coordinates of a reduced system of curves, which may have trivial components. Removing all trivial components results in a curve system representing a lamination whose reduced coordinates are exactly $(a_1, b_1, ..., a_n, b_n)$.

The fact that our piecewise linear actions have their rational counterpart—see the proof of Proposition 1.3—is also not accidental. As discovered by W. Thurston in [**190**], general (nonintegral) laminations appear naturally as infinity points of Teichmüller spaces. The action of the corresponding mapping class group on a Teichmüller space can be written in terms of rational functions by using special coordinate systems introduced by R.C. Penner in [**170**]. We recommend [**86**] and [**87**] as nice references on laminations, their connections with Teichmüller spaces, and coordinate systems on these spaces.

3. The Mosher normal form

We now turn to a new approach, based on the work of Lee Mosher in [**158**]. This approach has in common with the previous one that it uses triangulations in an essential way, but the principle is different. It consists in starting with a fixed triangulation T_* and associating with every braid β a distinguished description of the triangulation $\beta(T_*)$ obtained by letting β act on T_*. Here, no transversal curve is involved. The main result is that the description of the triangulation $\beta(T_*)$ can be made by using a finite alphabet consisting of elementary transformations, in a way that can be modelled by a finite state automaton.

As a result, we obtain a new automatic structure for the braid group B_n—different from the one provided by Garside theory as described in Chapter VI. In terms of the braid ordering, this construction results in a new characterization of σ-positive braids, and in a new decision algorithm which is quadratic in the length of the braids to be compared. Even stronger, we provide an *order automatic* structure on the braid group, in the sense that the relative order of two elements can be decided directly from their automatic normal forms: given two braids, we can construct their automatic normal forms in quadratic time, and from there the decision about which of the two elements is larger in the σ-ordering can be performed by a finite state automaton by comparing initial segments of the normal forms up to their first divergence, and a bounded number of steps beyond. Nevertheless, the algorithm is not very efficient in practice, as it comes naturally implemented on a finite state automaton whose number of states is exponential in the number of strands.

3.1. The combinatorial type of a triangulation. If T is a singular triangulation of S^2 with vertex set $\mathcal{P}$ and if φ is a homeomorphism of S^2 that preserves $\mathcal{P}$ globally, we recall that $\varphi(T)$ denotes the image of T under φ, *i.e.*, assuming that $e_1, \ldots, e_{3p-6}$ are the edges of T, the triangulation T' whose edges are $\varphi(e_1), \ldots, \varphi(e_{3p-6})$.

DEFINITION 3.1. The *combinatorial type* $[T]$ of a singular triangulation T is defined to be the orbit of T under the action of the group of orientation preserving homeomorphisms of S^2 that preserve $\mathcal{P}$.

We shall also use the term "combinatorial type" and similar notation for more complicated objects like an (ordered) pair of triangulations, etc.

LEMMA 3.2. *For each finite puncture set $\mathcal{P}$, there are only finitely many pairwise different combinatorial types of singular triangulations of $S^2 \setminus \mathcal{P}$.*

PROOF. In order to specify the combinatorial type of a singular triangulation, it suffices to do the following. Assign distinct labels to all edges and then for every face list the labels assigned to its sides in the clockwise order. This information can be encoded by a word of bounded length, hence the result. □

EXAMPLE 3.3. The standard triangulation of the boundary of the 3-simplex can be specified as $\{(a, b, c), (a, d, e), (d, c, f), (b, e, f)\}$; see Figure 9.

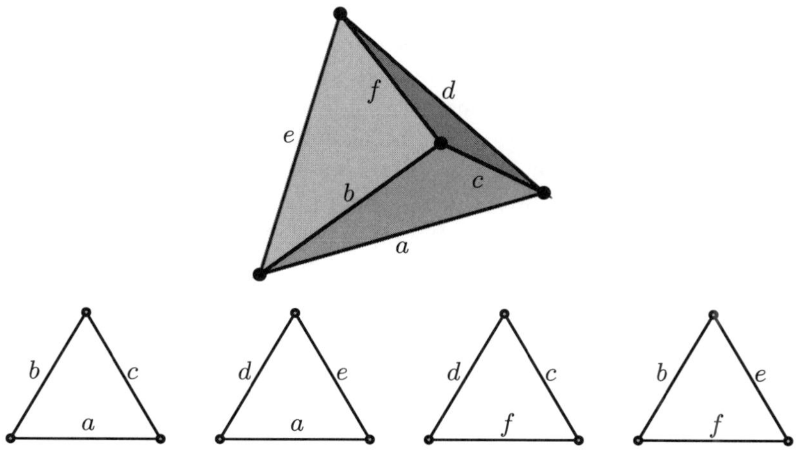

FIGURE 9. By putting a label at every edge and then cutting the triangulated surface into triangles, we keep the information how to assemble them back into a surface.

3.2. Pulling triangulations tight. Hereafter, we shall consider pairs of triangulations, and it will be important that their edges are chosen so as to avoid digons and similar trivial components. As in the case of normal curves in Proposition 2.8, we can use the pulling tight procedure of Chapter X.

DEFINITION 3.4. Let T and T' be two singular triangulations of S^2 with vertex set $\mathcal{P}$. We say that T and T' are *transverse* to each other if, for all edges e in T and e' in T', the edges e and e' either coincide or intersect transversely (possibly in several points).

DEFINITION 3.5. By a *D-disk* of a transverse pair of singular triangulations T, T' we shall mean a 2-disk whose interior is disjoint from T and T' and whose boundary consists of two arcs $\alpha \subseteq e$, $\alpha' \subseteq e'$, where e and e' are edges of T and T', respectively. If there is no D-disk of transverse singular triangulations T, T', then they are said to be *tight*.

LEMMA 3.6. *For any singular triangulations T, T' of S^2 with vertex set $\mathcal{P}$ there exists a homeomorphism φ of S^2 identical at punctures and isotopic to the identity relative to $\mathcal{P}$ such that the singular triangulations T and $\varphi(T')$ are tight.*

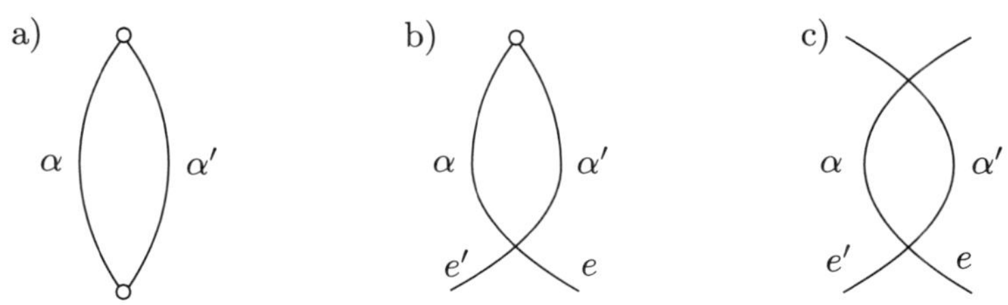

FIGURE 10. Three types of D-disks.

PROOF. Without loss of generality, we may assume that T and T' are transverse. Suppose they have a D-disk bounded by arcs α and α' described in Definition 3.5. There are three possible cases, indicated in Figure 10: a) the arcs α and α' can be whole edges of T and T'; b) only one common end of α and α' is a puncture; c) no one common end of α and α' is a puncture. In all these cases there exists a homeomorphism φ sending α to α' and preserving the rest of the singular triangulation T'. In cases b) and c), by a small perturbation of φ we make the edge $\varphi(e')$ transverse to T, see Figure 11. In this way, we obtain a singular triangulation $\varphi(T')$ which is transverse to T and has a smaller number of transverse intersection points with T. After finitely many applications of this procedure we obtain the desired homeomorphism. This process is called *pulling tight.* □

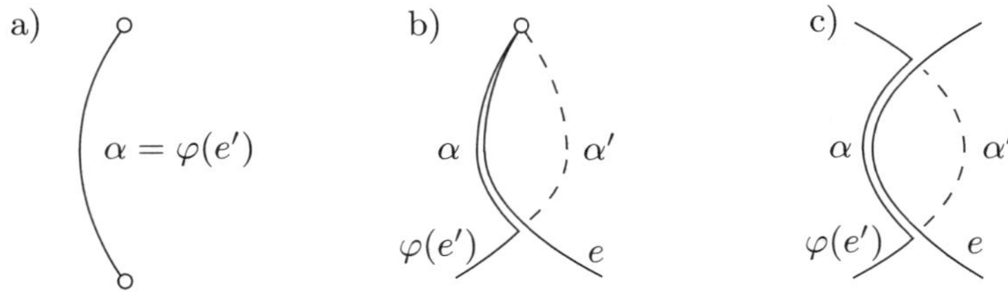

FIGURE 11. Pulling triangulations tight.

LEMMA 3.7. *If two singular triangulations T, T' are isotopic and tight, then they coincide.*

PROOF. For two transverse singular triangulations T_1, T_2, let $\rho_1(T_1, T_2)$ be the number of transverse intersections of edges of T_1 with those of T_2, $\rho_2(T_1, T_2)$ the number of edges of T_1 that are not edges of T_2, and $\rho(T_1, T_2)$ the sum $\rho_1(T_1, T_2) + \rho_2(T_1, T_2)$. The idea of the pulling tight process is to decrease $\rho(T, T')$ as much as possible by deformation of T'.

The assumption of the lemma and the general position argument imply that there exists a sequence of singular triangulations $T' = T_0, T_1, \dots, T_N = T$ transverse to T such that each passage $T_i \mapsto T_{i+1}$ is either eliminating a D-disk of T, T_i, in which case we have $\rho(T, T_{i+1}) < \rho(T, T_i)$ as described above, or the inverse operation. Among all such sequences, let us choose the one for which $\sum_i \rho(T, T_i)$ takes the minimum value. Let j be such that $\rho(T, T_j) = \max_i \rho(T, T_i)$.

Suppose that $j \neq 0$. Then both passages $T_j \mapsto T_{j-1}$ and $T_j \mapsto T_{j+1}$ consist in eliminating a D-disk. Let D_1 be the D-disk for the first transform and D_2 for the second one. Then the pair of operations $T_{j-1} \mapsto T_j \mapsto T_{j+1}$ can be replaced by another pair $T_{j-1} \mapsto T'_j \mapsto T_{j+1}$, where the first passage $T_{j-1} \mapsto T'_j$ eliminates D_2 and the second one $T'_j \mapsto T_{j+1}$ creates D_1. We will have $\rho(T, T'_j) < \rho(T, T_j)$, which contradicts the minimality of the sequence.

Thus, we have $j = 0$. The first passage $T_0 \mapsto T_1$ cannot be an elimination of a D-disk because T_0 and T are tight, and no D-disk may be created because $\rho(T, T_0) = \max_i \rho(T, T_i)$ by construction. Hence, $N = 0$ and $T' = T$. □

Now suppose that, in addition to the hypothesis of Lemma 3.6, we have one more singular triangulation T'' which is tight with both T and T'. Then, at each step of the pulling tight process described in the proof of Lemma 3.6, the singular triangulation T'' remains tight with $\varphi(T')$. Indeed, the intersection of any edge of T'' with a D-disk of T and T' must be an arc connecting a point at α with a point at α'; see Figure 12. Therefore, the homeomorphism φ in the proof of Lemma 3.6 can be chosen so as to preserve the singular triangulation T''.

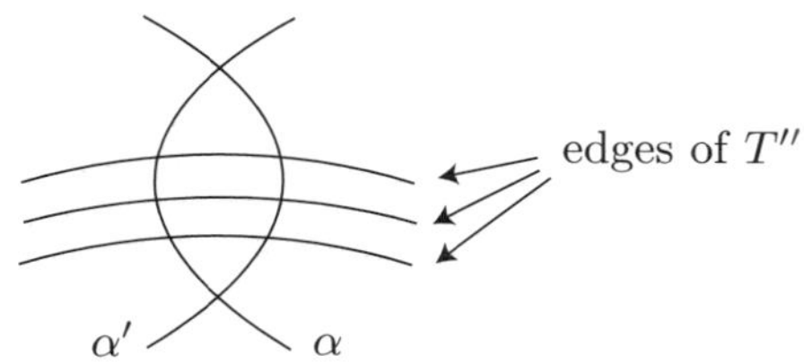

FIGURE 12. Triple tightening.

In the particular case when T and T' are isotopic, we get the following result, which is an analogue of Proposition X.1.6 for the case of triangulations. It says that the result of the pulling-tight process does not depend on the order in which we reduce D-disks.

PROPOSITION 3.8. *For three singular triangulations T, T' and T'' of S^2 with vertex set $\mathcal{P}$ such that T, T'' are tight, T', T'' are also tight, and T and T' are isotopic, there exists a homeomorphism φ isotopic to the identity relative to $\mathcal{P}$ that preserves T'' and sends T to T'.*

It follows from this proposition that singular triangulations T and T' are tight if and only if they are transverse and, with the notation of the proof of Lemma 3.7, $\rho(T, T'')$ takes the minimal value at $T'' = T'$ among all singular triangulations T'' isotopic to T'.

PROPOSITION 3.9. *Arbitrarily many singular triangulations $T_1, \dots, T_q$ of S^2 with vertex set $\mathcal{P}$ can be pulled tight pairwise. More precisely, there exist homeomorphisms $\varphi_1, \dots, \varphi_q$ of S^2 isotopic to identity relative to $\mathcal{P}$ such that the singular triangulations $\varphi_1(T_1), \dots, \varphi_q(T_q)$ are pairwise tight.*

PROOF. We apply the pulling tight process successively to pairs of singular triangulations: (T_1, T_2), (T_1, T_3), ..., (T_1, T_q), (T_2, T_3), (T_2, T_4), ..., $(T_2, T_q), \dots,$ (T_{q-1}, T_q). As we have seen above, at each step of this process, singular triangulations that are already tight remain tight. □

In what follows, we shall not distinguish a singular triangulation T with vertices at $\mathcal{P}$ from any isotopic singular triangulation, *i.e.*, a singular triangulation of the form $\varphi(T)$, where φ is a homeomorphism isotopic to the identity relative to $\mathcal{P}$. All simultaneously considered singular triangulations will be assumed to be pairwise tight.

3.3. Combing sequence. We shall now construct an automatic structure on the braid group B_n. We refer to Section XI.1.2—and to [**77**]—for basic definitions. In addition to what has been said in Chapter XI, we need one more remark, namely that it makes sense to say that a subset of $A^* \times A^*$ is a regular language. This means that elements of $A^* \times A^*$ are thought of as words in the alphabet $(A \cup \{\$\}) \times (A \cup \{\$\})$, where \$ is an artificially added symbol. If we are given a pair of words (w, w') in the alphabet A, we turn it into a word in $(A \cup \{\$\}) \times (A \cup \{\$\})$ as follows.

Let $w = a_1 \dots a_k$, $w' = a'_1 \dots a'_{k'}$, and assume that w is shorter than w'. Then the corresponding word in $(A \cup \{\$\}) \times (A \cup \{\$\})$ is

$$(a_1, a'_1) \dots (a_k, a'_k)(\$, a'_{k+1}) \dots (\$, a'_{k'}).$$

The case when the second word is shorter or they are of equal length is similar.

What we shall do is adapt the general construction of [**158**] to our specific case of the braid group B_n. The general idea is as follows. If β is a nontrivial braid, then it maps some fixed initial triangulation T_* to a new triangulation $\beta(T_*)$ that is not isotopic to T_*. Then, by Proposition 2.5, one can go from T_* to $\beta(T_*)$ by a finite sequence of flips. The idea is to select such a sequence of flips, and to use it as a distinguished specification of the braid β.

DEFINITION 3.10. By an *ordered oriented* singular triangulation we shall mean a singular triangulation whose edges are ordered and they are given an orientation.

Let T be a singular triangulation, T' an ordered oriented singular triangulation with edges $e_0, e_1, e_2, \dots$. If the triangulations T and T' do not coincide (as unordered nonoriented ones), let r be the least i satisfying $e_i \notin T$, and let α be the part of e_r between the starting point of e_r and the first intersection point with an edge of T.

DEFINITION 3.11. In the situation described above, we call the collection of arcs $\{e_i \mid 0 \leqslant i < r\} \cup \{\alpha\}$, which is supposed to keep ordering and orientation from T', the *leading part* of T' relative to T; see Figure 13. The edge f of T that cuts α will be called the *next-to-be-flipped* edge of T with target triangulation T'. The total number of transverse intersection points of edges of $T \setminus \{f\}$ with T' will be referred to as the *distance* from T to T' and will be denoted by $d(T, T')$—notice that d is not symmetric.

Let $\lambda = \{e_i \mid 0 \leqslant i < r\} \cup \{\alpha\}$ be the leading part of T' relative to T; then we denote by $\overline{\lambda}$ the union $\left(\bigcup_{i=0}^{r-1} e_i\right) \cup \alpha$ of the arcs from λ.

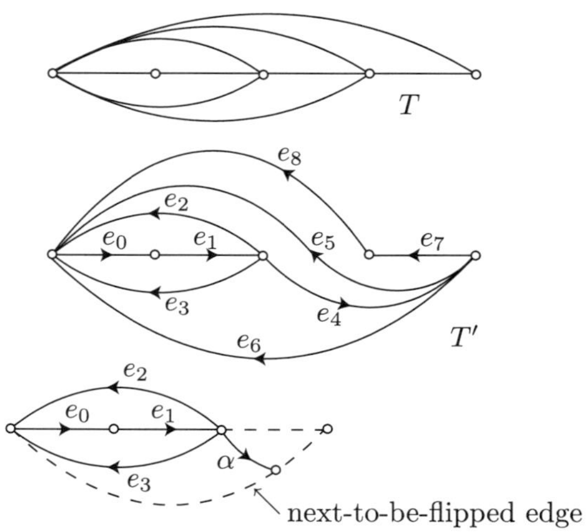

FIGURE 13. Leading part of a triangulation.

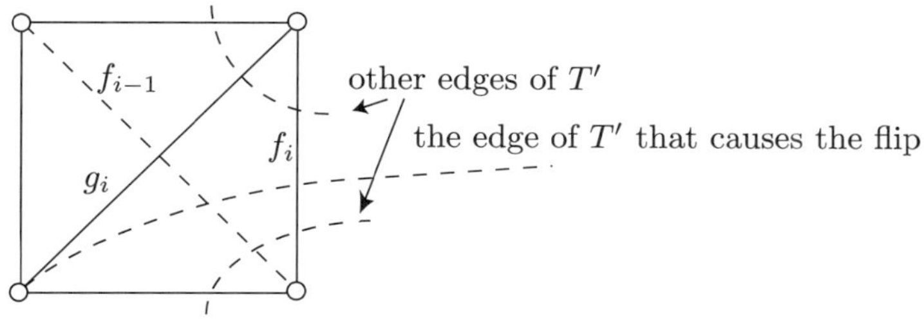

FIGURE 14. Combing terminates.

DEFINITION 3.12. Let T be a singular triangulation, and let T' be an ordered oriented triangulation. The *combing sequence* of T relative to T' is the sequence of singular triangulations $T_0, T_1, T_2, \ldots$ that is uniquely defined by the rules: $T_0 = T$, and T_{i+1} is obtained from T_i by flipping the next-to-be-flipped edge of T_i with target triangulation T'.

PROPOSITION 3.13. *The combing sequence $T_0, T_1, \ldots$ terminates, i.e., for some integer N, we have $T_N = T'$.*

PROOF. Let f_i be the next-to-be-flipped edge of the singular triangulation T_i and g_{i+1} be the edge of T_{i+1} that replaces f_i after the flip. It is easy to see that, for any $i = 1, 2, \ldots$, the edge g_i has a smaller number of transverse intersection points with the edges of T' than f_i does. Indeed, if the flips $T_{i-1} \to T_i$ and $T_i \to T_{i+1}$ are caused by the same edge of T', then f_i has one more intersection point with the edges of T' if compared with g_i; see Figure 14. If the edges causing those flips are different, then, by construction, g_i does not intersect T', but f_i does. In both cases, we have $d(T_{i-1}, T') > d(T_i, T')$. Indeed, we find $T_i \setminus \{f_i\} = (T_{i-1} \setminus \{f_{i-1}\}) \cup \{g_i\} \setminus \{f_i\}$. Since the distance from the starting triangulation T to the target triangulation T' is finite, for some N, we get $d(T_{N-1}, T') = 0$. For the next singular triangulation T_N, we will obviously have $T_N = T'$. □

3.4. Mosher normal form of a braid. Let us fix punctures $P_0, \ldots, P_{n+1}$ on the sphere S^2 and an arc e_* connecting P_0 with P_{n+1} and disjoint from the other punctures.

In this section we shall consider only singular triangulations with vertex set $\mathcal{P}$ and having e_* as an edge, which will be referred to as the *distinguished edge.* All homeomorphisms of S^2 considered in this section are supposed to be fixed on e_*. Recall that all homeomorphisms are also assumed to preserve orientation. The group $\mathcal{MCG}(S^2; \mathcal{P}, e_*)$ of isotopy classes of such homeomorphisms permuting the set $\mathcal{P}$ is clearly isomorphic to the mapping class group of an n-punctured disk, which, in turn, is isomorphic to B_n—by Proposition I.3.3. Indeed, we may cut the sphere S^2 along e_* and think of a self-homeomorphism fixed at e_* as a self-homeomorphism of the 2-disk thus obtained; see Figure 15.

In order to simplify notation, we use the same letter for a braid and any representative of the corresponding isotopy class from $\mathcal{MCG}(S^2_{n+3}, e_*)$. For instance, we shall write $\beta(T)$ for the image of a singular triangulation T under any homeomorphism presenting the braid β. This makes sense, since we have agreed above not to distinguish between isotopic singular triangulations. A singular triangulation of the cut sphere is a particular case of a diagram in the sense of Proposition I.3.3. It is clear that a homeomorphism which fixes a singular triangulation is isotopic to the identity. In conjunction with Proposition I.3.3, this implies

LEMMA 3.14. *In the settings above, the action of the braid group B_n on the set of singular triangulations—viewed up to isotopy—with $n+2$ vertices is free: only the trivial braid acts as the identity.*

In order to define the Mosher normal form, we must choose a *base ordered oriented singular triangulation.* We denote the base singular triangulation by T_* and its edges by $e_0, e_1, \dots, e_{3n-1}$. For the sake of definiteness, we assume that the distinguished edge is the last one, $e_* = e_{3n-1}$, and that the latter is oriented from P_0 toward P_{n+1}.

Let β be a braid in B_n. Recall that we regard β as a self-homeomorphism of S^2.

DEFINITION 3.15. Let $T_0, \dots, T_N$ be the combing sequence of $T_0 = T_*$ relative to $\beta(T_*)$, where the singular triangulation $\beta(T_*)$ is supposed to inherit an ordering and orientation from T_*. The sequence

$$[T_N \to T_{N-1}], [T_{N-1} \to T_{N-2}], \dots, [T_1 \to T_0], \tag{3.1}$$

which consists of combinatorial types of flips, is called the *Mosher normal form* of the braid β.

Recall that "combinatorial type" means up to a homeomorphism. In our case, this simply means that, for any braid β, we consider $[T \to T']$ and $[\beta(T) \to \beta(T')]$ to be equal. The singular triangulations appearing in the Mosher normal form of a braid are regarded as unordered and nonoriented. Since there are finitely many combinatorial types of singular triangulations and any singular triangulation can be flipped in finitely many different ways, there are only finitely many combinatorial types of flips. In other words, the Mosher normal form is a word in a *finite* alphabet.

PROPOSITION 3.16. *The set of Mosher normal words is a regular language.*

PROOF. We shall construct an automaton M that recognizes the set of Mosher normal words. This automaton will play a key role in constructing an algorithm detecting the order in the braid group.

First, we establish conditions a sequence of flips must satisfy in order to be the Mosher normal form of a braid. Let

$$[T'_N \to T''_{N-1}], [T'_{N-1} \to T''_{N-2}], \dots, [T'_1 \to T''_0] \tag{3.2}$$

be a sequence of combinatorial types of flips. If it is the Mosher form of a braid, then the following condition is necessarily satisfied.

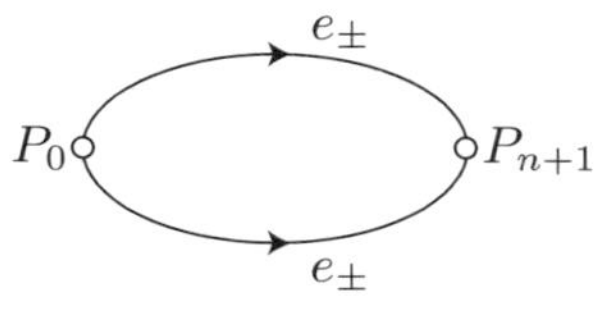

FIGURE 15. Cut sphere.

Condition 1. All subsequent flips in (3.2) are composable, and the starting and ending triangulation are combinatorially equivalent to T_*. In other words, we have

$$[T'_N] = [T''_0] = [T_*], \quad [T'_i] = [T''_i] \quad \text{for } i = 1, \ldots, N-1. \tag{3.3}$$

The latter means that there exist braids $\beta_0, \ldots, \beta_N$ satisfying

$$\beta_N(T_*) = T'_N \quad \beta_0(T''_0) = T_*, \quad \beta_i(T''_i) = T'_i. \tag{3.4}$$

If Condition 1 holds, then the braids $\beta_0, \ldots, \beta_N$ satisfying (3.4) are unique, since the action of braids on singular triangulations is free. Let

$$T_i = (\beta_0 \ldots \beta_{i-1})(T'_i) = (\beta_0 \ldots \beta_i)(T''_i),$$

and $\beta = (\beta_0 \ldots \beta_N)$. Then the sequence (3.2) can be rewritten as

$$[T_N \to T_{N-1}], [T_{N-1} \to T_{N-2}], \ldots, [T_1 \to T_0], \tag{3.5}$$

and we have

$$T_0 = T_*, \quad T_N = \beta(T_*).$$

So, the braid β is the only candidate to be a braid of which the sequence (3.2) is the Mosher normal form. Let λ_i be the leading part of $\beta(T_*)$ relative to T_i. The sequence (3.5), and hence the sequence (3.2), is the Mosher normal form of β if and only if the following holds.

Condition 2. We have $\lambda_{i-1} \neq \lambda_i$, and $\overline{\lambda_{i-1}} \subseteq \overline{\lambda_i}$ for $i = 1, \ldots, N$.

To verify Condition 2 for a given i, it is enough to know the combinatorial type of the pair (T_i, λ_i), which we denote by $[T_i, \lambda_i]$, and that of the flip $T_i \to T_{i-1}$. If it is satisfied, then the combinatorial type of the pair (T_{i-1}, λ_{i-1}) can be found from the knowledge of $[T_i \to T_{i-1}]$ and $[T_i, \lambda_i]$. It is now left to observe that there are only finitely many different combinatorial types of pairs (T, λ), where T is a singular triangulation and λ is an ordered collection of oriented arcs that *can* be a leading part of another singular triangulation T' relative to T.

We are now ready to construct an automaton M satisfying Conditions 1 and 2, which will complete the proof. □

DEFINITION 3.17. A *marked singular triangulation* is a pair (T, λ), where T is a singular triangulation and λ is either the same singular triangulation T equipped with an ordering and orientation or an ordered collection of pairwise disjoint oriented arcs $\lambda^0, \ldots, \lambda^r$ such that, for $0 \leqslant i < r$, the arc λ^i is an edge of T and λ^r is contained entirely in a face of T and joins a vertex of this face with a point at the opposite side. This side is called the *next-to-be-flipped* edge of (T, λ).

Clearly, if λ is the leading part of an ordered oriented singular triangulation T' relative to T, then the next-to-be-flipped edge of a marked triangulation (T, λ) coincides with the next-to-be-flipped edge of T with target triangulation T'.

States of M. All the states of M, except one dead-end state, are combinatorial types of marked triangulations. The pair $[T_*, T_*]$ (where the first entry is considered an unordered nonoriented singular triangulation) is the start state. The states of the form $[T_*, \lambda]$ are accept states, all the others are failure states.

Arrows of M. We set up arrows of M so that, for any two marked singular triangulations (T, λ), (T', λ') such that T' is obtained from T by a flip, the automaton M has an arrow from $[T, \lambda]$ to $[T', \lambda']$ marked with the flip $[T \to T']$ if and only if the inverse flip $T' \to T$ is performed on the next-to-be-flipped edge of (T', λ'),

$\overline{\lambda'} \subseteq \overline{\lambda}$ holds, and the enumeration of arcs in λ' is inherited from λ. All the other arrows point to the dead-end failure state.

In other words, let (T, λ) be a marked singular triangulation, and let $T' \to T''$ be a flip. The arrow originating from the state $[T, \lambda]$ and marked with the letter $[T' \to T'']$ points to the dead-end failure state unless the following two conditions are satisfied:

• The combinatorial types of T and T' coincide, $[T] = [T']$, *i.e.*, there exists a braid β satisfying $\beta(T) = T'$;

• The flip $T' \to T''$ cuts off a nontrivial part λ' of $\beta(\lambda)$.

If the conditions hold, the arrow will point to the state $[T'', \lambda']$.

It is shown in [**158**] that the Mosher normal form satisfies certain conditions which are expressed by saying that the mapping class groupoid has an automatic structure; see definitions in [**158**] or [**77**]. This implies the following result.

PROPOSITION 3.18. *The Mosher normal form of a braid can be computed in quadratic time in the length of the given braid word, provided that the number of strands is fixed.*

This means, in particular, that the Mosher form itself has length at most quadratic in the length of the corresponding braid.

3.5. Braid ordering via Mosher normal form. The following result is established in [**182**]:

PROPOSITION 3.19. *Under an appropriate choice of the base singular triangulation T_*, there exists an algorithm that, given the Mosher normal forms of two braids β, β' in B_n, detects their relative order in time linear in the length of the input, provided that n is fixed.*

REMARK 3.20. In [**182**] a more general statement is proved, where the braid group is replaced with a mapping class group of an arbitrary surface of finite type with nonempty boundary. The technique used in the general case is quite similar: the only difference is in some technical details like the choice of the base singular triangulation.

First, we give a simpler proof of Proposition 3.19 than that in [**182**]. Then we outline the original method, which, in fact, proves a stronger result. Proposition 3.19 will be a corollary to the following claim.

PROPOSITION 3.21. *Let M be the automaton constructed in Section 3.4 for some base singular triangulation T_* and let s_* be its start state. Under an appropriate choice of T_*, there exists a partial ordering $<$ on the set of states of M such that, for any two braids β, β' in B_n, the following holds.*

Let $a_N \dots a_1$, $a'_{N'} \dots a'_1$ be the Mosher normal forms of β and β', respectively, let j be the least integer satisfying $a'_j \neq a_j$, and assume that the words $a_N \dots a_j$ and $a'_{N'} \dots a'_j$ read to states s and s', respectively. Then we have $\beta < \beta'$ if and only if s and s' are comparable and we have $s < s'$.

PROOF. As a base singular triangulation, we choose any singular triangulation such that, for $i = 0, \dots, n$, the edge e_i starts at P_i and points to P_{i+1}, and for $i = 1, \dots, n-1$, the braid σ_i corresponds to a half-twist in a small neighbourhood of e_i, as shown in Figure 16. Then the union of $\bigcup_{0 \leqslant i \leqslant n} \beta(e_i)$ is nothing but the curve diagram of the braid β in the sense of Chapter X.

If the Mosher normal forms of two braids are different, then the braids themselves are different, and hence, their curve diagrams are different. The latter means that the first discrepancy in the combing sequences will be caused by the difference in $\beta(e_i)$ and $\beta'(e_i)$ with $i \leqslant n-1$. If this happens at the jth step of the combing process and we have $a_j^{-1} = [T_j \to T_{j+1}]$, $a_j'^{-1} = [T_j \to T'_{j+1}]$, then the leading parts λ_j and λ'_j of the target triangulations $\beta(T_*)$ and $\beta'(T_*)$ relative to T_j indicate the first divergence of the curve diagrams of β and β'.

Thus, the following partial ordering on the states of M will satisfy conditions of the claim. Comparable states are of the form $[T, \lambda]$, $[T, \lambda']$, where both λ and λ' consist of no more than n arcs, $\overline{\lambda}$ and $\overline{\lambda'}$ are simple curves coming from P_0, and no one of $\overline{\lambda}$, $\overline{\lambda'}$ is a part of the other. We set $[T, \lambda] < [T, \lambda']$ if and only if the first divergence of $\overline{\lambda'}$ from $\overline{\lambda}$ is to the left. □

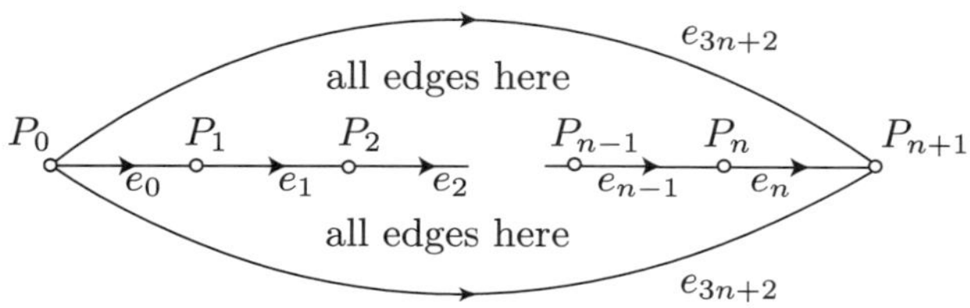

FIGURE 16. A base singular triangulation; only the important edges are displayed. The edge e_{3n+2} appears twice because this is a triangulation of the sphere S^2.

So, we can detect the relative order of two braids β, β' given the Mosher normal forms of their inverses as follows. First, we read the given Mosher normal forms from the end, find the first difference, and cut off the coinciding parts. Second, we input the truncated words to two automata identical to M and read the final states. Third, we compare the final states.

As shown in [**182**], if the base singular triangulation T_* is chosen in a more intelligent way, we shall not need to read the inverted Mosher normal forms completely. It will suffice to read them a little further after the first discrepancy. Namely, the following result has been proved.

PROPOSITION 3.22. *Assume that the base singular triangulation T_* is chosen as shown in Figure 17. Let T_0, T_1, ..., T_N be the combing sequence of T_* with target triangulation $\beta(T_*)$, where β is a braid and λ_i be the leading part of $\beta(T_*)$ relative to T_i for $i = 0, 1, \dots, N$. Then the combinatorial type of the marked singular triangulation (T_i, λ_i) can be found from the knowledge of the combinatorial types of the four succesive flips*

$$[T_i \to T_{i+1}], [T_{i+1} \to T_{i+2}], [T_{i+2} \to T_{i+3}], [T_{i+3} \to T_{i+4}]$$

in the general case $i \leqslant N-4$, and of the $(N-i)$ flips $[T_i \to T_{i+1}]$, ..., $[T_{N-1} \to T_N]$ in the special cases $i = N-3$, ..., $N-1$.

We skip the proof, which consists in searching finitely many possibilities. Notice that the choice of the base triangulation in Proposition 3.22 is also well adapted for detecting the order as in the proof of Proposition 3.21. As a consequence of Proposition 3.22 we get the following two statements:

COROLLARY 3.23. *Under the choice of* T_* *as in Proposition* 3.22, *the set of inverted Mosher normal forms of braids from* B_n *is a regular language.*

COROLLARY 3.24. *The braid group is order automatic, in the sense that there exists an algorithm which can be performed by a finite state automaton, which takes as its input the inverted Mosher normal forms of two braids—with an appropriate choice of base singular triangulation* T_**—and outputs the relative order of the braids. Moreover, the algorithm takes linear time. Under the choice of* T_* *as in Proposition* 3.22, *there exists a finite state automaton that detects the relative order of two braids by reading the inverted Mosher normal forms of their inverses no more than* 3 *steps further after the first discrepancy.*

We conclude the section by observing that under the choice of T_* as in Proposition 3.22, the relative order of β and 1 can be detected from the last element of the Mosher normal form of β, *i.e.*, the first flip in the combing sequence. The proof of the next assertion is then an easy exercise.

PROPOSITION 3.25. (i) *A braid* β *is* σ*-positive (resp.* σ*-negative) if and only if the first flip in the combing sequence of* T_* *relative to* $\beta(T_*)$ *is performed on the edge* e_{3i-1} *(resp.* e_{3i}*) for some* i *with* $1 \leqslant i \leqslant n$.

(ii) *For two braids* β, β', *we have* $\beta < \beta'$ *if and only if, under the choice of the base singular triangulation as in Figure* 17, *the last flip of the Mosher normal form of* $\beta^{-1}\beta'$ *occurs in the upper half-sphere.*

Thus we have obtained one more definition of the braid ordering from the list mentioned in the Introduction.

As a final remark, let us observe that the algorithms described in this section are not so easy to use in practice. The reason is the size of the automata, which is comparable with the number of singular triangulations. The latter grows exponentially with the number of punctures.

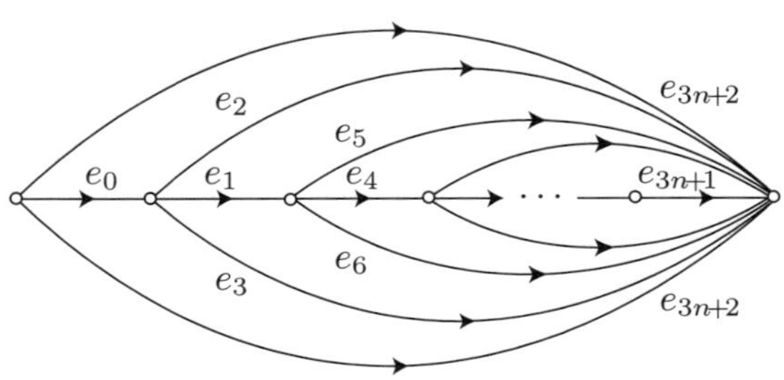

FIGURE 17. A better base singular triangulation.

CHAPTER XIII

Hyperbolic Geometry

In this chapter we present a simple and natural way to construct the σ-ordering and many other orderings of the braid group, and to study the properties of these orderings using techniques from hyperbolic geometry.

There are two key observations, which will be explained in detail, underlying the construction of the orderings.

The first one is that for a finitely generated group, like the braid group B_n, being left-orderable is equivalent to admitting a free action by orientation-preserving homeomorphisms on the real line. For instance, given an action of the group on $\mathbb{R}$ in which some point x of $\mathbb{R}$ has a trivial stabilizer, we can obtain an ordering on the group in a particularly simple way, namely by pulling back the standard ordering of the real line on the orbit of x.

The second key observation is that the n-times punctured disk D_n can be equipped with a hyperbolic metric, so that the universal cover of D_n is a subset of the hyperbolic plane; moreover, the boundary of the compactified universal cover is a circle, and there is a well-defined action of the braid group on this boundary circle, by homeomorphisms which all fix a common basepoint. Combining these two key observations, we immediately obtain that the braid groups are left-orderable.

This point of view on the orderability of braid groups is very attractive in a number of ways. First, it extends immediately to a proof that the mapping class groups of all compact surfaces with nonempty boundary are left-orderable. Second, it displays the σ-ordering as an example of a very large family of orderings which are all extremely natural, at least in the eyes of a geometer. Third, it leads to geometrically intuitive proofs of Property **S**—but not of Properties **A** and **C**. Finally, the classification of orderings arising in this way leads not only to an appealing classification of different ways of cutting a surface into pieces, it also ties in the theory of braid orderings with the vast field of dynamics of surface homeomorphisms [**34, 82**].

The fact that braid groups, and more generally mapping class groups of surfaces with nonempty boundary, act on the real line (which implies orderability) has certainly been known to Thurston and to members of his school for several decades. In fact, all the main ideas can even be traced back to Nielsen. For this reason we call the class of orderings studied in this chapter the *orderings of Nielsen–Thurston type.* The main reference for this chapter is the paper [**183**].

It seems certain that *all* the results in this chapter can be generalized to mapping class groups of any compact surface with nonempty boundary, with or without punctures. However, this has never been done explicitly.

The structure of this chapter is as follows. In Section 1, we construct a very natural B_n-action on $\mathbb{R}$ and give some examples of orderings of B_n induced by this action—among them the σ-ordering. In explaining why these orderings are essentially different, we introduce all the essential ideas for the systematic classification of orderings, which is carried out in Section 2. In Section 3 we prove that all orderings of B_n arising from our action on $\mathbb{R}$ have the subword property.

Throughout this chapter, groups act on the left.

1. Uncountably many orderings of the braid group

In this section we first explain the general correspondence between group actions on the real line and orderings of groups. Next we construct a very natural B_n-action on $\mathbb{R}$. Then we look at two sets of examples of orderings arising from this construction: the first one in order to demonstrate that different orderings can have essentially different properties, and the second one to illustrate how, on the other hand, two orderings whose constructions look very different may actually be equal. This zoo of examples, which includes the σ-ordering, introduces all the essential ideas for the systematic classification of orderings in the next section.

1.1. Orderability and group actions on $\mathbb{R}$. It can be shown that a non-trivial group is left-orderable if and only if it acts on some linearly ordered set by order-preserving bijections in such a way that only 1 acts trivially. For our purposes, actions on the real line suffice, according to the following folklore theorem, whose proof appears in [**98**].

PROPOSITION 1.1. *A countable group G is left-orderable if and only if G acts on $\mathbb{R}$ by orientation-preserving homeomorphisms in such a way that only 1_G acts by the identity map, i.e., if and only if there exists a monomorphism from G to* $\mathrm{Homeo}_+(\mathbb{R})$.

PROOF (SKETCH). Given an action of G on $\mathbb{R}$, we define an ordering $\prec$ as follows. We fix an enumeration $q_1, q_2, ...$ of the rationals. Consider distinct elements g, g' in G. We let k be the smallest integer satisfying $g(q_k) \neq g'(q_k)$. Now we define $g \prec g'$ if $g(q_k) < g'(q_k)$ and $g \succ g'$ if $g(q_k) > g'(q_k)$. This is a *linear* ordering: If we have $g(q_k) = g'(q_k)$ for all k in $\mathbb{N}$, then g and g' act by the same homeomorphism, because $\mathbb{Q}$ is dense in $\mathbb{R}$. The ordering is left-invariant, since $g(q_k) > g'(q_k)$ implies $h \circ g(q_k) > h \circ g'(q_k)$ for every h in G, because h acts orientation preservingly and k is also the least integer satisfying $h \circ g(q_k) \neq h \circ g'(q_k)$.

Conversely, if $\prec$ is a left-invariant ordering of G, then one can construct an action of G on $\mathbb{R}$ as follows: one can construct an order-preserving injection I of G to $\mathbb{R}$ such that all the image points are isolated. This uses the existence of a countable order-dense *discrete* subset of $\mathbb{R}$ with no first or last element; think of the midpoints of the deleted intervals in construction of the Cantor set, for example. Since G acts on itself in an order-preserving way by left multiplication, this yields an order-preserving action of G on the image of I. Finally, we interpolate, in order to extend this to an action on all of $\mathbb{R}$. □

We shall be interested primarily in a special case of the above result.

DEFINITION 1.2. Suppose G acts on $\mathbb{R}$, and suppose in addition that there exists a point x of $\mathbb{R}$ with $\mathrm{Stab}_G(x) = \{1\}$, *i.e.*, the points in the orbit of x under the G-action are in bijective correspondence with the elements of G. Then we define a left-invariant ordering $<_x$ of G by declaring that $g <_x g'$ holds if and only if we have $g(x) < g'(x)$.

Note that for a different point y in $\mathbb{R}$, the same construction method and the same G-action may yield a very different ordering; we shall see many instances of this behaviour.

1.2. An action of B_n on $\mathbb{R}$. We shall think of the braid group B_n as the mapping class group of a disk with n punctures: $B_n = \mathcal{MCG}(D_n)$. The following result is essentially due to Nielsen [**164, 165**]. Together with Proposition 1.1, it implies that B_n is left-orderable.

PROPOSITION 1.3. *There is a natural action by orientation-preserving homeomorphisms of the braid group B_n on a topological space which is homeomorphic to the real line. Moreover, only the trivial braid acts as the identity homeomorphism.*

PROOF. The n-punctured disk D_n can be equipped with a complete hyperbolic metric of finite volume such that the boundary of the disk is geodesic. Actually, there is a choice of a whole $\mathbb{R}^{2n-3}$-family—the Teichmüller space of D_n—of isotopy classes of hyperbolic metrics, but the choice is irrelevant for the $\mathbb{R}$-action we are aiming for. The metric on D_n lifts to a metric on the universal cover $\widetilde{D_n}$ of D_n, and then $\widetilde{D_n}$ can be isometrically embedded in the hyperbolic plane $\mathbb{H}^2$. We can compactify $\mathbb{H}^2$ by adding a circle at infinity $S^1_\infty = \partial\mathbb{H}^2$. We can then go on to compactify $\widetilde{D_n}$ by attaching its limit points on S^1_∞. The resulting space is homeomorphic to a closed disk, and by abuse of notation we shall still denote it $\widetilde{D_n}$.

The circle $\partial\widetilde{D_n}$ has two types of points: first the points at infinity, which form a Cantor set in the circle, and second their complement, $\partial\widetilde{D_n} \cap \mathbb{H}^2 = p^{-1}(\partial D_n)$ (where p denotes the covering projection), which consists of a countable number of open arcs; see Figure 1. We now choose, once and for all, a basepoint $*$ in one of these arcs. We are finally ready to define the real line on which B_n shall act: it is $\partial\widetilde{D_n} \setminus \{*\}$.

Next we consider an element β of B_n, represented by some homeomorphism φ from D_n to D_n. Now φ can be lifted to a homeomorphism $\widetilde{\varphi}$ of the universal cover—saying that $\widetilde{\varphi}$ is a *lifting* of φ means that $\varphi \circ p = p \circ \widetilde{\varphi}$ holds. In fact, there are infinitely many such liftings—as many as there are elements of $\pi_1(D_n)$—but we have one preferred choice: we demand that $\widetilde{\varphi}$ fix our basepoint $*$. Thus we have constructed an action of B_n on the real line $\partial\widetilde{D_n} \setminus \{*\}$.

To see that this action is well defined, we suppose that ψ is another representative of the same element β of B_n distinct from φ. Then ψ is related to φ by a homotopy which is fixed on ∂D_n. This homotopy lifts to a homotopy between $\widetilde{\varphi}$ and $\widetilde{\psi}$ which is fixed on the arcs $\partial\widetilde{D_n} \cap \mathbb{H}^2$. Since these arcs are dense in $\partial\widetilde{D_n}$, the homeomorphisms φ and ψ lift to exactly the same action on $\partial\widetilde{D_n}$.

Finally, in order to prove the second statement, we suppose that a homeomorphism φ acts trivially on $\partial\widetilde{D_n}$. Then, in particular, this homeomorphism fixes all liftings of the basepoint $*$ of D_n, and thus induces the identity homeomorphism on $\pi_1(D_n)$. This implies that φ represents the trivial element of $\mathcal{MCG}(D_n)$. □

Our definition of the action may look rather abstract. We shall actually think of it in the following way. If x is a point in $\partial\widetilde{D_n} \setminus \{*\}$, then a path $\widetilde{\Gamma}_x$ (typically a geodesic without self-intersections) in $\widetilde{D_n}$ from the basepoint $*$ to x projects to a path Γ_x (typically a geodesic with many self-intersections) in D_n which starts

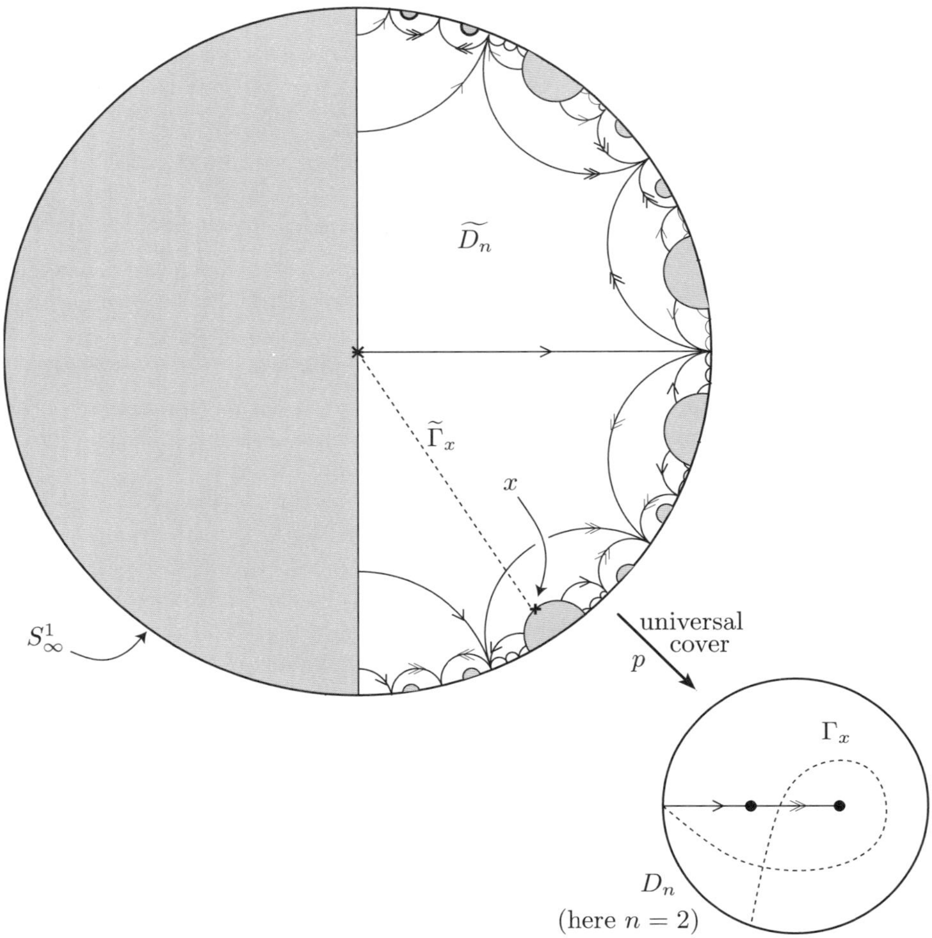

FIGURE 1. The disk D_n, and its universal cover $\widetilde{D_n}$ embedded in $\mathbb{H}^2$—indicated as the unshaded part.

at $p(*)$ in ∂D_n. Now a homeomorphism φ acts on this path Γ_x—just like it acted on curve diagrams—and the lifting of $\varphi(\Gamma_x)$ is a path in $\widetilde{D_n}$ from $*$ to some point in $\partial\widetilde{D_n}$. This is the point that we define to be $\varphi(x)$. Note that, if x was a point in S^1_∞, then all the paths mentioned were necessarily infinite. Note also that the path $\varphi(\Gamma_x)$ is, in general, not a geodesic even if Γ_x is geodesic.

Let us summarize what we have achieved so far. We have described an action by homeomorphisms of B_n on a topological space $\partial\widetilde{D_n} \setminus \{*\}$, which we managed to identify with $\mathbb{R}$. We shall say that an ordering of B_n is of Nielsen–Thurston type if it arises from a point x in $\mathbb{R}$ with $\mathrm{Stab}(x) = \{1\}$ in this action of B_n on $\mathbb{R}$. More precisely:

DEFINITION 1.4. We say that a left-ordering $\prec$ of B_n is of *Nielsen–Thurston type* if there exists an element x of $\mathbb{R}$ such that, for all β, β' in B_n, the relation $\beta \prec \beta'$ is equivalent to $\beta(x) <_{\mathbb{R}} \beta'(x)$.

Our aim will now be to classify the order types of B_n of Nielsen–Thurston type. For instance, we shall prove that for $n \geqslant 3$ there are uncountably many conjugacy classes of dense orderings of B_n, but only a finite number (for which a formula will be given) of conjugacy classes of discrete orderings of Nielsen–Thurston type. However, the main classification results, Propositions 2.5 and 2.8, are quite technical; in order to get some intuition for the most important phenomena that can occur, we first describe some examples.

1.3. Examples of different Nielsen–Thurston orderings. Let us start with a first set of examples which is meant to illustrate how some fundamentally different order types can result from our construction. A second set of examples which will be given in Section 1.4 will have the opposite aim: showing that many very different-looking geodesics give rise to the same orderings of B_n—thus the amount of different order types arising from our construction is quite limited, and this makes a classification possible.

DEFINITION 1.5. Two left-invariant orderings $\prec_1$ and $\prec_2$ of a group G are said to be *conjugate* if there exists an h in G such that $g \prec_1 g'$ is equivalent to $gh \prec_2 g'h$.

For instance, if a group G acts on $\mathbb{R}$, and some x in $\mathbb{R}$ satisfies $\mathrm{Stab}_G(x) = \{1\}$, then, for any h in G, the orderings of G induced by x and $h(x)$ are conjugate.

We shall see examples of orderings that are all pairwise nonconjugate. Some of them are dense, other are discrete. We recall from Section II.3.2 that an ordering $\prec$ of G is said to be *dense* if, for any two elements g, g' of G satisfying $g \prec g'$, there exist infinitely many elements h in G with $g \prec h \prec g'$. By contrast, an ordering is *discrete* if every element has a well-defined predecessor and successor, *i.e.*, if for g in G there exists a largest g and a smallest g' satisfying $g \prec g \prec g'$. It is an easy exercise to show that any left-ordering of a group is either dense or discrete. For example, we have seen in Section II.3.2 that the σ-ordering of B_n is discrete: the predecessor of the braid β is $\beta\sigma_{n-1}^{-1}$ and its successor is $\beta\sigma_{n-1}$.

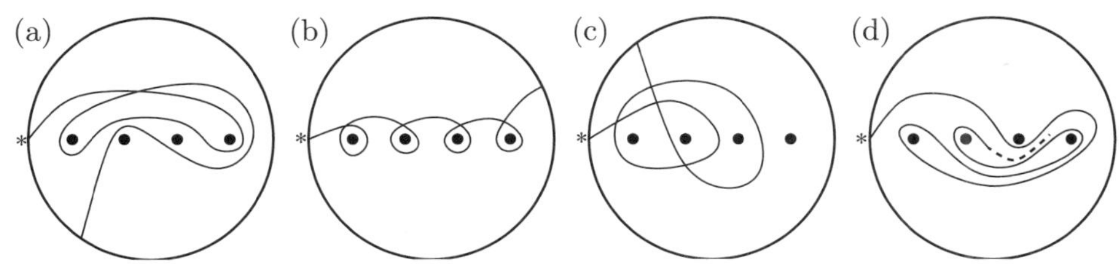

FIGURE 2. Four curves $\Gamma_a, \Gamma_b, \Gamma_c$, and Γ_d in D_4, for four points in $\partial\widetilde{D_4} \setminus \{*\}$. Figure (d) is meant to represent an infinite geodesic ray without self-intersections, whose closure would look like a generic geodesic lamination of D_4.

The first set of examples begins with the four curves shown in Figure 2. As explained above, each of the curves in this figure represents a point in $\partial\widetilde{D_4} \setminus \{*\}$, namely the endpoint of the lifting of the curve to $\widetilde{D_4}$ which starts at the point $*$ in $\partial\widetilde{D_4}$. We shall denote these four points a, b, c, and d.

LEMMA 1.6. *The point a in $\partial\widetilde{D_4} \setminus \{*\}$ has a nontrivial stabilizer under the B_4-action, and thus does not induce a linear ordering of B_4.*

PROOF. We observe that the first and the fourth puncture of D_4 are in the same path component of $D_4 \setminus \Gamma_a$. The element $\sigma_1^{-1}\sigma_2^{-1}\sigma_3\sigma_2\sigma_1$ of B_4 has support in this path component and interchanges the first and fourth puncture, *i.e.*, it can be represented by a homeomorphism that is the identity outside this path component. It is thus a nontrivial element of B_4 which acts trivially on the point a. □

More generally, we observe that a necessary condition for a geodesic to induce a linear ordering of B_4 is that it separates the punctures of D_n, in the sense that no two punctures of D_n should be in the same connected component of D_n with the geodesic removed. We shall see later on that this necessary condition is almost sufficient—a geodesic which separates the punctures may fail to induce a total ordering if it lies in a complementary component of the stable geodesic lamination of some pseudo-Anosov homeomorphism, but apart from this class of exceptions the necessary condition is sufficient.

From now on, we shall only be interested in geodesics that separate the punctures, such as for instance the geodesics Γ_b, Γ_c and Γ_d of Figure 2. It is a fact which for the moment we shall not prove, that the three points b, c and d do indeed have trivial stabilizer—with the proviso that for Γ_c we need an extra technical hypothesis, which is generically satisfied.

Thus, using Definition 1.2, each of these three points b, c and d in $\partial\widetilde{D_4} \setminus \{*\}$ induces a linear ordering of B_4, denoted $<_b$, $<_c$ and $<_d$, respectively. For instance, saying that a homeomorphism φ of D_4 represents an element of $\mathcal{MCG}(D_4)$ that is larger than 1 in the $<_b$-ordering means that the endpoint of the lifting of $\varphi(\Gamma_b)$ is to the left of the endpoint of the lifting of Γ_b, as seen from the basepoint of $\widetilde{D_4}$.

We now claim that the orderings $<_b$, $<_c$ and $<_d$ of B_4 are fundamentally different from each other. We recall from Section II.3.4 that a convex subgroup of a left-ordered group $(G, \prec)$ is a subgroup H of G such that, for h, h' in H, and g in G, the relation $h \prec g \prec h'$ implies $g \in H$. For instance, the subgroup of B_n generated by $\sigma_2, \dots, \sigma_{n-1}$ is convex in the σ-ordering.

PROPOSITION 1.7. (i) *The orderings $<_b$ and $<_c$ are discrete, whereas $<_d$ is dense. In particular, $<_d$ is not conjugate to either $<_b$ or $<_c$.*

(ii) *The ordering $<_b$ has a convex subgroup $\langle\sigma_2, \sigma_3\rangle$ isomorphic to B_3. By contrast, the ordering $<_c$ has a convex subgroup $\langle\sigma_1, \sigma_3\rangle$ isomorphic to $\mathbb{Z}^2$. The orderings $<_b$ and $<_c$ are not conjugate.*

PROOF (SKETCH). We shall only give some plausibility arguments. The general philosophy is the following. If for some path Γ and for some homeomorphism φ of D_n the support of φ is disjoint from a fairly long initial segment of Γ, then the element of B_n represented by φ should be fairly close to 1 in the ordering induced by Γ.

For instance, in Figure 3(c), we consider the liftings to $\widetilde{D_n}$ with initial point $*$ of the curves Γ_c and $\sigma_1^p(\Gamma_c)$, for every p in $\mathbb{Z} \setminus \{0\}$. These liftings coincide for quite a long initial segment of Γ_c, namely along the segment drawn in thin line in Figure 3(b). By contrast, only a relatively short tail consisting of the lifting of the latter parts of Γ_c, drawn bold in Figure 3(b), is at all affected by the σ_1-action. This means that the endpoint c of $\partial\widetilde{D_n} \setminus \{*\}$ of the lifting of Γ_c is very close to the endpoint $\sigma_1^p(c)$ of the lifting of $\sigma_1^p(\Gamma_c)$. By further contrast, under the action of σ_3, the immobile initial segment of Γ_c is much shorter, and the moving tail, bold in

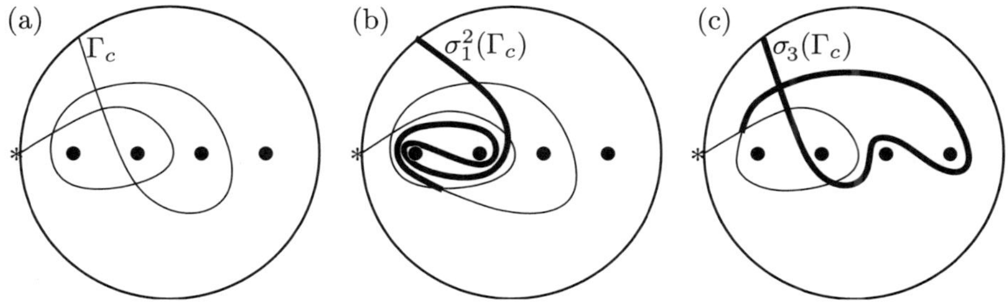

FIGURE 3. We have $1 <_c \sigma_1^2 <_c \sigma_3$, where $<_c$ denotes the ordering associated to the geodesic Γ_c.

Figure 3(c), is much longer. Indeed, it is easy to see that $\sigma_3(c)$ is further removed from c than $\sigma_1^p(c)$ for every p in $\mathbb{N}$. This is saying that we have $\sigma_1^p <_c \sigma_3$ for all p in $\mathbb{N}$—and in a similar fashion one can see that $\sigma_3^q <_c \sigma_2$ holds for all q in $\mathbb{N}$.

On the other hand, we have $\sigma_3^p <_b \sigma_2^q <_b \sigma_1$ for all p, q in $\mathbb{N}$. Indeed, the action of σ_3^p can be said to affect only the tail of Γ_b after the second self-intersection, whereas the effect of a σ_2^q-action is already noticable after the first self-intersection, and σ_1^r right from the start, for p, q, r in $\mathbb{Z}$.

Finally, for $<_d$, one can find homeomorphisms of D_n representing nontrivial elements of B_n which leave arbitrarily long initial segments of Γ_d untouched, which implies that the ordering $<_d$ is dense.

The only part of the proposition that remains to be proven is that $<_b$ and $<_c$ are not conjugate. Let us suppose, for a contradiction, that they are. Since the convex subgroups of a group are linearly ordered by inclusion, this would imply that there exists a subgroup Γ of B_4 which is conjugate in B_4 to the subgroup $\langle \sigma_1, \sigma_3 \rangle$, such that $\langle \sigma_2, \sigma_3 \rangle$ either is included in Γ or includes Γ. The first case is impossible since an Abelian group cannot contain a non-Abelian one. The second case is impossible because the subgroup $\langle \sigma_2, \sigma_3 \rangle$ has support in a subdisk of D_4 which contains only three of the punctures, whereas any subset of D_4 supporting the subgroup $\langle \sigma_1, \sigma_3 \rangle$—or any subgroup conjugate to this one—necessarily contains all four punctures. □

To summarize, the ordering depends critically on how the geodesic cuts up the disk: if two of the punctures are first separated from the two others, and then each of the pairs is split, we obtain an ordering which is qualitatively different from the ordering induced by a geodesic which splits off one puncture at a time.

There is, however, one more effect that can lead to two different geodesics inducing different orderings, even though the two sequences of more and more finely chopped subdisks of D_n induced by cutting along the geodesics are topologically indistinguishable. This effect is illustrated in Figure 4.

There are initial parts of the geodesics Γ_c and $\Gamma_{c'}$ (drawn in solid lines) which separate the first two punctures from the third and the fourth, and the pieces that result from the two cuts are topologically indistinguishable. The only difference between the curves is that the *direction* of the cut is opposite. This difference, however, suffices to make the orderings different: we can observe immediately in the picture that $\sigma_2^{-1}\sigma_1^{-1}\sigma_3\sigma_2 >_c 1$ holds, whereas $\sigma_2^{-1}\sigma_1^{-1}\sigma_3\sigma_2 <_{c'} 1$ does.

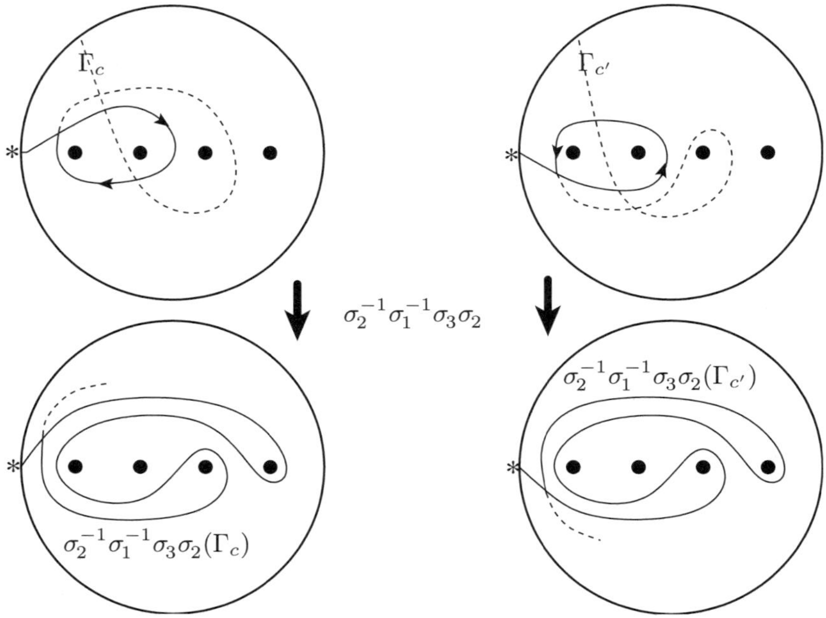

FIGURE 4. The geodesics Γ_c and $\Gamma_{c'}$ cut the disk D_n into pieces in very similar ways, yet $\sigma_2^{-1}\sigma_1^{-1}\sigma_3\sigma_2 >_c 1$ whereas $\sigma_2^{-1}\sigma_1^{-1}\sigma_3\sigma_2 <_{c'} 1$.

Indeed, using results that we shall present later on, one can prove that the geodesics Γ_c and $\Gamma_{c'}$—including the dotted parts—induce orderings which are not even conjugate. This completes our first set of examples: we have seen why different geodesics can induce different orderings of B_n.

1.4. Examples of coinciding Nielsen–Thurston orderings. We are now ready for the second set of examples. We shall see that geodesics which on superficial inspection appear to have no resemblance can give rise to the exact same orderings. This happens if they cut the disk D_n into pieces in essentially the same way.

At the same time we shall see that the σ-ordering considered throughout the book is a special case of an ordering arising from our Nielsen–Thurston type construction. This yields the seventh equivalent definition of the σ-positive ordering mentioned in the Introduction: we consider the ordering associated with the geodesic Γ_b, and its obvious generalization to disks with *any* finite number of punctures. We saw earlier that $\sigma_3^p <_b \sigma_2^q <_b \sigma_1$ holds for all natural numbers p and q. In fact, we have the following stronger result.

PROPOSITION 1.8. (i) *Let Γ_b be the geodesic in D_n indicated in Figure* 5(a). *Then the ordering $<_b$ coincides precisely with the σ-ordering, i.e., for all braids β, β' in B_n we have $\beta <_b \beta'$ if and only if the endpoint of the lifting of $\beta(\Gamma_b)$ is smaller (as a real number) than the endpoint of the lifting of $\beta'(\Gamma_b)$.*

(ii) *The same statement holds for the geodesic $\Gamma_{b'}$ in Figure* 5(b).

PROOF. Throughout the proof we shall refer to Figure 5. The proof is exactly the same for the two geodesics Γ_b and $\Gamma_{b'}$. In both cases, the geodesic has an initial segment up to the first self-intersection which separates the leftmost puncture from

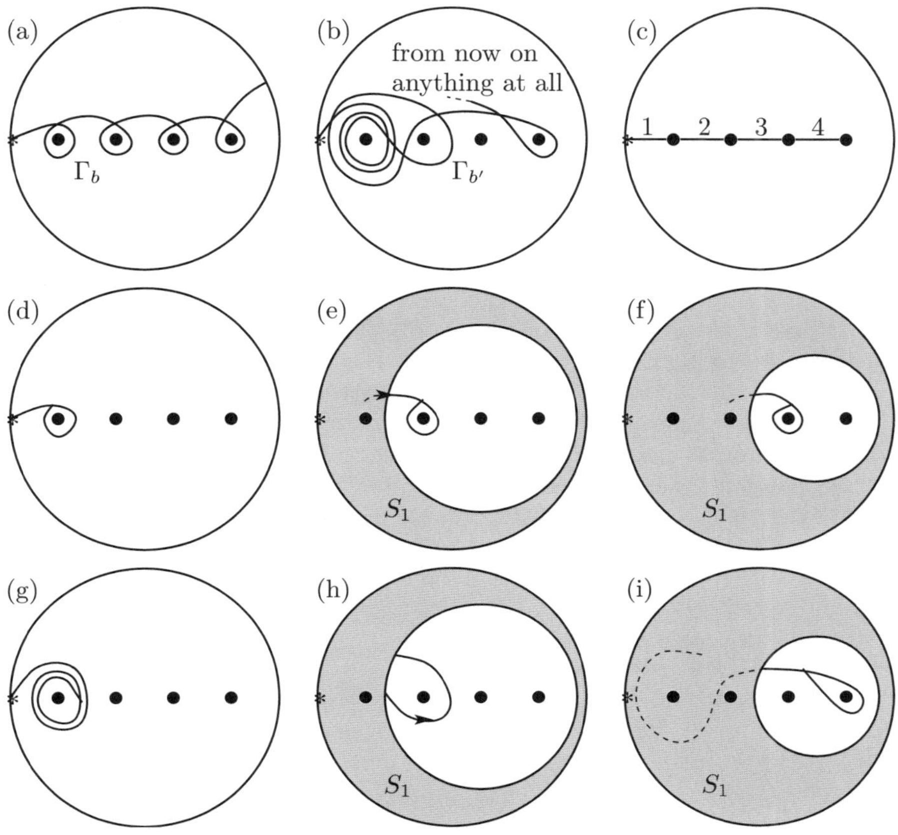

FIGURE 5. Two different geodesics Γ_b and $\Gamma_{b'}$ giving rise to the same ordering—in fact, this is exactly the σ-ordering considered thoughout this book. Figures (d)–(f) analyze Γ_b; Figures (g)–(i) analyze $\Gamma_{b'}$.

the other three. Moreover, if we pull tight the loop around the first puncture, we obtain exactly the first curve in the trivial curve diagram of Figure 5(c).

Considering the action of a braid β on these initial segments, we observe that the braid sends the segments further to the left if and only if it sends the first curve of the curve diagram in Figure 5(c) to the left. In other words, the relation $\beta >_b 1$ or $\beta >_{b'} 1$ holds if and only if β admits a σ_1-positive braid word representative—here we are using the characterisation of the σ-ordering given in Chapter X. In the same way, for braids β which possess a σ_1-negative representative word, we have $\beta <_b 1$ and $\beta <_{b'} 1$. Finally, the inital segments of Γ_b and $\Gamma_{b'}$ are fixed up to isotopy if and only if the braid admits a σ_1-free representative word, *i.e.*, one without a letter σ_1 or σ_1^{-1}. In summary, the action on the initial segments in Figures 5(d) or 5(g) already tell us whether a braid admits a σ_1-positive or a σ_1-negative representative word, and we only need to consider the later parts of the curves if the braid admits a σ_1-free representative word.

So from now on we suppose that the braid β admits a σ_1-free representative. Then the braid β is guaranteed to have a representative in the mapping class group whose support is disjoint from the region with geodesic boundary drawn shaded in Figures 5(e) and 5(h). Thus it is completely irrelevant for the induced ordering what the geodesic does between the first self-intersection and the moment when it leaves the shaded region.

When the geodesic enters the region that contains the three rightmost punctures, we consider its segment up to the next self-intersection or the moment it reenters the shaded region—whichever happens first. These segments for Γ_b and $\Gamma_{b'}$, drawn with bold line in Figures 5(e) and 5(h), both separate the second puncture from the third and fourth, and they do so in two ways which are essentially equivalent to each other. The precise sense in which the bold lines in Figures 5(e) and 5(h) are equivalent is the following: if in each of the figures we add a neighbourhood of the bold line to the shaded region and then delete the bold lines themselves, we obtain two isotopic figures.

Since the braid β admits a σ_1-free representative braid word, it induces a braid on three strings, which has a representative that acts only on the nonshaded disk containing punctures number 2, 3 and 4. We can now use exactly the same argument as in the first step. The bold line is sent further to the left or right, respectively, if and only if curve number 2 in the trivial curve diagram of Figure 5(c) is sent to the left or right, respectively. We remark that the cut is clockwise around the second puncture in Figure 5(e) but counterclockwise in 5(h)—however, we observe that this difference in orientation makes no difference for the induced ordering if only one puncture is being cut off.

If the bold lines in Figures 5(e) and 5(h) are fixed up to isotopy, *i.e.*, if the braid β admits a representative braid word which is σ_2-free as well, then we consider the only region of D_n which can still support a nontrivial element of B_n, namely the disk with geodesic boundary containing punctures number 3 and 4, as shown in Figures 5(f) and 5(i). Again, it is irrelevant for the induced ordering what the homotopy class of the geodesic is between the last point of the bold arc in the previous step and the first intersection point with the disk around the last two punctures. For instance, the part of the curve $\Gamma_{b'}$ which is drawn as a dotted arc in Figure 5(i) may be replaced by an arbitrarily complicated arc in the shaded region of Figure 5(i) without any effect on the induced ordering.

Finally, the bold arcs in Figures 5(f) and 5(i) separate the remaining two punctures. We recall that we are now considering the case where the braid β can be represented by a word σ_3^p with p in $\mathbb{Z}$. We observe that the arcs drawn in bold in Figures 5(f) and 5(i), as well as the arc number (3) in the curve diagram (c), are sent further to the left in the case $p > 0$, are stabilized in the case $p = 0$, and are sent further to the right in the case $p < 0$. □

REMARK 1.9. There are orderings of B_n which cannot be obtained by our Nielsen–Thurston type construction. For instance, one can consider the exponent sum homomorphism $\epsilon\colon B_n \to \mathbb{Z}$, and define a left-invariant ordering $<_\epsilon$ by declaring that $\beta <_\epsilon \beta'$ is true if we have either $\epsilon(\beta) < \epsilon(\beta')$, or $\epsilon(\beta) = \epsilon(\beta')$ and $\beta < \beta'$ (in the σ-ordering). One can prove that the ordering $<_\epsilon$ is not conjugate to any ordering of Nielsen–Thurston type.

2. The classification of orderings induced by the action on $\mathbb{R}$

In this section we state the main classification theorems for orderings arising from our action of B_n on the real line, and outline the proof. The explanation given here, using certain sequences of subsurfaces of D_n, is not quite the same as the one in [**183**], which uses curve diagrams. However, the two approaches are essentially equivalent.

2.1. The outline of the classification. We start with a very rough classification of Nielsen–Thurston type orderings into two classes: those arising from geodesics of *finite* and *infinite type.*

DEFINITION 2.1. A geodesic Γ_x in D_n *fills* the disk D_n, or is *filling,* if $D_n \setminus \Gamma_x$ has no path-connected component that contains two or more of the punctures of D_n.

For instance, the geodesics in Figure 2(b), (c), and (d) are filling whereas the one in (a) is not. We shall only be interested in geodesics filling D_n, because we know from Lemma 1.6 that those that are not filling do not give rise to linear orderings.

DEFINITION 2.2. A filling geodesic Γ is of *finite type* if one of the following two conditions is satisfied: either a finite initial segment of the geodesic already separates the punctures, or the geodesic *falls into a puncture*, in the sense that one of the punctures has the property that all but a finite initial segement of Γ lies in its cusp neighbourhood. A filling geodesic which is not of finite type is of *infinite type.*

For instance, the geodesics in Figure 2(b) and (c) are of finite type, and the one in Figure 2(d) is of infinite type. We stress that a finite type geodesic is not necessarily finite in length; for instance, a geodesic that falls into a puncture is infinite. Also, the geodesic $\Gamma_{b'}$ in Figure 5(b) is of finite type, regardless of whether it terminates after finite time on ∂D_n or continues forever.

So a first, very rough classification of orderings arising from our action of B_n on $\mathbb{R}$ is as follows: they are all induced by filling geodesics, and each such geodesic is either of finite or infinite type. We shall see that orderings arising from the two types of geodesics have very different properties, and we will treat the two cases separately.

2.2. Finite type geodesics. We start by looking at finite type geodesics. We are going to construct a geometrical invariant of such geodesics which contains just enough information about the geodesic in order to specify the induced ordering, but no more.

DEFINITION 2.3. A *subsurface sequence* is a finite sequence $\mathcal{S}_0, \mathcal{S}_1, \dots, \mathcal{S}_{n-1}$ of open connected submanifolds of D_n satisfying $\mathcal{S}_i \subsetneqq \mathcal{S}_{i+1}$ for $i = 0, \dots, n-2$. Moreover, we require that $\mathcal{S}_0$ is a regular neighbourhood of ∂D_n, and that $\mathcal{S}_{n-1}$ is D_n. We also require that, for $i > 0$, all components of $\partial\overline{\mathcal{S}_i}$, where $\overline{\mathcal{S}_i}$ is the closure of $\mathcal{S}_i$, are simple closed geodesics, one of which is ∂D_n. Finally, some of the surfaces must carry a certain extra structure, which will be specified below.

We shall not indicate the submanifold $\mathcal{S}_0$, which is just an annular neighbourhood of ∂D_n, in our pictures.

In the above situation, the surface $\mathcal{S}_i$ must be homeomorphic to a disk with $i+1$ holes (here the punctures are considered to be holes)—in particular $\mathcal{S}_{i+1}$ is obtained from $\mathcal{S}_i$ by adding one boundary component of the closure $\overline{\mathcal{S}_i}$ and, along this boundary component, an open surface homeomorphic to a disk with two holes. Thus if we compare the number of boundary components of $\overline{\mathcal{S}_i}$ to those of $\overline{\mathcal{S}_{i+1}}$, we can see three possible effects:

(i) either $\overline{\mathcal{S}_{i+1}}$ has one less boundary component than $\overline{\mathcal{S}_i}$, as in the transition from (c) to (d) in Figure 6; or

(ii) the number of boundary components remains constant, as in the transition from (e) to (f) in Figure 5; or

(iii) it may increase by one, similar to the transition from (b) to (c) in Figure 6.

DEFINITION 2.3 (CONTINUED). In case (iii), *i.e.*, when $\overline{\mathcal{S}_{i+1}}$ has two boundary components which were not present in $\overline{\mathcal{S}_i}$ while one boundary component of $\overline{\mathcal{S}_i}$ has disappeared, there must be one more piece of information present, namely a transverse orientation to a geodesic segment that connects the two new boundary components of $\overline{\mathcal{S}_{i+1}}$ inside $\mathcal{S}_{i+1} \setminus \mathcal{S}_i$.

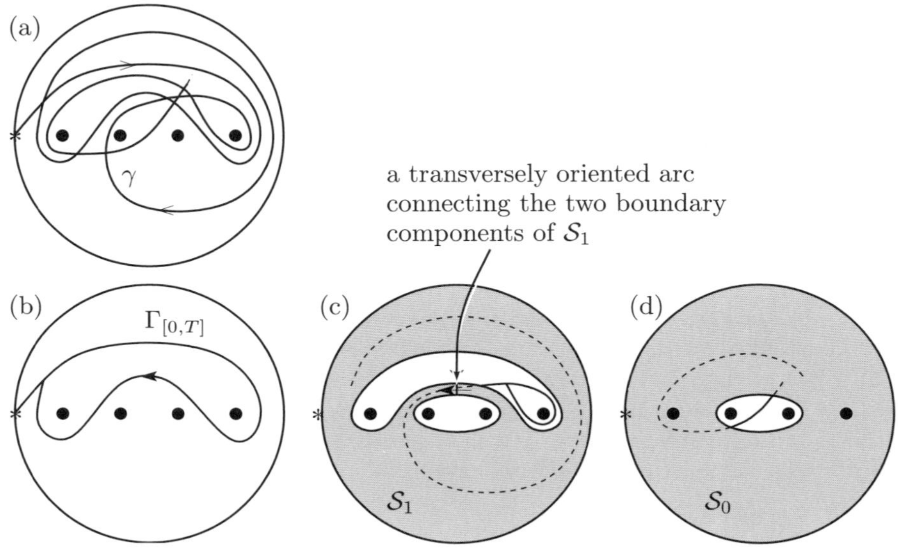

FIGURE 6. An example of the subsurface sequence associated to a geodesic Γ.

Examples of subsurface sequences can be found in Figures 5(d)–(f), 5(g)–(i) and 6(b)–(d).

DEFINITION 2.4. Two subsurface sequences $\mathcal{S}_0, \dots, \mathcal{S}_{n-1}$ and $\mathcal{S}'_0, \dots, \mathcal{S}'_{n-1}$ are *conjugate* if there exists a homeomorphism φ of D_n such that $\varphi(\mathcal{S}_i)$ is isotopic to $\mathcal{S}'_i$ for $i = 1, \dots, n-1$. Moreover, if $\mathcal{S}_i$ comes equipped with a transversely oriented geodesic arc, then the same must be true for $\mathcal{S}'_i$, and φ must carry (up to isotopy) the transversely oriented arc of $\mathcal{S}_i$ to the one of $\mathcal{S}'_i$, preserving the transverse orientation.

Next, using Figure 6, we explain how a finite or infinite geodesic $\Gamma\colon [0, M] \to D_n$ or $\Gamma\colon [0, \infty) \to D_n$ gives rise to a subsurface sequence. For t, t' in the domain of Γ, we denote by $\Gamma_{[t,t']}$ the restriction of Γ to the interval $[t, t']$. Let T be the smallest real number so that the initial segment $\Gamma_{[0,T]}$ of Γ has a self-intersection. Then $D_n \setminus \Gamma_{[0,T]}$ has two path components, each of which contains at least one of the punctures of D_n. We take an open regular neighbourhood $\widetilde{S}_1$ of $\partial D_n \cup \Gamma_{[0,T]}$, and define the surface $\mathcal{S}_1$ to be the unique subsurface of D_n which is homotopic to $\widetilde{S}_1$ and whose closure has geodesic boundary components. Note that, if $D_n \setminus \Gamma_{[0,T]}$ has a component which contains only one puncture, then $\mathcal{S}_1$ contains that entire component; see for instance Figure 5(e) and (h), where the boundary components of $\widetilde{S}_1$ that encircled the leftmost puncture disappeared in $\mathcal{S}_1$. Thus, if both components of $D_n \setminus \Gamma_{[0,T]}$ contain only one puncture, then we are in case (i) above. If one component contains one puncture and the other contains more, then we are in case (ii), and if both of components of $D_n \setminus \Gamma_{[0,T]}$ contain two or more punctures of D_n, then we are in case (iii). In the last case, $\mathcal{S}_1$ has two boundary components, which we connect by a geodesic segment in $\mathcal{S}_1$—this segment is unique up to isotopy. We equip this geodesic segment with the transverse orientation induced from the orientation of $\Gamma_{[0,T]}$. For instance, in Figure 6 the orientation of the curve in (b) induces the transverse orientation on the geodesic arc in (c). This completes the construction of $\mathcal{S}_1$.

The construction of the whole sequence of subsurfaces is now inductive. Let T' be the smallest positive real number with the property that $\Gamma(T') \notin \mathcal{S}_1$. The segment $\Gamma_{[T,T']}$ of Γ can be completely ignored. Furthermore, the point $\Gamma(T')$ lies in the boundary of one of the components of $D_n \setminus \mathcal{S}_1$, which are punctured disks. We look at the shortest initial segment of $\Gamma_{[T',\infty)}$ which either shows a self-intersection or intersects $\overline{\mathcal{S}_1}$ again. We add an open neighbourhood of this segment to $\mathcal{S}_1$, and apply a homotopy to the resulting subsurface so that its closure has geodesic boundary. This homotopy may kill some boundary components of $\overline{\mathcal{S}_1}$, as in the previous paragraph. This procedure yields the surface $\mathcal{S}_2$, etc.

Now, we have the following classification result for the finite type orderings:

PROPOSITION 2.5. (i) *If Γ_x is a filling geodesic of finite type, then it induces a linear ordering, i.e.,* Stab(x) *is trivial. Two filling geodesics of finite type Γ_x and Γ_y induce the same ordering if and only if they give rise to the same subsurface sequence. Moreover, Γ_x and Γ_y induce conjugate orderings if and only if their subsurface sequences are conjugate.*

(ii) *All orderings arising from filling geodesics of finite type are discrete.*

PROOF (SKETCH). (See [**183**] for full details.) The essential observation is that one can reconstruct the ordering $<_x$ from the subsurface sequence associated to Γ_x. Indeed, suppose that β is an element of B_n, and we want to decide which of $\beta >_x 1$, $\beta = 1$, or $\beta <_x 1$ is true. We start by choosing the maximal integer i such that β has a representative homeomorphism with support disjoint from $\mathcal{S}_i$. Then we let Γ be an embedded geodesic segment in $\mathcal{S}_{i+1}$ with endpoints in $\partial\overline{\mathcal{S}_i}$, which cannot be transformed by an homotopy into $\overline{\mathcal{S}_i}$: up to a movement of the endpoints in $\partial\overline{\mathcal{S}_i}$, there is only one such segment. Now we restrict our attention to the component D' of $D_n \setminus \mathcal{S}_i$ which contains $\mathcal{S}_{i+1} \setminus \mathcal{S}_i$. The segment Γ cuts the punctured disk D' into two components. If one of these components contains only one of the punctures of D_n, then we choose an orientation for Γ arbitrarily. If both components contain at least two punctures, then our geodesic segment Γ intersects exactly once the transversely oriented geodesic segment in $\mathcal{S}_{i+1}$ that came with the subsurface sequence. This transverse orientation induces an orientation of Γ. Now we recall that a representative of β acts on the punctured disk D' in a boundary-fixing way. We compare the geodesic Γ with a geodesic representative of $\beta(\Gamma)$ in D', and, more precisely, we compare their initial segments with respect to the chosen orientation on Γ. The key observation, which is proved in [**183**], is that $\beta >_x 1$ is true if and only if $\beta(\Gamma)$ branches off to the left of Γ, and $\beta <_x 1$ is true if and only if $\beta(\Gamma)$ branches off to the right of Γ. The case $\beta(\Gamma) = \Gamma$ cannot occur, because, if it did, β would have a representative homeomorphism with support disjoint from all of $\mathcal{S}_{i+1}$.

In particular, if we denote by $B_{\mathcal{S}}$ the subgroup of B_n consisting of elements which have representative homeomorphisms with support in a submanifold S of D_n, then we have a hierarchy of convex subgroups

$$\{1\} = B_{D_n \setminus \mathcal{S}_{n-1}} \subseteq B_{D_n \setminus \mathcal{S}_{n-2}} \subseteq \ldots \subseteq B_{D_n \setminus \mathcal{S}_1} \subseteq B_{D_n} = B_n.$$

The ordering we have just constructed is linear. With a little more work one can prove [**183**] that two different subsurface sequences give rise to different orderings. Thus there is indeed a one-to-one correspondence between subsurface sequences and order types. This completes the proof of (i).

In order to see that the ordering induced by a finite type geodesic Γ_x is discrete, as claimed in (ii), we observe that $D_n \setminus \mathcal{S}_{n-2}$, the complement of the last proper subsurface in the sequence, consists exactly of one twice-punctured disk. We consider the element β of B_n which can be represented by a homeomorphism with support in $D_n \setminus \mathcal{S}_{n-2}$, which exchanges these two punctures, namely the Dehn half-twist along an arc in $D_n \setminus \mathcal{S}_{n-2}$ connecting the punctures—see Definition X.2.5. This is the smallest element of B_n satisfying $\beta >_x 1$. □

We deduce the number of conjugacy classes of orderings of finite type on B_n:

PROPOSITION 2.6. *For $n \geqslant 2$, the number N_n of conjugacy classes of orderings of finite type of B_n is given by the recursive formula*

$$N_1 = 1, \quad N_2 = 1, \quad \text{and} \quad N_n = \sum_{k=1}^{n-2} \binom{n-2}{k-1} N_k \, N_{n-k}. \tag{2.1}$$

PROOF. It is easy to show constructively that every subsurface sequence can be obtained as the subsurface sequence associated to some finite type geodesic. Therefore, it suffices to prove that the number of conjugacy classes of subsurface sequences is given by (2.1). For our proof, it is more convenient to rewrite the formula as

$$N_2 = 1, \quad \text{and} \quad N_n = N_{n-1} + \sum_{k=2}^{n-2} \binom{n-2}{k-1} N_k \, N_{n-k}. \tag{2.2}$$

It is this formula that we shall prove.

The proof is by induction. For $n = 2$, (2.2) is obvious, since all our orderings must satisfy $\sigma_1 > 1$, and there is only one ordering of $B_2 \cong \langle \sigma_1 \rangle \cong \mathbb{Z}$ satisfying this condition.

Assuming that (2.2) holds for fewer than n punctures, we shall try to count conjugacy classes of subsurface sequences in D_n. There are two possibilities to be considered: either $\mathcal{S}_1$ has one boundary component in the interior of D_n or two, corresponding to cases (ii) and (iii) in Definition 2.3. The two cases will correspond also to the two summands in (2.2).

In the first case we can, after a suitable conjugation, suppose that $\mathcal{S}_1$ surrounds the leftmost puncture, and $\partial\overline{\mathcal{S}_1}$ contains no point left of the leftmost puncture. Then there are, up to conjugacy, N_{n-1} ways left to complete the subsurface sequence in the remaining disk $D_n \setminus \mathcal{S}_1$, which contains the punctures number $2, \ldots, n$.

In the second case, we can by a suitable conjugation achieve that $\mathcal{S}_1$ is isotopic to a neighbourhood of ∂D_n, together with a neighbourhood of a vertical line between punctures number k and $k+1$ for $2 \leqslant k \leqslant n-2$, and moreover that the transverse orientation on the horizontal geodesic segment that connects the two boundary components of $\overline{\mathcal{S}_1}$ in the interior of D_n points upwards. The number k is uniquely determined by these requirements. Thus we have found a way to conjugate a given subsurface sequence such that $\mathcal{S}_1$ is of some canonical type, and we have to classify the possible ways of continuing the subsurface sequence. A subsurface sequence must contain $n-2$ more elements $\mathcal{S}_2, \dots, \mathcal{S}_{n-1}$. Among the pieces $\mathcal{S}_i \setminus \mathcal{S}_{i-1}$, exactly $k-1$ must lie in the left half, *i.e.*, in the component of $D_n \setminus \mathcal{S}_1$ which contains punctures number $1, \dots, k$, and $n-k-1$ in the right half, *i.e.*, in the component of $D_n \setminus \mathcal{S}_1$ which contains punctures number $k+1, \dots, n$. There are $\binom{n-2}{k-1}$ ways to distribute the $k-1$ steps in the left half over the $n-2$ steps that are left to be made. Moreover, in the left half there are N_k, and in the right half N_{n-k} different subsurface sequences. Thus in the second case there are $\binom{n-2}{k-1}\, N_k\, N_{n-k}$ different subsurface sequences once the choice of $\mathcal{S}_1$ has been made. □

For instance, we get $N_2 = 1, N_3 = 1$, $N_4 = 3$, $N_5 = 9$, $N_6 = 39$, $N_7 = 189$ and $N_8 = 1107$—this last value is misprinted in [**183**]. The three different conjugacy classes of orderings of B_4 are represented by the geodesics Γ_b of Figure 5 (the associated ordering is the σ-ordering), and Γ_c and $\Gamma_{c'}$ of Figure 4.

It is an interesting fact that this sequence is known in combinatorics [**160**] in the following form. If we denote $\nu_n = N_{n+1}$, then ν_n is the number of permutations on n letters without double descent and without initial descent, *i.e.*, those permutations π such that $\pi(i) > \pi(i+1)$ implies $\pi(i) > \pi(i-1)$. Its generating function—see [**195**]—$f(x) = \sum_{n\geqslant 0} \nu_n \frac{x^n}{n!}$ satisfies the differential equation $f'(x) = 1 - f(x) + (f(x))^2$, and can even by given explicitly by

$$f(x) = \frac{1 + \frac{1}{\sqrt{3}} \tan(\frac{\sqrt{3}}{2} x)}{1 - \frac{1}{\sqrt{3}} \tan(\frac{\sqrt{3}}{2} x)}.$$

This finishes our discussion of finite type geodesics.

2.3. Infinite type geodesics. We now turn our attention to filling geodesics of infinite type. Such geodesics are necessarily infinite, because they fill the disk D_n while no finite initial segment does. Contrary to what may be suggested by Figure 2(d), such geodesics may have some self-intersections, but only finitely many. More precisely, if $\Gamma\colon [0,\infty) \to D_n$ is a filling geodesic of infinite type, then there exists a T in $\mathbb{R}_+$ such that all the self-intersections of Γ occur in the initial segment $\Gamma_{[0,T]}$. Cutting the disk D_n along this initial segment, one obtains some simply connected pieces, some pieces containing exactly one puncture, and exactly one piece that contains two or more punctures—in fact we shall see that this last component necessarily contains at least three punctures. The geodesic $\Gamma([T,\infty))$ separates the punctures in this last component, but no finite initial segment does; actually, it looks like an infinite type geodesic without self-intersections. Therefore, we can restrict our attention to such geodesics without self-intersection.

There is a large body of literature that helps us understand what a filling infinite type geodesic Γ without self-intersections must look like. Indeed, let us consider the bi-infinite, oriented path in D_n obtained by running along the geodesic Γ in the opposite direction (terminating on ∂D_n), followed by one turn around the circle ∂D_n, followed by the path Γ again, this time in the same direction as Γ. The closure of the unique geodesic isotopic to this path is a *geodesic lamination* of D_n in the sense of Nielsen–Thurston theory [**34**].

DEFINITION 2.7. A *subsurface sequence of infinite type* is a finite sequence $\mathcal{S}_1, \ldots, \mathcal{S}_k$, with $k < n-1$, of open connected submanifolds of D_n such that $\mathcal{S}_1, \ldots, \mathcal{S}_{k-1}$ satisfy the same conditions as the elements of a subsurface sequence according to Definition 2.3. Moreover, the surface $D_n \backslash \overline{\mathcal{S}_{k-1}}$ must be connected, *i.e.*, homeomorphic to a disk with $n-k+1$ punctures. The surface $\mathcal{S}_k$ must include $\mathcal{S}_{k-1}$, and $\mathcal{S}_k \setminus \mathcal{S}_{k-1}$ must be one complementary region of a geodesic lamination in the punctured disk $D_n \setminus \mathcal{S}_{k-1}$. In particular, the frontier of $\mathcal{S}_k$ must form a geodesic lamination in $D_n \setminus \mathcal{S}_{k-1}$.

From any filling geodesic of infinite type, one can construct a subsurface sequence of infinite type: the construction procedure is entirely analogous to the finite type case.

Much is known about the nature of geodesic laminations and their behaviour under the action of B_n [**34, 171**]. All we need to know for our purposes are the following facts: There are uncountably many geodesic laminations for D_n for $n \geqslant 3$, whereas for $n = 2$ there are only the simple closed curves. For any element β of B_n, the action of β on D_n stabilizes either zero or two geodesic laminations—and if there are two, then β is said to be pseudo-Anosov, and the two laminations are called the stable and unstable laminations of β.

Then we have the following classification result for infinite type orderings.

PROPOSITION 2.8. (i) *All orderings arising from infinite type geodesics are dense.*

(ii) *All but countably many of the uncountably many geodesics of infinite type induce linear orderings of B_n, i.e., all but countably many infinite type geodesics Γ have the property that* $\mathrm{Stab}(\Gamma)$ *is trivial.*

(iii) *Two geodesics Γ_x, Γ_y of infinite type induce the same ordering if and only if they give rise to the same subsurface sequence. Two geodesics Γ_x, Γ_y of infinite type induce conjugate orderings if and only if their subsurface sequences are conjugate.*

(iv) *There exist uncountably many different orderings of B_n which arise from infinite type geodesics, and also uncountably many conjugacy classes of such orderings.*

PROOF (SKETCH). Let Γ_x be a filling geodesic of infinite type. For the proof of (i), we notice that for an arbitrarily long initial segment of Γ_x one can find a geodesic arc connecting two punctures which is disjoint from the initial segment of Γ_x. So in order to find an element β of B_n such that $\beta(x)$ is arbitrarily close to x, it suffices to take a homeomorphism representing a nontrivial element of B_n with support in a neighborhood of the geodesic arc, for instance a Dehn half-twist along that arc.

For the proof of (ii), we recall that the group B_n is countable, and that each element of B_n stabilizes at most two geodesic laminations. Moreover, only countably many geodesics of infinite type can give rise to the same subsurface sequence. It follows that there can only be countably many geodesics which are stabilized by a nontrivial element of B_n.

Point (iii) is proved in [**183**]. The idea of the proof is quite similar to the finite type case. No proof will be given here.

As for (iv), the fact that there are uncountably many orderings induced by geodesics of infinite type follows immediately from statements (ii) and (iii). Indeed, there are even uncountably many conjugacy classes of such orderings, because we have the countable group B_n acting by conjugation on our uncountable set of orderings. The set of orbits is still uncountable. □

3. The subword property for all Nielsen–Thurston type orderings

The approach taken in this chapter yields a very natural proof of the left-orderability of B_n. However, it is not well adapted to proving Properties **A** and **C** in isolation, which refer specifically to σ_1-positive and σ_1-negative braid words; in this chapter, we shall not pursue these two properties further. Nevertheless, it is interesting and satisfying to see that all orderings of Nielsen–Thurston type do satisfy the subword property, *i.e.*, the counterpart of what was called Property **S** in the case of the σ-ordering:

PROPOSITION 3.1. *If $<_x$ is the ordering arising from our action of B_n on a point x of $\partial\widetilde{D}_n$ with* $\mathrm{Stab}(x) = \{1\}$, *then we have $\sigma_1\beta >_x \beta$.*

As the σ-ordering is a particular Nielsen–Thurston ordering, and every σ_i is a conjugate of σ_1, we deduce

COROLLARY 3.2 (Property **S**). *Every braid of the form $\beta^{-1}\sigma_i\beta$ is σ-positive.*

Proposition 3.1 is an immediate consequence of the following stronger result:

LEMMA 3.3. *If x is any point in $\partial\widetilde{D}_n \setminus \{*\}$ (which is homeomorphic to $\mathbb{R}$), then we have $\sigma_1(x) \geqslant x$ for the ordering induced by the ordering of $\mathbb{R}$.*

Roughly speaking, this means the following. Let us suppose that we sit at the point $*$ of $\partial\widetilde{D}_n$ and look toward $\partial\widetilde{D}_n \setminus \{*\}$. During a σ_1-action, we will see a homeomorphism of $\partial\widetilde{D}_n \setminus \{*\}$ which moves some points further to the left and leaves others fixed, but no points will jump to the right.

PROOF OF LEMMA 3.3 (SKETCH). The homeomorphism σ_1 of D_n is, up to isotopy, a Dehn half-twist along a geodesic arc e connecting the first two punctures. The preimage $\widetilde{e}$ of e in $\widetilde{D}_n$ consists of an infinite number of geodesics, each connecting two points of $\widetilde{D}_n \cap S^1_\infty$, which are preimages of the punctures under the projection p. We include those preimages in $\widetilde{e}$.

As before, let $\widetilde{\Gamma}_x$ be a geodesic in $\widetilde{D}_n$ connecting $*$ to x, and let Γ_x be its projection. Three cases are possible.

Case 1. The geodesic $\widetilde{\Gamma}_x$ is disjoint from $\widetilde{e}$. This means that Γ_x may be assumed to be disjoint from the support of the Dehn half-twist σ_1, hence we have $\sigma_1(x) = x$.

Case 2. The geodesic $\widetilde{\Gamma}_x$ intersects $\widetilde{e}$ transversely. We pay attention to the preimage e_0 of e that is met first when we go along $\widetilde{\Gamma}_x$ starting from $*$. Let x' and x'' be the endpoints of e_0, with $x' < x''$. Since e_0 intersects $\widetilde{\Gamma}_x$ we have $x' < x < x''$. From the definition of the Dehn half-twist, one can see that the lifting of σ_1 maps x' to x''. Indeed, the image of the geodesic $\widetilde{\Gamma}_{x'}$ connecting $*$ to x' will be a curve that goes along $\widetilde{\Gamma}_{x'}$ almost to the end, then turns left and goes along e_0 to x''. Since the σ_1-action on $\mathbb{R}$ is order-preserving, we have $\sigma_1(x) > \sigma_1(x') = x'' > x$.

Case 3. The geodesic $\widetilde{\Gamma}_x$ does not intersect $\widetilde{e}$ transversely but goes from $*$ to an endpoint of a geodesic contained in e. This means that Γ_x terminates at one of the first two punctures without intersecting the interior of the arc e. There will be exactly one geodesic e_0 in $\widetilde{e}$ connecting x', *i.e.*, x to some x'' satisfying $x'' > x'$ such that one can pass inside $\widetilde{D_n}$ from $*$ to any interior point of e_0 without intersecting $\widetilde{e}$ meanwhile. As in the previous case, the lifting of σ_1 maps $x = x'$ to x'', and we have $\sigma_1(x) = x'' > x' = x$. □

REMARK 3.4. We end this chapter with an extended general remark. At first glance, the geometrical approach to braid ordering developed in this chapter has two disadvantages. First, it is unnatural in the sense that it requires a choice of hyperbolic metric on the punctured disk D_n—and in particular it uses a nontrivial result from analysis, namely the uniformization theorem—but the set of orderings obtained by our construction is independent of this choice. Second, the geometrical approach seems to work only for B_n with finite n, not for B_∞. This makes it natural to wonder if these disadvantages could not be avoided by translating our geometric approach back into a combinatorial setting.

The answer to this question is that such a combinatorial generalisation is indeed possible, and we have already seen it: this is exactly the material of Section IX.3 concerning work of J. Funk [**93**].

Let us summarize this section from the point of view of the current chapter. One can define an ordering of the free group, interpreted as the fundamental group of the punctured disk $\pi_1(D_n)$. In this ordering, one element of the fundamental group is larger than a second one if a loop representing the first element goes more to the left than a loop representing the second. In other words, the relative order of two elements of $\pi_1(D_n)$ can be decided by lifting them to the universal cover and looking at the relative order of their endpoints in the real line $\partial\widetilde{D_n} \setminus \{*\}$. Now the braid group acts on this free group $\pi_1(D_n)$, as seen in Chapter IX, and it does so in an order-preserving fashion. So we can pull back the ordering on an orbit to obtain an ordering on the braid group. For instance, in Section IX.3 we considered the orbit of a loop resembling the curve Γ_b in Figure 2 from this chapter. This alternative approach to Nielsen–Thurston type orderings is interesting for at least two more reasons: it makes the geometric ideas described in the current chapter amenable to explicit computation, and it points towards connections of braid orderings with the theory of toposes [**93**].

CHAPTER XIV

The Space of all Braid Orderings

In this chapter, we consider the totality of all possible orderings of a given group, with special emphasis on the braid groups. The idea of putting a topology on this set has been considered for several years by such experts as E. Ghys, A. Sikora and others; its first appearance in the literature seems to be the paper [**187**] by Sikora. This approach has already led to significant advances in the study of orderable groups, for example in [**156**].

The space of (left-invariant) orderings is, for many groups, homeomorphic with the classical Cantor set, but we will see that the braid groups are an exception, although their spaces of orderings do contain certain natural Cantor sets. One advantage of this point of view is that it gives us another way, different from the method of Chapter XIII, to prove the existence of uncountably many distinct—in fact nonconjugate—orderings of the braid groups, all of which have the subword property, and therefore well-order the braid monoid B_n^+.

This short chapter is organized as follows. We gather a few general results about the space of orderings of a group in Section 1, whereas, in Section 2, we concentrate on the specific case of braid groups.

1. The spaces of orderings on a group

Here we will define a natural topology for the set of all left-invariant orderings on a left-orderable group. We describe the action of the group and its automorphisms by homeomorphisms on that space.

1.1. A topology on the powerset.

We begin by recalling some basic point-set topology. If X is a set, we use the notation $\{0,1\}^X$ to denote the set of all subsets of X, which is in one-to-one correspondence with the set of all functions from X to the set $\{0,1\}$. A subset Y of X corresponds to the function that maps x to 1 if and only if x lies in Y.

Being a special case of a product space, in which $\{0,1\}$ has discrete topology, we equip $\{0,1\}^X$ with the product topology: a basic open set consists of all functions in $\{0,1\}^X$ with specified values on a specified *finite* subset of X and arbitrary values for all other x in X. In the language of subsets, a basic open set is obtained by choosing a finite subset $\{x_1, \dots, x_p, y_1, \dots, y_q\}$ of X; the corresponding open set is then $\{S \in \{0,1\}^X \mid x_1 {\in} S, \dots, x_p {\in} S, y_1 {\notin} S, \dots, y_q {\notin} S\}$. We allow the possibility that the set of x_i or y_j, or both, are empty.

By a famous theorem of Tychonoff, since $\{0,1\}$ is a compact space, $\{0,1\}^X$ is also compact. In addition, $\{0,1\}^X$ is totally disconnected, meaning that any two points lie in disjoint open sets whose union is the whole space. To see this, consider subsets S_1 and S_2 of X. If S_1 and S_2 are distinct, then there exists x_0 in S_1 such that $x_0 \notin S_2$—or *vice-versa*. Then the two open sets $\{S \subseteq X \mid x_0 \in S\}$ and $\{S \subseteq X \mid x_0 \notin S\}$ of $\{0,1\}^X$ separate S_1 and S_2, and their union is all of $\{0,1\}^X$.

If X is a countably infinite set, we can fix an enumeration $x_1, x_2, \ldots$ of X and define the distance between two subsets A, B of G to be the sum of terms 2^{-k} for all k such that x_k belongs to the symmetric difference of A and B. In this way, $\{0,1\}^X$ becomes a metric space.

We recall that every nonempty compact metric space which is totally disconnected and has no isolated point is homeomorphic to the Cantor set; see, for example, [**110**, Corollary 2.98]. Therefore, we have the following criterion.

LEMMA 1.1. *If X is a countably infinite set and S is a closed nonempty subset of $\{0,1\}^X$, then S is homeomorphic to the Cantor set if and only if S has no isolated point.*

1.2. A topology for the set of all orderings of a group. We turn to the specific case of left-invariant orderings of a group.

DEFINITION 1.2. If G is a group, the set of all left-invariant orderings of G is denoted $LO(G)$, and the set of all bi-invariant orderings of G is denoted $O(G)$.

In what follows, we identify a left-invariant ordering $\prec$ of G with its positive cone P, and we will pass freely from one point of view to the other, sometimes referring to the cone P as an ordering of the group. Recall that a subset P of G is the positive cone of a left-invariant ordering precisely when the following hold:

(i) $P \cdot P \subseteq P$;

(ii) $P \cap P^{-1} = \emptyset$; and

(iii) $G \setminus \{1\} = P \cup P^{-1}$.

Thus we can consider $LO(G)$ as a subset of the collection $\{0,1\}^G$ of subsets of G. Then, it is natural to equip $LO(G)$ with the topology induced by the product topology on $\{0,1\}^G$ considered above. Note that this topology is the smallest topology so that, for any given $g, h \in G$, the set $U_{g,h}$ of left-invariant orders $\prec$ of G satisfying $g \prec h$, is an open set. This open set is identified with the set of all positive cones P for G satisfying $g^{-1}h \in P$. Then the following is clear [**187**]—an alternative argument can be found in [**41**].

PROPOSITION 1.3. *For every group G, the spaces $LO(G)$ and $O(G)$ are compact, totally disconnected spaces and $O(G) \subseteq LO(G) \subseteq \{0,1\}^G$ are inclusions of closed subsets.*

PROOF. As the space $\{0,1\}^G$ is compact and totally disconnected, it suffices to see that $LO(G)$ and $O(G)$ are closed in $\{0,1\}^G$, *i.e.*, that *not* being a positive cone is an open condition. Now P fails to satisfy condition (i) if there exist g, h in P such that gh does not belong to P, *i.e.*, if P belongs to the open set

$$\bigcup_{g,h \in G} \{P \mid g \in P,\ h \in P,\ gh \notin P\}.$$

Similarly, P fails to satisfy conditions (ii) or (iii) if it belongs to the open sets

$$\bigcup_{g \in G} \{P \mid g \in P,\ g^{-1} \in P\} \quad \text{or} \quad \bigcup_{g \in G \setminus \{1\}} \{P \mid g \notin P,\ g^{-1} \notin P\},$$

respectively. So $\{0,1\}^G \setminus LO(G)$ is the union of three open sets, and it follows that $LO(G)$ is closed in $\{0,1\}^G$. Similarly, the condition that the left-invariant ordering associated with P fails to be right-invariant means that there exist g, h in G satisfying $g \in P$ and $hgh^{-1} \notin P$, again an open condition. □

By Lemma 1.1, if G is infinite and countable, $\{0,1\}^G$ is a metric space homeomorphic to the standard Cantor set, and, therefore, the spaces $LO(G)$ and $O(G)$ may be viewed as closed subspaces of the Cantor set.

Note that, to specify a neighbourhood of a positive cone P, we do not need to consider conditions of the form $g \notin P$, as this is equivalent to the statement $g^{-1} \in P$.

DEFINITION 1.4 ([**6**]). An ordering $\prec$ in $LO(G)$ is said to be *finitely determined* if there is a finite subset $\{g_1, \dots, g_k\}$ of G such that $\prec$ is the *unique* left-invariant ordering of G satisfying $1 \prec g_i$ for $i = 1, \dots, k$.

For instance, if the positive cone of $\prec$ is finitely generated as a semigroup, then $\prec$ is finitely determined. Clearly, a left-invariant ordering is finitely determined if and only if it is not a limit point of $LO(G)$, so we conclude

PROPOSITION 1.5. *For a countable group G, the space $LO(G)$ is homeomorphic to the Cantor set if and only if it is nonempty and no left-invariant ordering of G is finitely determined.*

Left-orderable groups having only finitely many left-invariant orderings have been classified by Tararin—see for instance [**127**]. The simplest example is the infinite cyclic group, which has two left-invariant orderings. A less trivial example is the Klein group $\langle a, b; aba^{-1} = b^{-1}\rangle$, which has four left-invariant orderings. Notice that, if a group has only a finite number of left-invariant orderings, then each of them is finitely determined. On the other hand, according to Linnell [**139**] (see also [**162**]), the number of left-invariant orderings on a group cannot be countably infinite. For the case of Abelian groups, Sikora [**187**] proved

PROPOSITION 1.6. *If G is a countable free Abelian group of rank greater than 1 that is torsion-free, the space $LO(G)$ is homeomorphic to the Cantor set—and it is of course equal to $O(G)$ in this case.*

Notice that Abelian groups are left-orderable if and only if they are torsion-free. This is also the case for nilpotent groups. Actually, for countable torsion-free nilpotent groups, the result of Proposition 1.6 still holds; see [**162**].

Sikora made the following conjecture.

CONJECTURE 1.7. *Let F_n denote the free group of rank n. For $n \geqslant 2$, both $LO(F_n)$ and $O(F_n)$ are homeomorphic to the Cantor set.*

There exist two independent proofs showing that $LO(F_n)$ is homeomorphic to the Cantor set. The first [**154**] uses the theory of lattice-ordered groups, while the second [**162**] uses analysis of the dynamics of the real line associated with an ordering. The case of $O(F_n)$ seems to be open at the time of this writing.

We will prove the analogue of the conjecture for infinitely generated free groups in the next section by a rather trivial argument, but it is one which illustrates some key concepts of this point of view.

1.3. Actions on $LO(G)$. Assume that φ is a group automorphism of G. Then for every left-invariant ordering $\prec$ of G, there is a corresponding left-invariant ordering $\prec_\varphi$, defined by $x \prec_\varphi y$ if and only if $\varphi(x) \prec \varphi(y)$. It is easy to see that the mapping $\prec \mapsto \prec_\varphi$, and its inverse, are continuous. Moreover, we have $(\prec_\varphi)_\psi = \prec_{\varphi\psi}$, so this defines a natural action of $\mathrm{Aut}(G)$ on $LO(G)$ by homeomorphisms. Restricted to the cyclic subgroup generated by a single nontrivial automorphism φ of G, for example, we have an action of $\mathbb{Z}$ on $LO(G)$. The fixed set of this action consists of all orderings which are φ-invariant, *i.e.*, satisfy $\varphi(x) \prec \varphi(y) \Leftrightarrow x \prec y$.

This applies in particular to the case of inner automorphisms of G, leading to the notion of conjugate orderings already introduced in Definition XIII.1.5. If $g \in G$ is fixed and $\prec$ is an element of $LO(G)$, define $\prec_g$ by $x \prec_g y$ if and only if $xg^{-1} \prec yg^{-1}$ for all x, y in G. One easily checks that for fixed $g \in G$, the function $\prec \mapsto \prec_g$ is a homeomorphism of $LO(G)$, and the equality $(\prec_g)_h = \prec_{gh}$ holds. In other words, this defines a right-action of G on $LO(G)$. It is really a conjugation, because $x \prec_g y$ is equivalent to $gxg^{-1} \prec gyg^{-1}$. The subspace of $LO(G)$ fixed by all g in G under this action is exactly $O(G)$.

This action was used by D. Witte Morris [**156**] in a beautiful proof that, for *amenable* groups, left-orderability is equivalent to local indicability; see Definition XV.5.1 and the accompanying discussion. We will not pursue amenability here, except to mention that braid groups (on three or more strands) are not amenable, as they contain non-Abelian free subgroups. For five or more strands, they are not locally indicable; see Section XV.5.1.

In terms of positive cones, the reader may easily check that, if P is the positive cone for $\prec$, then the positive cone for $\prec_\varphi$ is $\varphi^{-1}(P)$. In particular, for the conjugated ordering $\prec_g$, the positive cone is the conjugate $g^{-1}Pg$.

Moreover, an ordering $\prec$ is bi-invariant if and only if $\prec_\varphi$ is also bi-invariant, so the automorphism action restricts to an action on $O(G)$.

As an illustration of these ideas, we will prove the following easy proposition. Let F_∞ denote the free group with a countably infinite basis.

PROPOSITION 1.8. *The spaces $LO(F_\infty)$ and $O(F_\infty)$ have no isolated points, and hence are homeomorphic to the Cantor set.*

PROOF. Denote the generators of F_∞ by $x_1, x_2, \ldots$. Suppose P is the positive cone for a left-invariant ordering $\prec$ of F_∞. Let U be a basic neighbourhood of P in $LO(F_\infty)$. That is, choose a finite subset S of P and take U to be the set of all positive cones which also include this subset. Each element of S involves only finitely many generators, hence we may choose n so that x_n does not appear in any element of S. There is an automorphism φ of F_∞ satisfying $\varphi(x_n) = x_n^{-1}$, and $\varphi(x_k) = x_k$ for $k \neq n$. Then $\prec_\varphi$ is a left-invariant ordering of F_∞ whose positive cone $\varphi^{-1}(P)$ also includes S. Yet it is distinct from $\prec$, since x_n is positive in one ordering if and only if it is negative in the other. We see that any neighbourhood of $\prec$ contains a different left-invariant ordering, so $\prec$ is a limit point. The same argument works for $O(F_\infty)$. □

2. The space of left-orderings of the braid groups

At the moment of this writing, our understanding of the structure of the space $LO(B_n)$ of left-invariant orderings of the braid groups is only fragmentary. It is interesting in that, unlike the free or free Abelian case, it is *not* homeomorphic to the Cantor set for finite n. However, it *is* homeomorphic to the Cantor set for $n = \infty$.

2.1. The σ-ordering is not isolated. Our goal will be to prove that the σ-ordering of B_n is not isolated in $LO(B_n)$; in fact we will show that it is a limit point of its conjugates. The result is established by Navas in **[162]** through a general approach for studying left-invariant orderings on general left-orderable groups. Here we provide a shorter argument which is more explicit, in that, for each basic open neighbourhood of the σ-ordering, it gives a method for finding a braid that will conjugate it into a different ordering which lies in that neighbourhood.

NOTATION 2.1. Hereafter, we denote by P_n the positive cone associated with the σ-ordering of B_n, *i.e.*, the set of all σ-positive n-strand braids.

We begin with the case of 3-strand braids for simplicity and because it implies the general case. We start with an easy preparatory result.

LEMMA 2.2. *If β_1 is a σ_1-positive braid in B_3, there exists a σ_1-positive braid β that satisfies $1 < \beta \leqslant \beta_1$ and does not commute with σ_2.*

PROOF. If β_1 does not commute with σ_2, we may take $\beta = \beta_1$. Otherwise, as noted in the proof of Proposition II.3.13, there must exist integers p, q satisfying $\beta_1 = (\sigma_1\sigma_2\sigma_1)^{2p}\sigma_2^q$, and the hypothesis that β_1 is σ_1-positive implies that p is positive. In this case, we may take, for example, $\beta = \sigma_1\sigma_2$. Indeed, we find $\beta^{-1}\beta_1 = \sigma_1(\sigma_1\sigma_2\sigma_1)^{2p-1}\sigma_2^q$. Hence $\beta^{-1}\beta_1$ is σ_1-positive, and $\beta < \beta_1$ is true. □

PROPOSITION 2.3. *The σ-ordering of the braid group B_3 is a limit point of its conjugates in $LO(B_3)$.*

PROOF. Given a finite subset S of P_3, we need to find a braid β satisfying
(i) $S \subseteq \beta P_3 \beta^{-1}$, and
(ii) $P_3 \neq \beta P_3 \beta^{-1}$.
Let β_1 be the $<$-smallest σ_1-positive element of $S \cup \{\sigma_1\}$. Applying Lemma 2.2, we find β that satisfies $1 < \beta \leqslant \beta_1$ and does not commute with σ_2. Let γ be any element of S. If γ is a power of σ_2, then $\beta^{-1}\gamma\beta$ is σ-positive by Property **S**. Otherwise, by hypothesis, we have $\beta \leqslant \beta_1 \leqslant \gamma$, hence $1 \leqslant \beta^{-1}\gamma$ and, *a fortiori*, $1 < \beta^{-1}\gamma\beta$ since β is σ-positive. So we have $\beta^{-1}S\beta \subseteq P_3$, *i.e.*, $S \subseteq \beta P_3 \beta^{-1}$, which proves (i).

For (ii), note that the least positive element of $\beta P_3 \beta^{-1}$ is $\beta\sigma_2\beta^{-1}$. If $\beta P_3 \beta^{-1}$ were equal to P_3, we would deduce $\beta\sigma_2\beta^{-1} = \sigma_2$, contradicting the choice of β. □

As Navas observed, the proof of Proposition 2.3 implies the general case.

PROPOSITION 2.4. *For $n \geqslant 3$, the σ-ordering of B_n is a limit point of its conjugates in $LO(B_n)$.*

PROOF. Let H be the subgroup of B_n generated by σ_{n-2} and σ_{n-1}. Then H is isomorphic to B_3 and the σ-ordering of B_n restricted to H corresponds with the σ-ordering of B_3. Moreover, the positive cone for the σ-ordering of H is $P_n \cap H$. Let S be a finite subset of P_n. By the proof of Proposition 2.3, there exists β in H such that $\beta^{-1}\gamma\beta$ is σ-positive for every γ in $S \cap H$, and $\beta(P_n \cap H)\beta^{-1}$ is distinct from $P_n \cap H$. Assume now $\gamma \in S \setminus H$. By hypothesis, γ must be σ_i-positive for some $i < n-2$. Its conjugate $\beta^{-1}\gamma\beta$ is σ_i-positive as well, hence σ-positive. We deduce $\beta^{-1}S\beta \subseteq P_n$, hence $S \subseteq \beta P_n \beta^{-1}$. Noting that $\beta P_n \beta^{-1}$ and P_n are distinct, because their intersections with H are distinct, we conclude as in Proposition 2.3 that P_n is the limit of all its conjugates $\beta P_n \beta^{-1}$. □

COROLLARY 2.5. *The positive cone P_n of B_n is not finitely generated as a semigroup.*

2.2. Isolated orderings of B_n. We shall now see that the space of left-invariant orderings of a braid group is quite different from that of a free group.

PROPOSITION 2.6. *The space $LO(B_n)$ has isolated points, and hence is not homeomorphic to the Cantor set.*

We saw above that the σ-ordering is a limit point in the space $LO(B_n)$, hence it cannot be used to prove Proposition 2.6. To do it, we shall consider other orderings of the braid groups introduced by Dubrovina and Dubrovin in [**71**]. The positive cones of these orderings are finitely generated as semigroups, and are therefore isolated points in $LO(B_n)$. For the sake of clarity, we will consider the case $n = 3$ in some detail.

PROPOSITION 2.7. *Let P_{DD} be the set of all 3-strand braids that are either σ_1-positive or σ_2-negative. Then P_{DD} is a positive cone and, as a semigroup, it is generated by $\sigma_1\sigma_2$ and σ_2^{-1}.*

PROOF. Let $\beta_1 = \sigma_1\sigma_2$, $\beta_2 = \sigma_2^{-1}$, and let Q be the subsemigroup of B_3 generated by β_1 and β_2. Clearly, β_1 and β_2 belong to P_{DD}, and so Q is included in P_{DD}. We need to show that if a 3-strand braid β is not 1, then exactly one of β or β^{-1} belongs to Q, according as $\beta \in P_{DD}$ or not.

Case 1. One of β or β^{-1} is σ_2-positive. In this case β equals σ_2^p, for some nonzero p and so β belongs to Q for $p < 0$, whereas β^{-1} belongs to Q for $p > 0$.

Case 2. The braid β is σ_1-positive. This means that there are integers $m_1, \dots, m_k$ satisfying

$$\beta = \sigma_2^{m_1}\, \sigma_1\, \sigma_2^{m_2} \sigma_1 \dots \sigma_1\, \sigma_2^{m_k}.$$

The identity $\sigma_1 = \beta_1\beta_2$ allows us to rewrite this as

$$\beta = \beta_2^{p_1}\, \beta_1\, \beta_2^{p_2}\, \beta_1 \dots \beta_1\, \beta_2^{p_k},$$

for some integers p_i. Now using the equality $\beta_1 = \beta_2\, \beta_1^2\, \beta_2$, which is easily checked, we may express β as a product

$$\beta = \beta_2^{q_1}\, \beta_1^{r_1}\, \beta_2^{q_2}\, \beta_1^{r_2} \dots \beta_1^{r_{k-1}}\, \beta_2^{q_k}$$

with all q_i and r_i nonnegative integers. This shows that β belongs to Q.

Case 3. The braid β is σ_1-negative. Then β^{-1} is σ_1-positive and we proceed as in Case 2 to show that β^{-1} belongs to Q. ☐

To complete the picture of $LO(B_3)$, it can be mentioned that the σ-ordering of B_3 is a limit of the conjugates of the P_{DD}-ordering of Proposition 2.7 [**162**]. Proving this amounts to showing that, for each finite subset S of B_3, there exists a braid β such that every braid in $\beta^{-1}S\beta$ has the same sign in the σ-ordering and the P_{DD}-ordering. As many conjugates of σ_2^{-1} are σ_1-negative, this is easily achieved.

Dubrovin and Dubrovina similarly define for each $n \geqslant 3$ an ordering of B_n whose positive cone is finitely generated as a semigroup. Indeed, they prove that the subsemigroup of B_n generated by

$$(\sigma_1\sigma_2 \dots \sigma_{n-1}), (\sigma_2\sigma_3 \dots \sigma_{n-1})^{-1}, (\sigma_3\sigma_4 \dots \sigma_{n-1}), \dots, (\sigma_{n-2}\sigma_{n-1})^{(-1)^{n-1}}, (\sigma_{n-1})^{(-1)^n},$$

is a positive cone and, therefore, the associated ordering is an isolated point in the space $LO(B_n)$.

2.3. A Cantor set within $LO(B_n)$. We shall now prove that, although the space $LO(B_n)$ is not itself a Cantor set, nevertheless, for $n \geqslant 3$, it contains Cantor sets in a natural way. We recall that P_n denotes the positive cone of the σ-ordering of B_n.

NOTATION 2.8. For $n \geqslant 3$, we put $Z_n = \{\beta^{-1}P_n\beta \mid \beta \in B_n\}$.

By Proposition 2.4, P_n is a limit point in Z_n, and so Z_n is an infinite set. As B_n is countable, so is Z_n. Note that, because conjugation is a homeomorphism of $LO(B_n)$, every point of Z_n is a limit point of Z_n. We now consider the closure of Z_n in $LO(B_n)$.

PROPOSITION 2.9. *For $n \geqslant 3$, let $\bar{Z}_n$ denote the closure of Z_n in $LO(B_n)$.*

(i) *The space $\bar{Z}_n$ is homeomorphic to the Cantor set.*

(ii) *Each of the uncountably many orderings in $\bar{Z}_n$ has the subword property, that is, the positive cone contains every braid of the form $\beta\sigma_i\beta^{-1}$—and hence all of B_n^+.*

(iii) *Each of these orderings is a well-ordering when restricted to B_n^+.*

(iv) *The space $\bar{Z}_n$ contains uncountably many conjugacy classes of orderings.*

PROOF. (i) Each point of $\bar{Z}_n$ is a limit point, and $\bar{Z}_n$ is a totally disconnected nonempty compact metric space, so it is homeomorphic to the Cantor set by the characterization mentioned earlier. To check (ii), we note that P_n has the subword property—this is Property **S**—and so does each point of Z_n as, by definition, the subword property is preserved under conjugation. Then, it is enough to check that satisfying the subword property is a closed condition. Now the set of all orderings of B_n that fail to satisfy the subword property is the open set

$$\bigcup_{\beta\in B_n}\bigcup_{i=1}^{n-1}\{P \in LO(B_n) \mid \beta\sigma_i\beta^{-1} \notin P\}.$$

Part (iii) is proved from the subword property as in Proposition II.4.1. For (iv), simply note that each conjugacy class is countable, and the countable union of countable sets cannot be uncountable, as $\bar{Z}_n$ happens to be. □

2.4. The case of B_∞. We will now show that the space $LO(B_\infty)$ is homeomorphic to the Cantor set, contrary to the spaces $LO(B_n)$ for finite n.

PROPOSITION 2.10. *The space $LO(B_\infty)$ has no isolated points, and hence is homeomorphic to the Cantor set. Moreover, each element of $LO(B_\infty)$ is a limit point of its conjugates.*

PROOF. Consider an arbitrary positive cone P for a left-invariant ordering of B_∞ and suppose S is a finite subset of P. We will show there is a positive cone $\sigma_i P\sigma_i^{-1}$ in B_∞ that also includes S and is distinct from P.

Choose n such that S is included in B_n. Then, for each $i > n$, every braid in S commutes with σ_i, so we have $S = \sigma_i S\sigma_i^{-1} \subseteq \sigma_i P\sigma_i^{-1}$.

On the other hand, we claim that there exists $i > n$ such that the sets P and $\sigma_i P\sigma_i^{-1}$ are different. For otherwise, we consider the subgroup $\mathrm{sh}^n(B_\infty)$, which is isomorphic to B_∞. The sets $P \cap \mathrm{sh}^n(B_\infty)$ and $(\sigma_i P\sigma_i^{-1}) \cap \mathrm{sh}^n(B_\infty)$ are positive cones for orderings of $\mathrm{sh}^n(B_\infty)$. If $P = \sigma_i P\sigma_i^{-1}$ is true for each $i > n$, then the cone $P \cap \mathrm{sh}^n(B_\infty)$ of $\mathrm{sh}^n(B_\infty)$ is invariant under conjugation by all elements of $\mathrm{sh}^n(B_\infty)$.

This would imply that $\mathrm{sh}^n(B_\infty)$ and, therefore, B_∞ are bi-orderable, which is not true. □

CHAPTER XV

Bi-ordering the Pure Braid Groups

We saw in Chapter II that the full braid group B_n is left-orderable, but not bi-orderable for $n \geqslant 3$. In this chapter, we will see that the *pure* braid group PB_n, a normal subgroup of B_n of index $n!$, can be given an ordering invariant under multiplication on both sides. The key is that free groups are bi-orderable, and PB_n is a semidirect product of free groups, according to Artin's combing technique.

With appropriate choice of conventions, the ordering has the property that positive pure braids—expressible in the generators σ_i using only positive exponents—are all greater than 1. We will also see that the set PB_n^+ of all positive pure n-strand braids is well-ordered under this ordering, and that its order type is the ordinal ω^{n-1}.

The ordering we will describe for PB_n is radically different from those defined for B_n in earlier chapters. It is natural to ask if there is a possible uniform ordering: a left-ordering of B_n which restricts to a bi-invariant ordering of PB_n. Perhaps surprisingly, the answer is that this is impossible.

The chapter is organized as follows. In a very short first section, we observe that pure braid groups must be bi-orderable owing to a general criterion involving the lower central series. In Section 2, we describe the Artin combing of pure braids and the Magnus expansion of a free group, which are then used in Section 3 to construct the concrete bi-invariant ordering of PB_n which is the main subject of the chapter. In Section 4, we investigate the restriction of the pure braid ordering to positive braids. Finally, the negative results about extending orderings from PB_n to B_n are explained in Section 5.

1. Lower central series

First we recall the definition of the pure braid groups.

DEFINITION 1.1. A braid β is said to be *pure* if the permutation associated with β is the identity. The set of all pure n-strand braids is denoted by PB_n.

Thus PB_n is the kernel of the canonical morphism of B_n onto $\mathfrak{S}_n$. Hence PB_n is a normal subgroup of B_n, and there is an exact sequence

$$1 \longrightarrow PB_n \longrightarrow B_n \longrightarrow \mathfrak{S}_n \longrightarrow 1.$$

The bi-orderability of PB_n follows from the work of Falk and Randell in [**80**], which shows that pure braid groups satisfy the hypothesis of the following proposition. We recall that the definition of the lower central series associated with a group G,

$$G = G_0 \supseteq G_1 \supseteq G_2 \supseteq \dots,$$

is given inductively by $G_{n+1} = [G_n, G]$, the group generated by commutators $hgh^{-1}g^{-1}$, with h in G_n and g in G. These are normal subgroups of G, and the quotient groups G_n/G_{n+1} are Abelian.

PROPOSITION 1.2. *Suppose G is a group which is residually nilpotent, meaning $\bigcap G_i = \{1\}$, and such that each G_n/G_{n+1} is torsion-free. Then G is bi-orderable.*

PROOF (SKETCH). It is straightforward to bi-order countable torsion-free Abelian groups, so take $\prec_n$ to be an arbitrary bi-invariant ordering of G_n/G_{n+1}. For any distinct elements g, h in G, let $N(g,h)$ be the greatest n such that $g^{-1}h$ belongs to G_n, so it represents a nontrivial class $[g^{-1}h]$ of G_N/G_{N+1}. Note that $N(h,g) = N(g,h)$ always holds. Define $g \prec h$ if and only if $1 \prec_{N(g,h)} [g^{-1}h]$. Then $\prec$ is a bi-invariant ordering on G. □

COROLLARY 1.3 ([**80**]). *For each n, the pure braid group PB_n is bi-orderable.*

We omit the proof, as the result will be proved later in this chapter.

2. Artin coordinates and Magnus expansion

We prefer another approach to constructing a bi-invariant ordering on PB_n, which has the advantage of being more explicit and of defining a well-ordering of PB_n^+. The current section lays the groundwork for this construction, following [**124**].

2.1. Artin combing. The standard inclusion $B_{n-1} \subseteq B_n$ restricts to the pure braid groups: $PB_{n-1} \subseteq PB_n$. However, in the case of pure braid groups, the inclusion has a left inverse.

DEFINITION 2.1. (See Figure 1.) For $n \geqslant 3$, we denote by r_n the mapping of PB_n to PB_{n-1} that corresponds to erasing the last strand.

FIGURE 1. The retraction r_3: erasing the third strand in $\sigma_1^2\sigma_2^2\sigma_1^2$ yields σ_1^4. As we consider *pure* braids, this is a homomorphism of PB_n onto PB_{n-1}.

LEMMA 2.2. *The mapping r_n is a homomorphism of PB_n onto PB_{n-1}. The kernel F_{n-1} of r_n is the set of pure n-strand braids that can be represented by a diagram in which the first $n-1$ strands go straight across, and it is a free group of rank $n-1$.*

PROOF. The set of pure n-strand braids representable so that the first $n-1$ strands go straight across is isomorphic to the fundamental group of a plane with $n-1$ points removed. As an induction on n and the van Kampen theorem show, the latter is a free group of rank $n-1$ (see Figure 3). □

We easily deduce the following structure result which connects pure braid groups with free groups. In our context, it will provide an inductive step for deducing an ordering of pure braid groups from an ordering of free groups.

PROPOSITION 2.3. *For each $n \geqslant 2$, the pure braid group PB_n is a semi-direct product of F_{n-1} and PB_{n-1}.*

PROOF. By Lemma 2.2, we have the exact sequence

$$1 \longrightarrow F_{n-1} \xrightarrow{\subset} PB_n \xrightarrow{r_n} PB_{n-1} \longrightarrow 1. \tag{2.1}$$

The inclusion of PB_{n-1} into PB_n is a right inverse for the homomorphism r_n, so the exact sequence of (2.1) is split and the group PB_n is a semidirect product of PB_{n-1} with the free group F_{n-1}. □

The process may be iterated to present PB_n as a semidirect product of the free subgroups $F_1, \dots, F_{n-1}$.

COROLLARY 2.4. *Each pure braid β in PB_n has a unique expression*

$$\beta = \beta_1 \beta_2 \dots \beta_{n-1}, \tag{2.2}$$

where β_i is a braid that admits a representation with all strands straight, except the $(i+1)$st, which can interact only with strands of lower index.

DEFINITION 2.5. (See Figure 2.) For β in PB_n, the braids $\beta_1, \dots, \beta_{n-1}$ of (2.2) are called the *Artin coordinates* of β.

FIGURE 2. Artin coordinates of the pure braid $\sigma_1^2\sigma_2^2\sigma_1^2$: the first coordinate β_1 is what remains when all strands but the first two are forgotten, the second coordinate β_2 is what remains in the remainder when all strands but the first three are forgotten, etc.

For the sequel we need to fix a basis of the free subgroup F_{n-1} of PB_n. Several choices are possible. Here, we take the squares of the generators $a_{i,j}$ used for the dual braid monoid B_n^{+*} in Chapter VIII.

LEMMA 2.6. (See Figure 3.) *For $1 \leqslant i < j \leqslant n$, put*

$$x_{i,j} = a_{i,j}^2 = \sigma_{j-1}^{-1} \dots \sigma_{i+1}^{-1} \sigma_i^2 \sigma_{i+1} \dots \sigma_{j-1}. \tag{2.3}$$

Then, for each j, the braids $x_{1,j}, \dots, x_{j-1,j}$ form a basis of the free subgroup F_j of PB_n.

As F_j is a free group, each Artin coordinate of a pure braid admits a unique reduced expression in terms of the generators $x_{i,j}$—actually this expression is what should be called the coordinate.

The operation of finding the Artin coordinates of a pure braid β of PB_n and expressing them in a fixed basis of each free group is known as the *combing* of β. Figure 4 shows an example. Geometrically, the process can be described as follows. We start with any diagram representing β and view combing as an ambient

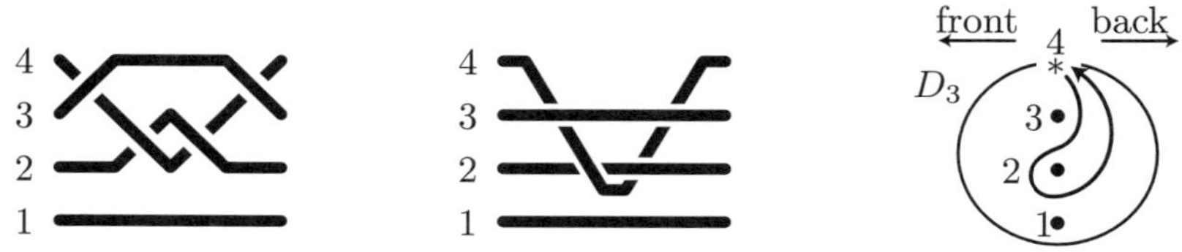

FIGURE 3. The generator $x_{2,4}$—or $a_{2,4}^2$—of PB_4. The 4th strand makes a loop around the second one, passing behind the intermediate third strand: on the left, the traditional representation; in the middle, the isotopic diagram where all strands except the one that makes the loop are pulled straight; on the right, the corresponding loop in the π_1 of a disk with 3 punctures—which corresponds to looking at the picture from the side.

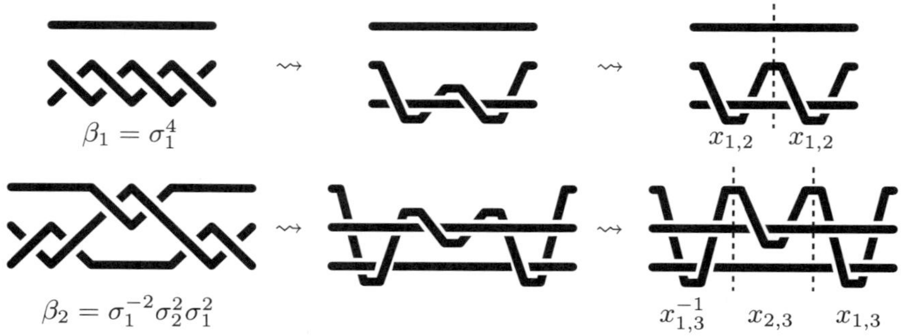

FIGURE 4. Combing of the pure braid $\sigma_1^2\sigma_2^2\sigma_1^2$: expressing the Artin coordinates β_1, β_2 in terms of the generators $x_{i,j}$.

isotopy that slides the crossings to the right. First, by definition of β_1, there is an isotopy which brings the first two strands into the position of β_1. Leaving those two strands fixed and sliding the other strings to the right, we obtain $\beta = \beta_1\beta'$, with β' represented by a diagram in which the first two strings are straight. Then we obtain β_2 and $\beta' = \beta_2\beta''$, with β'' represented by a diagram in which the first three strings are straight, etc. In describing this process Artin admitted that "*any attempt to carry this out on a living person would only lead to violent protests and discrimation against mathematics*".

The previous informal description can be turned into an algorithm that, starting with an expression of a pure braid β in terms of the generators σ_i, returns expressions for the Artin coordinates of β in terms of the generators $x_{i,j}$, but we shall not describe it here; see [**36**].

2.2. The Magnus expansion. If F is a free group, it is not obvious, at first glance, that F is bi-orderable. Actually it is known that F satisfies the hypotheses of Proposition 1.2. But a device of W. Magnus gives a uniform, and pretty, way of defining an ordering.

We denote by $\mathbb{Z}\langle\langle X_1, ..., X_n\rangle\rangle$ the ring of formal power series in n noncommuting indeterminates X_i. Such series are infinite sums of monomials, each of which is a word on the letters X_i, so they have the generic form

$$f = \sum_{W \in \{X_1,\dots,X_n\}^*} f_W W,$$

where $\{X_1, \ldots, X_n\}^*$ denotes the set of all finite length words on the alphabet $\{X_1, \ldots, X_n\}$. The length of the word W is called the *degree* of the monomial $f_W W$. As we consider n noncommutative variables, there exist n^d monomials of degree d. Addition of $\mathbb{Z}\langle\langle X_1, \ldots, X_n\rangle\rangle$ is defined by summing the coefficients, while multiplication is given by

$$\Big(\sum f_W W\Big)\Big(\sum g_W W\Big) = \sum_W \Big(\sum_{UV=W} f_U g_V\Big) W.$$

We use $O(X^k)$ to denote the ideal of $\mathbb{Z}\langle\langle X_1, \ldots, X_n\rangle\rangle$ made of the series involving only monomials of degree $\geqslant k$.

DEFINITION 2.7. Assume that F is a free group and $(x_1, \ldots, x_n)$ is a basis of F. The *Magnus expansion* of F relative to $(x_1, \ldots, x_n)$ is the map

$$\mu : F \longrightarrow \mathbb{Z}\langle\langle X_1, \ldots, X_n\rangle\rangle$$

defined by

$$\mu(x_i) = 1 + X_i, \quad \mu(x_i^{-1}) = 1 - X_i + X_i^2 - X_i^3 + \cdots.$$

EXAMPLE 2.8. For $w = x_1^{-1} x_2 x_1$, we find

$$\begin{aligned}\mu(w) &= (1 - X_1 + X_1^2 - X_1^3 + \cdots)(1 + X_2)(1 + X_1),\\ &= 1 + X_2 - X_1X_2 + X_2X_1 + X_1^2X_2 - X_1X_2X_1 \quad \bmod O(X^4).\end{aligned}$$

PROPOSITION 2.9 ([**144**]). *Assume that F is a free group, and μ is a Magnus expansion of F.*

(i) *The map μ is an injective map of F into $1 + O(X)$.*

(ii) *For each nonnegative k, the Magnus image of the kth term in the lower central series of F is included in $1 + O(X^{k+1})$.*

PROOF. (i) Let $(x_1, \ldots, x_n)$ be the basis of F involved in the definition of μ. Let w be a nontrivial element of F. Then we can write $w = x_{i_1}^{e_1} x_{i_2}^{e_2} \ldots x_{i_\ell}^{e_\ell}$ with $\ell \geqslant 1$, $i_r \neq i_{r+1}$ for r in $\{1, \ldots, \ell - 1\}$ and each e_r not equal to zero. When expanding the series $\mu(w)$, we find that it involves a unique monomial $X_{i_1}X_{i_2}\ldots X_{i_\ell}$, and its coefficient is the product $e_1e_2\ldots e_\ell$, which is not zero. It follows that $\mu(w)$ is not 1.

The proof of (ii) is an easy induction on k, and we leave it to the reader. □

Proposition 2.9(ii) is the key to Magnus' proof that free groups are residually nilpotent, and it will also be the technical property needed in our construction of an ordering on PB_n.

REMARK 2.10. For F a free group based on $(x_1, \ldots, x_n)$, Fox defined linear mappings $\partial/\partial x_i : \mathbb{Z}F \to \mathbb{Z}F$ for $i = 1, \ldots, n$, which are derivations; see [**14**] or [**40**] for details. There is, moreover, an augmentation map $\epsilon : \mathbb{Z}F \to \mathbb{Z}$. One of the utilities of these maps is that they give the coefficients of the Magnus expansion. For w in F, the coefficient of $X_{i_1} \ldots X_{i_r}$ in $\mu(w)$ is given by the appropriate rth partial derivative, followed by the augmentation; that is,

$$\epsilon\Big(\frac{\partial^r w}{\partial x_{i_1} \ldots \partial x_{i_r}}\Big).$$

2.3. Ordering free groups. We can use Magnus expansions to order free groups. First, we order $\mathbb{Z}\langle\langle X_1, ..., X_n\rangle\rangle$ as follows. For each d, the natural ordering $X_1 < ... < X_n$ induces a lexicographical ordering on monomials of total degree d. Therefore, we have a natural increasing enumeration of these monomials. For instance, for $n = d = 2$, the increasing enumeration of the degree 2 monomials is the sequence $(X_1^2, X_1X_2, X_2X_1, X_2^2)$.

DEFINITION 2.11. (i) For $d \geqslant 0$ and f in $\mathbb{Z}\langle\langle X_1, ..., X_n\rangle\rangle$, say $f = \sum f_W W$, we denote by $C_d(f)$ the sequence $(f_{W_1}, \dots, f_{W_N})$, where $W_1, \dots, W_N$ is the increasing enumeration of all degree d monomials. We denote by $c_d(f)$ the sum of all coefficients f_{W_i} in $C_d(f)$.

(ii) For f, g in $\mathbb{Z}\langle\langle X_1, ..., X_n\rangle\rangle$, we declare that $f <^{\text{SumLex}} g$ is true if there exists d such that the sequences $C_{d'}(f)$ and $C_{d'}(g)$ coincide for $d' < d$, and

- we have $c_d(f) < c_d(g)$, or
- we have $c_d(f) = c_d(g)$ and the sequence $C_d(f)$ is lexicographically smaller than the sequence $C_d(g)$, *i.e.*, there is an index k such that the first $k - 1$ entries are the same, and the kth entry in $C_d(f)$ is smaller than the kth entry in $C_d(g)$.

The above comparison procedure is a variant of the so-called `DegLex`-ordering, where one first considers the degree, and, then, a lexicographical ordering inside entries of a given degree. The specificity here is that we give priority to the sum of all coefficients corresponding to a given degree before starting the lexicographic comparison, which explains our terminology.

EXAMPLE 2.12. Let us compare the series f of Example 2.8 with the polynomial $g = 1 + X_2$. In degree 0, there is only the constant monomial, and we find $C_0(f) = C_0(g) = (1)$. In degree 1, the increasing enumeration of the two monomials is X_1, X_2, and we find $C_1(f) = C_1(g) = (0, 1)$. In degree 2, the increasing enumeration of the four monomials is $X_1^2, X_1X_2, X_2X_1, X_2^2$, and we have now

$$C_2(f) = (0, -1, 1, 0), \quad \text{and} \quad C_2(g) = (0, 0, 0, 0).$$

We find $c_2(f) = c_2(g) = 0$, so we compare the sequences $C_2(f)$ and $C_2(g)$ starting from the left. The second entry of f is smaller than that of g: so $f <^{\text{SumLex}} g$ is true.

LEMMA 2.13. *The relation $<^{\text{SumLex}}$ is a linear ordering of $\mathbb{Z}\langle\langle X_1, ..., X_n\rangle\rangle$ that is invariant under addition, and under multiplication on either side by an element of the multiplicative subgroup $1 + O(X)$.*

PROOF. First we claim that $f <^{\text{SumLex}} g$ is equivalent to $g - f >^{\text{SumLex}} 0$. Indeed, with obvious notation, we have $(g - f)_W = g_W - f_W$ for each monomial W, hence $c_d(g - f) = c_d(g) - c_d(f)$ for each degree d. Let d be the smallest degree for which there is a degree d monomial that does not have the same coefficient in f and g. Then d is also the smallest degree for which there is a degree d monomial with a nonzero coefficient in $g - f$. Assume $f <^{\text{SumLex}} g$. Then we have either $c_d(f) < c_d(g)$, hence $c_d(g - f) > 0$, or $c_d(f) = c_d(g)$ and $C_d(f)$ is lexicographically smaller than $C_d(g)$, hence $C_d(g - f)$ is lexicographically larger than the constant sequence $(0, \dots, 0)$. In both cases, we have $g - f >^{\text{SumLex}} 0$, which proves the claim.

Then, as $(g + h) - (f + h) = g - f$ is true, $f <^{\text{SumLex}} g$ implies $f + h <^{\text{SumLex}} g + h$.

For the product, we wish to show that $f <^{\text{SumLex}} g$ implies $fh <^{\text{SumLex}} gh$ for h in $1 + O(X)$. By the above observation, it is enough to prove that $f >^{\text{SumLex}} 0$ implies $fh >^{\text{SumLex}} 0$. Let d be the first degree for which $C_d(f)$ contains at least one nonzero coefficient. For each degree d monomial W, we have $(fh)_W = f_W$, as $f_U = 0$

holds for each proper prefix U of W. So we have $C_d(fh) = C_d(f)$, and, therefore, $fh >^{\text{SumLex}} 0$ is true.

The argument is similar for left invariance. □

Note that the ordering $<^{\text{SumLex}}$ on $\mathbb{Z}\langle\langle X_1, ..., X_n\rangle\rangle$ is not invariant under an arbitrary multiplication, typically by -1.

Using the Magnus expansion, we define an ordering of every finitely generated free group with a prescribed basis—naturally called the *Magnus ordering.*

DEFINITION 2.14. Assume that F is a free group and $(x_1, \dots, x_n)$ is a basis of F. For w, w' in F, we declare that $w <_\mu w'$ is true if we have $\mu(w) <^{\text{SumLex}} \mu(w')$, where μ is the Magnus expansion relative to $(x_1, \dots, x_n)$.

PROPOSITION 2.15. *For each finite rank free group F and each basis $(x_1, \dots, x_n)$ of F, the Magnus ordering of F relative to $(x_1, \dots, x_n)$ is a linear ordering that is invariant under multiplication on both sides.*

PROOF. By Proposition 2.9(i), the Magnus expansion is injective, so the relation $<_\mu$ is a linear ordering on F. Its invariance under multiplication on both sides follows from Lemma 2.13, since, by construction, the image of F_n under the Magnus expansion is included in the multiplicative subgroup $1 + O(X)$. □

EXAMPLE 2.16. Let us compare x_2 and $x_1^{-1}x_2x_1$. The Magnus expansions are

$$\mu(x_2) = 1 + X_2, \quad \mu(x_1^{-1}x_2x_1) = 1 + X_2 - X_1X_2 + X_2X_1 \mod O(X^2).$$

These series have been compared in Example 2.12: the latter is $<^{\text{SumLex}}$-smaller than the former. So we have $x_1^{-1}x_2x_1 <_\mu x_2$.

3. The Magnus ordering of PB_n

Here comes our main construction. We saw in Proposition 2.3 that the pure braid group is a semidirect product of free groups. Having ordered free groups via the Magnus expansion, we shall naturally deduce an ordering of PB_n.

3.1. Bi-ordering extensions. Left-orderability is inherited under extensions, but this is not necessarily true for bi-orderability. However, we have the following useful criterion.

LEMMA 3.1. *Assume we have an exact sequence of groups*

$$1 \longrightarrow N \stackrel{\subset}{\longrightarrow} G \stackrel{p}{\longrightarrow} H \longrightarrow 1,$$

and, moreover, $\prec_N$ is a left-invariant ordering of N and $\prec_H$ is a left-invariant ordering of H. For g, g' in G, declare that $g \prec g'$ is true if we have either $p(g) \prec_H p(g')$ or else $p(g) = p(g')$ and $1 \prec_N g^{-1}g'$.

(i) *The relation $\prec$ is a left-invariant ordering of G.*

(ii) *If $\prec_N$ and $\prec_H$ are bi-invariant orderings, then $\prec$ is a bi-invariant ordering of G if and only if conjugation of N by G is order-preserving, i.e., $f \prec_N f'$ implies $g^{-1}fg \prec_N g^{-1}f'g$ for all f, f' in N and g in G.*

The proof is straightforward and left to the reader.

EXAMPLE 3.2. The Klein bottle group $K = \langle x, y\,;\, x^{-1}yx = y^{-1}\rangle$ fits in an exact sequence $1 \to \mathbb{Z} \to K \to \mathbb{Z} \to 1$, where the infinite cyclic subgroup is generated by y. The group K is therefore left-orderable. However, K cannot be bi-ordered, for such an ordering must be invariant under conjugation and the defining relation would lead to the contradiction that $1 \prec y$ is equivalent to $1 \prec y^{-1}$. The problem here is that the map $y \mapsto y^{-1}$ cannot possibly be order-preserving.

3.2. Preparatory results. We recall the exact sequence

$$1 \longrightarrow F_{n-1} \xrightarrow{\subset} PB_n \xrightarrow{r_n} PB_{n-1} \longrightarrow 1.$$

In view of Lemma 3.1, in order to use the Magnus ordering of F_{n-1} to define a bi-invatiant ordering on PB_n, we need to show that this ordering is invariant under conjugation by elements of PB_n. To this end, we shall use the following result.

LEMMA 3.3. *Assume that φ is an automorphism of a free group F, and let φ_{ab} be the induced automorphism on the Abelianization $F/[F,F]$ of F. If φ_{ab} is the identity, then the Magnus ordering on F (with respect to any fixed basis) is invariant under φ.*

PROOF. It suffices to show that $1 <_{\mu} w$ implies $1 <_{\mu} \varphi(w)$. By Proposition 2.9(ii), $\mu([F,F])$ lies in the subgroup $1 + O(X^2)$ of $\mathbb{Z}\langle\langle X_1, ..., X_n\rangle\rangle$. The hypothesis that φ_{ab} is trivial implies that, for each i, the element $x_i^{-1}\,\varphi(x_i)$ belongs to $[F,F]$, and therefore we have

$$\mu(\varphi(x_i)) = \mu(x_i) \mod O(X^2).$$

Now assume $w \neq 1$. Let d be the smallest positive degree for which there exists a degree d monomial with nonzero coefficient in $\mu(w)$. The series $\mu(\varphi(w))$ is obtained from $\mu(w)$ by replacing each occurrence of X_i by some element of $X_i + O(X^2)$. Then, for each degree d monomial W, we have $\mu(\varphi(w))_W = \mu(w)_W$, and therefore $c_d(\mu(\varphi(w))) = c_d(\mu(w))$ and $C_d(\mu(\varphi(w))) = C_d(\mu(w))$. So $1 <^{\mathrm{SumLex}} \mu(w)$ is equivalent to $1 <^{\mathrm{SumLex}} \mu(\varphi(w))$, *i.e.*, in the Magnus ordering, $1 <_{\mu} w$ is equivalent to $1 <_{\mu} \varphi(w)$. □

Thus we are left with showing that the Abelianization of the conjugacy action of PB_n on F_{n-1} is trivial. This is what the next lemma does.

LEMMA 3.4. *Assume that β lies in PB_n and φ is the automorphism of F_{n-1} defined by $\varphi(x) = \beta x \beta^{-1}$. Then the Abelianized map φ_{ab} is the identity.*

PROOF. We claim that, for each i in $\{1, \dots, n-2\}$, there exists an element w_i in F_{n-1} (depending on β) satisfying

$$\varphi(x_{i,n}) = w_i x_{i,n} w_i^{-1}, \tag{3.1}$$

which implies that φ_{ab} is the identity. As PB_n is generated by F_{n-1} and PB_{n-1}, it is enough to prove the claim when β either belongs to F_{n-1} or belongs to PB_{n-1}. In the first case, we can take $w_i = \beta$ for every i.

For the second case, we shall prove that, for every β in B_{n-1}—and not only in PB_{n-1}—there exists an element w_i in F_{n-1} that satisfies

$$\varphi(x_{i,n}) = w_i x_{\pi(i),n} w_i^{-1}, \tag{3.2}$$

where π is the permutation associated with β. For $\beta = \sigma_i$ with $i \leqslant n-2$, the images of Figure 5 give the relations

$$\varphi(x_{i,n}) = x_{i,n}^{-1} x_{i+1,n} x_{i,n}, \quad \varphi(x_{i+1,n}) = x_{i,n}, \quad \varphi(x_{j,n}) = x_{j,n} \text{ for } j \neq i, i+1,$$

which have the expected form. The case of σ_i^{-1} follows easily, and the case of an arbitrary β then follows from an induction on the length of an expression of β in terms of the generators $\sigma_i^{\pm 1}$. So (3.2) is established. In particular, for β in PB_{n-1}, the permutation π is the identity, and we obtain (3.1). So the lemma is proved. □

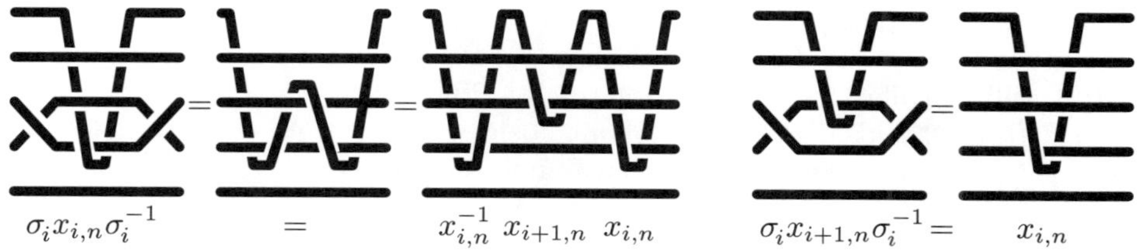

FIGURE 5. Conjugation action of σ_i on the generators $x_{i,n}$ and $x_{i+1,n}$ of the subgroup F_{n-1} of PB_n.

REMARK 3.5. Up to a sign change, the action of B_{n-1} on $\mathrm{Aut}(F_{n-1})$ corresponding to (3.2) is the classical Artin representation of B_{n-1} into $\mathrm{Aut}(F_{n-1})$, which was described in Chapter IX. Also note that, although F_{n-1} is normal in PB_n and PB_n is normal in B_n, the subgroup F_{n-1} is *not* normal in B_n. For example, F_{n-1} is not closed under conjugation by σ_{n-1}.

By gathering Lemmas 3.3 and 3.4, we obtain

PROPOSITION 3.6. *For each n, the Magnus ordering of F_{n-1} is invariant under conjugation by PB_n.*

CONVENTION 3.7. Above, and everywhere hereafter, when we speak of *the* Magnus expansion of the free subgroup F_j of PB_n, and of *the* Magnus ordering, we always refer to the preferred basis $(x_{1,j}, \ldots, x_{j-1,j})$.

3.3. Ordering the pure braid groups. We now have the ingredients for bi-invariant ordering PB_n inductively.

DEFINITION 3.8. For β, β' in PB_2, we declare that $\beta <_{\mathrm{M},2} \beta'$ is true if we have $\beta = \sigma_1^{2e}$ and $\beta' = \sigma_2^{2e'}$ with $e < e'$. For $n \geqslant 3$, and β, β' in PB_n, we declare that $\beta <_{\mathrm{M},n} \beta'$ is true if we have
- either $r_n(\beta) <_{\mathrm{M},n-1} r_n(\beta')$,
- or $r_n(\beta) = r_n(\beta')$ and $r_n(\beta)^{-1}\beta <_\mu r_n(\beta)^{-1}\beta'$.

The relation $<_{\mathrm{M},n}$ is called the *Magnus relation* on PB_n.

Note that the above recursive definition is formally similar to those established in Chapters VII and VIII for the σ^Φ-ordering of B_n^+ and B_n^{+*}.

Then we have the expected result:

PROPOSITION 3.9. *For each $n \geqslant 2$, the Magnus relation on PB_n is a bi-invariant ordering.*

PROOF. We use induction on $n \geqslant 2$. For $n = 2$, the result is obvious, as B_2^+ is a copy of $\mathbb{Z}$ and our ordering corresponds to the standard ordering of integers. Assume $n \geqslant 3$. Then we have the exact sequence

$$1 \longrightarrow F_{n-1} \longrightarrow PB_n \longrightarrow PB_{n-1} \longrightarrow 1,$$

and, by definition, the relation $<_{M,n}$ is obtained from the Magnus ordering on F_{n-1} and the relation $<_{M,n-1}$ on PB_{n-1} using the scheme of Lemma 3.1.

By Proposition 2.15, the Magnus ordering on F_{n-1} is a bi-invariant ordering, and, by Proposition 3.6, it is invariant under the action of PB_n. On the other hand, by induction hypothesis, the relation $<_{M,n-1}$ is a bi-invariant ordering of PB_{n-1}. Then Lemma 3.1 implies that the relation $<_{M,n}$ is a bi-invariant ordering of PB_n. □

We have thus obtained a new proof of Corollary 1.3, *i.e.*, of the bi-orderability of the pure braid group PB_n for each finite n. From now on, the Magnus relation on PB_n will be called the *Magnus ordering* of PB_n.

If β, β' are braids of PB_{n-1}, they also belong to PB_n. As we have $r_n(\beta) = \beta$ and $r_n(\beta')$, it immediately follows from Definition 3.8 that $\beta <_{M,n} \beta'$ is true if and only if $\beta <_{M,n-1} \beta'$ is. So we can drop the subscript n and simply write $<_M$ for the Magnus ordering of pure braids.

3.4. Magnus ordering *vs.* Artin coordinates. Instead of appealing to the recursive construction of Definition 3.8, we can compare pure braids directly by using their Artin coordinates.

PROPOSITION 3.10. *For β, β' in PB_n, the relation $\beta <_M \beta'$ is true if and only if the sequence of Artin coordinates of β is smaller than the sequence of Artin coordinates of β' with respect to the lexicographical extension of the Magnus orderings of each subgroup F_j.*

PROOF. Use induction on $n \geqslant 2$. For $n = 2$, the result is obvious. Assume $n \geqslant 3$. Assume $\beta, \beta' \in PB_n$ and $\beta <_M \beta'$. Let $(\beta_1, \ldots, \beta_{n-1})$ and $(\beta'_1, \ldots, \beta'_{n-1})$ be the Artin coordinates of β and β'. We observe that the Artin coordinates of $r_n(\beta)$ are $(\beta_1, \ldots, \beta_{n-2})$, and that $r_n(\beta)^{-1}\beta$ is equal to β_{n-1}. So, for $r_n(\beta) <_M r_n(\beta')$, we obtain $(\beta_1, \ldots, \beta_{n-2}) <^{\text{Lex}} (\beta'_1, \ldots, \beta'_{n-2})$ by induction hypothesis, and we deduce $(\beta_1, \ldots, \beta_{n-1}) <^{\text{Lex}} (\beta'_1, \ldots, \beta'_{n-1})$. On the other hand, if we have $r_n(\beta) = r_n(\beta')$ and $r_n(\beta)^{-1}\beta <_M r_n(\beta')^{-1}\beta'$, we obtain now $(\beta_1, \ldots, \beta_{n-2}) = (\beta'_1, \ldots, \beta'_{n-2})$ and $\beta_{n-1} <_\mu \beta'_{n-1}$, again implying $(\beta_1, \ldots, \beta_{n-1}) <^{\text{Lex}} (\beta'_1, \ldots, \beta'_{n-1})$. □

As the Magnus ordering $<_\mu$ is defined in terms of the ordering on $\mathbb{Z}\langle\langle X_1, \ldots, X_n\rangle\rangle$, we can equivalently characterize the ordering of pure braids in terms of power series.

COROLLARY 3.11. *For β in PB_n, define the $\mathbb{Z}\langle\langle X_1, \ldots, X_n\rangle\rangle$-coordinates of β to be the Magnus expansions of its Artin coordinates. Then, for β, β' in PB_n, the relation $\beta <_M \beta'$ is true if and only if the $\mathbb{Z}\langle\langle X_1, \ldots, X_n\rangle\rangle$-coordinates of β are smaller than the $\mathbb{Z}\langle\langle X_1, \ldots, X_n\rangle\rangle$-coordinates of β' with respect to the lexicographical extension of the* SumLex*-ordering of $\mathbb{Z}\langle\langle X_1, \ldots, X_n\rangle\rangle$.*

EXAMPLE 3.12. Let us compare $\beta = \sigma_1^2\sigma_2^2\sigma_1^2$ and $\beta' = \Delta_3^2 = (\sigma_1\sigma_2\sigma_1)^2$. First, we find $\beta_1 = \sigma_1^4 = x_{1,2}^2$ and $\beta'_1 = \sigma_1^2 = x_{1,2}$. We have $x_{1,2}^2 >_\mu x_{1,2}$ in the Magnus ordering of F_1—*i.e.*, equivalently, $X_1^2 >^{\text{SumLex}} X_1$ in the ordering of $\mathbb{Z}\langle\langle X_1, \ldots, X_n\rangle\rangle$—so $\beta >_M \beta'$ is true.

For a second example, let us keep β, and consider $\beta'' = \sigma_1^4\sigma_2^2$. The first coordinate of β'' is $\beta''_1 = \sigma_1^4$. The first Artin coordinates of β and β'' coincide, and we go to the second coordinates. The latter are $\beta_2 = \sigma_1^{-2}\sigma_2^2\sigma_1^2$ and $\beta''_2 = \sigma_2^2$, which gives $\beta_2 = x_{1,3}^{-1}x_{2,3}x_{1,3}$ and $\beta''_2 = x_{2,3}$ in terms of the generators $x_{i,3}$. As seen in Example 2.16, we have $x_{1,3}^{-1}x_{2,3}x_{1,3} <_\mu x_{2,3}$ in F_2—equivalently, $(1 - X_1 + X_1^2 - X_1^3 + \ldots)(1 + X_2)(1 + X_1) <^{\text{SumLex}} 1 + X_2$ is true—so we find $\beta'' <_M \beta$.

3.5. More properties. Let us now consider the group PB_∞. Because the Magnus ordering of PB_n extends that of PB_{n-1} for each n, we can extend it to the pure braid group PB_∞: for β, β' in PB_∞, we say that β is smaller than β' in PB_∞ if β is smaller than β' is true in any group PB_n that contains both β and β'. We immediately find that the pure braid group PB_∞ is also bi-orderable:

COROLLARY 3.13. *The Magnus ordering of PB_∞ is a linear ordering that is invariant under multiplication on both sides.*

We conclude with two easy general properties of the pure braid ordering.

PROPOSITION 3.14. (i) *For $n \geqslant 3$, the Magnus ordering of PB_n is a dense ordering.*
(ii) *The Magnus ordering of PB_∞ is invariant under the shift endomorphism.*

PROOF. For (i), it is enough to find a sequence of pure braids which converge to the identity. This can be accomplished by taking terms deeper and deeper in the lower central series of the free group F_{n-1} and applying Proposition 2.9(ii).

(ii) We prove using induction on n that, for β, β' in PB_n, the relation $\beta <_{\mathrm{M}} \beta'$ implies $\mathrm{sh}(\beta) <_{\mathrm{M}} \mathrm{sh}(\beta')$. Indeed, if $r_n(\beta) <_{\mathrm{M}} r_n(\beta')$ is true, the induction hypothesis implies $\mathrm{sh}(r_n(\beta)) <_{\mathrm{M}} \mathrm{sh}(r_n(\beta'))$, hence $r_{n+1}(\mathrm{sh}(\beta)) <_{\mathrm{M}} r_{n+1}(\mathrm{sh}(\beta'))$, which gives $\mathrm{sh}(\beta) <_{\mathrm{M}} \mathrm{sh}(\beta')$.

On the other hand, if $r_n(\beta)$ and $r_n(\beta')$ are equal, so are $r_{n+1}(\mathrm{sh}(\beta))$ and $r_{n+1}(\mathrm{sh}(\beta'))$. Then the Artin combing of $\mathrm{sh}(\beta)$ is obtained from that of β by translating all indices by $+1$, and similarly for β'. As the ordering on monomials in $\mathbb{Z}\langle\langle X_1, ..., X_n\rangle\rangle$ is invariant under such a translation, $\beta <_{\mathrm{M}} \beta'$ implies $\mathrm{sh}(\beta) <_{\mathrm{M}} \mathrm{sh}(\beta')$. □

4. The ordering of positive pure braids

We shall now investigate the restriction of the Magnus ordering of PB_n to positive pure braids and, more generally, to submonoids of PB_n obtained by taking the intersection of PB_n with a submonoid of B_n of a convenient type, typically the dual braid monoid B_n^{+*} of Chapter VIII. The main result is that, in all cases, the restriction of the Magnus ordering is a well-ordering, with an ordinal type that can be easily determined.

4.1. The main observation. As each braid relation preserves the length, there exists a well-defined homomorphism

$$\epsilon : B_n \to \mathbb{Z} \tag{4.1}$$

that maps every generator σ_i to 1. The integer $\epsilon(\beta)$, called the *exponent sum* of β, is the difference between the number of positive and negative letters in every braid word representing β, *i.e.*, it is the algebraic sum of the exponents.

On the other hand, we recall that, for β in F_j, we denote by $\mu(\beta)$ the Magnus expansion of β relative to $(x_{1,j}, \dots, x_{j-1,j})$, while $c_1(\mu(\beta))$ denotes the sum of all coefficients of degree 1 monomials in $\mu(\beta)$.

LEMMA 4.1. *Assume that β is a pure n-strand braid and $\beta_1, \dots, \beta_{n-1}$ are the Artin coordinates of β. Then we have*

$$\epsilon(\beta) = 2 \cdot \sum_{i=1}^{n-1} c_1(\mu(\beta_i)). \tag{4.2}$$

PROOF. By definition, we have $\beta = \beta_1 ... \beta_{n-1}$, and $\epsilon(\beta) = \epsilon(\beta_1) + \cdots + \epsilon(\beta_{n-1})$, so it is enough to check (4.2) for each braid β_{j-1} with $2 \leqslant j \leqslant n$. In this case, β_{j-1} admits a (unique) reduced decomposition

$$\beta_{j-1} = x_{i_1,j}^{e_1} \cdots x_{i_\ell,j}^{e_\ell}.$$

Then the degree 1 part in the Magnus expansion $\mu(\beta_{j-1})$ is $e_1 X_{i_1} + \cdots + e_\ell X_{i_\ell}$, and, therefore, we have $c_1(\mu((\beta_{j-1})) = e_1 + \cdots + e_\ell$. On the other hand, for each i, we have $\epsilon(x_{i,j}) = 2$, so we find $\epsilon(\beta_{j-1}) = 2e_1 + \cdots + 2e_\ell$, and (4.2) follows. □

This simple observation implies the following.

LEMMA 4.2. *Assume $\beta, \beta' \in PB_n$ and $\beta' <_{\rm M} \beta$. Then we have $r_n(\beta') <_{\rm M} r_n(\beta)$ or $\epsilon(\beta') \leqslant \epsilon(\beta)$ (or both).*

PROOF. Assume $\beta' <_{\rm M} \beta$. Then $r_n(\beta') >_{\rm M} r_n(\beta)$ is impossible as, by definition, it would imply $\beta' >_{\rm M} \beta$. So, if (i) is not true, we must have $r_n(\beta') = r_n(\beta)$. Then the Artin coordinates of β and β' take the form $\beta_1, \dots, \beta_{n-1}$ and $\beta_1, \dots, \beta_{n-2}, \beta'_{n-1}$, and the hypothesis implies $\beta'_{n-1} <_\mu \beta_{n-1}$, *i.e.*, $\mu(\beta'_{n-1}) <^{\rm SumLex} \mu(\beta_{n-1})$. By definition of the ordering of $<^{\rm SumLex}$, this is possible only if we have $c_1(\mu(\beta'_{n-1})) \leqslant c_1(\mu(\beta_{n-1}))$. Now, (4.2) gives

$$\frac{\epsilon(\beta')}{2} = \sum_{i=1}^{n-2} c_1(\mu(\beta_i)) + c_1(\mu(\beta'_{n-1})) \leqslant \sum_{i=1}^{n-2} c_1(\mu(\beta'_i)) + c_1(\mu(\beta_{n-1})) = \frac{\epsilon(\beta)}{2},$$

and we have $\epsilon(\beta') \leqslant \epsilon(\beta)$. □

4.2. The well-order property. We are ready to study the restriction of the pure braid ordering to positive pure braids.

NOTATION 4.3. For each n, we put $PB_n^+ = PB_n \cap B_n^+$.

By construction, PB_n^+ is a submonoid of PB_n for each n—but we do not claim that, for $n \geqslant 3$, it is generated as a monoid by the braids $x_{i,j}$. For instance, the positive pure 3-strand braid $\sigma_1\sigma_2^2\sigma_1$ cannot be expressed as the product of two braids $x_{i,j}$, as an exhaustive search easily shows.

PROPOSITION 4.4. *For $n \geqslant 3$, every braid β in PB_n^+ satisfies $\beta \geqslant_{\rm M} r_n(\beta)$.*

PROOF. Assume $\beta <_{\rm M} r_n(\beta)$. We have $r_n(r_n(\beta)) = r_n(\beta)$, so it is impossible to have $r_n(\beta) <_{\rm M} r_n(r_n(\beta))$. Hence, by Lemma 4.2, we must have $\epsilon(\beta) \leqslant \epsilon(r_n(\beta))$. Now, owing to the construction of r_n as the operation of removing the nth strand, the relation $\epsilon(\beta) \leqslant \epsilon(r_n(\beta))$ is possible for a positive braid β only if $\beta = r_n(\beta)$ holds, contradicting $\beta <_{\rm M} r_n(\beta)$. So $\beta <_{\rm M} r_n(\beta)$ is impossible. □

An immediate induction on n gives the following consequence.

COROLLARY 4.5. *For $n \geqslant 2$, every braid β in PB_n^+ satisfies $\beta \geqslant_{\rm M} 1$.*

We now turn to the well-order property. We shall deduce it from the following consequence of Lemma 4.2.

LEMMA 4.6. *For each braid β in PB_n^+, the interval $[r_n(\beta), \beta]$ contains only finitely many braids in PB_n^+.*

PROOF. Assume that γ satisfies $r_n(\beta) \leqslant \gamma \leqslant \beta$. By Lemma 4.2, we must have $\epsilon(\gamma) \leqslant \epsilon(\beta)$, since $\gamma <_{\text{M}} r_n(\beta)$ is excluded by hypothesis. Infinitely many braids γ satisfy $\epsilon(\gamma) \leqslant \epsilon(\beta)$, but only finitely of them may belong to B_n^+—actually not more than $(n-1)^{\epsilon(\beta)}$ such braids may exist since every braid in B_n^+ can be represented by a braid word involving only positive letters σ_i. □

We are ready to establish our main result about the Magnus ordering of PB_n^+. If $(A, \prec)$ and $(B, \prec)$ are two (linearly) ordered sets, their *product* is defined to be the cartesian product $A \times B$ equipped with the lexicographical order: $(a, b) \prec (a', b')$ is true if and only if either $a \prec a'$ is true, or we have $a = a'$ and $b \prec b'$.

PROPOSITION 4.7. *For each $n \geqslant 3$, the ordered set $(PB_n^+, <_{\text{M}})$ is isomorphic to the product of $(PB_{n-1}^+, <_{\text{M}})$ and $(\mathbb{N}, <)$.*

PROOF. For each braid β in PB_n^+, we put

$$I(\beta) = (\, r_n(\beta)\, ,\, \#[r_n(\beta), \beta)\,).$$

By Lemma 4.6, the interval $[r_n(\beta), \beta)$ is finite, so I sends PB_n^+ to $PB_{n-1}^+ \times \mathbb{N}$.

We claim that I is a strictly increasing mapping of $(PB_n^+, <_{\text{M}})$ into the product of $(PB_{n-1}^+, <_{\text{M}})$ and $(\mathbb{N}, <)$. Indeed, assume that β, β' belong to PB_n^+ and $\beta <_{\text{M}} \beta'$ is satisfied. Two cases may occur. If $r_n(\beta)$ and $r_n(\beta')$ are distinct, then necessarily $r_n(\beta) <_{\text{M}} r_n(\beta')$ holds, and, by definition, we have $I(\beta) < I(\beta')$. Otherwise, we have $r_n(\beta) = r_n(\beta')$. Then saying that $\beta <_{\text{M}} \beta'$ is true means that the interval $[r_n(\beta), \beta)$ is a proper initial segment of the interval $[r_n(\beta), \beta')$, and we then have $\#[r_n(\beta), \beta) < \#[r_n(\beta), \beta')$, hence $I(\beta) < I(\beta')$ again.

Next, we claim that I is surjective. Indeed, we first observe that the image of I is an initial segment of $(PB_{n-1}^+, <_{\text{M}}) \times (\mathbb{N}, <)$, *i.e.*, that a lower bound of an element of the image still belongs to the image. The only problem is with the second coordinate. Now, by construction, if there exists β such that $\#[r_n(\beta), \beta)$ is p, then, for each $q \leqslant p$, there exists γ satisfying $r_n(\gamma) = r_n(\beta)$ and $\#[r_n(\beta), \beta) = q$, namely the $(q+1)$st braid in the interval $\#[r_n(\beta), \beta)$.

Second, we observe that, for each braid β in PB_{n-1}^+, the braid $\beta\sigma_{n-1}^{2p}$ is a braid in PB_n^+ that satisfies

$$r_n(\beta\sigma_{n-1}^{2p}) = \beta \quad \text{and} \quad \#[\beta, \beta\sigma_{n-1}^{2p}) \geqslant p,$$

which shows that the image of I is all of $PB_{n-1}^+ \times \mathbb{N}$.

So I is the expected isomorphism. □

Proposition 4.7 leads to a complete characterization of the ordered set $(PB_n^+, <_{\text{M}})$. The construction is illustrated in Figure 6.

PROPOSITION 4.8. *For each $n \geqslant 2$, the restriction of the pure braid ordering of PB_n to PB_n^+ is a well-ordering, and its ordinal type is ω^{n-1}.*

PROOF. We use induction on $n \geqslant 2$. For $n = 2$, the ordered set $(PB_2^+, <_{\text{M}})$ is a copy of $(\mathbb{N}, <)$, whose ordinal type is ω, *i.e.*, ω^1 by definition. Assume $n \geqslant 3$. By Proposition 4.8 and by the induction hypothesis, $(PB_n^+, <_{\text{M}})$ is the product of the two well-orderings $(PB_{n-1}^+, <_{\text{M}})$ and $(\mathbb{N}, <)$, so it is a well-ordering, and its ordinal type is the reversed product of the ordinal types, *i.e.*, $\omega \cdot \omega^{n-2}$, which is ω^{n-1}. □

4.3. An extension. The results of the previous section extend to more general submonoids of the pure braid group PB_n.

PROPOSITION 4.9. *Assume that M is a submonoid of the pure braid group PB_n that is generated by a finite family of braids, each of which has a positive exponent sum with respect to the σ_i generators. Then the restriction of the pure braid ordering of PB_n to M is a well-ordering, and its ordinal type is at most ω^{n-1}.*

PROOF. The argument is the same as for PB_n^+. The point that guarantees the well-order property and the upper bound on the ordinal type is the result that, for each braid β in M, only finitely many braids of M may lie in the interval $[r_n(\beta), \beta)$. By Lemma 4.2, a sufficient condition is that, for each natural number p, only finitely many elements of M have their exponent sum bounded by p. If every generator of M has a positive exponent sum, then an element β of M satisfying $\epsilon(\beta) \leqslant p$ must admit a decomposition of length at most p in the generators. So, if the latter are finite in number, only finitely many such elements may exist. □

The result applies in particular to the dual braid monoids B_n^{+*} of Chapter VIII.

COROLLARY 4.10. *For each $n \geqslant 2$, the restriction of the pure braid ordering of PB_n to $PB_n \cap B_n^{+*}$ is a well-ordering, and its ordinal type is ω^{n-1}.*

PROOF. The monoid B_n^{+*} is generated by the elements $a_{i,j}$, each of which has an exponent sum $+1$, so Proposition 4.9 applies. As for the ordinal type, it cannot be higher than ω^{n-1} by Proposition 4.9, and it cannot be less than ω^{n-1}, because B_n^{+*} includes B_n^+, and the ordinal type of PB_n^+ is ω^{n-1}. So the ordinal type must be ω^{n-1} exactly. □

5. Incompatibility of the orderings

The ordering of pure braids constructed in this chapter is quite different from the σ-ordering of B_n, despite the fact that positive braids are greater than 1 in both the orderings. In particular, the Magnus ordering of PB_n is *not* the restriction of the σ-ordering of B_n to pure braids, and, on the other hand, it cannot be extended into a left-invariant ordering of B_n. The purpose of this section is to show that, much more generally, no bi-invariant ordering of PB_n can be extended into a left-invariant ordering of B_n, a result of [**180**], also proved independently in [**71**].

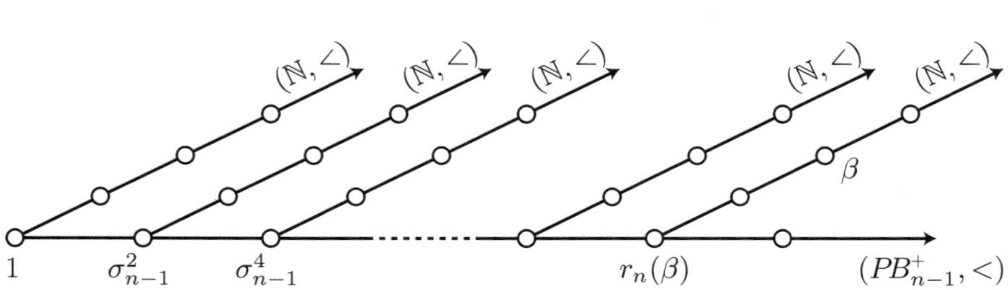

FIGURE 6. The well-ordering of PB_n^+. In each point of the well-ordering PB_{n-1}^+ we attach a copy of $(\mathbb{N}, <)$. Each braid β lies on the line attached at $r_n(\beta)$. If the base line has ordinal type ω^{n-2}, then the new well-ordering has ordinal type $\omega \cdot \omega^{n-2}$, *i.e.*, ω^{n-1}.

5.1. Local indicability. Our main tool will be the notion of local indicability. It was introduced by Higman [**107**], who was motivated in part by the zero-divisor conjecture mentioned in Section III.1.2.

DEFINITION 5.1. A group is said to be *indicable* if it has a quotient isomorphic to $\mathbb{Z}$. A group is said to be *locally indicable* if each finitely generated subgroup not equal to $\{1\}$ is indicable.

Examples of locally indicable groups are free Abelian groups and free groups. More generally we have the following.

PROPOSITION 5.2. *Bi-orderable groups are locally indicable and locally indicable groups are left-orderable. Neither of these implications is reversible.*

SKETCH OF PROOF. Assume that $\prec$ is a bi-invariant ordering on a group G. Consider a finitely generated subgroup $H = \langle h_1, \dots, h_r \rangle$, with notation chosen so that $1 \prec h_1 \prec \dots \prec h_r$ holds, and also assume that this collection of generators is minimal. The family of convex subgroups of a given ordered group is linearly ordered by inclusion and is closed under arbitrary unions and intersections. Let K be the union of all convex subgroups of H which do not contain h_r. Then K is convex and one argues it is also normal in H, which implies that the quotient group H/K inherits a bi-invariant ordering. Moreover, H/K is Archimedian, so by a theorem of Hölder [**111**], it is isomorphic with a subgroup of $(\mathbb{R}, +)$. Being finitely generated, H/K is therefore isomorphic with a sum of infinite cyclic groups, and so there is a nontrivial homomorphism $H \to H/K \to \mathbb{Z}$, completing the first half of the proposition.

Burns and Hale [**31**] proved that a group is left-orderable if and only if every nontrivial finitely generated subgroup has a nontrivial quotient which is left-orderable. Since $\mathbb{Z}$ is left-orderable, the second half of the proposition follows.

The irreversibility of both implications can be shown by using the braid groups as examples. We will see in Section 5.3 that B_3 and B_4 are locally indicable, whereas they are not bi-orderable, and, below, that B_n, for $n \geqslant 5$, is left-orderable but not locally indicable. □

We recall calculations of the commutator subgroups of braid groups due to Gorin and Lin [**102**]; see also [**159**] for details and specific presentations.

PROPOSITION 5.3. *The commutator subgroups $[B_n, B_n]$ of B_n have the following properties:*

- *$[B_2, B_2]$ equals $\{1\}$,*
- *$[B_3, B_3]$ is a free group of rank 2,*
- *$[B_4, B_4]$ is a semidirect product of two free groups, each of rank 2,*
- *For $n \geqslant 5$, $[B_n, B_n]$ is finitely generated and perfect, meaning that it equals its own commutator subgroup.*

COROLLARY 5.4. *For $n \geqslant 5$, the braid group B_n is not locally indicable—but it is indicable.*

PROOF. The subgroup $[B_n, B_n]$, which is finitely generated, cannot have $\mathbb{Z}$ as a quotient, as any homomorphism of a perfect group to an Abelian group must have a trivial image.

As for indicability, the augmentation mapping ϵ provides, for each $n \geqslant 2$, a surjective homomorphism of B_n onto the integers. □

5.2. The Conrad property. In order to prove that no bi-invariant ordering of PB_n can extend into a left-invariant ordering of B_n, we shall use the Conrad property introduced in Definition II.2.6. We recall that a left-invariant ordering $\prec$ of a group G is said to be *Conradian* if for all g, h in G that are greater than 1, there exists a positive integer p satisfying $h \prec gh^p$.

Conrad used this property in **[38]** to show that left-ordered groups which are Conradian share many of the properties of bi-orderable groups. It is easy to see that every bi-invariant ordering is Conradian.

PROPOSITION 5.5. *Every group that admits a Conradian left-invariant ordering is locally indicable.*

The proof is essentially the argument outlined in Proposition 5.2. It is also known that, if a group has a left-invariant ordering which is both Archimedian and Conradian, then it must be Abelian.

We can now address the question of extending orderings from PB_n to B_n. The key observation is the following result.

PROPOSITION 5.6. *Let $(G, \prec)$ be a left-ordered group and suppose H is a subgroup of G of finite index. If $(H, \prec)$ is Conradian, then so is $(G, \prec)$.*

PROOF. Assume the hypothesis, but that $(G, \prec)$ is not Conradian. Then there exist $g, h \in G$ with $1 \prec g$ and $1 \prec h$ and $gh^p \prec h$ for all positive p. First note that $h \succ 1$ implies $gh \succ g$ and therefore $h \succ gh \succ g$. Next note that $h \succ gh^p$ implies $h \succ gh \succ g^2h^p$, and a simple induction shows $h \succ g^qh^p$ for all integers p, q satisfying $p \geqslant 0$ and $q \geqslant 1$. Since H is of finite index, there is a positive integer r such that g^r and h^r belong to H. But then we have $g^r(h^r)^p \prec h \prec h^r$ for all positive integers p, which contradicts the assumption that the ordering is Conradian on H. □

COROLLARY 5.7. *If $(G, \prec)$ is a left-ordered group such that the ordering is also right-invariant on a subgroup of finite index, then $(G, \prec)$ is Conradian.*

By Proposition 5.5, this in turn implies

COROLLARY 5.8. *If $(G, \prec)$ is a left-ordered group such that the ordering is also right-invariant on a subgroup of finite index, then G is locally indicable.*

Gathering the results, we deduce the main result of the section.

PROPOSITION 5.9. *For $n \geqslant 5$, there is no left-invariant ordering of B_n which restricts to a bi-invariant ordering on PB_n.*

PROOF. If such an ordering existed, then Corollary 5.8 would imply that B_n is locally indicable. This would contradict Corollary 5.4. □

5.3. The cases of three and four strands. The cases of B_3 and B_4 are special. We shall see here that, contrary to B_n for $n \geqslant 5$, they are locally indicable and admit Conradian orderings.

The basic observation is that local indicability is inherited by subgroups, but also by extensions: if

$$1 \to K \to G \to H \to 1$$

is an exact sequence of groups in which K and H are locally indicable, then G is also locally indicable. The proof is straightforward.

PROPOSITION 5.10. *For $n \leqslant 4$, the braid group B_n is locally indicable.*

PROOF. The result is obvious for $n = 2$. For $n = 3$, we note the exact sequence

$$1 \to [B_3, B_3] \to B_3 \to \mathbb{Z} \to 1 \tag{5.1}$$

associated to the commutator subgroup, a free group. Both $[B_3, B_3]$ and $\mathbb{Z}$ are locally indicable, so it follows that B_3 is as well. For $n = 4$, consider the homomorphism $B_4 \to B_3$ that preserves σ_1 and σ_2 and maps σ_3 to σ_1. The kernel K is the normal closure of $\sigma_3\sigma_1^{-1}$ in B_4, and so lies in the commutator subgroup, a semidirect product of free groups, and is therefore locally indicable—actually it is known that K is a free group generated by $\sigma_1\sigma_3^{-1}$ and $\sigma_2\sigma_1\sigma_3^{-1}\sigma_2^{-1}$; see [**96, 102**], or [**122**] for a simple argument. The exact sequence

$$1 \to K \to B_4 \to B_3 \to 1 \tag{5.2}$$

and local indicability of K and B_3 finish the argument. □

A theorem of Brodskii [**25**] (see also [**180**] and [**162**]) asserts that the sufficient condition of Proposition 5.5 is also necessary: a group is locally indicable if and only if it admits a Conradian left-invariant ordering. So the groups B_3 and B_4 must admit left-invariant orderings that are Conradian. Actually, this can be proved directly using a constructive argument.

PROPOSITION 5.11. *The groups B_3 and B_4 admit left-invariant orderings that are Conradian.*

PROOF. Free groups have bi-orderings, namely the Magnus orderings constructed earlier in this chapter, which are therefore Conradian. It is straightforward to check that, if

$$1 \to K \to G \to H \to 1$$

is an exact sequence, and K and H have Conradian left-orderings, then the left-ordering of G given in Lemma 3.1 is also Conradian. By the first exact sequence (5.1), we construct a Conradian left-invariant ordering for B_3. Similarly, using (5.2) and noting that semidirect products of free groups also have Conradian left-orderings, we can construct a Conradian left-ordering of B_4. □

CHAPTER XVI

Open Questions and Extensions

In this chapter we mention further results and discuss open questions connected with the various aspects of braid orderings considered in this book.

We should start, however, with a very general remark. There are many approaches to braid groups that have not been considered in this book. In fact, braid groups play a role in many areas of mathematics that have not even been mentioned here—*e.g.*, algebraic geometry or mathematical physics. We can therefore still hope that new, illuminating perspectives on braid orderings will emerge in the future.

The chapter is organized as follows. In Section 1, we list some general questions about the σ-ordering and related topics. Then in Section 2, we discuss more specific questions that arise in the context of the successive chapters of this book, taken in the order in which they appear. Finally, we address in Section 3 some of the many extensions of braid groups from the point of view of order properties.

1. General questions

We begin with three types of questions involving the σ-ordering in general, namely its uses, its structure, and the problem of finding σ-positive representatives.

1.1. Uses of the braid ordering. In Chapter III we listed several applications of the orderability of braid groups and of the more specific properties of the σ-ordering of braids. However, up to now, the applications are not so plentiful and not so strong. This situation contrasts with the seemingly deep and, at the least, sophisticated properties of the σ-ordering explained in this text, which may appear as a promising sign for potentially powerful applications. So, although vague, the first open question is the following.

QUESTION 1.1. *How to use the braid ordering?*

In particular, one of the deepest properties of the σ-ordering of B_n known so far is the fact that its restriction to the braid monoid B_n^+, and even to the dual braid monoid B_n^{+*}, is a well-ordering. As emphasized in Section III.3.1, the well-order property is a very strong condition which enables one to distinguish one element in each nonempty subset—so, typically, in each conjugacy class or each Markov class. But, so far, this observation was of no use because we had no effective way to identify such minimal elements in practice, for instance in the case of the conjugacy problem. Thus, a special case of Question 1.1 is

QUESTION 1.2. *How to take advantage of the fact that the σ-ordering restricted to B_n^+ and B_n^{+*} is a well-ordering?*

To raise less fuzzy questions, we may think more specifically of the conjugacy problem. Let us say that two positive braids β, β' are positively conjugate if there exists a positive braid γ satisfying $\beta\gamma = \gamma\beta'$. As Δ_n^2 is central and multiplying any n-strand braid with a sufficient power of Δ_n^2 yields a positive braid, solving the conjugacy problem of the group B_n is algorithmically equivalent to solving the positive conjugacy problem of the monoid B_n^+. Now, for each positive braid β, the positive conjugacy class of β is a nonempty subset of B_n^+, hence, by the well-order property, it admits a $<$-least element.

QUESTION 1.3. *Can one effectively compute the $<$-least element in a positive conjugacy class?*

A similar question can be raised with "Markov equivalence class" replacing "conjugacy class"; a solution would typically associate a computable, well-defined ordinal number with each knot.

The recent developments described in Chapters VII and VIII around the alternating and cycling normal forms of braids have not been exploited thus far, and they might be useful here.

1.2. Structure of the braid ordering. To a large extent, the structure of the σ-ordering of braids remains mysterious. Even in the case of B_3, the examples of Section II.2.1 show that the order $<$ is a complicated object. By contrast, the results of Chapters VII and VIII give a much simpler description for the restriction of $<$ to the submonoids B_n^+ and B_n^{+*} of B_n. The reason why the description is more satisfactory for B_n^+ than for B_n is that we have a simple recursive definition describing how the ordering of B_n^+ can be obtained from that of B_{n-1}^+. It is natural to raise the question of finding similar constructions for B_n, *i.e.*, more precisely, to raise

QUESTION 1.4. *Does there exist a simple recursive definition of the σ-ordering on B_n from the σ-ordering on B_{n-1}?*

QUESTION 1.5. *Does there exists a (computable) unique normal form on B_n so that, for any two braids β, β', whether $\beta < \beta'$ holds can be read directly from the normal forms of β and β'?*

The handle reduction algorithm of Chapter V does not answer Question 1.5, because it does not lead to a unique normal form, and because the result can be used to compare a braid with 1, but not directly to compare two braids. It is natural to wonder whether Bressaud's normal form of Section XI.1 might be useful here. Note that the algorithm based on the Mosher normal form presented in Section XII.3 yields a positive answer to the above question, except that the normal forms are not braid words but sequences of edge flips of singular triangulations.

1.3. Sigma-positive representatives. We mentioned in Proposition II.1.21 that the algorithmic complexity of the σ-ordering is at most quadratic. Several proofs have been given—in particular the stronger version of Chapter XII involving random access machine (RAM) complexity. It seems unlikely that there exist subquadratic algorithms, and the current result might be close to be optimal.

The situation is different with the stronger question of finding σ-positive representatives. Property **C** says that every nontrivial braid admits at least one representative braid word that is σ-positive or σ-negative. The exponential upper bound of Proposition II.1.22 is certainly far from optimal. Actually, all proofs of Property **C** sketched in this text lead to algorithmic methods for finding σ-positive representatives. Some solutions are inefficient: the only upper bound proved for the method of Chapter IV is a tower of exponentials of exponential height. Similarly, the proof of Chapter VII relies on a transfinite induction, and deriving a complexity statement is not easy. On the other hand, some methods, like handle reduction of Chapter V or transmission-relaxation of Chapter XII, seem quite efficient for producing short σ-positive expressions, but no result has been proved so far.

CONJECTURE 1.6. *For every $n > 3$, there exist numbers C_n, C'_n such that every nontrivial n-strand braid represented by a word of length ℓ has a σ-positive or σ-negative representative of length at most $C_n \cdot \ell$. Moreover, such a representative word can be found by an algorithm whose running time is bounded by $C'_n \cdot \ell^2$.*

Very recently, J. Fromentin announced a proof of Conjecture 1.6, with the explicit bounds $C_n = C'_n \leqslant 12n^2$. His method consists in turning the main proof of Section VIII.3 into an algorithm that, running on an arbitrary n-strand braid word w, translates it into a fractionary expression in the $a_{i,j}^{\pm 1}$ letters, then puts the numerator and the denominator in cycling normal form, and returns an explicit σ^Φ-positive or σ^Φ-negative expression for the braid represented by w.

REMARK 1.7. The braid $\sigma_1\sigma_2\sigma_3^{-1}\sigma_2\sigma_1^{-1}$ has no σ-positive representative of length 6 or less [**83**, Theorem 5.1], so $C_4 \geqslant 7/5$ must hold. Moreover, the word

$$\sigma_1\sigma_2^{-2}\sigma_3^2\sigma_4^{-2}\ldots\sigma_{n-1}^{2e}\sigma_{n-2}^{2e}\sigma_{n-3}^{-2e}\ldots\sigma_2^2\sigma_1^{-1}, \tag{1.1}$$

with $e = \pm 1$ according to the parity of n, has no σ_1-positive or σ_1-negative representative with fewer than $(n-2)(n+1)$ crossings. As the above n-strand braid word has length $4(n-2)$, it seems that C_n needs to satisfy $C_n \geqslant (n+1)/4$.

Finally, let us mention possible connections with the problem of finding geodesics in braid groups. For β a braid, let us denote by $\ell_\sigma(\beta)$ the minimal length of a braid word representing β. The *geodesic problem* is the question of effectively finding, for each braid word w, an equivalent braid word w' satisfying $\ell(w') = \ell_\sigma(\beta)$, *i.e.*, finding a shortest representative of β.

It is shown in [**169**] that the B_∞-version of the geodesic problem is co-*NP*-complete. However, this result says nothing about the problem in a fixed group B_n, nor about the problem of finding *quasi-geodesics*, *i.e.*, about algorithms that, starting with a braid word w, would produce an equivalent word of length $O(\ell_\sigma(\overline{w}))$. As for the latter problem, the symmetric version of the greedy normal form provides, for each n, a quadratic algorithm returning for each n-strand braid word w an equivalent braid word of length at most $n^2\ell_\sigma(\overline{w})$. *A priori*, the problem of finding short representatives is unconnected with the problem of finding σ-positive representatives. In particular, the examples of (1.1) show that, when n is unbounded, the ratio between the length of the shorterst σ-positive representative and the length of the shortest representative may be at least $n/4$. However, the algorithms solving the latter problem often turn out to also solve the former, at least in part.

2. More specific questions

We turn to more specific questions involving the σ-ordering of braids and the various approaches that have been developed in the text. For simplicity, we organize the questions according to the chapters they refer to.

2.1. Self-distributivity. Many questions about self-distributivity in general, and about the self-distributive structure of braids in particular, remain open. We shall mention one such question here, and refer to [**52**] and [**51**] for many more.

We saw in Section IV.2 that, under the hypothesis that $(S, *)$ is a left-cancellative LD-system, there exists a partial action of B_n on S^n. The action is partial in that $\boldsymbol{x} \bullet \beta$ need not exist for each $\boldsymbol{x}$ in S^n and each braid β. In Proposition IV.2.5, we proved that, for every braid β in B_n, there exist $\boldsymbol{x}$ in S^n such that $\boldsymbol{x} \bullet \beta$ is defined. Reversing the point of view, let us introduce, for $\boldsymbol{x}$ in S^n,

$$D_S(\boldsymbol{x}) = \{\beta \in B_n \mid \boldsymbol{x} \bullet \beta \text{ is defined}\}.$$

As the action of positive braids is always defined, we have $B_n^+ \subseteq D_S(\boldsymbol{x}) \subseteq B_n$. If $(S, *)$ is a rack, the braid action is defined everywhere, and so, for each $\boldsymbol{x}$, we have $D_S(\boldsymbol{x}) = B_n$. On the other hand, if S is the LD-system $(B_\infty, *)$ of Definition IV.1.7, it is easy to see that $D_{B_\infty}(1, \dots, 1)$ never contains σ_i^{-1}, and, therefore, it is a proper subset of B_n. In some cases studied in [**132**], $D_S(\beta_1, \dots, \beta_n)$ coincides with B_n^+, and, then, the restriction of the braid order $<$ to $D_S(\beta_1, \dots, \beta_n)$ is a well-ordering.

CONJECTURE 2.1 (Laver). *For all braids $\beta_1, \dots, \beta_n$, the subset $D_{B_\infty}(\beta_1, \dots, \beta_n)$ of B_n is well ordered by the σ-ordering.*

Note that the question is a pure problem of braids, in that it involves no objects other than braids.

2.2. Handle reduction. We have seen that handle reduction, as described in Chapter V, is a very efficient solution to the braid word problem in practice—actually, the most efficient known so far; see, for instance, [**35**] for a comparison with the Tetris algorithm of Chapter XI. However, there remains a large gap between the complexity bound established in Proposition V.1.5 and the experimental values of Tables V.1 and V.2. This suggests that the argument of Section V.2 is far from optimal. One may hope that this is the manifestation of some deep and as yet unknown aspect of the geometry of braids.

CONJECTURE 2.2. *For each n, the handle reduction algorithm for B_n has a quadratic time complexity, and a linear space complexity: starting from a braid word of length ℓ, the running time lies in $O(\ell^2)$ and all words produced during the algorithm have length in $O(\ell)$—so does, in particular, the final reduced word.*

Clearly, Conjecture 2.2 implies Conjecture 1.6. The second statement in Conjecture 2.2 would be a consequence of a positive solution to the following more general conjecture about the subword reversing method—which extends without change to many group presentations; see [**53**] and [**64**].

CONJECTURE 2.3. *If w is an n-strand braid word length ℓ, and if w' is a freely reduced braid word obtained from w by a sequence of special transformations in the sense of Definition* V.2.7, *each immediately followed by a free reduction, then the length of w' is at most $C_n \cdot \ell$, where C_n is some constant that depends on n.*

It has been demonstrated experimentally in [**161**] that, by combining two handle reductions, namely, starting from a braid word w, first reducing w to w', and then reducing $\Phi_n(w')$ to $\Phi_n(w'')$, one obtains a final word w'' that is a short representative of $\overline{w}$. A. Myasnikov conjectured a positive answer to

QUESTION 2.4. *Does the above double handle reduction yield quasi-geodesics in B_n? In particular, does there exist a constant C_n such that, for w, w'' as above, one obtains $\ell(w'') \leqslant C_n \cdot \ell_\sigma(\overline{w})$?*

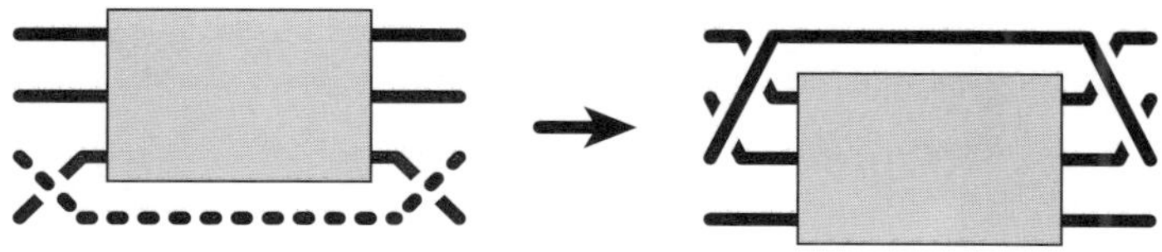

FIGURE 1. Coarse reduction of a σ_1-handle: instead of skirting around the next crossings, we push the strand responsible for the handle over the whole intermediate part.

To conclude with a perhaps easier question, let us come back to the *coarse handle reduction* briefly alluded to at the end of Chapter V. This variant of handle reduction consists in replacing a handle of the form $\sigma_i^e \cdot \mathrm{sh}^i(v) \cdot \sigma_i^{-e}$ with $\sigma_{i+1}^{-e} \dots \sigma_{n-1}^{-e} \cdot \mathrm{sh}^{i-1}(v) \cdot \sigma_{n-1}^{e} \dots \sigma_{i+1}^{e}$, as illustrated in Figure 1.

QUESTION 2.5. *Does coarse handle reduction converge?*

The arguments of Sections V.2.4 and V.2.5 are still valid, but those of Section V.2.2 are not, as the words obtained using coarse reduction from a word that is drawn in some set $\mathrm{Div}(\beta)$ may escape from $\mathrm{Div}(\beta)$. Experiments suggest that coarse reduction always converges, but the proof is still to be found.

2.3. Connection with the Garside structure. The results of Section VI.3 remain partial, and it is an obvious question to ask for a complete description of the $<$-increasing enumeration $S_{n,d}$ of the divisors of Δ_n^d similar to the one given in Section VI.2 for the case $n = 3$. The general case is probably difficult, but the case of 4-strand braids should be doable. Owing to the recursive rule of Definition VI.1.6, one can expect the generic entry of $S_{4,d-1}$ to have six copies in $S_{4,d}$, but some entries from $S_{4,d-2}$ have three copies in $S_{4,d}$ only.

More promising might be the questions of braid combinatorics to which the approach of Chapter VI leads. Counting problems involving braids have been little investigated, and a number of questions remain open. We saw in Section VI.1 that a crucial role in counting problems connected with the greedy normal form is played by a certain $n! \times n!$ matrix M_n, whose rows and columns are indexed by permutations of $\{1, \dots, n\}$, and the (π, π')-entry of M_n is 1 if and only if all descents of π'^{-1} are descents of π, and is 0 otherwise. In particular, the number of positive n-strand braids that divide Δ_n^d is directly connected with the eigenvalues of M_n—and of an equivalent smaller matrix $\widehat{M_n}$ whose size is the number of partitions of n. Table 1 below shows the associated characteristic polynomials for small values of n, immediately leading to

CONJECTURE 2.6. *For each n, the characteristic polynomial of M_{n-1} divides that of M_n. More precisely, the spectrum of M_n is the spectrum of M_{n-1}, plus $p(n) - p(n-1)$ nonzero eigenvalues.*

Very recently, a proof of the first part of the conjecture has been announced by F. Hivert, J.C. Novelli, and J.Y. Thibon in [**109**]. They use the framework of noncommutative symmetric and quasi-symmetric functions connected with combinatorial Hopf algebras, and they construct an explicit derivation that connects M_{n-1} and M_n.

TABLE 1. Characteristic polynomial of M_n up to a power of x—together with the corresponding spectral radius ρ_n and its relative growth.

$$P_{M_1}(x) = x - 1$$
$$P_{M_2}(x) = P_{M_1}(x) \cdot (x - 1)$$
$$P_{M_3}(x) = P_{M_2}(x) \cdot (x - 2)$$
$$P_{M_4}(x) = P_{M_3}(x) \cdot (x^2 - 6x + 3)$$
$$P_{M_5}(x) = P_{M_4}(x) \cdot (x^2 - 20x + 24)$$
$$P_{M_6}(x) = P_{M_5}(x) \cdot (x^4 - 82x^3 + 359x^2 - 260x + 60)$$
$$P_{M_7}(x) = P_{M_6}(x) \cdot (x^4 - 390x^3 + 6,024x^2 - 13,680x + 8,640)$$

n	1	2	3	4	5	6	7	8
ρ_n	1	1	2	5.449	18.717	77.405	373.990	2,066.575
$\rho_n/(n\rho_{n-1})$	-	0.5	0.667	0.681	0.687	0.689	0.690	0.691

Furthermore, for small values of n, all numbers $b_{n,d}(\beta)$—except $b_{n,d}(\Delta_n)$, which is 1—grow like ρ_n^d, where ρ_n is the spectral radius of M_n. Whether this is always true is unknown, but it makes it natural to investigate ρ_n. The trivial upper bound $\#\mathrm{Div}(\Delta_n^d) \leqslant (n!)^d$ suggests that we compare ρ_n with $n\rho_{n-1}$. The values listed in Table 1 may suggest that this ratio tends to $\log 2$.

Finally, it should be clear that all the above questions involving the symmetric groups can be extended to other finite Coxeter groups and to the corresponding braid groups, *i.e.*, the spherical Artin–Tits groups of Section 3.1.

2.4. Alternating decompositions. The recursive characterization of the σ-ordering of B_n^+ by means of the Φ_n-splitting provides a very simple description of this ordering. However, in the current exposition, this description, as well as all results of Section VII.4, is deduced from Burckel's delicate combinatorial methods, which involve in particular transfinite inductions.

QUESTION 2.7. *Does there exist for the recursive characterization of the σ-ordering of B_n^+, and for the other results of Section* VII.4, *a direct proof in the vein of the one described in Chapter* VIII*?*

The two approaches developed in Chapters VII and VIII are quite similar, and answering Question 2.7 in the positive should not be impossible. However, the Artin relations differ from the Birman–Ko–Lee relations in that some of them involve words of length 3, and this small technical difference might make the solution more difficult in the case of B_n^+.

Another natural question—that is problably connected with the previous one—involves the computation of the ordinal rank. With Corollary VII.2.22, we have a simple closed formula that expresses the rank of any positive 3-strand braid in the well-ordering of B_3^+ in terms of its Φ-normal form, which itself is very easily computed. In this way, we arguably obtain an optimal description of the ordering, as we identify the position of any element in an absolute way.

QUESTION 2.8. *Does there exist a similar method for determining the ordinal rank of an arbitrary braid in $(B_n^+, <^\Phi)$?*

A general solution is proposed in [**28**]. It relies on Burckel's notion of reducible words, and consists in counting how many irreducible braid words precede a given one in the tree ordering. The method is algorithmically efficient only in the case of 3 strands, and further investigation is certainly needed in the general case.

2.5. Dual braid monoids. The results of Chapter VIII are quite recent and many open questions remain, in particular the counterpart of Question 2.8. Another natural question would be to determine the ranks of the elements of B_n^+ inside $(B_n^{+*}, <^\Phi)$, hence to compare the ranks of a positive braid in $(B_n^+, <^\Phi)$ and in $(B_n^{+*}, <^\phi)$.

Another problem is to study the action of conjugacy on B_n^{+*}, in particular in view of Question 1.3. The definition of the cycling normal form suggests the introduction of a cycling operation similar to that used in the Garside-based solution to the conjugacy problem, and one may hope for progress in this direction. A similar approach is of course possible with the alternating decompositions of Chapter VII, but the fact that the family of generators $a_{i,j}$ is closed under conjugacy might make the context of Chapter VIII more suitable.

Other types of questions connected with the Φ- and the ϕ-normal forms involve random walks on the monoids B_n^+ or B_n^{+*} and possible stabilization phenomena, as studied for instance in [**149**]. To state a simple question, we may ask

QUESTION 2.9. *Assume that X is a random walk on B_n^+ (resp. B_n^{+*}). What is the expectation for the Φ_n-breadth (resp. the ϕ_n-breadth) of X?*

In other words, what is the average Φ_n-breadth of a random positive n-strand braid of length ℓ? Experiments suggest a connection with $\sqrt{\ell}$ that is not explained so far.

Another natural question is whether the Φ- and ϕ-normal forms might be connected with an automatic structure on B_n. It is known that the languages of normal words are regular languages, but it is unclear whether any form of the fellow traveler property might be satisfied.

Finally, we saw in Proposition II.4.2 that the restriction of the σ-ordering to every submonoid of B_n generated by finitely many conjugates of the generators σ_i is a well-ordering. So, the results about B_n^+ and B_n^{+*} might extend to more general monoids.

QUESTION 2.10. *Let B_n^{++} be the submonoid of B_n generated by all braids of the form $\beta\sigma_i\beta^{-1}$ with β a simple n-strand braid. What is the order type of the restriction of the σ-ordering to B_n^{++}?*

More generally, the algebraic study of the monoid B_n^{++} is a natural question that has not yet been addressed. It is known that this monoid is not a Garside monoid in the usual sense, but it seems nevertheless to satisfy much of the interesting properties of a Garside monoid, and, in particular, it might be associated with a new automatic structure on B_n.

2.6. Automorphisms of a free group. The study of automorphism groups and outer automorphism groups of free groups is currently an area of intense activity; see for instance [**191**] for an excellent survey. The analogy between $\mathrm{Aut}(F_n)$ and $\mathrm{Out}(F_n)$ on the one hand and mapping class groups on the other is one of the driving forces behind this research. Now, it is very well known, and explained in Chapter IX, that the braid group B_n is a subgroup of $\mathrm{Aut}(F_n)$, so it is natural to ask the following questions.

QUESTION 2.11. *Which subgroups of* Aut(F_n) *and* Out(F_n) *are left-orderable? Which ones are bi-orderable?*

There is certainly no shortage of torsion-free subgroups, *i.e.*, of candidates for being (left-)orderable. Indeed, let us consider the natural homomorphisms of Aut(F_n) to GL($n, \mathbb{Z}$) and of Out(F_n) to GL($n, \mathbb{Z}$). Using a result of Baumslag and Taylor, one can show that the preimage of any torsion-free finite-index subgroup of GL($n, \mathbb{Z}$) under either of these homomorphisms is a torsion-free finite-index subgroup of Aut(F_n) and of Out(F_n).

2.7. Curve diagrams. In Chapter X we gave a proof of Property **C** by using a relaxation algorithm for curve diagrams, in the sense explained in Chapter XI. More precisely, the algorithm works by repeatedly sliding a puncture along a so-called useful arc, and relaxing the diagram after each slide. We saw that the length of the σ-consistent output braid could grow exponentially with the length of the input braid word. The reason for this is that the length of each of the useful arcs can grow exponentially with the length of the input, whereas the length of the relaxing braid (the puncture slide) is proportional to the length of the useful arc. So the algorithm in question is inefficient, but it is so for an obvious reason, and it is easy to invent improvements of the algorithm.

In fact, it seems that the idea of relaxing curve diagrams, as explained in Chapter XI, rather tends to lead to algorithms which are very efficient, but whose efficiency is difficult to prove. Hence we have the following very vague problem.

QUESTION 2.12. *Is there a precise, provable statement which expresses the idea that any relaxation type algorithm which does not have an obvious obstruction to being of polynomial complexity has a quadratic time complexity and a linear space complexity?*

2.8. Relaxation algorithms. Question 2.12 applies in particular to the two types of relaxation algorithms discussed in detail in Chapter XI, namely Bressaud's relaxation algorithm and the transmission-relaxation schemes from [**74**].

A more concrete conjecture specifically involves the Tetris algorithm of Section XI.1. Note that the truth of the following conjecture would imply the truth of Conjecture 1.6:

CONJECTURE 2.13. *For each n, the Tetris algorithm for B_n has a quadratic time complexity, and a linear space complexity. Starting from a braid word of length ℓ in the generators $\sigma_{i,j,p}$, the running time lies in $O(\ell^2)$, and all words produced during the algorithm have length in $O(\ell)$. Moreover, the linear constants in these bounds depend linearly on the braid index n.*

As mentioned in Section XI.1, the language of braid words in normal form is recognized by a finite-state automaton, but it fails to be (synchronously) automatic.

QUESTION 2.14. *Is there an automatic structure on the braid group that is conceptually close to the approach of Section XI.1?*

Bressaud's original motivation was related to the study of random walks on the braid group B_n and of its Poisson boundary. This boundary has been identified by Kaimanovich and Masur [**117**] as the space of uniquely ergodic measured foliations on the disk D_n. Bressaud's relaxation procedure from Section XI.1 may be applied to such a foliation just as well as to a curve diagram, yielding an infinite braid word—this is like a continued fraction expansion of the measured foliation [**157**]. Thus, in the context of trying to find a more combinatorial description of the Poisson boundary, one can ask

QUESTION 2.15. *Is Bressaud's normal form stable for random walks on B_n?*

In the context of Section XI.2, the braid group B_n is equipped with the metric which gives to the braid $\Delta_{i,j}^d$ the length $\log_2(|d|+1)$. This approach provides a combinatorial model of the thick part of the Teichmüller space $\mathcal{T}$, equipped with the Teichmüller metric d_{Teich}. Further squashing this metric by giving length one to any nonzero power of a Garside-like braid $\Delta_{i,j}$ yields a combinatorial model of the Teichmüller space, equipped with the Weil–Peterson metric d_{WP}—for both of the above statements; see Rafi [**178**]. Now any word whose letters are of the form $\Delta_{i,j}^d$ represents a path in all three spaces: the Cayley graph of the braid group, the combinatorial model of $(\mathcal{T}, d_{\text{Teich}})$, and the combinatorial model of $(\mathcal{T}, d_{\text{WP}})$.

CONJECTURE 2.16. *The set of braid words produced by the transmission-relaxation algorithm forms a family of parametrized uniform quasi-geodesics in all three spaces.*

If this were true, then this normal form could serve as a very concrete and algorithmically efficient tool in the exploration of these spaces.

2.9. Triangulations. In Chapter XII we studied the Mosher normal form of a braid, which is a sequence of combinatorial types of triangulations on the surface D_n, where each element of the sequence is obtained from the preceding one by an edge flip. The connection between braid groups and triangulation sequences comes from the fact that the complex of triangulations of D_n, where triangulations are adjacent in the complex if they differ by an edge flip, is a refinement of the Cayley graph of the braid group, and, more precisely, is quasi-isometric to it. We saw in Chapter XII that the Mosher normal form is a useful tool for understanding the σ-ordering.

Since Mosher's discovery of the (automatic) normal form, other complexes which are quasi-isometric to the Cayley graph of B_n have greatly contributed to our understanding of braid groups, for instance the train track complex [**104**] and the marking complex [**151**]. Other complexes, like the pants complex [**24**] and the curve complex [**151**], have also been studied in great depth.

QUESTION 2.17. *Can any of the above-mentioned complexes contribute to our understanding of braid orderings?*

2.10. Hyperbolic geometry. We have seen in Chapter XIII how to define an infinite family of distinct left-invariant orderings on B_n whose restriction to B_n^+ is a well-ordering, but we did not address the determination of the length of that well-ordering. The following question may well be quite easy to answer.

QUESTION 2.18. *What are the possible ordinal types for the restriction of $\prec$ to B_n^+ when $\prec$ is an ordering of Nielsen–Thurston type?*

2.11. The space of all orderings of B_n. We have seen in Chapter XIV that there exist many orders on B_n. In connection with Questions 2.18 above and 2.24 below, it is natural to raise:

QUESTION 2.19. *Assume that $\prec$ is a left-invariant ordering of B_n that satisfies the subword property. What are the possible ordinal types for the restriction of $\prec$ to B_n^+?*

As B_n is not bi-orderable, one could imagine that only long orders may exist on it. This is *not* the case, as is shown in the following ordering, which was already considered in Remark XIII.1.9. We recall that ϵ denotes the exponent sum, *i.e.*, the homomorphism of B_∞ to $\mathbb{Z}$ that maps every σ_i to 1.

PROPOSITION 2.20. *For β, β' in B_n, declare that $\beta <_\epsilon \beta'$ is true if we have either $\epsilon(\beta) < \epsilon(\beta')$, or $\epsilon(\beta) = \epsilon(\beta')$ and $\beta < \beta'$. Then $<_\epsilon$ is a left-invariant ordering of B_n whose restriction to B_n^+ is a well-ordering of ordinal type ω.*

PROOF. For each (positive) braid β, each braid γ satisfying $\gamma <_\epsilon \beta$ must satisfy $\epsilon(\gamma) \leqslant \epsilon(\beta)$. For fixed n, there exist only finitely many positive braids γ satisfying this condition. □

Note that the ordering $<_\epsilon$ is the lexicographical ordering deduced from the exact sequence

$$1 \to [B_n, B_n] \to B_n \to \mathbb{Z} \to 1$$

using the σ-ordering of the commutator subgroup and the usual ordering of $\mathbb{Z}$.

Other natural questions involve convex subgroups. As already mentioned, the family of convex subgroups of a left-ordered group is linearly ordered under inclusion and closed under unions and intersections. According to Section II.3.4, the σ-ordering of B_n has exactly n convex subgroups, including $\{1\}$ and B_n itself, and only the latter two subgroups are normal.

On the other hand, with respect to the ordering of Proposition 2.20, the commutator subgroup $[B_n, B_n]$ is both convex and normal. Other convex subgroups are $H_k \cap [B_n, B_n]$, where H_k denotes the subgroup of B_n generated by $\sigma_k, \sigma_{k+1}, \dots, \sigma_{n-1}$, but they are not normal.

For a third example, consider the special case $n = 3$. The commutator subgroup $[B_3, B_3]$ is free on two generators, hence it is bi-orderable. In fact, using the Magnus ordering of this free group, one obtains infinitely many convex subgroups, namely the inverse images of the ideals $1 + O(X^k)$. Using this ordering of $[B_3, B_3]$ and the lexicographic ordering as described in the previous paragraph, one can construct a left-ordering of B_3 which has infinitely many distinct convex subgroups.

QUESTION 2.21. *What convex subgroups must a left-invariant ordering of B_n admit? Is there a left-invariant ordering of B_n which has no convex subgroups at all, other than $\{1\}$ and B_n? What about bi-invariant orderings of PB_n?*

2.12. Pure braid groups. The Magnus ordering of the pure braid group PB_n shares with the σ-ordering of B_n the property that its restriction to the monoid B_n^+ of positive braids—and, similarly, to the dual braid monoid B_n^{+*}—is well-ordered. This leads to several problems.

First, every positive pure braid receives a unique ordinal rank that describes its position in the well-ordered set $(PB_n^+, <_{\mathrm{M}})$. As in the case of B_n^+ and the σ-ordering, we can raise

QUESTION 2.22. *Does there exist a practical method for determining the rank of a pure braid in $(PB_n^+, <_{\mathrm{M}})$?*

We observed that the Magnus ordering extends to the pure braid group PB_∞, and we can consider its restriction to the positive monoid PB_∞^+. It is easy to see that $(PB_\infty^+, <_{\mathrm{M}})$ is not a well-ordering as it admits the infinite descending sequence $\sigma_1^2 >_{\mathrm{M}} \sigma_2^2 >_{\mathrm{M}} \dots$. The situation resembles that of the σ-ordering. In the latter case, we obtained a well-ordering of B_∞^+ by considering a flipped version so as to reverse the problematic inequalities. The point is that $\mathrm{sh}(B_{n-1}^+)$ is the initial segment of $(B_n^+, <)$ determined by σ_1, implying that, after the flip, B_{n-1}^+ is the initial segment of $(B_n^+, <^\Phi)$ determined by σ_{n-1}. The counterpart of that property fails for the Magnus ordering of PB_n: every pure braid in $\mathrm{sh}(PB_\infty^+)$ is smaller than σ_1^2, but the converse is false, as we have for instance $1 <_{\mathrm{M}} \sigma_2\sigma_1^2\sigma_2 <_{\mathrm{M}} \sigma_1^2$. The example of $1 <_{\mathrm{M}} \sigma_3\sigma_2\sigma_1^2\sigma_2\sigma_3 <_{\mathrm{M}} \sigma_3\sigma_2^2\sigma_3$ shows that $\mathrm{sh}(PB_{n-1}^+)$ is not even convex in PB_n^+. This does not discard the possibility of defining a flipped version of the Magnus ordering on PB_n^+, and then on PB_∞^+, but the structure of the latter is unclear.

QUESTION 2.23. *Let $<_{\mathrm{M},n}^\Phi$ denote the image of the Magnus ordering of PB_n under the flip automorphism Φ_n. Is the induced ordering of PB_∞^+ a well-ordering? If it is, what is its order type? Would some variant of the Magnus ordering be more suitable for such constructions?*

There seems to be a large difference of complexity between the σ-ordering of B_n and the Magnus ordering of PB_n. In particular, this difference is visible in the gap between the order types of the restrictions to B_n^+, namely the relatively large ordinal $\omega^{\omega^{n-2}}$ for the former, to be compared with the modest ordinal ω^{n-1} for the latter. It is natural to wonder whether this difference is essential.

QUESTION 2.24. *Can there exist a bi-invariant ordering of PB_n whose restriction to PB_n^+ is a well-ordering of order type larger than ω^{n-1}?*

The point here is that we consider *bi*-invariant orderings of PB_n: the restriction of the σ-ordering to PB_n is a left-invariant ordering of PB_n, whose restriction to PB_n^+ is a well-ordering whose order type is easily checked to be $\omega^{\omega^{n-2}}$.

More generally, one could wonder whether a bi-invariant ordering on a group G can be as complicated—in a sense to be made precise—as a left-invariant ordering of G.

3. Generalizations and extensions

The braid groups can be generalized in many respects, so extending the results mentioned in this text to other groups is an obvious task. Of course, several types of extensions may be considered: extending orderability, extending the specific σ-ordering of braids, extending the various approaches that lead to that ordering, extending the associated algorithms, etc. Here, we shall briefly review a few results and conjectures involving such extensions, but we shall not try to be exhaustive.

3.1. Artin–Tits groups. Starting from the presentation of B_n, rather than from any geometric description, we can situate braid groups in a larger framework of a completely different nature: they are special cases of Artin–Tits groups, and more specifically spherical Artin–Tits groups, as introduced in [**65, 21**].

An *Artin–Tits* group is, by definition, a group admitting a presentation with finitely many generators $s_1, \dots, s_n$ and relations of the form $s_is_js_is_j\dots = s_js_is_js_i\dots$, where the words on both sides of the equality sign have the same length (finite and at least 2) depending on i and j, and there is at most one relation for each pair $\{i, j\}$.

For instance, finitely generated free groups (no relations) and free Abelian groups (commutation relations between all pairs of generators) are Artin–Tits groups. An Artin–Tits group is said to be *spherical* if the associated Coxeter group, namely the group obtained by adding the relations $s_i^2 = 1$ for $i = 1, \dots, n$, is finite [**112**]. The braid group B_n is then the spherical Artin–Tits group associated with the symmetric group $\mathfrak{S}_n$, thus corresponding with the so-called Coxeter type A_{n-1}.

QUESTION 3.1. *Which Artin–Tits groups are left-orderable or bi-orderable?*

Currently, the only Artin–Tits groups known to be left-orderable are those that embed in mapping class groups. Among the spherical ones, these are all but those of type $\mathrm{E}_6, \mathrm{E}_7$, and E_8 [**193, 173**]. Let us mention that, if the Artin–Tits group of type E_8 is left-orderable, then, due to embedding properties, all spherical Artin–Tits groups are; see [**159**]. Among the nonspherical ones, there is one well-known family of groups that are bi-orderable [**72**], namely the right-angled Artin–Tits groups (also called partially commutative groups), which have only commutation relations. Indeed, these groups embed in pure surface braid groups; see Section 3.2.

The more specific question of extending the σ-ordering of braid groups to other Artin–Tits groups seems rather artificial and is not very promising in general. It is well known that sending s_1 to σ_1^2 and s_i to σ_i for $i \geqslant 2$ defines an embedding of the type B_n Artin–Tits group into the corresponding type A_n group, *i.e.*, into the braid group B_{n+1}. In this way, one obtains an exact counterpart of the σ-ordering for each type B_n Artin–Tits group and, more generally, for every Artin–Tits group that is a product of type A and type B Artin–Tits groups.

In [**186**], Hervé Sibert proves

PROPOSITION 3.2. *The counterpart of Property* **A** *is true in every Artin–Tits group. The counterpart of Property* **C** *is true only for those groups that are products of type* A *and type* B *groups.*

Thus, except in the special cases of types A and B, extending the definition of the σ-ordering leads to a partial ordering only.

Many algebraic properties of spherical Artin–Tits groups extend to a larger class of groups called Garside groups [**53, 55, 63, 175, 174, 185**]. In particular, the latter are known to be torsion-free.

QUESTION 3.3. *Is every Garside group left-orderable?*

3.2. Mapping class groups and surface braid groups. We defined in Section I.3 the mapping class group $\mathcal{MCG}(\mathcal{S}, \mathcal{P})$ of any compact surface $\mathcal{S}$ relative to a finite set of punctures $\mathcal{P}$. Closely related is the *n-strand braid group* $B_n(\mathcal{S})$ of a surface $\mathcal{S}$. It can be defined as the fundamental group of the configuration space of n unlabelled points in $\mathcal{S}$. More geometrically, we can fix arbitrarily n distinguished points $P_1, \dots, P_n$ in the interior of $\mathcal{S}$. Then $B_n(\mathcal{S})$ is the group of isotopy classes of braids in $[0,1] \times \mathcal{S}$, where each strand starts at one of the points $\{0\} \times P_i$ and ends at one of the points $\{1\} \times P_j$. For instance, we have $B_1(\mathcal{S}) = \pi_1(\mathcal{S})$ for every surface $\mathcal{S}$. It is a simple fact [**13**] that for all compact surfaces $\mathcal{S}$, the braid group $B_n(\mathcal{S})$ is in a natural way a subgroup of $\mathcal{MCG}(\mathcal{S}, \{P_1, \dots, P_n\})$, except if $\mathcal{S}$ is one of the following: the sphere S^2, the sphere with one or with two points removed, the torus, or the Klein bottle.

PROPOSITION 3.4. *Let* $\mathcal{S}$ *be any compact surface with nonempty boundary. Then* $\mathcal{MCG}(\mathcal{S})$ *is left-orderable.*

In this statement the surface may or may not have punctures, and may or may not be orientable. A proof of this fact appeared in [**182**]. It uses a simple generalisation of the curve diagram construction from Section X.1.2. Since subgroups of left-orderable groups are also left-orderable, we deduce the following.

COROLLARY 3.5. *Let $\mathcal{S}$ be any compact surface with or without punctures, orientable or nonorientable, but necessarily with $\partial\mathcal{S} \neq \emptyset$. Then $B_n(\mathcal{S})$ is left-orderable.*

However, no interesting analogue of the notion of σ-positivity is known in this case. Nevertheless, it would be interesting to generalize the classification of Nielsen–Thurston type orderings encountered in Chapter XIII to this setting.

The situation is much more subtle if $\mathcal{S}$ is a compact surface without boundary. The mapping class groups of such surfaces have torsion, and consequently are not left-orderable.

QUESTION 3.6. *If $\mathcal{S}$ is a compact orientable surface without boundary, is the surface braid group $B_n(\mathcal{S})$ left-orderable?*

Other interesting questions occur when we consider the pure braid groups of a surface. By definition, the pure n-strand braid group in a surface $\mathcal{S}$, denoted $PB_n(\mathcal{S})$, is the fundamental group of the configuration space of n labelled points in the surface $\mathcal{S}$—or, equivalently the group of pure braids in $\mathcal{S} \times [0,1]$ where each strand has one endpoint in $\mathcal{S} \times \{0\}$ and the other in $\mathcal{S} \times \{1\}$.

Some of these braid groups are quite obviously not bi-orderable. For instance, for $n \geqslant 3$, the pure n-strand braid group of the sphere $PB_n(S^2)$ has torsion. Indeed, if Δ^2 denotes the usual full-twist braid inside an embedded disk in S^2 containing all the punctures, then Δ^2 is nontrivial whereas its square Δ^4 is trivial—this is the famous belt trick.

Similarly, if the surface $\mathcal{S}$ is nonorientable, then, for $n \geqslant 2$, a generator σ_i^2 of $PB_n(\mathcal{S})$ is conjugate to its own inverse—the conjugating element being a pure braid which pushes the two strands involved in σ_i once around an embedded Möbius band. So $PB_n(\mathcal{S})$ has generalized torsion.

J. González-Meneses proved [**100**] that these obvious obstructions to bi-orderability are the only ones:

PROPOSITION 3.7. *If $\mathcal{S}$ is an orientable closed surface of genus $g \geqslant 1$, then, for $n \geqslant 1$, the pure braid group $PB_n(\mathcal{S})$ is bi-orderable.*

The proof works by developing the ideas of Section XV.3, and combining them with some delicate combinatorics in surface braid groups [**101**].

As an immediate consequence of the theorem, we have that all right-angled Artin groups are bi-orderable, because, according to [**39**], they embed in pure surface braid groups. As a further corollary we have that all subgroups of right-angled Artin groups are bi-orderable, and this class of groups is surprisingly rich: it contains, for instance, all graph braid groups, all surface groups except the three simplest nonorientable ones, and certain 3-manifold groups.

Beyond the question of orderability, one may also wish to extend other techniques developed in this book to the more general context of mapping class groups. In this respect, it is especially tempting to try and generalize the ideas of Section XI.2 to other situations. This leads to the following ambitious claim.

CONJECTURE 3.8. *All results and techniques mentioned in Section* XI.2 *can be generalized to mapping class groups of higher genus surfaces.*

One could even speculate whether similar techniques might be applied to the outer automorphism group of free groups, possibly with applications to geometry of the outer space.

3.3. Torelli groups. The *Torelli group* of a surface $\mathcal{S}$ is defined to be the subgroup of $\mathcal{MCG}(\mathcal{S})$ consisting of those elements which act trivially on the homology $H_1(\mathcal{S}, \mathbb{Z})$, *i.e.*, on the Abelianization of $\pi_1(\mathcal{S})$. For a good recent survey on what is known and not known about Torelli groups, see [**81**].

PROPOSITION 3.9. *For each compact surface* $\mathcal{S}$*, the Torelli group of* $\mathcal{S}$ *is residually nilpotent, and hence bi-orderable.*

The proof follows from the deep structure theory of Torelli groups, whose fundamental results are due to Dennis Johnson [**114**]. A crucial role in this theory is played by the so-called Johnson filtration, a certain infinite sequence of subgroups different from the lower central series of the Torelli group, such that the quotient of two successive terms is always torsion-free Abelian. The structure of the Johnson filtration is in fact a manifestation of a more general phenomenon; see [**9**].

QUESTION 3.10. *Is the mapping class group* $\mathcal{MCG}(\mathcal{S})$ *virtually orderable or even virtually bi-orderable? In particular, what are the orderability properties of the kernel of the action on* $H_1(\mathcal{S}, \mathbb{Z}/p\mathbb{Z})$*, where* p *is a prime?*

This subgroup of elements acting trivially on homology with $\mathbb{Z}/p\mathbb{Z}$-coefficients is torsion-free [**113**, Chapter 1], but by a result of Hain [**103, 153**] its Abelianization is finite, at least when the genus of $\mathcal{S}$ is 3 or more. These results are related to the well-known question whether the mapping class group of a closed surface virtually surjects to $\mathbb{Z}$, *i.e.*, whether it has a finite index subgroup which has an infinite Abelian quotient.

3.4. Surface groups and 3-manifold groups. It is shown in [**181**] that the fundamental group—or, equivalently, the one-string braid group—of every compact surface, except for the projective plane $\mathbb{R}P^2$, is left-orderable. Moreover, with the further exception of the Klein bottle, all surface fundamental groups are actually bi-orderable.

The situation is more subtle when considering the case of fundamental groups of compact 3-manifolds, which we will refer to simply as 3-manifold groups. A study of these groups is initiated in the paper [**18**], where necessary and sufficient conditions are derived for the left-orderability and bi-orderability of fundamental groups of the important class of Seifert-fibred 3-manifolds (manifolds which are foliated by topological circles). It is also shown there that for each of the eight 3-dimensional geometries, there exist manifolds modelled on that geometry which have left-orderable groups and also there exist examples whose groups are not left-orderable.

Recall that a 3-manifold is called irreducible if every smooth 2-sphere bounds a 3-ball in the manifold. An important general result of [**18**] is that all compact irreducible orientable 3-manifolds with a positive first Betti number have left-orderable groups. In particular, all knot and link groups are left-orderable.

QUESTION 3.11. *Which knot groups are bi-orderable?*

The group of the figure eight knot 4_1 is bi-orderable. The first unknown case is the knot 5_2 in knot tables.

QUESTION 3.12. *Given an automorphism φ of a surface group G (or more generally of any bi-orderable group), under what conditions does there exist a bi-invariant ordering of G which is φ-invariant, meaning $x \prec y$ implies $\varphi(x) \prec \varphi(y)$?*

This is relevant to the study of 3-manifolds which are bundles over S^1, with surface fibres. If φ is the monodromy associated with such a fibration, then a φ-invariant bi-invariant ordering of the fibre's group naturally leads to a bi-invariant ordering of the fundamental group of the total space, and vice versa. In [**172**] this observation, as well as the techniques described in Chapter XV, are used to prove that certain fibred knots with pseudo-Anosov monodromy have bi-orderable groups. By contrast, the group of any torus knot cannot be bi-ordered, because it contains elements which do not commute, while a power of one of those elements commutes with the other, which cannot occur in a bi-orderable group.

CONJECTURE 3.13. *If G is the fundamental group of a closed orientable (irreducible) 3-manifold, then G is virtually bi-orderable, i.e., there exists a subgroup of finite index which is bi-orderable.*

It is shown in [**18**] that Conjecture 3.13 holds for Seifert-fibred 3-manifolds, and more generally for all manifolds with a geometric structure, except possibly hyperbolic manifolds. We do not even know if hyperbolic manifold groups are virtually left-orderable.

To put the difficulty of these questions into perspective, we point out that from general properties of orderable groups and covering space theory one can show that any 3-manifold satisfying Conjecture 3.13 also satisfies a certain well-known conjecture in 3-manifold theory; this conjecture states that any closed, orientable, irreducible 3-manifold $\mathcal{M}$ with an infinite fundamental group has a finite-sheeted cover $\widetilde{\mathcal{M}}$ with positive first Betti number. This conjecture remains open despite Perelman's recent proof of the geometrization conjecture.

In another direction, the pure braid group can be regarded as the fundamental group of the complement of the family of hyperplanes $z_i = z_j$ in the space $\mathbb{C}^n$ with coordinates $z_1, \ldots, z_n$. The analysis of orderability for PB_n applies to many other (but not all) complex hyperplane arrangements.

PROPOSITION 3.14. *The fundamental group of the complement of every hyperplane arrangement of fibre type is bi-orderable.*

For further details and a recent discussion of the fundamental groups of hyperplane arrangements, see [**167**].

3.5. A topological completion. Let us now come back to the specific case of braids and their σ-ordering. Another line of research consists in looking for extensions of that particular ordering to larger spaces. Here we start with a topological completion.

As mentioned in Section II.3.2, the topology on B_∞ associated with the σ-ordering is metrizable, the radius 2^{-n} ball centered at 1 being the shifted subgroup $\mathrm{sh}^n(B_\infty)$. With respect to that topology, a sequence $\beta_1, \beta_2, \ldots$ converges to the trivial braid if for each integer n there exists an integer p such that, for $q > p$, all braids β_q belong to $\mathrm{sh}^n(B_\infty)$. The order topology renders B_∞ homeomorphic to $\mathbb{Q}$—in particular, not completely metrizable.

Very recently, P. Fabel announced in [**79**] the following construction of a completion of B_∞. Let D_∞ be the closed unit disk in $\mathbb{C}$ centered at 0 with punctures on the real line at $0, \frac{1}{2}, \frac{2}{3}, \frac{3}{4}, \ldots$, and let H_∞ be the group of homeomorphisms of D_∞ fixing the boundary pointwise. Let $M(H_\infty)$ be the group of isotopy classes of H_∞. Viewing it as the mapping class group of a disk containing the punctures at $0, \frac{1}{2}, \ldots, 1-\frac{1}{n}$, one embeds B_n in $M(H_\infty)$. One can equip $M(H_\infty)$ with a metric d by declaring

$$d(\beta_1, \beta_2) = \inf_{h_1,h_2} \sup_{z \in D_n} d_{\mathbb{C}}(h_1(z), h_2(z)) + \inf_{h_1,h_2} \sup_{z \in D_n} d_{\mathbb{C}}(h_1^{-1}(z), h_2^{-1}(z)),$$

where h_i is a homeomorphism of D_∞ representing the isotopy class β_i. The result announced by Fabel is

PROPOSITION 3.15. *The group $(M(H_\infty), d)$ is complete as a metric space, it contains B_∞ as a dense subgroup, and it left-ordered by an ordering that extends the σ-ordering of B_∞.*

QUESTION 3.16. *Are all completions extending the σ-ordering of B_∞ essentially equivalent?*

3.6. Parenthesized braids. We shall conclude with another seemingly promising extension of the braids and their σ-ordering—one that will at least enable us to end with a nice figure.

Thompson's group F is a finitely presented group which, in many respects, is a cousin of the braid groups. Like the latter, it can be introduced in many different ways, and it has very rich properties involving geometric group theory and dynamical systems; see [**32**] for an introduction. The open question of its possible amenability has provided a strong motivation for studying the group F in recent years. Let us mention that F is nothing but the counterpart of the group G_{LD} of Section IV.3 when the associativity law replaces the self-distributivity law.

For our current purposes, it is enough to know that F admits the presentation

$$\langle a_1, a_2, \ldots \mid a_i a_{j-1} = a_j a_i \text{ for } j \geqslant i+2 \rangle. \tag{3.1}$$

It has recently been observed that the groups B_∞ and F can be married in a natural way. This was done independently by M. Brin in [**22, 23**] by constructing what was seen as a braided version of the Thompson group, and in [**58, 59**] by constructing what was seen as a Thompson version of B_∞. The groups so obtained are essentially similar; variants also appeared in [**120**] and [**92**].

From our current point of view, the most natural description is probably the one involving *parenthesized braids.*

DEFINITION 3.17. The *group of parenthesized braids* $B_\bullet$ is defined by two infinite series of generators $\sigma_1, \sigma_2, \ldots, a_1, a_2, \ldots$, subject to the following relations for $i \geqslant 1$ and $j \geqslant i+2$:

$$\left\{ \begin{array}{llll} \sigma_i\sigma_j = \sigma_j\sigma_i, & \sigma_i a_j = a_j \sigma_i, & a_i a_{j-1} = a_j a_i, & a_i \sigma_{j-1} = \sigma_j a_i, \\ \sigma_i\sigma_{i+1}\sigma_i = \sigma_{i+1}\sigma_i\sigma_{i+1}, & & \sigma_{i+1}\sigma_i a_{i+1} = a_i \sigma_i, & \sigma_i\sigma_{i+1} a_i = a_{i+1}\sigma_i. \end{array} \right. \tag{3.2}$$

The elements of $B_\bullet$ can be visualized using braid diagrams in which the distances between strands are not uniform. An ordinary braid diagram connects an initial sequence of equidistant positions to a similar final sequence. A parenthesized

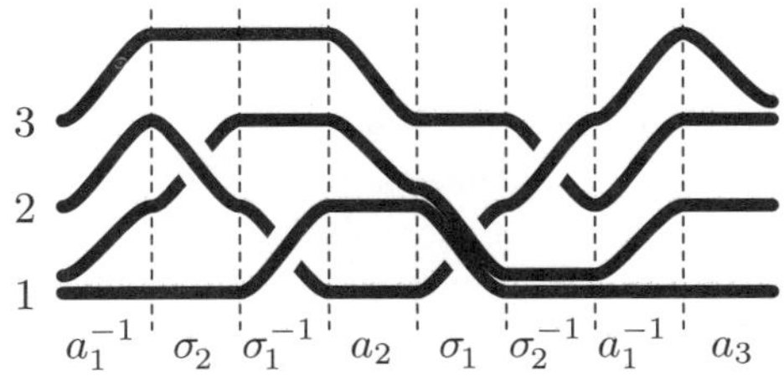

FIGURE 2. The parenthesized braid diagram encoded by the word $a_1^{-1}\sigma_2\sigma_1^{-1}a_2\sigma_1\sigma_2^{-1}a_1^{-1}a_3$. As two strands start from close to 1, the initial positions of the strands can be represented by (••)••, similarly, the final positions correspond to ••(••). The generator σ_i corresponds to the usual crossing between positions i and $i+1$ (but there may be several strands close to these positions), whereas a_i corresponds to shrinking positions from $i+1$ to i.

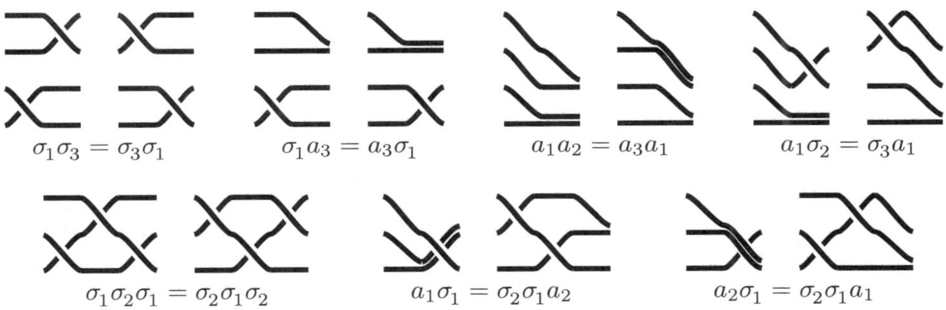

FIGURE 3. Relations of $B_\bullet$ and the corresponding diagrams isotopies: those of the top line are commutations and quasi-commutations; those of the bottom line are braid relations.

braid diagram connects a parenthesized sequence of positions to another possibly different parenthesized sequence of positions, the intuition being that grouped positions are (infinitely) closer than ungrouped ones. A typical example is shown in Figure 2. The generator σ_i corresponds to the usual crossing operator, with the difference that it involves all strands that start in the vicinity of i and $i+1$. The generator a_i corresponds to shrinking all strands that start in the vicinity of i and translating the next ones so as to avoid a gap.

The relations of (3.2) correspond to the isotopies displayed in Figure 3. As can be expected, the elements σ_i generate a copy of B_∞, while the elements a_i generate a copy of Thompson's group F. More precisely, $B_\bullet$ is a group of fractions for a monoid that is a bi-crossed product of B_∞^+ and of the monoid F^+ defined by the presentation of (3.1).

PROPOSITION 3.18. *The group $B_\bullet$ is left-orderable, by an ordering that extends the σ-ordering of B_∞.*

An application of that result—more exactly, of the specific form of the elements larger than 1 in terms of σ-positive expressions—is that the Artin representation of braids extends to $B_\bullet$, and that the latter embeds in the mapping class group of a sphere with a Cantor set of punctures (Figure 4).

Let us also mention that the natural subgroup of $B_\bullet$ corresponding to pure braids was recently shown to be bi-orderable, by an ordering that extends the ordering of PB_∞ constructed in Chapter XV [**30**].

These results do not prove that the parenthesized braid group is an extraordinary object. After all, that two groups can be glued together in a somewhat tricky way is nothing exceptional, so $B_\bullet$ might very well be just an amusing example. However, we think that $B_\bullet$ is really an important object. Once again, the variety of approaches that lead to $B_\bullet$ or to close variants, and, mainly, the unexpected way in which the technical properties fit together, suggest that something interesting is hidden there. In particular, the self-distributive structure of B_∞ described in Chapter IV extends to $B_\bullet$, a surprising result that certainly reflects deep properties. We hope for—and even predict—future applications.

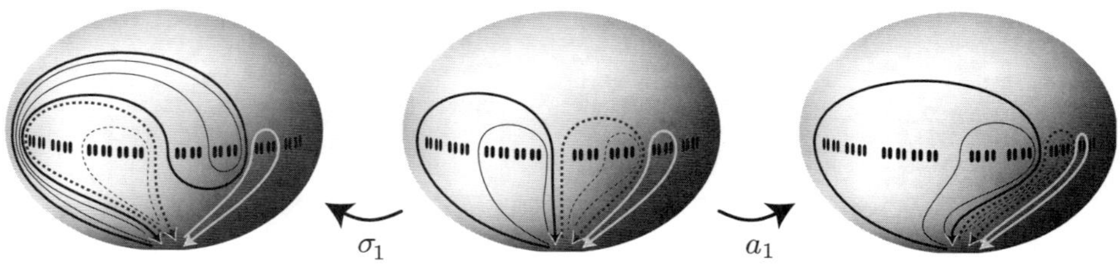

FIGURE 4. Embedding parenthesized braids in the mapping class group of a sphere with a Cantor set of punctures: σ_i acts by the usual half-twist, while a_i acts as a dilatation-translation along the equator.

Key Definitions

Sigma-ordering:

- For β, β' in B_∞, the relation $\beta < \beta'$ is true if $\beta^{-1}\beta'$ is σ-positive.
- For β, β' in B_∞, the relation $\beta <^{\Phi} \beta'$ is true if $\beta^{-1}\beta'$ is σ^{Φ}-positive.

Sigma-positive braid word:

- A braid word is *σ-positive* if the σ_i with lowest index occurs positively only.
- A braid word is *σ^{Φ}-positive* if the σ_i with highest index occurs positively only.

Sigma-positive braid:

- A braid is *σ-positive* if it admits a σ-positive representative word.
- A braid is *σ^{Φ}-positive* if it admits a σ^{Φ}-positive representative word.

Property A (Acyclicity):

- A σ-positive braid is nontrivial.

Property C (Comparison):

- Every nontrivial braid of B_n can be represented by an n-strand braid word that is σ-positive or σ-negative.

Property S (Subword):

- Every braid of the form $\beta^{-1}\sigma_i\beta$ is σ-positive.

Complementary Definitions

- A braid word is *σ_i-positive* if it contains at least one σ_i, no σ_i^{-1}, no $\sigma_j^{\pm 1}$ with $j < i$.
- ... *id.* ... *σ_i-negative* if ... at least one σ_i^{-1}, no σ_i, no $\sigma_j^{\pm 1}$ with $j < i$.
- ... *id.* ... *σ_i-free* if ... no $\sigma_j^{\pm 1}$with $j \leqslant i$.

- A braid is called *σ_i-positive* if it admits a σ_i-positive expression, etc.

Property A (second, equivalent form): A σ_1-positive braid is nontrivial.

Property C (second, equivalent form): Every braid of B_n can be represented by an n-strand braid word that is σ_1-positive, σ_1-negative, or σ_1-free.

Bibliography

1. S.I. Adyan, *Fragments of the word Delta in a braid group*, Mat. Zametki Acad. Sci. SSSR **36** (1984), no. 1, 25–34, (Russian); English translation in *Math. Notes of the Acad. Sci. USSR* **36** (1984), no. 1, p. 505–510.
2. I. Agol, J. Hass, and W.P. Thurston, *The computational complexity of knot genus and spanning area*, Trans. Amer. Math. Soc. **358** (2006), no. 9, 3821–3850, (electronic).
3. I. Anshel, M. Anshel, and D. Goldfeld, *An algebraic method for public-key cryptography*, Math. Res. Lett. **6** (1999), 287–291.
4. E. Artin, *Theorie der Zöpfe*, Abh. Math. Sem. Univ. Hamburg **4** (1925), 47–72.
5. ———, *Theory of braids*, Ann. of Math. **48** (1947), 101–126.
6. S. Arworn and Y. Kim, *On finitely-determined total orders*, preprint.
7. L. Bacardit and W. Dicks, *Actions of the braid group, and new algebraic proofs of results of Dehornoy and Larue*, arXiv: math.GR/0705.0587.
8. V.G. Bardakov, *On the theory of braid groups*, Mat. Sb. **183** (1992), no. 6, 3–42, (Russian. English summary); English translation in *Acad. Sci. Sb. Math.* **76** (1993), no. 1, p. 123–153.
9. H. Bass and A. Lubotzky, *Linear-central filtrations on groups*, The mathematical legacy of Wilhelm Magnus: groups, geometry and special functions (Brooklyn, NY, 1992), Contemp. Math., vol. 169, American Mathematical Society, 1994, pp. 45–98.
10. D. Bessis, *The dual braid monoid*, Ann. Sci. École Norm. Sup. **36** (2003), 647–683.
11. M. Bestvina and K. Fujiwara, *Quasi-homomorphisms on mapping class groups*, preprint; arXiv:math.GR/0702273, 2007.
12. S. Bigelow, *Braid groups are linear*, J. Amer. Math. Soc. **14** (2001), no. 2, 471–486.
13. J. Birman, *On braid groups*, Comm. Pure Appl. Math. **22** (1969), 41–72.
14. ———, *Braids, Links, and Mapping Class Groups*, Ann. of Math. Stud., vol. 82, Princeton Univ. Press, 1974.
15. J. Birman, K.H. Ko, and S.J. Lee, *A new approach to the word problem in the braid groups*, Adv. Math. **139** (1998), no. 2, 322–353.
16. J. Birman and W. Menasco, *Studying links via closed braids III: classifying links which are closed 3-braids*, Pacific J. Math. **161** (1993), 23–113.
17. N. Bourbaki, *Algèbre, chapitres I–III*, Hermann, Paris, 1970.
18. S. Boyer, D. Rolfsen, and B. Wiest, *Orderable 3-manifold groups*, Ann. Inst. Fourier (Grenoble) **55** (2005), 243–288.
19. X. Bressaud, *A normal form for braids*, J. Knot Theory Ramifications **17** (2008), no. 6, 697–732.
20. E. Brieskorn, *Automorphic sets and braids and singularities*, Braids, Contemp. Math., vol. 78, American Mathematical Society, 1988, pp. 45–117.
21. E. Brieskorn and K. Saito, *Artin-Gruppen und Coxeter-Gruppen*, Invent. Math. **17** (1972), 245–271.
22. M. Brin, *The algebra of strand splitting I. A braided version of Thompson's group V*, J. Group Theory, to appear; arXiv math.GR/040642.
23. ———, *The algebra of strand splitting II. A Presentation for the braid group on one strand*, Int. J. Algebra and Computation **16** (2006), 203–219.
24. J. Brock, *The Weil-Petersson metric and volumes of 3-dimensional hyperbolic convex cores*, J. Amer. Math. Soc. **16** (2003), 495–535.
25. S. D. Brodskii, *Equations over groups, and groups with one defining relation*, Sibirsk. Mat. Zh. **25** (1984), 84–103.
26. S. Burckel, *L'ordre total sur les tresses positives*, PhD. Thesis, Université de Caen, 1994.
27. ———, *The well-ordering on positive braids*, J. Pure Appl. Algebra **120** (1997), no. 1, 1–17.

28. ———, *Computation of the ordinal of braids*, Order **16** (1999), 291–304.
29. ———, *Syntactical methods for braids of three strands*, J. Symbolic Comput. **31** (2001), 557–564.
30. J. Burillo and J. González-Meneses, *Biorderings on pure braided Thompson's groups*, Quarterly J. Math., to appear.
31. R. Burns and V. Hale, *A note on group rings of certain torsion-free groups*, Canad. Math. Bull. **15** (1972), 441–445.
32. J.W. Cannon, W.J. Floyd, and W.R. Parry, *Introductory notes on Richard Thompson's groups*, Enseign. Math. **42** (1996), 215–257.
33. L. Carlucci, P. Dehornoy, and A. Weiermann, *Unprovability statements involving braids*, preprint; arXiv:math.LO/0711.3785, 2007.
34. A. Casson and S. Bleiler, *Automorphisms of surfaces after Nielsen and Thurston*, London Math. Soc. Student Texts, vol. 9, Cambridge University Press, 1988.
35. J. Chamboredon, *Tresses, relaxation de lacet et forme normale de Bressaud*, Master Memoir, University de Caen, 2007, http://www.eleves.ens.fr/home/chambore/maths.en.html.
36. W.L. Chow, *On the algebraic braid group*, Ann. of Math. **49** (1948), 654–658.
37. A. Clay and D. Rolfsen, *Densely ordered braid subgroups*, J. Knot Theory Ramifications **16** (2007), no. 7, 869–878.
38. P.F. Conrad, *Right-ordered groups*, Michigan Math. J. **6** (1959), 267–275.
39. J. Crisp and B. Wiest, *Quasi-isometrically embedded subgroups of braid and diffeomorphism groups*, Trans. Amer. Math. Soc. **359** (2007), no. 11, 5485–5503, (electronic).
40. R.H. Crowell and R.H. Fox, *Introduction to Knot Theory*, Grad. Texts in Math., vol. 57, Springer-Verlag, 1977.
41. M. Dabkovska, M. Dabkowski, V. Harizanov, J. Przytycki, and M. Veve, *Compactness of the space of left orders*, J. Knot Theory Ramifications **16** (2007), 267–256.
42. P. Dehornoy, *Infinite products in monoids*, Semigroup Forum **34** (1986), 21–68.
43. ———, *Free distributive groupoids*, J. Pure Appl. Algebra **61** (1989), 123–146.
44. ———, *Sur la structure des gerbes libres*, C. R. Acad. Sci. Paris Sér. I Math. **309** (1989), 143–148.
45. ———, *Deux propriétés des groupes de tresses*, C. R. Acad. Sci. Paris Sér. I Math. **315** (1992), 633–638.
46. ———, *Structural monoids associated to equational varieties*, Proc. Amer. Math. Soc. **117** (1993), no. 2, 293–304.
47. ———, *Braid groups and left distributive operations*, Trans. Amer. Math. Soc. **345** (1994), no. 1, 115–151.
48. ———, *The structure group for the associativity identity*, J. Pure Appl. Algebra **111** (1996), 59–82.
49. ———, *A fast method for comparing braids*, Adv. Math. **125** (1997), 200–235.
50. ———, *Groups with a complemented presentation*, J. Pure Appl. Algebra **116** (1997), 115–137.
51. ———, *Strange questions about braids*, J. Knot Theory Ramifications **8** (1999), no. 5, 589–620.
52. ———, *Braids and Self-Distributivity*, Progr. Math., vol. 192, Birkhäuser, 2000.
53. ———, *Groupes de Garside*, Ann. Sci. École Norm. Sup. (4) **35** (2002), 267–306.
54. ———, *Study of an identity*, Algebra Universalis **48** (2002), 223–248.
55. ———, *Thin groups of fractions*, Combinatorial and Geometric Group Theory, Contemp. Math., vol. 296, American Mathematical Society, 2002, pp. 95–128.
56. ———, *Braid-based cryptography*, Group Theory, Statistics, and Cryptography, Contemp. Math., vol. 360, American Mathematical Society, 2004, pp. 5–33.
57. ———, *The group of fractions of a torsion free lcm monoid is torsion free*, J. Algebra **281** (2004), 303–305;.
58. ———, *Geometric presentations of Thompson's groups*, J. Pure Appl. Algebra **203** (2005), 1–44.
59. ———, *The group of parenthesized braids*, Adv. Math. **205** (2006), 354–409.
60. ———, *Combinatorics of normal sequences of braids*, J. Combin. Theory Ser. A **114** (2007), 389–409.
61. ———, *Still another approach to the braid ordering*, Pacific J. Math. **232** (2007), no. 1, 139–176.

62. ———, *Alternating normal forms for braids and locally Garside monoids*, J. Pure Appl. Algebra **212** (2008), no. 1, 2416–2439.
63. P. Dehornoy and L. Paris, *Gaussian groups and Garside groups, two generalisations of Artin groups*, Proc. London Math. Soc. **79** (1999), no. 3, 569–604.
64. P. Dehornoy and B. Wiest, *On word reversing in braid groups*, Internat. J. Algebra Comput. **16** (2006), no. 5, 941–957.
65. P. Deligne, *Les immeubles des groupes de tresses généralisés*, Invent. Math. **17** (1972), 273–302.
66. F. Deloup, *Palindromes and orderings in Artin groups*, Algebr. Geom. Topol. **5** (2005), 419–442.
67. F. Digne and J. Michel, *Garside and locally Garside categories*, arXiv: math.GR/0612652.
68. R. Dougherty, *Critical points in an algebra of elementary embeddings*, Ann. Pure Appl. Logic **65** (1993), 211–241.
69. R. Dougherty and T. Jech, *Finite left-distributive algebras and embedding algebras*, Adv. Math. **130** (1997), 201–241.
70. A. Drápal, *Persistence of cyclic left-distributive algebras*, J. Pure Appl. Algebra **105** (1995), 137–165.
71. T. Dubrovina and N. Dubrovin, *On braid groups*, Sb. Math. **192** (2001), 693–703.
72. G. Duchamp and J.-Y. Thibon, *Simple orderings for free partially commutative groups*, Internat. J. Algebra Comput. **2** (1992), no. 3, 351–355.
73. I. Dynnikov, *On a Yang-Baxter mapping and the Dehornoy ordering*, Uspekhi Mat. Nauk **57** (2002), no. 3, 151–152, (Russian); English translation in *Russian Math. Surveys* **57** (2002), no. 3.
74. I. Dynnikov and B. Wiest, *On the complexity of braids*, J. Europ. Math. Soc. **9** (2007), no. 4, 801–840.
75. E.A. El-Rifai and H.R. Morton, *Algorithms for positive braids*, Quart. J. Math. Oxford Ser. **45** (1994), no. 2, 479–497.
76. D. Epstein, *Curves on 2-manifolds and isotopies*, Acta Math. **115** (1966), 83–107.
77. D. Epstein, J.W. Cannon, D.F. Holt, S.V.F. Levy, M.S. Paterson, and W.P. Thurston, *Word Processing in Groups*, Jones and Bartlett Publ., 1992.
78. P. Etingof, T. Schedler, and A. Soloviev, *Set-theoretical solutions to the quantum Yang–Baxter equation*, Duke Math. J. **100** (1999), no. 2, 169–209.
79. P. Fabel, *The mapping class group of a disk with infinitely many holes*, preprint, 2007.
80. M. Falk and R. Randell, *The lower central series of a fiber-type arrangement*, Invent. Math. **82** (1985), 77–88.
81. B. Farb, *Some problems on mapping class groups and moduli space*, Problems on Mapping Class Groups and Related Topics, Proc. Sympos. Pure Math., vol. 74, 2006, pp. 11–55.
82. A. Fathi, F. Laudenbach, and V. Poenatu, *Travaux de Thurston sur les surfaces*, Astérisque, vol. 66-67, Soc. Math. de France, 1979.
83. R. Fenn, M.T. Greene, D. Rolfsen, C. Rourke, and B. Wiest, *Ordering the braid groups*, Pacific J. Math. **191** (1999), 49–74.
84. R. Fenn, D. Rolfsen, and J. Zhu, *Centralisers in braid groups and singular braid monoids*, Enseign. Math. **42** (1996), 75–96.
85. R. Fenn and C.P. Rourke, *Racks and links in codimension 2*, J. Knot Theory Ramifications **1** (1992), 343–406.
86. V.V. Fock, *Dual Teichmüller spaces*, http://front.math.ucdavis.edu/dg-ga/9702018.
87. V.V. Fock and A.B. Goncharov, *Dual Teichmuller and lamination spaces*, to appear in the Handbook on Teichmuller theory; arXiv:math/0510312.
88. H. Friedman, *Higher set theory and mathematical practice*, Ann. Math. Logic **2** (1971), 325–357.
89. ———, *On the necessary use of abstract set theory*, Adv. Math. **41** (1981), 209–280.
90. J. Fromentin, *The cycling normal form on dual braid monoids*, arXiv: math.GR/0712.3836.
91. ———, *A well-ordering of dual braid monoids*, Comptes Rendus Mathematique **346** (2008), no. 13-14, 729–734.
92. L. Funar and C. Kapoudjian, *On a universal mapping class group in genus zero*, Geom. Funct. Anal. **14** (2004), 965–1012.
93. J. Funk, *The Hurwitz action and braid group orderings*, Theory Appl. Categ. **9** (2001), no. 7, 121–150.

94. F.A. Gambaudo and E. Ghys, *Braids and signatures*, Bull. Soc. Math. France **133** (2005), no. 4, 541–579.
95. F.A. Garside, *The braid group and other groups*, Quart. J. Math. Oxford Ser. **20** (1969), 235–254.
96. B.J. Gassner, *On braid groups*, Abh. Math. Sem. Univ. Hamburg **25** (1961), 10–22.
97. K.F. Gauss, *Handbuch 7*, Univ. Göttingen collection.
98. E. Ghys, *Groups acting on the circle*, Enseign. Math. **47** (2001), no. 2, 329–407.
99. D. Goldsmith, *Homotopy of braids—in answer to a question of E. Artin*, Topology Conference, Lecture Notes in Math., vol. 375, Springer, Berlin, 1974, Virginia Polytech. Inst. and State Univ., Blacksburg, Va., 1973, pp. 91–96.
100. J. González-Meneses, *Ordering pure braid groups on compact, connected surfaces*, Pacific J. Math. **203** (2002), 369–378.
101. J. González-Meneses and L. Paris, *Vassiliev invariants for braids on surfaces*, Trans. Amer. Math. Soc. **356** (2004), no. 1, 219–243.
102. E.A. Gorin and V.Ya. Lin, *Algebraic equations with continuous coefficients, and certain questions of the algebraic theory of braids*, Sb. Math. **7** (1969), 569–596.
103. R. Hain, *Torelli groups and geometry of moduli spaces of curves*, Current topics in complex algebraic geometry, MSRI Publ., Berkeley, vol. 28, 1995, pp. 97–143.
104. U. Hamenstädt, *Geometry of the mapping class groups II: (Quasi)-geodesics*, preprint, arXiv: math.GR/0511.349, 2005.
105. A. Hatcher and W. Thurston, *A presentation for the mapping class group of a closed orientable surface*, Topology **19** (1980), no. 3, 221–237.
106. M. Hertweck, *A counterexample to the isomorphism problem for integral group rings*, Ann. of Math. **154** (2001), 115–136.
107. G. Higman, *The units of group rings*, Proc. London Math. Soc. **46** (1940), no. 2, 231–248.
108. ______, *Ordering by divisibility in abstract algebras*, Proc. London Math. Soc. **2** (1952), 326–336.
109. F. Hivert, J.-C. Novelli, and J.-Y. Thibon, *Sur une conjecture de Dehornoy*, Comptes Rendus Mathematique **346** (2008), no. 7-8, 375–378.
110. J.G. Hocking and G.S. Young, *Topology*, Addison–Wesley, Reading MA, 1961.
111. O. Hölder, *Die Axiome der Quantität und die Lehre vom Mass*, Math.-Phys. Kl **53** (1901), 1–64.
112. J.E. Humphreys, *Reflection groups and Coxeter groups*, Cambridge Stud. Adv. Math., vol. 29, Cambridge University Press, Cambridge, 1990.
113. S. Ivanov, *Subgroups of Teichmüller modular groups*, Transl. Math. Monogr., vol. 115, American Mathematical Society, Providence, RI, 1992.
114. D. Johnson, *A survey of the Torelli group*, Contemp. Math., vol. 20, American Mathematical Society, 1983, pp. 165–178.
115. V. Jones, *Hecke algebra representations of braid groups and link polynomials*, Ann. of Math. **126** (1987), 335–388.
116. D. Joyce, *A classifying invariant of knots: the knot quandle*, J. Pure Appl. Algebra **23** (1982), 37–65.
117. V. Kaimanovich and H. Masur, *The Poisson boundary of the mapping class group*, Invent. Math. **125** (1996), no. 2, 221–264.
118. S. Kamada, *Braid and Knot Theory in Dimension Four*, Math. Surveys Monogr., vol. 95, American Mathematical Society, 2002.
119. A. Kanamori, *The Higher Infinite*, Perspect. Math. Logic, Springer Verlag, 1994.
120. C. Kapoudjian and V. Sergiescu, *An extension of the Burau representation to a mapping class group associated to Thompson's group T*, Geometry and Dynamics, Contemp. Math., vol. 389, American Mathematical Society, 2005, pp. 141–164.
121. C. Kassel, *L'ordre de Dehornoy sur les tresses*, Séminaire Bourbaki, Astérisque, vol. 276, Soc. Math. France, 2002, exposé 865 (novembre 1999), pp. 7–28.
122. C. Kassel and C. Reutenauer, *Sturmian morphisms, the braid group B_4, Christoffel words and bases of F_2*, Ann. Mat. Pura Appl. (4) **186** (2007), no. 2, 317–339.
123. C. Kassel and V. Turaev, *Braid groups*, Springer Verlag, 2007.
124. D.M. Kim and D. Rolfsen, *An ordering for groups of pure braids and fibre-type hyperplane arrangements*, Canad. J. Math. **55** (2002), 822–838.

125. L. Kirby and J. Paris, *Accessible independence results for Peano Arithmetic*, Bull. London Math. Soc. **14** (1982), 285–293.
126. K.H. Ko, S. Lee, J.H. Cheon, J.W. Han, J. Kang, and C. Park, *New public-key cryptosystem using braid groups*, Proc. Crypto 2000, Lecture Notes in Comput. Sci., vol. 1880, Springer Verlag, 2000, pp. 166–184.
127. A.I. Kokorin, V.M. Kopyutov, and N.Ya. Medvedev, *Right-Ordered Groups*, Plenum Publishing Corporation, 1996.
128. D. Krammer, *The braid group B_4 is linear*, Invent. Math. **142** (2000), 451–486.
129. ———, *A class of Garside groupoid structures on the pure braid group*, Trans. Amer. Math. Soc. **360** (2008), 4029–4061.
130. R.H. La Grange and A.H. Rhemtulla, *A remark on the group rings of order preserving permutation groups*, Canad. Math. Bull. **11** (1968), 679–680.
131. D.M. Larue, *Left-distributive and left-distributive idempotent algebras*, PhD. Thesis, University of Colorado, Boulder, 1994.
132. ———, *On braid words and irreflexivity*, Algebra Universalis **31** (1994), 104–112.
133. R. Laver, *Elementary embeddings of a rank into itself*, Abstracts Amer. Math. Soc. **7** (1986), 6.
134. ———, *The left distributive law and the freeness of an algebra of elementary embeddings*, Adv. Math. **91** (1992), no. 2, 209–231.
135. ———, *A division algorithm for the free left distributive algebra*, Logic Colloquium '90 (Oikkonen and *al*, eds.), Lect. Notes in Logic, vol. 2, Springer Verlag, 1993, pp. 155–162.
136. ———, *On the algebra of elementary embeddings of a rank into itself*, Adv. Math. **110** (1995), 334–346.
137. ———, *Braid group actions on left distributive structures and well-orderings in the braid group*, J. Pure Appl. Algebra **108** (1996), no. 1, 81–98.
138. A. Levy, *Basic Set Theory*, Springer Verlag, 1979.
139. P. Linnell, *The topology on the space of left orderings of a group*, preprint; arXiv math.GR/0607470.
140. ———, *Zero divisors and $L^2(G)$*, C. R. Acad. Sci. Paris Sér. I Math. **315** (1992), no. 1, 49–53.
141. P.A. Linnell and T. Schick, *Finite group extensions and the Atiyah conjecture*, J. Amer. Math. Soc. **20** (2007), 1003–1051.
142. J. Longrigg and A. Ushakov, *Cryptanalysis of shifted conjugacy authentication protocol*, preprint; arXiv:math.GR/0708.1768.
143. J.H. Lu, M. Yan, and Y.C. Zhu, *On the set-theoretical Yang–Baxter equation*, Duke Math. J. **104** (2000), no. 1, 1–18.
144. W. Magnus, A. Karrass, and D. Solitar, *Combinatorial Group Theory*, J. Wiley and Sons, New York, 1966.
145. A.I. Malcev, *On the embedding of group algebras in division algebras*, Dokl. Akad. Nauk SSSR (N.S.) **60** (1948), 1499–1501.
146. A.V. Malyutin, *Fast algorithms for the recognition and comparison of braids*, Zap. Nauchn. Sem. POMI **279** (2001), 197–217, (Russian).
147. ———, *Twist number of (closed) braids*, St. Peterburg Math. J. **16** (2005), no. 5, 791–813.
148. ———, *Pseudo-characters of braid groups and primeness of links*, preprint, 2006.
149. A.V. Malyutin and A.M.Vershik, *Poisson-Furstenberg boundary of the braid groups and Markov–Ivanovsky normal form*, arXiv:math.GT/0707.1109.
150. A.V. Malyutin and N.Yu. Netstvetaev, *Dehornoy's ordering on the braid group and braid moves*, St. Peterburg Math. J. **15** (2004), no. 3, 437–448.
151. H. Masur and Y. Minsky, *Geometry of the complex of curves II: hierarchical structure*, GAFA, Geom. Funct. Anal. **10** (2000), 902–974.
152. S.V. Matveev, *Distributive groupoids in knot theory*, Sb. Math. **119** (1982), no. 1-2, 78–88.
153. J.D. McCarthy, *On the first cohomology group of cofinite subgroups in surface mapping class groups*, Topology **40** (2001), no. 2, 401–418.
154. S. McCleary, *Free lattice ordered groups represented as o-2-transitive l-permutation groups*, Trans. Amer. Math. Soc. **290** (1985), no. 1, 81–100.
155. R. McKenzie and R.J. Thompson, *An elementary construction of unsolvable word problems in group theory*, Word Problems (Boone and *al*, eds.), Stud. Logic Found. Math., vol. 71, North Holland, 1973, pp. 457–478.

156. D. Morris, *Amenable groups that act on the line*, Algebr. Geom. Topol. **6** (2006), 2509–2518.
157. L. Mosher, *Train track expansions of measured foliations*, unpublished notes available on http://andromeda.rutgers.edu/~mosher.
158. ———, *Mapping class groups are automatic*, Ann. of Math. **142** (1995), 303–384.
159. J. Mulholland and D. Rolfsen, *Local indicability and commutator subgroups of Artin groups*, preprint; arXiv: math.GR/0606116, 2006.
160. E. Munarini, *Sequence number A080635 in Sloane's "On-Line Encyclopedia of Integer Sequences"*, http://www.research.att.com/projects/OEIS?Anum=3DA080635.
161. A.G. Myasnikov, V. Shpilrain, and A.Ushakov, *A practical attack on some braid group based cryptographic protocols*, CRYPTO 2005, Lecture Notes in Comput. Sci., vol. 3621, Springer, 2005, pp. 86–96.
162. A. Navas, *On the dynamics of (left) orderable groups*, preprint; arXiv: math.GR/0710.2466, 2007.
163. B.H. Neumann, *On ordered division rings*, Trans. Amer. Math. Soc. **66** (1949), 202–252.
164. J. Nielsen, *Untersuchungen zur Topologie des geschlossenen zweiseitigen Flächen*, Acta Math. **50** (1927), 189–358.
165. ———, *Collected Mathematical Papers, edited by V.L. Hansen*, Birkhäuser, Boston-Basel-Stuttgart, 1986.
166. S.Yu. Orevkov, *Strong positivity in the right-invariant order on a braid group and quasipositivity*, Mat. Zametki **68** (2000), no. 5, 692–698, (Russian); English translation in *Math. Notes* **68** (2000), no. 5-6, 588-593.
167. L. Paris, *On the fundamental group of the complement of a complex hyperplane arrangement*, Singularities and Arrangements, Sapporo and Tokyo, 1998, Adv. Stud. Pure Math., vol. 27, Kinokuniya, 2000, pp. 257–272.
168. D.S. Passman, *The Algebraic Structure of Group Rings*, Pure Appl. Math, Wiley Interscience, 1977.
169. M.S. Paterson and A.A. Razborov, *The set of minimal braids is co-NP-complete*, J. Algorithms **12** (1991), 393–408.
170. R.C. Penner, *The decorated Teichmüller space of punctured surfaces*, Comm. Math. Phys. **113** (1987), no. 2, 299–339.
171. R.C. Penner and J.L. Harer, *Combinatorics of train tracks*, Ann. Math. Stud., vol. 125, Princeton Univerity Press, 1992.
172. B. Perron and D. Rolfsen, *On orderability of fibred knot groups*, Math. Proc. Cambridge Philos. Soc. **135** (2003), 147–153.
173. B. Perron and J.P. Vannier, *Groupe de monodromie géométrique des singularités simples*, Math. Ann. **306** (1996), no. 2, 231–245.
174. M. Picantin, *The center of thin Gaussian groups*, J. Algebra **245** (2001), no. 1, 92–122.
175. ———, *The conjugacy problem in small Gaussian groups*, Comm. Algebra **29** (2001), no. 3, 1021–1038.
176. V.V. Prasolov and A.B. Sossinsky, *Knots, links, braids, and 3-manifolds*, Transl. Math. Monogr., vol. 154, American Mathematical Society, 1997.
177. J. Przytycki, *Classical roots of knot theory*, Chaos Solitons Fractals **9** (1998), no. 4, 5, 531–545.
178. K. Rafi, *A combinatorial model for the Teichmller metric*, Geom. Funct. Anal., to appear.
179. K. Reidemeister, *Knotentheorie*, Ergeb. Math. Grenzgeb., vol. 1, Julius Springer, Berlin, 1932, English translation: Knot theory, BCS Associates, Moscow, Idaho (1983).
180. A. Rhemtulla and D. Rolfsen, *Local indicability in ordered groups: braids and elementary amenable groups*, Proc. Amer. Math. Soc. **130** (2002), no. 9, 2569–2577.
181. D. Rolfsen and B. Wiest, *Free group automorphisms, invariant orderings and applications*, Algebr. Geom. Topol. **1** (2001), 311–320 (electronic).
182. C. Rourke and B. Wiest, *Order automatic mapping class groups*, Pacific J. Math. **194** (2000), no. 1, 209–227.
183. H. Short and B. Wiest, *Orderings of mapping class groups after Thurston*, Enseign. Math. **46** (2000), 279–312.
184. W. Shpilrain, *Representing braids by automorphisms*, Internat. J. Algebra Comput. **11** (2001), no. 6, 773–777.
185. H. Sibert, *Extraction of roots in Garside groups*, Comm. Algebra **30** (2002), no. 6, 2915–2927.
186. ———, *Algorithmique des tresses*, PhD. Thesis, Université de Caen, 2003.

187. A.S. Sikora, *Topology on the spaces of orderings of groups*, Bull. London Math. Soc. **36** (2004), 519–526.
188. L. Solomon, *A Mackey formula in the group ring of a Coxeter group*, J. Algebra **41** (1976), 255–268.
189. W. Thurston, *Finite state algorithms for the braid group*, circulated notes, 1988.
190. ———, *On the geometry and dynamics of diffeomorphisms of surfaces*, Bull. Amer. Math. Soc. **19** (1988), no. 2, 417–431.
191. K. Vogtmann, *Automorphisms of free groups and outer space*, Geom. Dedicata **94** (2002), 1–31.
192. M. Wada, *Group invariants of links*, Topology **31** (1992), no. 2, 399–406.
193. B. Wajnryb, *An elementary approach to the mapping class group of a surface*, Geom. Topol. **3** (1999), 405–466.
194. B. Wiest, *Dehornoy's ordering of the braid groups extends the subword ordering*, Pacific J. Math. **191** (1999), 183–188.
195. H. Wilf, *generatingfunctionology*, Academic Press, Inc., Boston, MA, 1990, available at http://www.math.upenn.edu/~wilf/DownldGF.html.

Index of Notation

Chapter VI

Chapter VII

Chapter VIII

Chapter IX

Chapter X

Chapter XI

Chapter XII

Chapter XIII

Chapter XIV

Chapter XV

Chapter XVI

Index

Titles in This Series

148 **Patrick Dehornoy with Ivan Dynnikov, Dale Rolfsen, and Bert Wiest,** Ordering braids, 2008
147 **David J. Benson and Stephen D. Smith,** Classifying spaces of sporadic groups, 2008
146 **Murray Marshall,** Positive polynomials and sums of squares, 2008
145 **Tuna Altinel, Alexandre V. Borovik, and Gregory Cherlin,** Simple groups of finite Morley rank, 2008
144 **Bennett Chow, Sun-Chin Chu, David Glickenstein, Christine Guenther, James Isenberg, Tom Ivey, Dan Knopf, Peng Lu, Feng Luo, and Lei Ni,** The Ricci flow: Techniques and applications, Part II: Analytic aspects, 2008
143 **Alexander Molev,** Yangians and classical Lie algebras, 2007
142 **Joseph A. Wolf,** Harmonic analysis on commutative spaces, 2007
141 **Vladimir Maz'ya and Gunther Schmidt,** Approximate approximations, 2007
140 **Elisabetta Barletta, Sorin Dragomir, and Krishan L. Duggal,** Foliations in Cauchy-Riemann geometry, 2007
139 **Michael Tsfasman, Serge Vlăduţ, and Dmitry Nogin,** Algebraic geometric codes: Basic notions, 2007
138 **Kehe Zhu,** Operator theory in function spaces, 2007
137 **Mikhail G. Katz,** Systolic geometry and topology, 2007
136 **Jean-Michel Coron,** Control and nonlinearity, 2007
135 **Bennett Chow, Sun-Chin Chu, David Glickenstein, Christine Guenther, James Isenberg, Tom Ivey, Dan Knopf, Peng Lu, Feng Luo, and Lei Ni,** The Ricci flow: Techniques and applications, Part I: Geometric aspects, 2007
134 **Dana P. Williams,** Crossed products of C^*-algebras, 2007
133 **Andrew Knightly and Charles Li,** Traces of Hecke operators, 2006
132 **J. P. May and J. Sigurdsson,** Parametrized homotopy theory, 2006
131 **Jin Feng and Thomas G. Kurtz,** Large deviations for stochastic processes, 2006
130 **Qing Han and Jia-Xing Hong,** Isometric embedding of Riemannian manifolds in Euclidean spaces, 2006
129 **William M. Singer,** Steenrod squares in spectral sequences, 2006
128 **Athanassios S. Fokas, Alexander R. Its, Andrei A. Kapaev, and Victor Yu. Novokshenov,** Painlevé transcendents, 2006
127 **Nikolai Chernov and Roberto Markarian,** Chaotic billiards, 2006
126 **Sen-Zhong Huang,** Gradient inequalities, 2006
125 **Joseph A. Cima, Alec L. Matheson, and William T. Ross,** The Cauchy Transform, 2006
124 **Ido Efrat, Editor,** Valuations, orderings, and Milnor K-Theory, 2006
123 **Barbara Fantechi, Lothar Göttsche, Luc Illusie, Steven L. Kleiman, Nitin Nitsure, and Angelo Vistoli,** Fundamental algebraic geometry: Grothendieck's FGA explained, 2005
122 **Antonio Giambruno and Mikhail Zaicev, Editors,** Polynomial identities and asymptotic methods, 2005
121 **Anton Zettl,** Sturm-Liouville theory, 2005
120 **Barry Simon,** Trace ideals and their applications, 2005
119 **Tian Ma and Shouhong Wang,** Geometric theory of incompressible flows with applications to fluid dynamics, 2005
118 **Alexandru Buium,** Arithmetic differential equations, 2005
117 **Volodymyr Nekrashevych,** Self-similar groups, 2005
116 **Alexander Koldobsky,** Fourier analysis in convex geometry, 2005

115 **Carlos Julio Moreno,** Advanced analytic number theory: L-functions, 2005
114 **Gregory F. Lawler,** Conformally invariant processes in the plane, 2005
113 **William G. Dwyer, Philip S. Hirschhorn, Daniel M. Kan, and Jeffrey H. Smith,** Homotopy limit functors on model categories and homotopical categories, 2004
112 **Michael Aschbacher and Stephen D. Smith,** The classification of quasithin groups II. Main theorems: The classification of simple QTKE-groups, 2004
111 **Michael Aschbacher and Stephen D. Smith,** The classification of quasithin groups I. Structure of strongly quasithin K-groups, 2004
110 **Bennett Chow and Dan Knopf,** The Ricci flow: An introduction, 2004
109 **Goro Shimura,** Arithmetic and analytic theories of quadratic forms and Clifford groups, 2004
108 **Michael Farber,** Topology of closed one-forms, 2004
107 **Jens Carsten Jantzen,** Representations of algebraic groups, 2003
106 **Hiroyuki Yoshida,** Absolute CM-periods, 2003
105 **Charalambos D. Aliprantis and Owen Burkinshaw,** Locally solid Riesz spaces with applications to economics, second edition, 2003
104 **Graham Everest, Alf van der Poorten, Igor Shparlinski, and Thomas Ward,** Recurrence sequences, 2003
103 **Octav Cornea, Gregory Lupton, John Oprea, and Daniel Tanré,** Lusternik-Schnirelmann category, 2003
102 **Linda Rass and John Radcliffe,** Spatial deterministic epidemics, 2003
101 **Eli Glasner,** Ergodic theory via joinings, 2003
100 **Peter Duren and Alexander Schuster,** Bergman spaces, 2004
99 **Philip S. Hirschhorn,** Model categories and their localizations, 2003
98 **Victor Guillemin, Viktor Ginzburg, and Yael Karshon,** Moment maps, cobordisms, and Hamiltonian group actions, 2002
97 **V. A. Vassiliev,** Applied Picard-Lefschetz theory, 2002
96 **Martin Markl, Steve Shnider, and Jim Stasheff,** Operads in algebra, topology and physics, 2002
95 **Seiichi Kamada,** Braid and knot theory in dimension four, 2002
94 **Mara D. Neusel and Larry Smith,** Invariant theory of finite groups, 2002
93 **Nikolai K. Nikolski,** Operators, functions, and systems: An easy reading. Volume 2: Model operators and systems, 2002
92 **Nikolai K. Nikolski,** Operators, functions, and systems: An easy reading. Volume 1: Hardy, Hankel, and Toeplitz, 2002
91 **Richard Montgomery,** A tour of subriemannian geometries, their geodesics and applications, 2002
90 **Christian Gérard and Izabella Łaba,** Multiparticle quantum scattering in constant magnetic fields, 2002
89 **Michel Ledoux,** The concentration of measure phenomenon, 2001
88 **Edward Frenkel and David Ben-Zvi,** Vertex algebras and algebraic curves, second edition, 2004
87 **Bruno Poizat,** Stable groups, 2001
86 **Stanley N. Burris,** Number theoretic density and logical limit laws, 2001

For a complete list of titles in this series, visit the
AMS Bookstore at **www.ams.org/bookstore/**.